T0348755

Industrial Waste Treatment Handbook
Second Edition

Industrial Waste Treatment Handbook
Second Edition

Woodard & Curran, Inc.

AMSTERDAM · BOSTON · HEIDELBERG · LONDON
NEW YORK · OXFORD · PARIS · SAN DIEGO
SAN FRANCISCO · SINGAPORE · SYDNEY · TOKYO
BUTTERWORTH–HEINEMANN IS AN IMPRINT OF ELSEVIER

Butterworth–Heinemann is an imprint of Elsevier
30 Corporate Drive, Suite 400, Burlington, MA 01803, USA
Linacre House, Jordan Hill, Oxford OX2 8DP, UK

 Recognizing the importance of preserving what has been written, Elsevier prints its
books on acid-free paper whenever possible.

Library of Congress Cataloging-in-Publication Data

Application submitted

British Library Cataloguing-in-Publication Data

A catalogue record for this book is available from the British Library.

ISBN 13: 978-0-7506-7963-3
ISBN 10: 0-7506-7963-8

For information on all Elsevier Butterworth–Heinemann publications,
visit our Web site at www.books.elsevier.com

Transferred to Digital Printing 2011

Dedication

To Franklin E. Woodard, Ph.D. Without Frank's tireless dedication to the first edition, this second edition would not be possible. His boundless enthusiasm and expertise in waste treatment practices are an inspiration to all. He is an engineer, a mentor, an educator, a peer, and a friend. Thank you, Frank.

Table of Contents

Preface
to First Edition

This book has been developed with the intention of providing an updated primary reference for environmental managers working in industry, environmental engineering consultants, graduate students in environmental engineering, and government agency employees concerned with wastes from industries. It presents an explanation of the fundamental mechanisms by which pollutants become dissolved or suspended in water or air, then builds on this knowledge to explain how different treatment processes work, how they can be optimized and how one would go about efficiently selecting candidate treatment processes.

Examples from the recent work history of Woodard & Curran, as well as other environmental engineering and science consultants, are presented to illustrate both the approach used in solving various environmental quality problems and the step by step design of facilities to implement the solutions. Where permission was granted, the industry involved in each of these examples is identified by name. Otherwise, no name was given to the industry, and the industry has been identified only as to type of industry and size. In all cases, the actual numbers and all pertinent information have been reproduced as they occurred, with the intent of providing accurate illustrations of how environmental quality problems have been solved by one of the leading consultants in the field of industrial wastes management.

This book is intended to fulfill the need for an updated source of information on the characteristics of wastes from numerous types of industries, how the different types of wastes are most efficiently treated, the mechanisms involved in treatment, and the design process itself. In many cases, "tricks" that enable lower cost treatment are presented. These "tricks" have been developed through many years of experience and have not been generally available except by word of mouth.

The chapter on Laws, and Regulations is presented as a summary as of the date stated in the chapter itself and/or the addendum that is issued periodically by the publisher. For information on the most recent addendum, please call the publisher or the Woodard & Curran office in Portland, Maine ((207) 774-2112).

Preface
to Second Edition

As the change in author's name implies, this book has been turned over, in a manner of speaking, to the firm (Woodard & Curran) who will perpetuate it in continually updated editions, for many years to come. In growing from 12 employees in 1979 to over 460 in 2005, our knowledge of industries and their wastes has increased in breadth and depth. We have brought this to bear on the updating of this, the second edition, and are confident that the reader will benefit greatly.

As was stated in the preface to the first edition, the readership that the authors had in mind included environmental managers working in industry, environmental engineering consultants, graduate students in environmental engineering, and federal, state, or regional employees of government agencies, who are concerned with wastes from industries.

The book maintains its approach of identifying the fundamental chemical and physical characteristics of each target pollutant, then identifying the mechanism by which that target pollutant is held in solution or suspension by the waste stream (liquid, gaseous, or solid). The most efficient method by which each target pollutant can be removed from the waste stream can then be determined.

The chapter on laws and regulations has been expanded significantly, especially in the area of air pollution control. Again, this chapter is up to date as of the end of 2005. The reader is invited to call Woodard & Curran's office in Portland, Maine at (207) 774-2112 for information on new laws or regulations.

Acknowledgments

This second edition was a collaborative effort involving a number of individuals. Being a second edition, however, we would be remiss if we did not acknowledge the individuals, corporations and business organizations that contributed to the first edition. No distinction has been made between first and second edition contributors. We have attempted to cite all contributors. If we have neglected to cite someone, it is unintentional and we extend our sincerest apologies. Thus, heartfelt gratitude and acknowledgements are extended to:

Adam H. Steinman, Esq.; Aeration Technologies, Inc.; R. Gary Gilbert; Albert M. Pregraves; Andy Miller; Claire P. Betze; Connie Bogard; Connie Gipson; Dennis Merrill; Dr. Steven E. Woodard; Geoffrey D. Pellechia; George Abide; George W. Bloom; Henri J. Vincent; Dr. Hugh J. Campbell; J. Alastair Lough; Janet Robinson; Dr. James E. Etzel; James D. Ekedahl; Karen L. Townsend; Katahdin Analytical Services; Keith A. Weisenberger; Kurt R. Marston; Michael Harlos; Michael J. Curato; Patricia A. Proux-Lough; Paul Bishop; Randy E. Tome; Eric P. King; Raymond G. Pepin; Robert W. Severance; Steven N. Whipple; Steven Smock; Susan G. Stevens; Terry Rinehart; Cambridge Water Technology, Inc.; Katherine K. Henderson; Mohsen Moussavi; Lee M. Cormier; Nimrata K. (Tina) Hunt; Peter J. Martin; Dixon P. Pike, Esq.; Bruce S. Nicholson, Esq.; Charlotte Perry; Thomas R. Eschner; Ethan Brush; Kimberly A. Pontau; James H. Fitch, Jr.; Paul M. Rodriguez; Kyle M. Coolidge; Gillian J. Wood; Sarah Hedrick; Chigako Wilson; Jonathan A. Doucette; Ralph Greco, Jr.; Todd A. Schwingle; Christian Roedlich, Ph.D.; and Sharon E. Ross.

Many of these individuals contributed text or verbal information from which Frank freely drew in the production of the first edition. While the second edition contains some new information, it is in large part a repeat of the first edition, and it took effort from dozens of people to recreate what Frank originally produced.

1 Evaluating and Selecting Industrial Waste Treatment Systems

The approach used to develop systems to treat and dispose of industrial wastes is distinctly different from the approach used for municipal wastes. There is a lot of similarity in the characteristics of wastes from one municipality, or one region, to another. Because of this, the best approach to designing a treatment system for municipal wastes is to analyze the performance characteristics of many existing municipal systems and deduce an optimal set of design parameters for the system under consideration. Emphasis is placed on the analysis of other systems, rather than on the waste stream under consideration. In the case of industrial waste, however, few industrial plants have a high degree of similarity between products produced and wastes generated. Therefore, emphasis is placed on analysis of the wastes under consideration, rather than on what is taking place at other industrial locations. This is not to say that there is little value in analyzing the performance of treatment systems at other more or less similar industrial locations. Quite the opposite is true. It is simply a matter of emphasis.

Wastes from industries are customarily produced as liquid wastes (such as process wastes, which go to an on-site or off-site wastewater treatment system), solid wastes (including hazardous wastes, which include some liquids), or air pollutants; often, the three are managed by different people or departments. These wastes are managed and regulated differently, depending on the characteristics of the wastes and the process producing them. They are regulated by separate and distinct bodies of laws and regulations, and, historically, public and governmental focus has shifted from one category (e.g., wastewater) to another (e.g., hazardous wastes) as the times change. However, the fact is that the three categories of wastes are closely interrelated, both as they impact the environment and as they are generated and managed by individual industrial facilities. For example, solid wastes disposed of in the ground can influence the quality of groundwater and surface waters by way of leachate entering the groundwater and traveling with it through the ground, then entering a surface water body with groundwater recharge. Volatile organics in that recharge water can contaminate the air. Air pollutants can fall out to become surface water or groundwater pollutants, and water pollutants can infiltrate the ground or volatilize into the air.

Additionally, waste treatment processes can transfer substances from one of the three waste categories to one or both of the others. Air pollutants can be removed from an air discharge by means of a water solution scrubber. The waste scrubber solution must then be managed in such a way that it can be discarded in compliance with applicable regulations. Airborne particulates can be removed from an air discharge using a bag house, thus creating solid waste to be managed. On still a third level, waste treatment or disposal systems themselves can directly impact the quality of the air, water, or ground. Activated sludge aeration tanks are very effective in causing volatilization of substances from wastewater. Failed landfills can be potent polluters of both groundwater and surface water. The goal of the manager or engineer is thus to design treatment processes that minimize the volume and toxicity of both process waste and the final treatment residue, since final disposal can incur significant cost and liability.

Industrial waste treatment thus encompasses a wide array of environmental, technical, and regulatory considerations. Regardless of the industry, the evaluation and selection of waste treatment technologies typically follows a logical series of steps that help to meet the goal of minimizing waste toxicity and volume. These steps start with a bird's-eye description and evaluation of the waste-producing processes and then move through a program of increasingly detailed evaluations that seek the optimal balance of efficiency and cost, where cost includes both treatment and disposal. The following sections present an illustration of this process, as applied to two very different waste streams: industrial wastewater and air emissions. The sections show, through specific examples, the basic engineering approach to evaluating and selecting waste treatment technologies. This approach is implicit in the more detailed descriptions provided in subsequent chapters.

Treatment Evaluation Process: Industrial Wastewater

Figure 1-1 illustrates the approach for developing a well-operating, cost-effective treatment system for industrial wastewater. The first step is to gain familiarity with the manufacturing processes themselves. This usually starts with a tour of the facility and then progresses through a review of the literature and interviews with knowledgeable people. The objective is to gain an understanding of how wastewater is produced. There are two reasons for understanding the origin of the water: the first is to enable an informed and therefore effective waste reduction, or minimization (pollution prevention), program; the second is to enable proper choice of candidate treatment technologies.

Subsequent steps shown in Figure 1-1 examine, in increasing detail, the technical

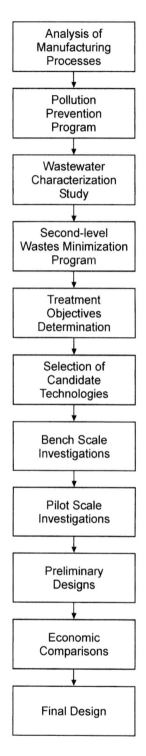

Figure 1-1 Approach for developing an industrial wastewater treatment system.

and economic merits of available technologies, thereby narrowing the field of candidates as the level of scrutiny increases. Understanding and correctly applying each of these steps are critical to successful identification of the best treatment approach. These steps are described in detail in the following text.

Step 1: Analysis of Manufacturing Processes

One of the first steps in the analysis of manufacturing processes is to develop a block diagram that shows how each manufacturing process contributes wastewater to the treatment facility. A block diagram for a typical industrial process, which in this case involves producing finished woven fabric from an intermediate product of the textile industry, is provided in Figure 1-2. Each block of the figure represents a step in the manufacturing process. The supply of water to each point of use is represented on the left side of the block diagram. Wastewater that flows away from each point of wastewater generation is shown on the right side.

In this example, the "raw material" (woven greige goods) for the process is first subjected to a process called "desizing," where the substances used to provide strength and water resistance to the raw fabric, referred to as "size," are removed. The process uses sulfuric acid; therefore, the liquid waste from this process would be expected to have a low pH, as well as containing the substances that were used as sizing. For instance, if starch were the substance used to size the fabric, the liquid waste from the desizing process would be expected to exhibit a high biochemical oxygen demand (BOD), since starch is readily biodegradable.

As a greater understanding of the process is gained, either from the industry's records (if possible) or from measurements taken as part of a wastewater characterization study, process parameters would be indicated on the block diagram. These process parameters may include any number of the following:

flow rates, total quantities for a typical processing day, upper and lower limits, and characteristics such as BOD, chemical oxygen demand (COD), total suspended solids (TSS), total dissolved solids (TDS), and any specific chemicals being used. Each individual step in the overall industrial process would be developed and shown on the block diagram, as illustrated in Figure 1-2.

Step 2: Wastes Minimization and Wastes Characterization Study

After becoming sufficiently familiar with the manufacturing processes as they relate to wastewater generation, the design team should institute a wastes minimization program (actually part of a pollution prevention program), as described in Chapter 4. Then, after the wastes reduction program has become fully implemented, a wastewater characterization study should be carried out, as described in Chapter 5.

The ultimate purpose of the wastewater characterization study is to provide the design team with accurate and complete information on which to base the design of the treatment system. Both quantitative and qualitative data are needed to properly size the facility and to select the most appropriate treatment technologies.

Often, enough new information about material usage, water use efficiency, and wastes generation is learned during the wastewater characterization study to warrant a second level of wastes minimization effort. This second part of the wastes minimization program should be fully implemented, and then its effectiveness should be verified by more sampling and analyses, which amount to an extension of the wastewater characterization study.

A cautionary note is appropriate here concerning maintenance of the wastes minimization program. If a treatment facility is designed and, more specifically, sized based on implementation of a wastes minimization program, and that program is not maintained, causing wastewater increases in volume,

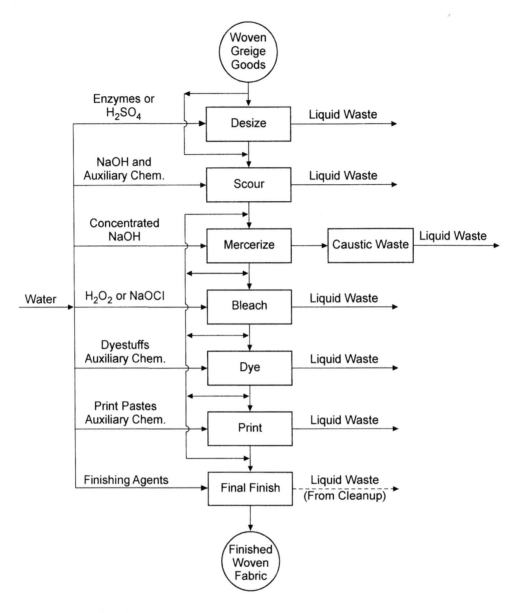

Figure 1-2 Typical industrial process block diagram for a woven fabric finishing process (from the EPA Development Document for the Textile Mills Industry).

strength, or both, the treatment facility will be underdesigned and overloaded at the start. It is extremely important that realistic goals be set and maintained for the wastes minimization program, and that the design team, as well as the industry's management team, is fully aware of the consequences of overloading the treatment system.

Step 3: Determine Treatment Objectives

After the volume, strength, and substance characteristics of the wastewater have been established, the treatment objectives must be determined. These objectives will depend on where the wastewater is to be sent after treatment. If the treated wastewater is discharged to another treatment facility, such as a regional facility or a Publicly Owned Treatment Works (POTW), it must comply with pretreatment requirements. As a minimum, compliance with the Federal Pretreatment Guidelines issued by the Environmental Protection Agency (EPA) and published in the Federal Register is required. Some municipal or regional treatment facilities have pretreatment standards that are more stringent than those required by the EPA.

If the treated effluent is discharged to an open body of water, permits issued by the National Pollutant Discharge Elimination System (NPDES) and the appropriate state agency must be obtained. In all cases, Categorical Standards issued by the EPA apply, and it is necessary to work closely with one or more government agencies while developing the treatment objectives.

Step 4: Select Candidate Technologies

Once the wastewater characteristics and the treatment objectives are known, candidate technologies for treatment can be selected. Rationale for selection is discussed in detail in Chapter 7. The selection should be based on one or more of the following:

- Successful application to a similar wastewater
- Knowledge of chemistry, biochemistry, and microbiology
- Knowledge of available technologies, as well as knowledge of their respective capabilities and limitations

Then, bench-scale investigations should be conducted to determine technical as well as financial feasibility.

Step 5: Bench-Scale Investigations

Bench-scale investigations have the purpose of quickly and efficiently determining the technical feasibility and a rough approximation of the financial feasibility of a given technology. Bench-scale studies range from rough experiments, in which substances are mixed in a beaker and results observed almost immediately, to rather sophisticated continuous flow studies, in which a refrigerated reservoir contains representative industrial wastewater, which is pumped through a series of miniature treatment devices that are models of the full-size equipment. Typical bench-scale equipment includes the six-place stirrer shown in Figure 1-3(a); small columns for ion exchange resins, activated carbon, or filtration media, shown in Figure 1-3(b); and "block aerators," shown in Figure 1-3(c), for performing microbiological treatability studies, as well as any number of custom-designed devices for testing the technical feasibility of given treatment technologies.

Because of scale-up problems, it is seldom advisable to proceed directly from the results of bench-scale investigations to the design of a full-scale wastewater treatment system. Only in cases in which there is extensive experience with both the type of wastewater being treated and the technology and types of equipment to be used can this approach be justified. Otherwise, pilot-scale investigations should be conducted for each technology that appears to be a legitimate candidate for reliable, cost-effective treatment.

The objective of pilot-scale investigations is to develop the data necessary to determine the minimum size and least-cost system of equipment that will enable a design of a treatment system that will reliably meet its intended purpose. In the absence of pilot-scale investigations, the design team is obliged to be conservative in estimating design criteria for the treatment system. The likely result is that a pilot test will pay for itself by allowing less conservative design criteria to be used.

Step 6: Pilot-Scale Investigations

A pilot-scale investigation is a study of the performance of a given treatment technology using the actual wastewater to be treated, usually on site and using a representative

Figure 1-3 (a) Photograph of a six-place stirrer.

model of the equipment that would be used in the full-scale treatment system. The term *representative model* refers to the capability of the pilot treatment system to closely duplicate the performance of the full-scale system. In some cases, accurate scale models of the full-scale system are used. In other cases, the pilot equipment bears no physical resemblance to the full-scale system. For example, fifty-five gallon drums have been successfully used for pilot-scale investigations.

It is not unusual for equipment manufacturers to have pilot-scale treatment systems that can be transported to the industrial site on a trailer. A rental fee is usually charged, and there is sometimes an option to include an operator in the rental fee. It is important, however, to keep all options open. Operation of a pilot-scale treatment system that is rented from one equipment manufacturer might produce results that indicate that another type of equipment, using or not using the same technology, would be the wiser choice. Figure 1-4 presents a photograph of a pilot-scale wastewater treatment system.

One of the difficulties in operating a pilot-scale treatment system is the susceptibility of system upsets, which may be caused by slug doses, wide swings in temperature, plugging of the relatively small diameter pipes, or a lack of familiarity on the part of the operator. Therefore, it is critical to operate a pilot-scale treatment system for a sufficiently long period of time to:

1. Evaluate its performance on all combinations of wastes that are reasonably expected to occur during the foreseeable life of the prototype system.

2. Provide sufficient opportunity to evaluate all reasonable combinations of operation parameters. When operation parameters are changed—for instance, the volumetric loading of an air scrubber, the chemical feed rate of a sludge press, or the recycle ratio for a reverse osmosis system—the system must operate for sufficient time to achieve a steady state before the data to be used for evaluation are taken. This can be particularly

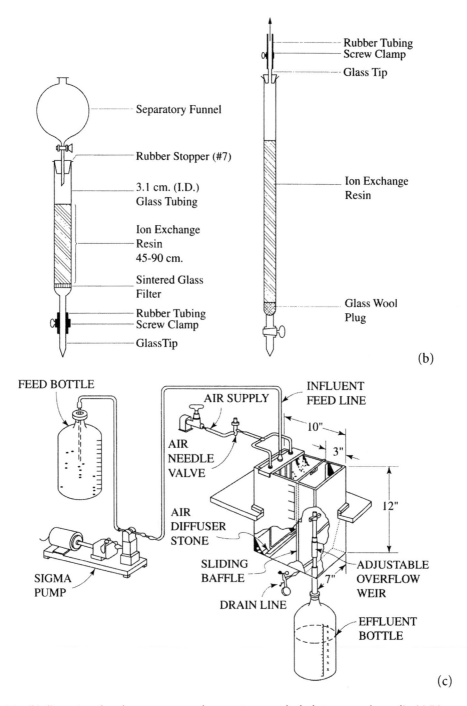

Figure 1-3 (b) Illustration of a column set up to evaluate treatment methods that use granular media. (c) Diagrammatic sketch of a column set up to evaluate treatment methods that use granular media.

Figure 1-4   Photograph of a pilot-scale wastewater treatment system.

problematic in anaerobic biological treatment systems, which can take months to equilibrate. Of course, it will be necessary to obtain data during the period just after operation parameters are changed, to determine when a steady state has been reached.

During the pilot plant operation period, observations should be made to determine whether or not performance predicted from the results of the bench-scale investigations is being confirmed. If performance is significantly different from that which had been predicted, it may be prudent to stop the pilot-scale investigation work and try to determine the cause for the performance difference.

Step 7: Prepare Preliminary Designs

The results of the pilot-scale investigations show which technologies are capable of meeting the treatment objectives, but do not enable an accurate estimation of capital and operating costs. A meaningful cost-effectiveness analysis can take place only after the completion of preliminary designs of the technologies that produced satisfactory effluent quality in the pilot-scale investigations. A preliminary design, then, is the design of an entire waste treatment facility, carried out in sufficient detail to enable accurate estimation of the costs for construction, operation, and maintenance. It must be complete to the extent that the sizes and descriptions of all of the pumps, pipes, valves, tanks, concrete work, buildings, site work, control systems, and manpower requirements are established. The difference between a preliminary design and a final design is principally in the completeness of detail in the drawings and in the specifications. It is almost as though the team that produces the preliminary design could use it directly to construct the plant. The extra detail that goes into the final design is principally to communicate all of the intentions of the design team to people not involved in the design process.

Step 8: Conduct Economic Comparisons

The choice of treatment technology and complete treatment system between two or more systems proven to be reliably capable of meeting the treatment objectives should be based on a thorough analysis of all costs over the expected life of the system. Because this evaluation often drives the final choice, accurate cost estimates, based on an appropriate level of detail, are essential. How much detail is necessary? This is illustrated in the following example, which shows an actual evaluation of treatment alternatives for a manufacturing facility considering a treatment system upgrade. The example illustrates both the types of charges to be considered, as well as the level of detail necessary to support technology selection at this stage of the evaluation. Actual costs (which were accurate at the time of the first edition of this book) are shown for illustrative purposes only, and should not be used as a basis for current evaluations.

Example 1-1: Estimating Costs for Treatment Technology Selection

This example illustrates an economic comparison of five alternatives for treating wastewater from an industrial plant producing microcrystalline cellulose from wood pulp. This plant discharged about 41,000 gallons per day (GPD) of wastewater with a BOD concentration of approximately 20,000 mg/L to the local POTW. The municipality that owned the POTW charged the industry a fee for treatment, and the charge was proportional to the strength, in terms of the biochemical oxygen demand (BOD); total suspended solids (TSS); fats, oils and greases (FOG); and total daily flow (Q).

In order to reduce the treatment charges from the POTW, the plant had the option of constructing and operating its own wastewater treatment system. Since there was not an alternative for discharging the treated wastewater to the municipal sewer system, there would continue to be a charge from the

POTW, but the charge would be reduced in proportion to the degree of treatment accomplished by the industry. Because the industry's treated wastewater would be further treated by the POTW, the industry's treatment system is referred to as a "pretreatment system," regardless of the degree of treatment accomplished.

Four alternatives for the treatment of this waste were evaluated:

1.   Sequencing batch reactors (SBR)
2.   Rotating biological contactors (RBC)
3.   Fluidized bed anaerobic reactors
4.   Expanded bed anaerobic reactors

Both capital and operation and maintenance (O&M) costs for each of these systems were evaluated.

Capital Costs

Tables 1-1 through 1-4 show the capital costs associated with each one of these alternatives. The number and type of every major piece of equipment is included, and a general cost estimate is provided for categories of costs (site work or design) that cannot be accurately estimated at this stage. Buildings, utilities, labor, and construction are all captured.

The estimated costs for the major items of equipment presented in this example, referred to as "cost opinions," were obtained by soliciting price quotations from actual vendors. Ancillary equipment costs were obtained from cost-estimating guides, such as Richardson's, as well as experience with similar projects. Elements of capital cost, such as equipment installation, electrical, process piping, and instrumentation, were estimated as a fixed percentage of the purchase price of major items of equipment. Costs for the building, including plumbing and heating, ventilation, and air conditioning (HVAC), were estimated as a cost per square foot of the buildings. At this level of cost opinion, it is appropriate to use a con-

tingency of 25% and to expect a level of accuracy of ± 30% for the total estimated cost.

This example also shows an interesting circumstance from which engineers should not shy away: the evaluation of a technology that is not yet commercially available. At the time of the writing of the first edition of this text, this was the case for the expanded bed anaerobic reactor. However, this technology showed promise and therefore was retained in the evaluation. The cost was estimated by using the major system components from the fluidized bed anaerobic reactor (Table 1-3), but deleting items that are not required for the expanded bed system, such as clarifiers, sludge-handling equipment, and other equipment.

As a result of these deletions, the estimated capital cost for the expanded bed anaerobic reactor system is $1,700,000.

O&M Costs

Operational costs presented for each treatment alternative include the following elements:

*   Chemicals
*   Power
*   Labor
*   Sludge disposal, if applicable
*   Sewer use charges
*   Maintenance

Because these costs are present for the life of the system, O&M costs are often much more important in the evaluation process than capital costs. In consequence, O&M costs must include as much detail as capital costs, if not more. For instance, in this example the bases for estimating the annual operating cost for each of the above elements were: (1) the quantity of chemicals required for the average design value; (2) power costs for running pumps, motors, blowers, etc.; (3) manpower required to operate the facility; (4) sludge disposal costs, assuming sludge would be disposed of at a local landfill;

Table 1-1   Capital Cost Opinion, Sequencing Batch Reactors—Alternative #1

Equipment	No. Units	Size	Installed Cost ($)
SBR Feed Pumps	3	220 GPM	20,000
Blowers	5	1,500 ACFM	120,000
Aeration Equipment	2	6,000 ACFM	77,000
Floating Mixer	2	15 HP	66,000
Floating Decanter	2	1,200 GPM	44,000
Decant Pump	2	1,200 GPM	26,000
Waste Sludge Pumps	2	450 GPM	13,000
Sludge Press	1	100 ft^3	186,000
Filter Feed Pumps	2	60 GPM	8,000
Thickener	1	100 GPM	100,000
Thickener Feed Pumps	2	50 GPM	8,000
Air Compressor	1	100 CFM	44,000
Ammonia Feed System	1	360 PPD	22,000
Phosphoric Acid Feed System	1	15 GPD	6,000
Potassium Chloride Feed System	1	50 PPD	11,000
Sludge Tank Mixer	1	15 HP	13,000
Filter Feed Tank Mixer	1	5 HP	7,000
		Subtotal:	771,000
Site Work @ 5%			39,000
Electrical & Instrumentation @ 10%			77,000
Process Pipes & Valves @ 10%			77,000
30' × 60' Building @ $65/\text{ft}^2$			117,000
2-SBR Tanks (390,000 gal)			300,000*
Sludge Holding Tank (160,000 gal)			95,000
Equalization Tank (50,000 gal)			58,000
		Subtotal:	1,534,000
OH & P @ 22%			337,000
		Subtotal:	1,871,000
Engineering @ 12%			225,000
		Subtotal:	2,096,000
Contingency @ 25%			524,000
Estimated Construction Cost:			2,620,000
		Say:	2,600,000

*Total for both tanks.

Table 1-2   Capital Cost Opinion, Rotating Biological Contactors—Alternative #2

Equipment	No. Units	Size	Installed Cost ($)
RBC Feed Pumps	3	220 GPM	20,000
Clarifiers	2	40° Diameter	195,000
Sludge Pumps	2	100 GPM	11,000
Sludge Press	1	100 ft^3	186,000
Filter Feed Pumps	2	60 GPM	8,000
Thickener	1	100 GPM	100,000
Thickener Feed Pumps	2	50 GPM	8,000
Air Compressor	1	100 CFM	44,000
Ammonia Feed System	1	360 PPD	22,000
Phosphoric Acid Feed System	1	15 GPD	6,000
Potassium Chloride Feed System	1	60 PPD	11,000
Sludge Tank Mixer	1	15 HP	13,000
Filter Feed Tank Mixer	1	5 HP	7,000
Blowers	5	500 CFM	63,000
Aeration System	1	2,000 CFM	44,000
		Subtotal:	738,000
Site Work @ 5%			37,000
Electrical & Instrumentation @10%			74,000
Process Pipes & Valves @ 10%			74,000
30' × 60' Building @ $65/ft^2			117,000
Sludge Holding Tank (160,000 gal)			95,000
RBC Tanks (Concrete)			350,000
		Subtotal:	1,485,000
OH & P @ 22%			326,000
RBC Shafts & Enclosures			1,000,000
		Subtotal:	2,811,000
Engineering @ 12%			337,000
		Subtotal:	3,148,000
Contingency @ 25%			787,000
Estimated Construction Cost:			3,935,000
		Say:	3,900,000

(5) the cost for sewer use charges based on present rates; and (6) maintenance costs as a fixed percentage of total capital costs.

The estimated sewer use charges for each treatment alternative are given in Table 1-5 and show the spread of estimated sewer costs alone among all alternatives. Tables 1-6 through 1-9 show the yearly O&M costs for the SBR, the RBC, the fluidized bed anaerobic reactors, and the expanded bed anaerobic reactors; they show the types of fees considered. For the fluidized bed system, additional information on gas recovery is also included to show the potential offsetting of O&M costs. O&M costs for the expanded bed system were estimated from the fluidized bed costs

Table 1-3 Capital Cost Opinion, Fluidized Bed Anaerobic Reactors—Alternative #3

Equipment	No. Units	Size	Installed Cost ($)
Reactor Feed Pumps	3	220 GPM	20,000
Secondary Clarifiers	2	40' Diameter	195,000
Sludge Pumps	2	20 GPM	3,000
Filter Press	1	40 ft^3	108,000
Filter Feed Pumps	2	60 GPM	8,000
Sludge Transfer Pumps	2	80 GPM	8,000
Sludge Tank Mixer	1	10 HP	1,000
Filter Feed Tank Mixer	1	5 HP	7,000
Compressor	1	100 CFM	44,000
Gas Recovery Blower	1	40 CFM	19,000
		Subtotal:	413,000
Site Work @ 5%			21,000
Electrical & Instrumentation @ 10%			41,000
Process Pipes & Valves @ 10%			41,000
30' × 30' Building @ $65/ft^2			59,000
Sludge Holding Tank (30,000 gal)			35,000
		Subtotal:	612,000
OH & P @ 22%			135,000
		Subtotal:	747,000
Upflow Fluidized Bed Reactor System			1,000,000
		Subtotal:	1,747,000
Engineering @ 12%			210,000
		Subtotal:	1,957,000
Contingency @ 25%			489,000
Estimated Construction Cost:			2,446,000
		Say:	2,450,000

Table 1-4 Capital Cost Opinion, Expanded Bed Anaerobic Reactors—Alternative #4

Equipment	No. Units	Size	Installed Cost ($)
Reactor Feed Pumps	3	220 GPM	20,000
Gas Recovery Blower	1	40 CFM	19,000
		Subtotal:	39,000
Site Work @ 5%			2,000
Electrical & Instrumentation @ 10%			4,000
Process Pipes & Valves @ 10%			4,000
30' × 20' Building @ $65/ft^2			39,000
		Subtotal:	88,000
OH & P @ 22%			19,000
		Subtotal:	107,000

Table 1-4 Capital Cost Opinion, Expanded Bed Anaerobic Reactors—Alternative #4 *(continued)*

Equipment	No. Units	Size	Installed Cost ($)
Upflow Fluidized Bed Reactor System			1,000,000
		Subtotal:	1,107,000
Engineering @ 12%			133,000
		Subtotal:	1,240,000
Contingency @ 35%			434,000
Estimated Construction Cost:			1,674,000
		Say:	1,700,000

Table 1-5 Estimated Sewer Use Charges

Scenario	Yearly Cost ($)[*]
No Treatment	928,000
SBR Alternative	325,000
RBC Alternative	325,000
Fluidized Bed Alternative	384,000
Expanded Bed Alternative	335,000

[*]Based on flow, TSS, and BOD5 charges currently incurred.

Table 1-6 Yearly O&M Cost Summary, Sequencing Batch Reactors—Alternative #1

Item	Unit	Quantity	Unit Cost ($)	Yearly Cost ($)[*]
Chemicals				
Ammonia (Anhydrous)	Ton	66	135	8,900
Phosphoric Acid (85%)	lb	83,000	0.22	18,300
Potassium Chloride (99%)	lb	41,000	0.67	27,500
Polymer	lb	9,000	1.00	9,000
Power	kw-hr	3,000,000	0.054	162,000
Labor	man-hr	4,380	38	166,400
Sludge Disposal[†]	Ton	3,600	70	252,000
Sewer Use Charges[‡]	—	—	—	325,000
Maintenance[**]	—	—	—	52,000
			Total:	1,021,000
			Say:	1,000,000

[*]Total Rounded to nearest $50,000.
[†]Sludge assumed to be nonhazardous; includes transportation.
[‡]Per Table 1-5.
[**]Assumed to be 2% of total capital cost.

Table 1-7 Yearly Operating Cost Summary, Rotating Biological Contactors—Alternative #2

Item	Unit	Quantity	Unit Cost ($)	Yearly Cost ($)[*]
Chemicals				
Ammonia (Anhydrous)	Ton	66	135	8,900
Phosphoric Acid (85%)	lb	83,000	0.22	18,300
Potassium Chloride (99%)	lb	41,000	0.67	27,500
Polymer	lb	9,000	1.00	9,000
Power	kw-hr	890,000	0.054	48,100
Labor	man-hr	4,380	38	166,400
Sludge Disposal[†]	Ton	3,600	70	252,000
Sewer Use Charges[‡]	—	—	—	325,000
Maintenance[**]	—	—	—	78,000
			Total:	933,200
			Say:	950,000

[*]Total rounded to nearest $50,000.
[†]Sludge assumed to be nonhazardous; includes transportation.
[‡]Per Table 1-5.
[**]Assumed to be 2% of total capital cost.

Table 1-8 Yearly Operating Cost Summary, Fluidized Bed Anaerobic Reactor—Alternative #2

Item	Unit	Quantity	Unit Cost ($)	Yearly Cost ($)[*]
Chemicals				
Ammonia (Anhydrous)	Ton	5	135	700
Phosphoric Acid (85%)	lb	7,000	0.22	1,500
Potassium Chloride (99%)	lb	3,000	0.67	2,000
Polymer	lb	6,000	1.00	6,000
Power	kw-hr	262,000	0.054	14,100
Labor	man-hr	4,380	38	166,400
Sludge Disposal[†]	Ton	1,642	70	114,900
Sewer Use Charges[‡]	—	—	—	384,000
Maintenance[**]				39,000
	Total:			738,600
	Say:			700,000
Gas Recovery[††]	MCF:	19,000	3.00	(57,000)
	Total with Gas Recovery:			643,000
	Say:			650,000

[*]Total rounded to nearest $50,000.
[†]Sludge assumed to be nonhazardous; includes transportation.
[‡]Per Table 1-5.
[**]Assumed to be 2% of total capital cost.
[††]Unit cost includes amortized cost of gas recovery equipment.

Table 1-9   Yearly Operating Cost Summary, Expanded Bed Anaerobic Reactor—Alternative #4

Item	Unit	Quantity	Unit Cost ($)	Yearly Cost ($)[*]
Chemicals				
Ammonia (Anhydrous)	Ton	5	135	700
Phosphoric Acid (85%)	lb	7,000	0.22	1,500
Potassium Chloride (99%)	lb	3,000	0.67	2,000
Polymer	lb	6,000	1.00	6,000
Power	kw-hr	262,000	0.054	14,000
Labor	man-hr	2,190	38	88,200
Sewer Use Charges[†]	—	—	—	335,000
Maintenance[‡]				34,000
Total:				481,500
Say:				500,000
Gas Recovery[**]	MCF:	19,000	3.00	(57,000)
Total with Gas Recovery:				443,000
Say:				450,000

[*]Total rounded to nearest $50,000.
[†]Per Table 1-4.
[‡]Assumed to be 2% of total capital cost.
[**]Unit cost includes amortized cost of gas recovery equipment.

and adjusted by reducing labor costs by 75% (since no sludge dewatering is required) and eliminating sludge disposal costs, since cellulose can be recycled.

Annualized Costs

Calculating annualized costs is the final component of an economic cost comparison and is a convenient method for making long-term economic comparisons between treatment alternatives. To obtain annualized costs, the capital cost for the alternative in question is amortized over the life of the system, which, for the purpose of this example, is assumed to be 20 years. The cost of money is assumed to be 10%. This evaluation shows the total capital and O&M costs of each system over a 20-year period.

The annualized cost for each alternative is shown in Table 1-10. While these totals show a fairly close spread of costs, the significant effect of energy recovery on total costs of the fluidized and expanded bed systems is apparent.

Based on this economic analysis, the expanded bed anaerobic reactor is the preferred alternative. Note that it is the only alternative that provides an annual cost that is less than installing no pretreatment system and continuing to pay high surcharge fees. In some cases, paying additional surcharge fees is not an acceptable alternative, because the pollutant loading to the POTW would be so high it would violate a pretreatment permit limit, not just a surcharge limit.

At this stage, sufficient information is typically available to choose a final design.

Step 9: Final Design

The final design process for the selected technology is both a formality, during which standardized documents (including plans and specifications) are produced, and a procedure, during which all of the subtle details of the facility that is to be constructed are worked out. The standardized documents have a dual purpose; the first is to provide a common basis for several contractors to

Table 1-10   Annualized Costs

Total Capital Alternative	Annual Capital Cost ($)	Cost ($)*	Total Annual O&M Cost ($)†	Cost ($)
#1 SBRs	2,600,000	300,000	1,000,000	1,300,000
#2 RBCs	3,900,000	450,000	950,000	1,400,000
#3 Fluidized Bed	2,450,000	300,000	700,000	1,000,000
			(650,000)	(950,000)
#4 Expanded Bed	1,700,000	200,000	500,000	700,000
			(450,000)	(650,000)
#5 No Treatment	—	900,000	—	900,000

*Assumes 20-yr. life, 10% cost of money.
†Assumes no increase in future O&M costs. Numbers in parentheses reflect energy recovery.

prepare competitive bids for constructing the facility, and the second is to provide complete instructions for building the facility, so that what gets built is exactly what the design team intended.

Step 10: Solicitation of Competitive Bids for Construction

The purpose of the competitive bidding process is to ensure that the facility developed by the design team will be built at the lowest achievable cost. However, the contractors invited to participate in the bidding process should be carefully selected on the basis of competence, experience, workmanship, and reliability, so that quality is ensured regardless of the bid price. In the end, the best construction job for the lowest possible price will not have a chance of being realized if the best contractor is not on the list of those invited to submit bids.

The foundation of the bidding process is the "plans and specifications." The first duty of the plans and specifications is to provide all information in sufficiently complete detail so that each of the contractors preparing bids will be preparing cost proposals for exactly the same, or truly equivalent, items. It is essential that each contractor's bid proposal be capable of being compared on an "apples to apples" basis; that is, regardless of which contractor builds the facility, it will be essentially identical in all respects relating to performance, reliability, operation and mainte-

nance requirements, and useful life. The key to obtaining this result is accuracy and completeness, down to the finest details, of the plans and specifications.

As it has developed in the United States, the bidding process follows the block diagram shown in Figure 1-5. Figure 1-5 illustrates that the first of six phases is to develop a list of potential bidders, as discussed previously. This list is developed based on past experience, references, and dialog with contractors regarding their capabilities. Other means for developing the list can involve advertising for potential bidders in local and regional newspapers, trade journals, or publications issued by trade associations. In the second phase, a formal request for bids is issued, along with plans, specifications, a bid form, and a timetable for bidding and construction.

The third phase, shown in Figure 1-5, the prebid conference, is key to the overall success of the project. This phase involves assembling all potential contractors and other interested parties, such as potential subcontractors, vendors, and suppliers, for a meeting at the project site. This site visit normally includes a guided and narrated tour, a presentation of the engineer's/owner's concept of the project, and a question-and-answer period. This meeting can result in identification of areas of the design that require additional information or changes. If this is the case, the additional information and/or changes are then addressed to all par-

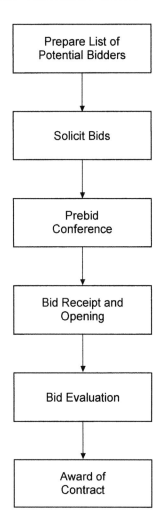

Figure 1-5   Illustration of the bidding process.

ties by issuance of formal addenda to the plans and specifications.

The final three phases—receipt and opening of bids, bid evaluation, and award of contract—are highly interrelated. Upon receipt, the bids are reviewed to determine accuracy and completeness and to identify the lowest responsible bidder. If all bids are higher than was expected, the industry's management and engineers have the opportunity to explore alternatives for redesign of the project. Finally, the project is awarded to the contractor submitting the lowest responsible bid. Construction or implementation can now begin.

These steps complete the normal sequence of events for the identification, selection, and construction of an industrial wastewater treatment system, either for pretreatment or final treatment of wastewater. These steps are common to most design approaches. In the following section, the use of this approach is illustrated for an entirely different waste stream: pollutants in air emissions. This comparison shows both the flexibility and overall utility of the process.

Treatment Evaluation Process: Air Emissions

The control and treatment of air emissions from a facility can be one of the more challenging aspects of a manager's or engineer's job because of the number, type, and often invisible nature of these emissions. Under federal regulations, the discharge of substances to the air, no matter how slight, is regarded as air pollution. A federal permit, as well as a state license or permit, must cover all discharges over a certain quantity per unit time. Local ordinances or regulations may also apply.

Discharges to the air can be direct, by means of a stack or by way of leaks from a building's windows, doors, or other openings. The latter are referred to as "fugitive emissions." Volatilization of organic compounds, such as solvents and gasoline from storage containers, transfer equipment, or even points of use, is an important sources of air discharges. Another source of discharge of volatile organics to the air is aerated wastewater treatment systems.

Management of discharges to the air is almost always interrelated with management of discharges to the water and/or the ground, since air pollution control devices usually remove substances from the air discharge (usually a stack) and transfer them to a liquid solution or suspension, as with a scrubber, or to a collector of solids, as with a bag house. For this reason, a total system approach to environmental pollution control is preferred, and this approach should include a pollution prevention program with vigorous waste minimization.

There are three phases to the air pollution cycle. The first is the discharge at the source; the second is the dispersal of pollutants in the atmosphere; and the third is the reception of pollutants by humans, animals, or inanimate objects. Management of the first phase is a matter of engineering, control, and operation of equipment. The second phase can be influenced by stack height, but meteorology dictates the path of travel of released pollutants. Since the motions of the atmosphere can be highly variable in all dimensions, management of the third phase, which is the ultimate objective of air pollution control, requires knowledge of meteorology and the influence of topography.

Chapter 3 presents a detailed synopsis of laws and regulations pertaining to protection of the nation's air resources. While these laws differ significantly from those governing wastewater, the goals are the same: to minimize the mass and toxicity of pollutants released to the environment. Likewise, the engineering process for identifying technologies suitable for achieving this goal follows the same general process as illustrated in detail for wastewater, with a few important twists and differences. The general steps, however, are the same:

- Analysis of the manufacturing process
- Wastes minimization and characterization study
- Identification of treatment objectives
- Selection of candidate technologies
- Bench-scale investigations
- Pilot-scale investigations
- Preliminary design
- Economic comparisons
- Final design

In the following sections, this process is described again (in less detail, since many steps are the same as described previously) for specific application to the selection of air treatment technologies. The treatment of emissions from a cement manufacturing facility is used as an example, since this sort of operation, like many industries, has both point source and fugitive emissions.

Analysis of Manufacturing Process

As with wastewater, successful and cost-effective air pollution control has its foundation in complete awareness of all of the individual sources, fugitive as well as point sources. The process of cataloging each and every individual air discharge within an industrial manufacturing or other facility is most efficiently done by first developing detailed diagrams of the facility as a whole. Depending on the size and complexity of the facility, it may be advantageous to develop separate diagrams for point sources and sources of fugitive emissions. Next, a separate block diagram for each air discharge source should be developed. The purpose of each block diagram is to illustrate how each manufacturing process and wastewater or solid wastes treatment or handling process contributes unwanted substances to the air. Figures 1-6 through 1-8 are examples that pertain to a facility that manufactures cement from limestone.

Figure 1-6 is a diagram of the facility as a whole, showing the cement manufacturing process as well as the physical plant, including the buildings, parking lots, and storage facilities.

At this particular facility, cement, manufactured for use in making concrete, is produced by grinding limestone, cement rock, oyster shell marl, or chalk, all of which are principally calcium carbonate, and mixing the ground material with ground sand, clay, shale, iron ore, and blast furnace slag, as necessary, to obtain the desired ingredients in proper proportions. This mixture is dried in a kiln, and then ground again while mixing with gypsum. The final product is then stored, bagged, and shipped. Each of the individual production operations generates or is otherwise associated with dust, or "particulates," and is a potential source of air pollutant emissions exceeding permit limits.

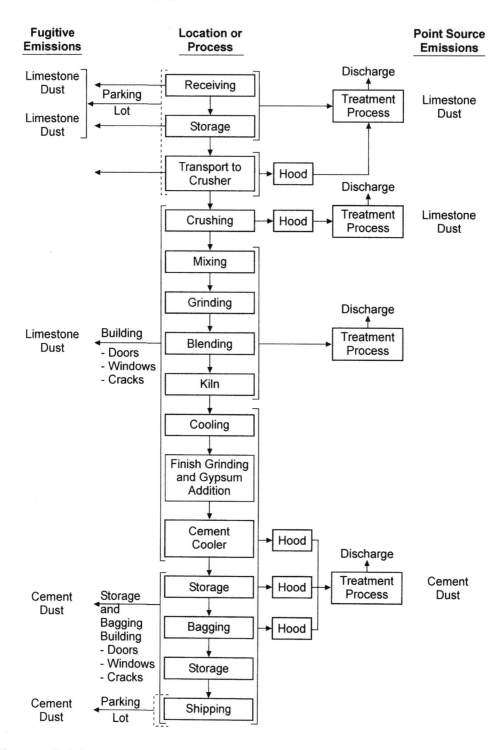

Figure 1-6   Block diagram of cement manufacturing plant.

Figure 1-7 illustrates that raw materials are received and stockpiled at the plant and are potential sources of particulate emissions due to the fine particles of "dust" generated during the mining, transportation, and loading and unloading processes. Their susceptibility to being blown around if they are out in the open is also a factor. To control fugitive emissions from these sources, it is necessary to conduct all loading, unloading, grinding, and handling operations within enclosures that are reasonably airtight but are also ventilated for the health and safety of employees. Ventilation requires a fresh air intake and a discharge. The discharge requires a treatment process. Candidate treatment processes for this application include bag houses, wet scrubbers, and electrostatic precipitators, possibly in combination with one or more inertial separators. Each of these treatment technologies is discussed in Chapter 8.

A very important aspect of air pollution control is to obtain and then maintain a high degree of integrity of the buildings and other enclosures designed to contain potential air pollutants. Doors, windows, and vents must be kept shut. The building or enclosure must be kept in good repair to avoid leaks. In many cases, it is necessary to maintain a negative pressure (pressure inside building below atmospheric pressure outside building) to prevent the escape of gases or particulates. Maintaining the integrity of the building or enclosure becomes very important, in this case, to minimizing costs for maintaining the negative pressure gradient.

As further illustrated in Figure 1-7, the next series of processing operations constitutes the cement manufacturing process itself, and starts with crushing, then proceeds through mixing, grinding, blending, and drying in a kiln. Each of these processes generates large amounts of particulates, which must be contained, transported, and collected by use of one or more treatment technologies, as explained in Chapter 8. In some cases, it may be most advantageous from the points of view of reliability or cost effectiveness, or both, to use one treatment system for all point sources. In other cases, it might prove best to treat one or more of the sources individually.

Continuing through the remaining processes illustrated in Figure 1-6, the finished product (cement) must be cooled, subjected to "finish grinding," cooled again, stored, and then bagged and sent off to sales distribution locations. Again, each of these operations is a potential source of airborne pollutants, in the form of "particulate matter," and it is necessary to contain, transport, and collect the particulates using hoods, fans, ductwork, and one or more treatment technologies, as explained in Chapter 8.

The next step in the process of identifying each and every source of air pollutant discharge from the cement manufacturing plant being used as an example is to develop a block diagram for each individual activity that is a major emission source. Figure 1-8 illustrates this step. Figure 1-8 is a block diagram of the process referred to as the "kiln," in which the unfinished cement is dried using heat. This diagram pertains to only the manufacturing process and does not include sources of emissions from the physical plant, most of which are sources of fugitive emissions.

Figure 1-8 shows that the inputs to the kiln include partially manufactured (wet) cement and hot air. The outputs include dry partially manufactured cement and exhaust air laden with cement dust, or particulates. The diagram then shows that there are four candidate technologies to treat the exhaust gas to remove the particulates before discharge to the ambient air. The four candidate technologies are:

- Electrostatic precipitator
- Cyclone
- Bag house
- Wet scrubber

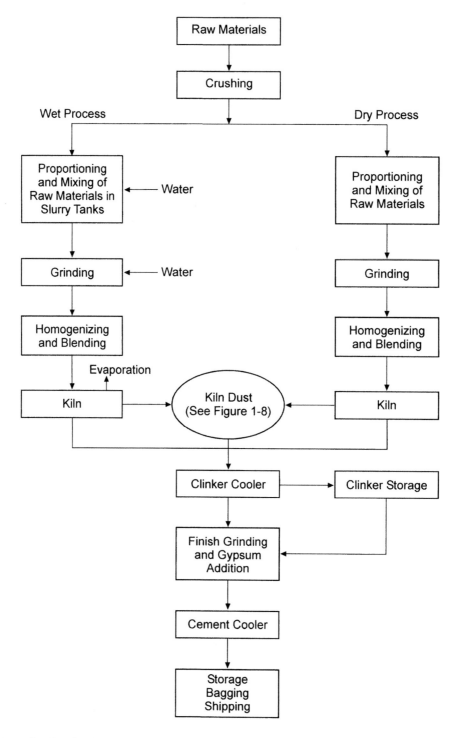

Figure 1-7   Flow sheet for the manufacture of Portland cement.

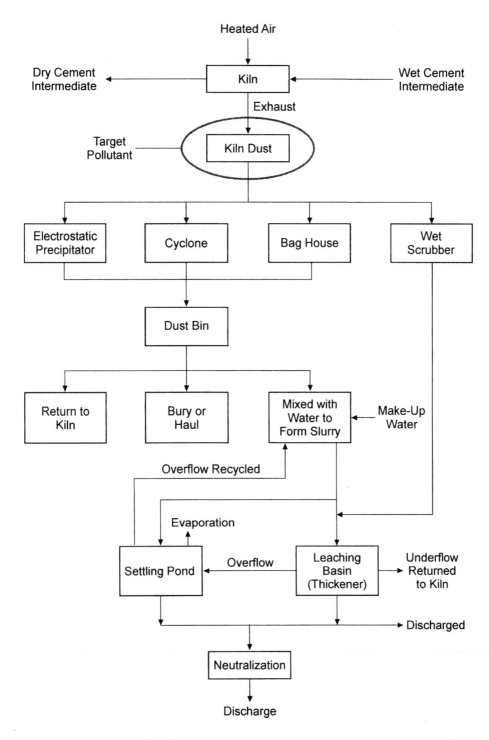

Figure 1-8   Kiln dust collection and handling.

Each of these technologies is worthy of further investigation, including investigation of technical feasibility and cost effectiveness. Also, each of these technologies results in a residual, which must be handled and disposed of.

For instance, the bag house technology produces a residual that can be described as a dry, fine dust—essentially, "raw" cement. This material can be stored in a "dust bin" (the dust bin must be managed as a potential air pollution source), and from there many options are possible. The dust could be:

- Returned to the kiln in an attempt to increase the yield of the manufacturing process
- Buried
- Mixed with water to form a slurry
- Hauled (as a by-product) to another point of use

The first of the above options is only a partial solution at best, since there must be some "blow down," if only to maintain quality specifications for the finished product. Burial is a final solution, but it must be accomplished within the parameters of good solid waste disposal practice. "Water slurry" is only an interim treatment step. Forming a water slurry transforms the air pollution potential problem to a water pollution potential problem (a "cross-media" effect). The slurry can be transported to another location without risk of air pollution, but once there, it must be dewatered by sedimentation before final disposal within the bounds of acceptable solid waste and wastewater disposal practices.

The foregoing example illustrates how an entire manufacturing facility must be analyzed and diagrammed to define each and every source of discharge of pollutants to the air as an early step in a technically feasible and cost-effective air pollution control program. The next steps are presented below.

Wastes Minimization and Characterization Study

After all potential sources of air pollutants have been identified, the objectives of the industry's pollution prevention program should be addressed. Wastes minimization is only one aspect of a pollution prevention program, but it is a critical one. Each source should first be analyzed to determine whether it could be eliminated. Next, material substitutions should be considered to determine whether less toxic or costly substances can be used in place of the current ones. Then it should be determined whether or not a change in present operations—for instance, improved preventative maintenance or equipment—could significantly reduce pollutant generation. Finally, it should be determined whether or not improvements in accident and spill prevention, as well as improved emergency response, are warranted.

After a prudent wastes minimization program has been carried out, a period of time should be allocated to determine whether the changes made appear to be permanent. This phase of the overall air pollution control program is important, because if the determination of air pollutant flow rates and concentrations is made on the basis of improved maintenance and operational procedures, and if the facility regresses to the way things were done previously, the handling and treatment equipment designed on the basis of the improved procedures will be overloaded and will fail.

Once all air pollution flows and loads have become stabilized, each of the sources should be subjected to a characterization program to determine flow rates and target pollutant concentrations (flows and loads) for the purpose of developing design criteria for handling and treatment facilities. Examples of handling facilities are hoods, fans, and ductwork. Examples of treatment equipment are electrostatic precipitators and fabric filters (bag houses, for instance). The characterization study amounts to developing estimates of emission rates based on historical records of the facility under consideration or those of a similar facility. For instance, material balances showing amounts of raw materials purchased and products sold can be used to estimate loss rates.

Treatment Objectives

Treatment objectives are needed to complete the development of design criteria for handling and treatment equipment. The air discharge permit, either in-hand or anticipated, is one of the principal factors used in this development, since it specifies permissible levels of discharge. Another principal factor is the strategy to be used regarding allowances—that is, whether or not to buy allowances from another source or to reduce emissions below permit limits and attempt to recover costs by selling allowances. This strategy and its legal basis are discussed in Chapter 3. Only after all treatment objectives have been developed can candidate treatment technologies be determined. However, it may be beneficial to employ an iterative process in which more than one set of treatment objectives and their appropriate candidate technologies are compared as competing alternatives in a financial analysis to determine the most cost-effective system.

Selection of Candidate Technologies

After the characteristics of air discharges, in terms of flows and loads, have been determined (based on stabilized processes after changes were made for wastes minimization), and treatment objectives have been agreed upon, candidate technologies for removal of pollutants can be selected. The principles discussed in Chapters 2 and 8 are used as the bases for this selection. The selection should be based on one or more of the following:

- Successful application in a similar set of conditions
- Knowledge of chemistry

- Knowledge of options available, as well as knowledge of capabilities and limitations of those alternative treatment technologies

The next step is to conduct bench-scale investigations to determine technical and financial feasibility.

Bench-Scale Investigations

Unless there is unequivocal proof that a given technology will be successful in a given application, it is imperative that a rigorous program of bench-scale, followed by pilot-scale, investigations be carried out. Such a program is necessary for standard treatment technologies as well as innovative technologies. The cost for this type of program will be recovered quickly as a result of the equipment being appropriately sized and operated. Underdesigned equipment will simply be unsuccessful. Overdesigned equipment will cost far more to purchase, install, and operate.

The results of a carefully executed bench-scale pollutant removal investigation will provide the design engineer with reliable data on which to determine the technical feasibility of a given pollutant removal technology, as well as a preliminary estimate of the costs for purchase, construction and installation, and operation and maintenance. Without such data, the design engineer is forced to use very conservative assumptions and design criteria. The result, barring outrageous serendipity, will be unnecessarily high costs for treatment throughout the life of the treatment process.

Pilot-Scale Investigations

Bench-scale investigations are the first step in a necessary procedure for determining the most cost-effective treatment technology. Depending on the technology, inherent scale-up problems may make it inadvisable to design a full-scale treatment system based only on data from bench-scale work. The next step after bench-scale investigations is the pilot-scale work. A pilot plant is simply a small version of the anticipated full-scale treatment system.

A good pilot plant should have the capability to vary operational parameters. It is not sufficient to merely confirm that successful treatment, in terms of compliance with discharge limitations, can be achieved using the same operating parameters as those determined by the bench-scale investigations. Again, it would be outrageous serendipity if the results of the bench-scale investigations truly identified the most cost-effective, as well as reliable, full-scale treatment system design and operating parameters.

The pilot-scale investigation should be carried out at the industrial site, using a portion of the actual gas stream to be treated. A pilot-scale treatment unit (e.g., a wet scrubber or an electrostatic precipitator) can, in many cases, be rented from a manufacturer and transported to the site.

The pilot plant should be operated continuously over a representative period of time, so as to include as many of the waste stream variations as reasonably possible that the full-scale unit is expected to experience. One difficulty in carrying out a pilot-scale study is that smaller units are more susceptible to upset, fouling, plugging, or other damage from slug doses caused by spills or malfunctions in processing equipment. Also, unfamiliarity on the part of the pilot plant operator, either with the gas stream being treated, the processing system from which the stream is generated, or the pilot plant itself, can result in the need for prolonged investigations.

As with wastewater treatment pilot plant investigations, it is critically important to operate a pilot-scale treatment system for a sufficiently long period of time to:

- Include as many combinations of wastes that are reasonably expected to occur during the foreseeable life of the prototype system, as is reasonably possible.

- Evaluate combinations of operation parameters, within reason. When operation parameters are changed (e.g., the volumetric loading of an air scrubber, the chemical feed rate of a pH neutralization system, or the effective pore size of a bag house–type fabric filter), the system must operate for a long enough time to achieve steady state before any data to be used for evaluation are taken. Of course, it will be necessary to obtain data during the period just after operation parameters are changed, to determine when steady state has been reached.

Observations should be made to determine whether or not the performance of the pilot plant, using a particular set of parameters, is in the range of what was predicted from the results of the bench-scale investigations. If the difference in performance is significant, it may be prudent to stop the pilot-scale investigation work and try to determine the cause.

Preliminary Design

The results of the pilot-scale investigations show which technologies are capable of meeting the treatment objectives, but do not enable an accurate estimation of capital and operating costs. A meaningful cost analysis can take place only after the completion of preliminary designs of those technologies that proved to produce satisfactory effluent quality in the pilot-scale investigations. A preliminary design is a design of an entire treatment facility, carried out in sufficient detail to enable accurate estimation of the costs for constructing and operating a treatment facility. It must be complete to the extent that the sizes and descriptions of all of the pumps, pipes, valves, tanks, concrete work, buildings, site work, control systems, and manpower requirements are established. The difference between a preliminary design and a final design is principally in the completeness of detail of the drawings and in the specifications. It is almost as though the team that produces the preliminary design could use it directly to construct the plant. The extra detail that goes into the final design is principally to communicate all of the intentions of the design team to people not involved in the design.

Economic Comparisons

The choice of treatment technology and complete treatment system between two or more systems proven to be reliably capable of meeting the treatment objectives should be based on a thorough analysis of all costs over the expected life of the system. This evaluation is conducted in a similar manner, and to a similar level of detail, as described in the previous section on wastewater.

Conclusion

The engineering approach described in the previous sections will provide a solid basis for the evaluation of almost any industrial process technology or treatment train. It can be applied, in abbreviated form, to the evaluation of individual equipment choices or, in very detailed form, to the design of an entire treatment facility. Additional detail on specific aspects of industrial waste treatment are provided in subsequent chapters and reflect the basic engineering tenets described in this chapter. This engineering approach, when coupled with a sound understanding of the cost and consequences of final disposal, will help ensure a sound and effective treatment system for even the most complex waste stream.

Treatment Evaluation Process: Solid Wastes

Industrial wastes that are discharged to neither air nor water are classified as solid, industrial, or hazardous waste. At the federal level, these wastes are regulated primarily by the Resource Conservation and Recovery Act

(RCRA), which contains specific design and management standards for both hazardous wastes (Subtitle C) and municipal solid wastes (Subtitle D).

In general, the process of evaluating solid waste treatment does not follow the same process as liquid waste or air discharges. A common goal of the treatment of liquid wastes and air discharges is to produce a solid waste for disposal. By the time solid waste is created, it is often implied that there are no remaining "treatment" options. That being said, a solid waste "treatment evaluation process" would likely involve evaluation of the processes in the facility that generates the solid wastes, as well as the liquid and air treatment processes that contribute to the solid wastes generated, with the goal of minimizing the solid waste that is handled, transported, and ultimately disposed of.

An important aspect of the solid waste treatment evaluation process is an understanding of how solid wastes are characterized and, subsequently, handled and disposed of so that a facility maintains compliance with the appropriate regulations. As is discussed in greater detail in Chapter 9, solid waste management and disposal often represent significant and constantly increasing costs for industry. To minimize these costs and reduce the likelihood of enforcement actions by regulators, environmental managers must ensure that a sound solid waste management program is in place and that all personnel, from laborers to top managers, are vigilant in carrying it out. Some key points to bear in mind when evaluating the solid waste generated by a facility and developing a solid waste management program are as follows.

1. *Know the facility waste streams.* As with industrial wastewaters, these are seldom the same for different plants. As a first step, facilities must know how much of each type of solid waste they are producing.

2. *Keep wastes segregated.* Heavy fines, as well as criminal sentences, are the penalties for improper waste disposal. Facilities must ensure that hazardous wastes are not put in the trash dumpster, that listed hazardous wastes are not mixed with other nonhazardous materials, and generally that wastes are handled as they're supposed to be.

3. *Choose waste disposal firms carefully.* Because facilities can be held responsible for cleanup costs of the waste facilities they use, waste transporters and facilities should be chosen carefully.

4. *Institute a pollution prevention program that includes a vigorous wastes minimization effort.* Where possible, reduce the quantity or toxicity of materials used in production.

5. *Keep areas clean.* Frequent spills or releases not only present safety hazards, but also increase the amount of facility decontamination necessary at closure.

6. *Keep good records.* Industry-wide, a great deal of money is wasted on testing and disposing of unknown materials or in investigating areas with insufficient historical data. Good recordkeeping is essential to keep both current and future waste management costs to a minimum.

Excellent texts, which discuss in detail the many aspects of solid, industrial, and hazardous waste management, are available; these references are listed in the bibliography at the end of this chapter and can be consulted for specific information.

Bibliography

American Society of Civil Engineers. *Manual of Practice: Quality in the Construction Project—A Guide for Owners, Designers, and Contractors.* New York: 1988.

CELDS: United States Army Corps of Engineers, Construction Engineering Research.

Chanlett, E. T. *Environmental Protection.* 2nd ed. New York: McGraw-Hill, 1979.

Dunne, T., and L. D. Leopold. *Water in Environmental Planning.* San Francisco: Freeman, 1978.

Pruett, J. M. "Using Statistical Analysis for Engineering Decisions." *Plant Engineering* (May 13, 1976): 155–157.

U.S. Environmental Protection Agency. *Design Criteria for Mechanical, Electric, and Fluid System and Component Reliability.* EPA-430-99-74-001. Washington, DC: U.S. Government Printing Office, 1974.

U.S. Environmental Protection Agency. *ISO 14000 Resource Directory, National Risk Management Research Laboratory.* EPA/625/R-97/003. Cincinnati, OH: 1997.

Vesilind, P. A. *Environmental Pollution and Control.* 2nd ed. Ann Arbor, MI: Ann Arbor Science Publishers, 1983.

Willis, J. T., ed. *Environmental TQM.* 2nd ed. New York: McGraw-Hill, 1994.

2   Fundamentals

Although the laws and regulations that require industrial wastewater treatment are constantly changing, the fundamental principles on which treatment technologies are based do not change. This chapter presents a summarized version of the basic chemistry and physics on which treatment technologies are based, with the objective of showing that a command of these principles can enable quick, efficient identification of very effective treatment schemes for almost any given type of wastewater.

The fundamental idea upon which the approach suggested in this chapter is based can be stated as follows: if the mechanisms by which individual pollutants become incorporated into a waste stream can be identified, analyzed, and described, the most efficient methodology of removal, or "treatment," will be obvious.

As an example of the usefulness of this approach to quickly develop an effective, efficient treatment scheme, consider the leachate from a landfill that is to be pretreated and then discharged to a municipal wastewater treatment facility (a publicly owned treatment works, or POTW). The waste sludge from the POTW is disposed of by land application, therefore, a restrictive limitation is placed on heavy metals in the pretreated leachate. Analysis of this leachate shows that the content of iron is higher than permissable by the landfill. Other metals, such as cadmium (probably from discarded batteries), zinc, copper, nickel, and lead, may also be present in excess of the concentrations allowed by the landfill's pretreatment permit but are substantially lower than the iron content.

Knowledge of the following enabled quick conceptualization of a treatment scheme:

- All metals are sparingly soluble in water.
- Iron in the divalent state (2+) is highly soluble in water, whereas iron in the trivalent state (3+) is not.
- Iron can be converted from the divalent state to the trivalent state by passing air through the aqueous solution containing the dissolved iron. (The oxygen in the air oxidizes the ferrous (divalent) ion to ferric [trivalent] ion.)
- Because substances such as cadmium, zinc, and lead are so sparingly soluble, they will tend to adsorb to the surface of almost any solid particle in an aqueous solution.

In this scheme, the leachate was conveyed to a simple, open concrete tank, where air was bubbled through it. The soluble ferrous compounds in the leachate were oxidized to insoluble iron oxide; the precipitated iron oxide coagulated and flocculated because of the gentle mixing action of the air bubbles; and the dissolved species of other metals adsorbed to the iron oxide particles and co-precipitated with them. Next, the aerated leachate was allowed to settle, which effectively removed all of the heavy metals to within the limits of the pretreatment permit.

The following sections of this chapter have been developed to explain the fundamental chemical, physical, and thermodynamic principles by which pollutants become dissolved, suspended, or otherwise incorporated into waste streams to form homogeneous or heterogeneous mixtures. At the end of this chapter, several simple examples are given to further illustrate the usefulness of the technique, whereby fundamental concepts of chemistry and physics can be applied to efficiently deduce appropriate treatment schemes.

Electron Configurations and Energy Levels

According to the theory of quantum mechanics, an electron is most likely to be found in a region in space known as an orbital. Several "rules" govern the orbitals in which electrons will be located within an atom or a molecule. The first has to do with energy level. As atoms increase in size, the additional protons always reside in the nucleus, but the additional electrons reside in successively larger orbitals, which, in turn, exist within successively larger concentric shells. The electrons within orbitals that are closer to the nucleus are of lower energy level than those in larger orbitals. One of the strict rules of electron location is that no electron can occupy an orbital of higher energy level until all orbitals of lower energy are "full." A second rule is that only two electrons can occupy any atomic orbital, and these electrons must have opposite spins. These electrons of opposite spin are called "electron pairs." Electrons of like spin tend to get as far away from each other as possible. This rule, the Pauli exclusion principle, is the most

important of all the influences that determine the properties and shapes of molecules.

Figure 2-1, (a) and (b), shows two ways to depict a spherical electron orbital. Figure 2-1(a) presents the orbital as a spherical cloud surrounding the nucleus. Figure 2-1(b) is simply a convenient, two-dimensional representation of the orbital. Figure 2-2 shows the shapes of the two orbitals of lowest energy level, which are the two smallest orbitals as well.

Figure 2.2(a) shows that the smallest and, therefore, the lowest energy level orbital is designated the "ls" orbital and is approximately spherical. The center of the 1s orbital coincides with the center of the nucleus. Figure 2.2(b) shows that the next larger orbital is called the "2s" orbital and is also spherical, with its center coinciding with the center of the nucleus. Next in size (and energy level) are three orbitals of equal energy that have two approximately spherical lobes each and can thus be described as having shapes similar to dumbbells. These "2p orbitals" are shown in Figure 2-3(a), (b), and (c). Figure 2-3(a), (b), and (c) show that the three two-lobed orbitals are arranged so as to be as far away from one another as possible and are thus arranged such that the center of each lobe lies on one of three axes that are perpendicular to one another. The center of the atomic nucleus coincides with the origin of the three axes. The axes are referred to as the x-, y-, and z-axes, and the three orbitals are called the 2px, 2py, and 2pz orbitals.

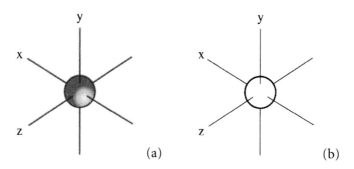

(a) (b)

Figure 2-1 (a) Spherical electron orbital as a spherical cloud surrounding the nucleus. (b) Two-dimensional representation of the electron orbital.

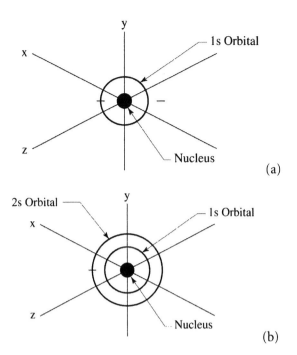

Figure 2-2 (a) 1s orbital. (b) 1s and 2s orbital.

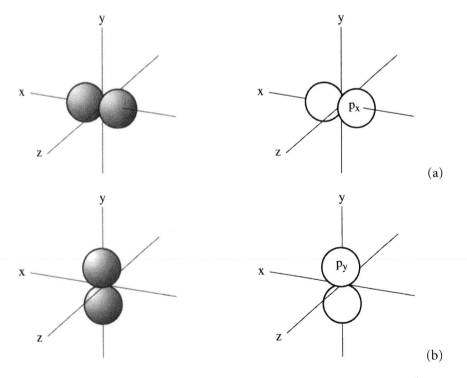

Figure 2-3 (a) 2px orbital. (b) 2py orbital.

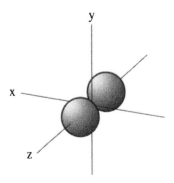

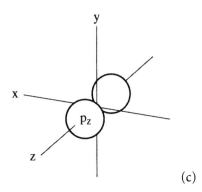

(c)

Figure 2-3   (c) 2pz orbital.

The electron orbitals exist within electron shells. The electron shells are concentric and are numbered 1, 2, 3, 4, and so on. The smaller shells, closer to the nucleus, must become full before electrons will be found in orbitals in higher (larger) shells. Shell 1 is full with one (1s) orbital. The total number of electrons in shell 1, when it is full, then, is two. Shell 2 is full when it contains four (one 2s and three 2p) orbitals. Thus, a full shell 2 contains eight electrons. Figure 2-4 shows an atom with the five orbitals of lowest energy level. The ls orbital resides within shell 1, and the 2s, 2px, 2py, and 2pz orbitals reside within shell 2. The center of the nucleus coincides with the origin of the three axes.

In the sections that follow, these orbitals will be discussed as they relate to the stability of molecules.

Electrical and Thermodynamic Stability

A fundamental and very important law of nature is that all elements tend toward ever-greater stability. Chemical elements are electrically stable when the number of electrons equals the number of protons. They are thermodynamically stable when their outermost, or largest, electron shell is full.

An element with more electrons than protons or more protons than electrons carries an

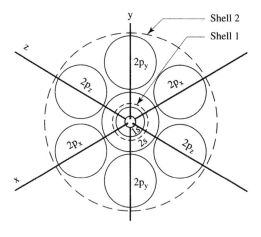

Figure 2-4   Arrangement of electron orbitals within shells 1 and 2.

electric charge and is referred to as an "ion." If electrons are in excess of protons, the charge is negative, and the element is said to be an anion. If protons are in excess of electrons, the charge is positive, and the element is said to be a cation. Both cations and anions are electrically active and are therefore said to be electrically unstable. Cations and anions are attracted to each other because of their opposite electric charges and thus form chemical compounds. When a cation and an anion join to form a chemical compound, the respective charges cancel each other out and they become electrically stable.

Because of a strong tendency to attain thermodynamic stability, atoms tend to either lose electrons until the next smaller shell (being full) becomes the outermost shell or gain electrons until the original outermost shell becomes full. Electrons in shells that are not full are called "valence electrons" and are active in chemical bonding. Electrons in full shells are essentially unreactive. The only role these electrons play in chemical bonding is that of insulating the valence electrons of other atoms from the attraction of the positively charged nucleus.

The elements at the far right of the periodic table have full outermost shells in their natural state. The number of electrons equals the number of protons. As they are both electrically and thermodynamically stable, these elements are unreactive. They are known as the "noble elements" or "noble gases."

Chemical Structure and Polarity of Water

Water molecules are polar in nature: similar to a magnet, one side has a positive "pole" and the opposite side has a negative "pole." This polarity arises from the spatial arrangement of protons and electrons in the individual hydrogen and oxygen atoms that make up each water molecule.

Before proceeding to an examination of the structure and electrically charged charac-

teristics of water, it is useful to examine the construction of hydrogen, oxygen, and the six elements that lie between them in size; it is also useful to consider, in a step-by-step way, how each successive proton and its associated electron influence the characteristics of each element.

Consider hydrogen first; it is the smallest of the elements. Hydrogen consists of one proton within a small, extremely dense nucleus and one electron contained within an orbital that is more or less spherical and surrounds the nucleus. Figure 2-5 is a two-dimensional portrayal of the three-dimensional hydrogen atom, but is sufficient to show that, at any given instant, the negatively charged electron is able to counteract the positively charged nucleus within only a small region of the space that the atom occupies.

Figure 2-5 illustrates that, if a charge detector could be placed near the hydrogen atom, it would detect a negative charge in the region near the electron and a positive charge everywhere else. The positive charge would register strongest in the region opposite in space to the region occupied by the electron. At any given instant then, a hydrogen atom is a polar object, having a negatively charged region and a positively charged region. In this sense, a hydrogen atom exhibits properties of a tiny magnet. However, since the electron is in continual motion, and at any given instant can be found anywhere within the approximately spherical orbital surrounding

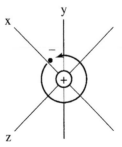

Figure 2-5   Diagram of hydrogen atom.

the nucleus, the net effect of an isolated hydrogen atom is to appear electrically neutral and not polar.

The element that is next larger in size to hydrogen is helium. Helium has two protons in its nucleus and two electrons in the single (1s) orbital that resides within shell 1. Helium is thus both electrically stable (two protons and two electrons) and thermodynamically stable (full outer shell); it is one of the noble elements. Helium also has two neutrons within its nucleus, which add mass to the atom but do not affect the reactivity of the atom.

The next element larger than helium is lithium. Lithium consists of three protons in the nucleus and three electrons. Two of the electrons reside within the 1s orbital within shell 1, and the third resides within the 2s orbital within shell 2. Lithium is an extremely reactive element because of its tendency to give up its third electron, thus leaving it thermodynamically stable with a full outer shell (the 1s shell) and a positive charge. When the lithium atom gives up its third electron, it is said to be a monovalent cation (1+).

Each of the next successively larger elements—beryllium, boron, carbon, nitrogen, and oxygen—has, respectively, one additional proton and one additional electron. While the protons are all in the nucleus, the electrons are in, successively, the 1s orbital within shell 1, and the 2s, then the 2p orbitals within shell 2. Beryllium has two electrons in shell 2, both residing within the 2s orbital. Although this outermost electron orbital is full, the outermost shell, shell 2, is not. Shell 2 is full when it contains eight electrons. Beryllium, therefore, strongly tends to lose two electrons, thus becoming a divalent cation (2+). Boron, the next larger element, must either lose three electrons or gain five to attain a full outer shell. Its tendency is to lose three rather than gain five. It therefore tends to become a cation but less strongly so than lithium or beryllium.

Carbon, the next larger element, has six protons and six electrons. Carbon must therefore lose four electrons or gain four to attain a full outer electron shell. In fact, carbon has almost no tendency to do either, but tends to attain a full outer shell by way of a process called electron sharing. Using the electron-sharing mechanism, carbon can enter into what is known as the covalent bonding process, as illustrated in Figure 2-6.

Figure 2-6 depicts the formation of methane, CH_4, which is the result of the covalent bonding of four hydrogen atoms

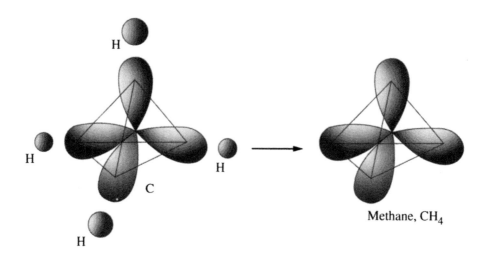

Methane, CH_4

Figure 2-6   Formation of a molecule of methane.

with a single carbon atom. Figure 2-6 shows a carbon atom with each of the four valence electrons arranged in space so that they are as far away from one another as they can be and still occupy the same electron shell. Some rearranging of orbitals is required to make this configuration possible. Because carbon has a total of six electrons, it is expected, on the basis of the foregoing discussion, that two reside as a pair within the 1s orbital, another pair resides in the 2s orbital, and one unpaired electron resides in each of two 2p orbitals; however, this arrangement would allow only two covalent bonds, yielding CH_2 if those bonds were with hydrogen atoms. This configuration would result in carbon having six electrons in its outer shell, whereas eight are required for a full outer shell. What actually happens, because of the very strong tendency for carbon to attain a full outer shell of eight electrons, is that the 2s orbital and the three 2p orbitals form four hybrid orbitals of equal energy level having one unpaired electron in each. These four hybrid orbitals are called sp3 orbitals and are shaped as shown in Figure 2-7. An sp3 orbital has two lobes of unequal size, and the nucleus of the atom resides between the two lobes. Figure 2-8 shows that the four hybrid sp3 orbitals of a carbon atom are arranged in such a way as to be as far from one another as possible. The carbon atom thus attains a shape such that, if straight lines were drawn to connect the outer limits of each lobe, a tetrahedron would be drawn. Thus, carbon atoms are said to exist in the shape of a regular (all sides equal in size) tetrahedron, with one valence electron available to form a covalent bond at each corner of the tetrahedron and with the center of the nucleus coinciding with the center of mass of the tetrahedron. Figure 2-9 shows a two-dimensional representation of Figure 2-6. Figure 2-10 depicts four hydrogen atoms, each of which needs one additional electron to fill its outer shell, combining with each of the four valence electrons of the carbon atom to form the

Figure 2-7   Configuration of a hybrid sp3 orbital.

methane molecule. Each of the four electron pairs is shared in a covalent bond, and the molecule exists as a cohesive unit. This is in contrast to compounds of lithium, for example, which are formed by ionic bonds. In a water solution, lithium gives up its one valence electron to another atom—chlorine, for instance—and exists as the discrete lithium ion, Li+, not physically attached to another ion. This is the principal difference between covalent bonds and ionic bonds.

Moving on to an examination of the characteristics of oxygen and then water, it is seen that oxygen has eight protons in its nucleus and eight electrons arranged such that two reside in the 1s orbital within shell 1, two are within the 2s orbital within shell 2, and four are within the 2p orbitals in shell 2, such that one 2p is full with a pair and the other two 2p orbitals have one unpaired electron each. Two additional electrons are needed to fill shell 2. To accomplish this, the oxygen atom forms a covalent bond with each of two hydrogen atoms, as illustrated in Figure 2-11.

Figure 2-11 shows the spatial arrangement of the six valence electrons of oxygen. Four of the six electrons are arranged into two pairs, and two are arranged so that each can participate in a covalent bond and thus be included in an electron pair. When the covalent bonds have been formed, the resulting four electron

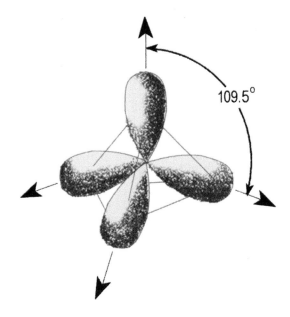

Figure 2-8   The four sp3 orbitals of an atom of carbon.

pairs arrange themselves so as to be as far away from one another as possible, resulting in another tetrahedral structure of four sp3 orbitals, similar to that of the carbon molecule. It is the mutual repulsion caused by like charges (negative) associated with each of the

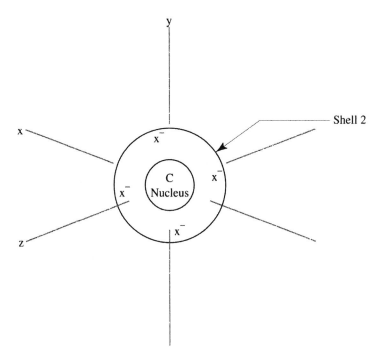

Figure 2-9   Two-dimensional representation of a carbon atom showing the four valence electrons.

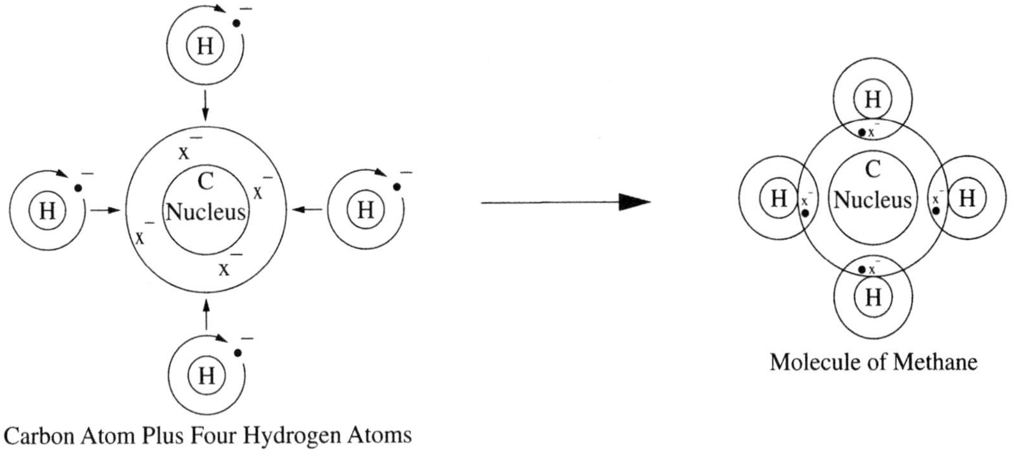

Carbon Atom Plus Four Hydrogen Atoms

Molecule of Methane

Figure 2-10   A carbon atom forms covalent bonds with each of four hydrogen atoms to produce a molecule of methane.

electron pairs that cause them to form the tetrahedral structure and thus be as far away from one another as possible while still occupying the same electron shell.

There is a difference between the tetrahedral structures of methane molecules and water molecules. In the case of methane, there are equivalent hydrogen atom–carbon valence electron structures at each of the four corners. In the case of water, two of the corners of the tetrahedral structure contain the

hydrogen-oxygen valence electron pair, and two contain simply an electron pair. There is equal repulsion between the electron pairs in the case of methane; therefore, a regular tetrahedron structure results. The structure of the methane molecule is almost perfectly symmetrical. In the case of water, however, there is not equal repulsion between the structures at the corners of the tetrahedral molecule, and a distorted tetrahedral structure results.

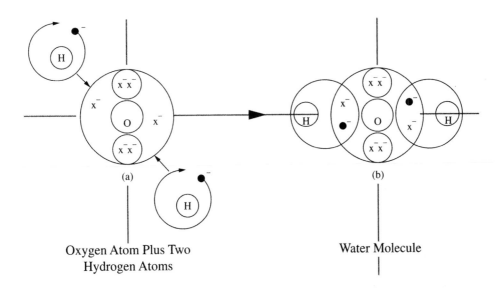

(a)

(b)

Oxygen Atom Plus Two
Hydrogen Atoms

Water Molecule

Figure 2-11   An oxygen atom forms covalent bonds with each of two hydrogen atoms to produce a molecule of water.

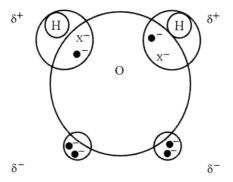

Figure 2-12   Relative position of electron pairs on a water molecule.

Figure 2-12 shows a slightly different two-dimensional representation of a water molecule. In the region of each of the two hydrogen atoms that are involved in the makeup of each water molecule, the electron pair that makes up the covalent bond spends most of the time in the space between the oxygen atom and the hydrogen's nucleus. The result is that the positive charge of each hydrogen nucleus is exposed. Figure 2-12 further shows that the two electron pairs that occupy the corners of the distorted tetrahedral structure of the water molecule not occupied by hydrogen atoms, display a negative charge. Therefore, each water molecule is seen to have regions of positive charge and regions of negative charge. Each water molecule thus exhibits characteristics of a tiny magnet, having a negative "pole" and a positive "pole," and is said to be a polar molecule. Methane, on the other hand, as described previously, has a tetrahedral structure where all four corners are identical. Methane is therefore said to be nonpolar.

The consequence of the polar versus nonpolar nature of molecules is well illustrated by the remarkable difference in physical state between water and methane. Water has a molecular weight of 18 and is liquid at room temperature. Methane has a molecular weight of 14, very close to that of water, but is a gas at room temperature. The reason that such a low-molecular-weight substance like water is liquid at room temperature is illustrated in Figure 2-13. Figure 2-13 shows that water molecules, acting like tiny magnets, attract each other, whereas methane molecules, having no polar properties, have no attraction for one another. The result in the case of water is that the space between molecules is relatively small, compared with the space between methane molecules at room temperature.

Hydrogen Bonding

The attraction between the positive areas on the hydrogen atoms of each water molecule and the negatively charged areas that result from the two unbonded electron pairs on the oxygen atom of each water molecule is referred to as "hydrogen bonding." Hydrogen bonding not only accounts for water existing as a liquid at room temperature, but also accounts for the capability of water to solvate, or dissolve, many substances; thus, its nickname is "the universal solvent."

Solutions and Mixtures

Industrial wastewater is the aqueous discard that results from substances having been dissolved or suspended in water, typically during the use of water in an industrial manufacturing process or the cleaning activities that take place along with that process. The objective of industrial wastewater treatment

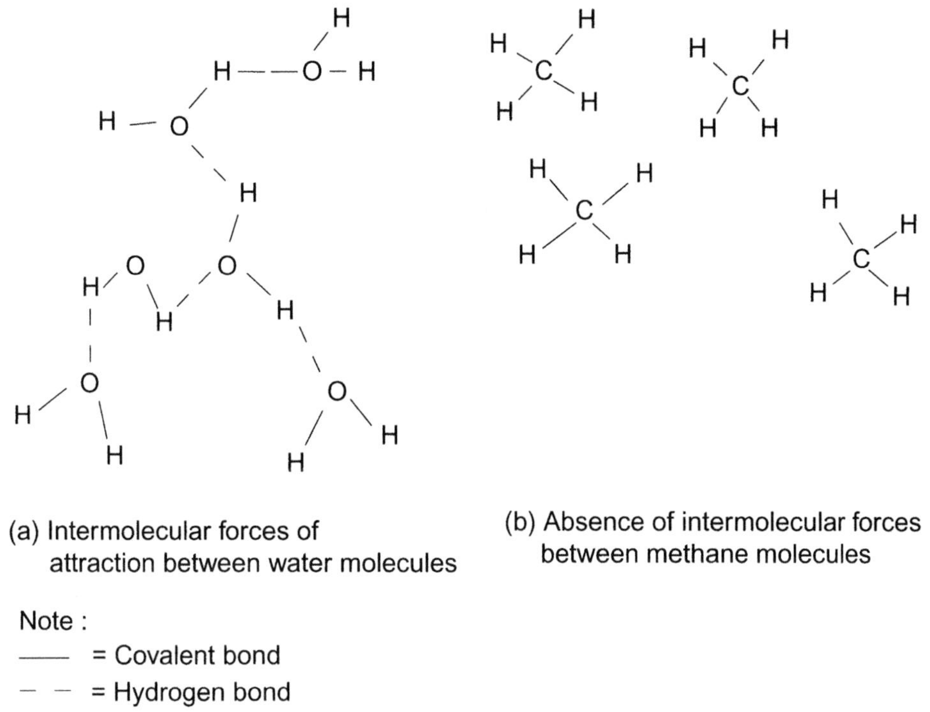

(a) Intermolecular forces of
attraction between water molecules

(b) Absence of intermolecular forces
between methane molecules

Note :
———— = Covalent bond
— — = Hydrogen bond

Figure 2-13 Strong attractive forces between molecules of water as compared with weak attractive forces between molecules of methane.

is to remove those dissolved or suspended substances. The best approach to working out an effective and efficient method of industrial wastewater treatment is to understand how substances are dissolved or suspended in water and then to deduce plausible chemical or physical actions that would reverse those processes. However, before we enter into a discussion on how solutions and mixtures are formed, it is important to understand the second law of thermodynamics, which is often the driving force behind these phenomena.

Second Law of Thermodynamics

The second law of thermodynamics states that, in nature, all systems tend toward a state in which free energy is minimized. The free energy of a system is given by the following expression, known as the Gibbs free energy equation:

$$G = H - TS \qquad (2-1)$$

where G is the free energy (often referred to as the Gibbs free energy) and is that portion of the heat content, or enthalpy, H, that is available to do work, isothermally; T is the absolute temperature; and S is entropy. Entropy can be described as the "degree of disorganization," that is, the more highly organized a system is, the lower its entropy. Consider a glass of water into which a teaspoon of sodium chloride (table salt) is placed. Sodium chloride exists as a crystalline solid in its pure form, and because it has a density greater than that of water, the crystals of sodium chloride will first sink to the bottom of the glass and sit there for a while. After a period of time, however, the salt crystals will "disappear," having gone into solution. The sodium chloride ionizes to sodium ions and chloride ions (cations and anions, respectively). Explanation of this phenomenon provides a convenient way to explain both the second law of thermodynamics and the Gibbs free energy equation.

When the salt is on the bottom of the glass, the system is highly organized. The salt is in one place and the water is in another, although they are both in the same container. The total enthalpy, H, of the system includes all of the bond energies within the system. This includes all of the ionic bonds within the sodium chloride crystals and all of the bonds within the water molecules and the hydrogen bonds between water molecules. Over a period of time, the negatively and positively charged sites on water molecules surround individual sodium and chloride ions, respectively, until they are all dissolved and evenly dispersed throughout the volume of water in the glass. When this process, illustrated in Figure 2-14, is complete, entropy is maximized (disorganization of the system is complete; everything in the glass is randomly distributed), and the sum of all the chemical bond energies within the system is greater than it was before dissolution took place. That is, the sum of all the bond energies between water molecules and chloride ions (hydrogen bonds), plus the bonds of electrical attraction between water molecules and sodium ions, plus the bonds within the water molecules themselves, are greater than the sum of all the bond energies within the system when the salt was a solid on the bottom of the glass.

Solutions

Solutions are formed when one or more substances, known as the solutes, dissolve in another medium, known as the solvent. When dissolved, the solutes, in accordance with the second law of thermodynamics, become distributed uniformly throughout the volume of the solvent. As illustrated in Figure 2-14, substances that ionize into cations and anions are soluble in polar solvents, such as water. Substances that do not ionize, such as oil, are poorly soluble (everything is soluble, to some degree, in water) in polar solvents, but are highly, and in many cases infinitely, soluble in nonpolar solvents. Substances that are polar in nature and therefore soluble in water are said to be "hydrophilic." The mechanism of solubility of sodium chloride in water, as described previously, illustrates how hydrophilic substances can become dissolved in water. Nonpolar substances are only sparingly soluble in water and are said to be "hydrophobic."

Thus, it is seen that water dissolves substances that are capable of ionizing or otherwise exhibiting an electric charge, either positive or negative. The resulting mixture of water and dispersed ions is referred to as a "true solution." Liquids such as light mineral oil, which are made up of molecules formed by the covalent bonding process, are hydrophobic and form true solutions most easily in nonpolar solvents.

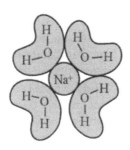
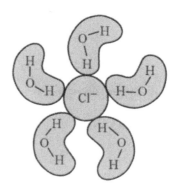

Figure 2-14   Depicts the soluble behavior of sodium chloride in water.

Emulsions

Hydrophobic substances can be induced to go into a state that is equivalent in many ways to a water solution, by a process known as emulsification. An emulsion is equivalent to a solution in that it consists of a stable mixture that will not separate under quiescent conditions (except in the case of temporary emulsions). A sample taken from anywhere within the volume of the mixture is representative of the mixture as a whole, and as the mixture flows from one place to another, it does not change in character.

Pure hexane is a liquid that is sparingly soluble in water. It is composed of a six-carbon chain with hydrogen atoms bonded by covalent bonds at all bonding sites other than those involved in the six-carbon chain. Figure 2-15 depicts a hexane molecule.

Hexane has a molecular weight of about 86 and exists as a liquid at room temperature. Very small electrical attractive forces exist between molecules, resulting from the fact that, at the site of any given hydrogen atom bonded by covalent bonds to a carbon atom, when both of the electrons involved in the bond are located between the hydrogen and the carbon, the nucleus of the hydrogen presents a positively charged site on the molecule. Figure 2-16 depicts this state.

In fact, the two electrons involved in each hydrogen-carbon bond are in continuous orbital motion around the hydrogen nucleus, so the time during which the positively charged nucleus is exposed is very brief and intermittent, and the strength of the charge is relatively weak. In exactly the same fashion, when one or both of the electrons involved in any given hydrogen-carbon bond are on the

$$C$$
$$..$$
$$H$$
$$\delta +$$

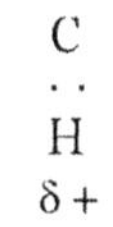

Figure 2-16   Carbon-hydrogen covalent bond exhibiting positive charge.

side of the hydrogen nucleus that is away from the carbon atom, a negative charge is presented to the surrounding environment. At any given instant, then, there is a probability that on each hexane molecule there will be one or more positively charged sites and one or more negatively charged sites, and there will be an electrical or magnetic-type attraction between hexane molecules. These charged sites are much weaker than those involved in the hydrogen bonding that is characteristic of water, however; and for this reason, hexane is not soluble in water.

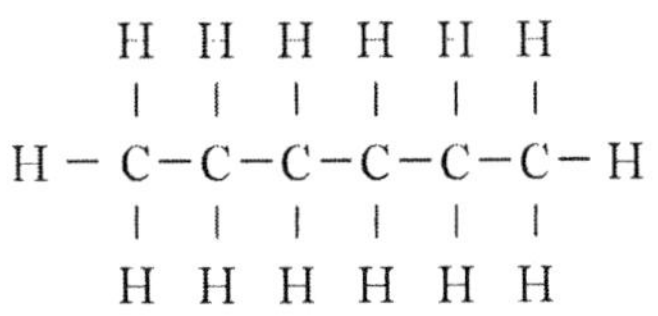

Hexane can be emulsified in water, however, and the result is, for practical purposes, equivalent to true solution as far as industrial waste is concerned. There are at least two ways to accomplish emulsification. One way is to use an emulsifying agent, such as a detergent. The second way is to mix hexane and water together vigorously. The mechanisms of these two methods for forming an emulsion are explained as follows.

Forming an Emulsion with an Emulsifying Agent

Figure 2-17 presents a representation of a typical detergent.

The synthetic detergent molecule shown in Figure 2-17 consists of two active components: a group that will ionize in water, in this case the sulfonate group (-S=O-); and a group that is nonpolar in nature and is therefore attracted to and soluble in organic material, such as fats, oils, and greases. Figure 2-18 depicts the process by which detergent molecules form a link between water and substances that are not soluble in water (note—dirt usually adheres to a thin film of oil on skin, clothing, etc.).

```
    H  H  H  H  H  H
    |  |  |  |  |  |
H — C — C — C — C — C — C — H
    |  |  |  |  |  |
    H  H  H  H  H  H
```

Figure 2-15   A single molecule of hexane.

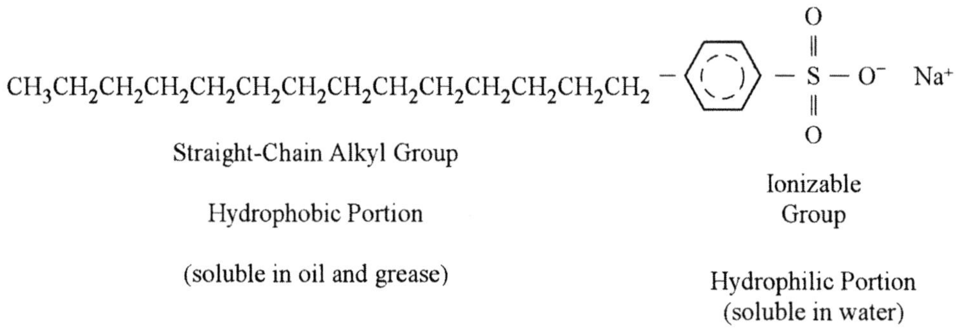

$$CH_3CH_2CH_2CH_2CH_2CH_2CH_2CH_2CH_2CH_2CH_2CH_2CH_2CH_2 - \langle \bigcirc \rangle - \overset{\overset{O}{\parallel}}{\underset{\underset{O}{\parallel}}{S}} - O^- \quad Na^+$$

Straight-Chain Alkyl Group

Hydrophobic Portion

(soluble in oil and grease)

Ionizable
Group

Hydrophilic Portion
(soluble in water)

Figure 2-17   A molecule of alkylbenzenesulfonate.

The organic, nonpolar portion of the detergent molecule clings to the oily particle. The particle of "dirt" also clings to the oily particle. The forces involved include the strong hydrophobic nature of the dirt, the oil, and the nonpolar portion of the detergent molecule. The strong hydrophobic nature of the oil and dirt particles results in the medium of water molecules forcefully excluding these substances from the bulk solution. The free energy of the system is reduced when hydrogen bonds between water molecules are formed. No bonds of any type will form between molecules of water and molecules that make up the dirt or oil particles.

In summary, the organic portion of the detergent sticks to the oil or dirt; the ionized inorganic portion dissolves in water; and the detergent thus forms a link between the two.

The second law of thermodynamics is thus seen to be the driving force in the process of cleaning with water and a detergent. Ordinary soap, of course, cleans in exactly the same way as synthetic detergents. The principal differences between soaps and detergents are that the hydrophobic, organic (nonpolar) portion of a soap is usually a more simple organic compound than that of a typical detergent; also, the hydrophilic (ionizable) portion of a typical detergent is characteristically the salt of a strong acid (sodium sulfonate, for instance), whereas it is usually the salt of a weak acid (sodium carbonate, for instance) in the case of soaps.

Figure 2-19 shows how oil and other nonpolar substances can be held in a solution-like suspension, referred to as an emulsion, by an emulsifying agent such as a synthetic detergent.

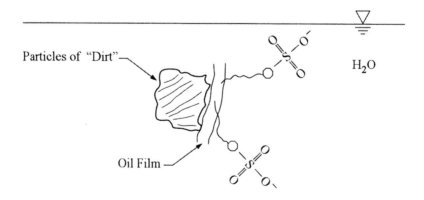

Figure 2-18   Detergent molecule forming a link between water and oily particle with attached "dirt."

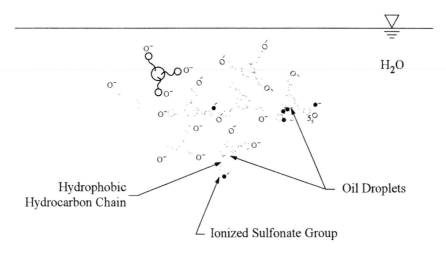

Figure 2-19   Emulsion by use of an emulsifying agent.

Forming an Emulsion by Vigorous Mixing

Figure 2-20 shows an emulsion of a light oil in water. By vigorous mixing, as with a kitchen blender, the oil has been separated into very fine droplets. There is an electric charge on the surface of each oil droplet. This charge can be either positive or negative, depending on the type of oil and whichever ions or other substances exist in the water that adhere to the surface of the oil droplets, but it is almost always negative.

The fact that the charge is of the same type on all droplets causes mutual repulsion between the droplets. It is this mutual repulsion that is the source of stability of the emulsion. There are at least two forces at work that tend to destabilize the emulsion and cause the oil and the water to separate into distinct phases. Gravity is one force. If the oil is less dense than water, gravity tends to cause the water to sink below all of the oil, resulting in the oil forming a separate layer on top of the water. Gentle mixing tends to cause the droplets of oil to collide, whereupon the droplets would coalesce until finally the oil would form a completely separate phase over or under the water, depending on its density. If the mutually repulsive forces caused by similar surface charges on the droplets are sufficiently strong, the destabilizing forces will not prevail, and the emulsion will remain stable.

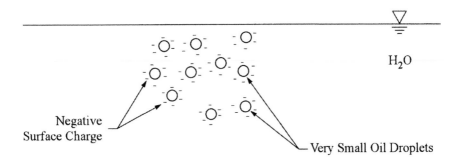

Figure 2-20   Emulsion by vigorous mixing.

Colloidal Suspensions

A third type of industrial wastewater mixture, in addition to true solutions (pollutants dissolved in water) and emulsions (pollutants suspended in water by emulsification), is the colloidal suspension. In all three types of wastewater mixtures, the pollutants are held in the water medium by electrical forces. The forces are those of attraction in the case of true solutions and are those of repulsion in the case of both emulsions and colloidal suspensions. In fact, colloidal suspensions are essentially identical to emulsions formed by vigorous mixing, described in the previous section, in that the source of stability for the mixture, mutual repulsion by like electric charges, results from dissociated bonds (i.e., the resulting sites of attraction and/or cohesion). Figure 2-21 illustrates the source of stability of colloidal suspensions of clay. Figure 2-21 shows the physical structure of clay as an indefinitely extended sheet of crystalline hydrous aluminum silicate. Many of the chemical bonds within the crystal lattice structure are covalent. This arrangement of silicon, aluminum, oxygen, and hydrogen atoms results in a relatively strong negative charge on each flat surface of the "indefinitely extended sheet." This charge attracts cations, such as magnesium, aluminum, ferrous and ferric iron, potassium, and so on, and these cations attract individual sheets together to make up an indefinitely extended three-dimensional structural mass of clay.

When the clay is pulverized into very small (colloidal-sized) particles, the pulverizing process amounts to the dissociation of countless covalent bonds, whereby each broken bond results in a site having a negative charge. Each of these sites contributes to an electric charge surrounding the surface of each particle, and if the particles are mixed into water, the particles repel each other. After the mixing has taken place, three important forces act on the suspension: (1) gravity acts to cause the particles to settle to the bottom of whatever contains the suspension; (2) Brownian and other forces, referred to as "thermal agitation," keep the particles in ceaseless motion, tending to make them collide (these collisions, if successful, would result in coalescence, reversing the dispersal process); and (3) repulsive forces caused by like electric charges on the surface of each particle tend to prevent collisions and even settling caused by the force of gravity. If the repulsive forces are strong enough to overcome the forces of gravity as well as the forces of thermal agitation, then the particles will be successfully held away from each other, and the colloidal suspension will be stable.

The principal difference, therefore, between an emulsion and a colloidal suspension as they relate to industrial wastes is that

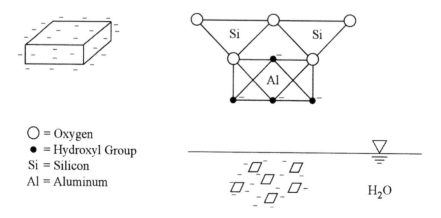

O = Oxygen
● = Hydroxyl Group
Si = Silicon
Al = Aluminum

H₂O

Figure 2-21   Source of the negative charge that is present on the surface of particles of clay.

the suspended substance, or "pollutant," in an emulsion is a liquid under ambient conditions, whereas the pollutant in the case of a colloidal suspension is a solid. Substances in addition to clays that form colloidal suspensions include those substances that do not readily dissolve into ions in water and are therefore "insoluble" (meaning only sparingly soluble, or very poorly soluble), but they can be pulverized into very small particles having a surface charge. Almost any substance that exists as a solid at room temperature, but will not dissolve to any great extent in water, can be made to form a colloidal suspension.

Mixtures Made Stable by Chelating Agents

There are cases when mixtures are cannot be emulsified, nor are they a colloidal suspension. A good example of this is when metal ions are present in water. While most metal ions are soluble in water within certain pH ranges, they are quite insoluble outside the appropriate pH range. Even within the optimum pH ranges for solubility, however, most metals are soluble to only a limited extent. Certain chemical agents, called "chelating agents," are able to "hold" metal ions in "solution" over broad ranges of pH, inside as well as outside the optimum pH ranges for solubility, at concentrations far in excess of their solubility limits. The name "chelating agents" is derived from the Greek word "chele," which means "claw." These agents have physical structures that accommodate or "fit" the metal cations like an object in the grip of a claw, and thus they "seize," or "sequester," the metal cations to prevent them from forming insoluble salts or hydroxides with anions or from entering into ion exchange reactions. Chelating agents can be inorganic in nature—for instance, polyphosphates (the active ingredient in some commercial scale and corrosion inhibitors, e.g., Calgon)—or they can be organic—for instance, EDTA (ethylenediaminetetraacetic

acid). Typically, a chelating agent consists of "ligand" atoms such as oxygen, nitrogen, or sulfur, which have two electrons available to form a "coordinate" bond with the metal ion. The ligand portion of the chelating agent bonds the cation, and the "claw" structure protects the cation from other chemical influences.

Summary

Wastewater can be described as a mixture of undesirable substances, or "pollutants," in water. If the mixture is stable, the pollutants will not settle out of the water under quiescent conditions under the influence of gravity; one or more treatment processes, other than simple sedimentation, must be used to render the water suitable to be returned to the environment. The key to determining an efficient, effective treatment process generally lies in the ability to recognize which forces are responsible for the stability of the mixture.

This chapter has described five general types of mixtures: (1) true solutions, where the stability arises from hydrogen bonding between water molecules and the electrical charge associated with each ion; (2) emulsions caused by emulsifying agents, where stability is provided by an agent, such as a detergent, that links small droplets of a liquid substance to water by having one portion of the agent dissolved in the water and another dissolved in the droplets of suspended liquid pollutant; (3) emulsions in which the stability of a mixture of small droplets of a liquid pollutant in water arises from the repulsion caused by like electric charges on the surface of each droplet; (4) colloidal suspensions in which small particles of a nonsoluble solid are held away from each other by the repulsive forces of like electric charges on the surface of each solid particle; and (5) solutions in which ions that would not normally be soluble in water under the prevailing conditions are held in solution by so-called chelating agents.

In each of the five cases, the most efficient way to develop an effective treatment scheme is to directly address the force responsible for the stability of the wastewater mixture.

Examples

The following descriptive examples are intended to show how a working knowledge of the foregoing fundamental concepts can be used to quickly deduce the proper technologies for efficiently treating different types of industrial waste streams.

Poultry Processing Wastewater

The processing of poultry involves receiving live birds and preparing them for retail sale. As shown in Figure 2-22, a typical poultry processing operation includes at least six operations in which water is used, contaminated, then discharged, including washdown for plant cleanup. Several of the operations involve water having direct contact with the birds as they are being processed (defeathered, washed, eviscerated, washed again, chilled, then cut up if desired), leading one to expect that any and all of the constituents of

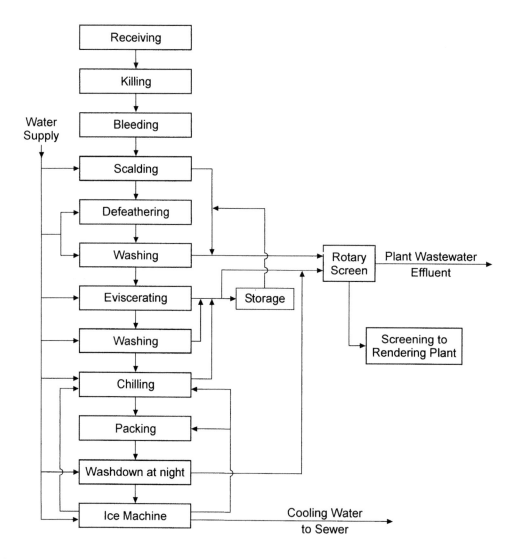

Figure 2-22 Schematic of a poultry processing system.

chicken flesh and blood will be found in the wastewater. These substances would be expected to include blood cells, fats and oils, protein materials, and suspended solids of various makeup, including dirt.

Most of the materials in the wastewater would be expected to be organic in nature and therefore biodegradable. For this reason, a biological treatment process, aerobic or anaerobic, would appear to be a good choice for treatment. Laboratory analysis of the combined wastewater from a typical poultry processing plant, however, as shown in Table 2-1, reveals that the oil and grease content is relatively high. Also, it is reasonable to expect that much of the biochemical oxygen demand (BOD) would result from the presence of blood. Poultry fat is known to be relatively soft and therefore more soluble, especially in hot water, than other animal fats; it would likely exist in "solution" as a colloidal suspension, or emulsion. The pertinent indication here is that the fats, oil, and grease (FOG) component of poultry processing wastewater would be relatively slowly degraded by a biological treatment process. On the other hand, a combined physical/chemical treatment process should work well, because surface chemistry holds the FOG material in suspension. A gravity sepa-ration process should work well, once the suspension is destabilized, by taking advantage of surface chemistry phenomena.

A second major component of poultry processing wastewater is blood. This is obvious from Figure 2-22, which shows that two of the major steps in the processing of poultry are killing followed by bleeding and that equipment used in these steps is cleaned by washing with water each day. Because blood contains red blood cells, which can be described as large particles that will not truly dissolve in water but will disperse uniformly throughout a volume of water in which they are placed, the conclusion is that blood must form a stable suspension in water. The stabilizing force in the suspension of blood is most likely electric surface charge; therefore, chemical coagulation would quickly come to mind as an efficient method for its removal.

The previous considerations, taken together with the knowledge that (1) poultry fat has a specific gravity less than 1.0, and (2) red blood cells have a specific gravity very close to that of water, indicate that chemical coagulation followed by dissolved air flotation (DAF) should be a very promising candidate technology for treatment of poultry processing wastewater.

Table 2-1   Characteristics of Raw Wastewater Poultry Processing[*]

Sampling Point	Flow (MGD)	BOD (lb/day)	SS (lb/day)	Grease (lb/day)
Main Drain	0.052	430	460	40
New York Room (excluding feather flush)				
Main Drain	0.448	1,680	1,140	850
Processing Room				
Evisceration	0.30	910	930	520
Carriage Water				
Chiller Overflow	0.05	240	150	90
Chiller Dump	0.09	440	270	115
Scalder Overflow	0.027	200	210	5
Scalder Dump	0.015	110	80	2
Washdown (night)	0.36	240	810	750
Total for Plant	0.860	2,850	2,410	1,640

[*]Data based on a 34-day average; production averaged 72,000 broilers/day.

Wastewater from Metal Galvanizing

Wastewater from a metal galvanizing process can be expected to contain dissolved zinc. A quick check of the properties of various compounds reveals that zinc hydroxide, zinc sulfide, and zinc phosphate are all very poorly soluble in water within certain ranges of pH. Therefore, treatment of the wastewater using caustic, a sulfide compound, or a soluble compound of phosphate would probably be viable candidates for treatment of the wastewater. However, zinc phosphate is seen to be the least soluble in the acid pH range; therefore, it would likely produce a sludge that would be more stable than either the hydroxide or the sulfide. For this reason, wastewater from a galvanizing process would seem to be best treated by addition of a soluble compound of phosphorus (phosphoric acid, for instance, or super phosphate fertilizer). A further check would reveal that the phosphate compounds of all metals likely to be present in wastewater from a galvanizing process (iron, lead, nickel, and tin) are all very poorly soluble in water of low or neutral pH ranges, further indicating treatment with phosphorus as a likely optimal treatment approach.

Removal of Heavy Metals from Industrial Wastewater Containing Dissolved Iron

Many industrial wastewaters contain iron as well as other heavy metals, all of which must be removed to one degree or another. It is often possible to take advantage of two characteristics of metal ions: the first being specific to iron, and the second being common to many metal ions.

Iron is soluble in water in its plus-two valence state (ferrous form) but is insoluble in its plus-three valence state (ferric form). Moreover, the ferrous form can readily be oxidized to the ferric form by contact with atmospheric oxygen (air). Therefore, it is easy to remove iron from wastewater by simply bubbling air through it (which accomplishes the conversion of the dissolved ferrous ions to insoluble ferric ions); then providing a slow mix time of 5 or 10 minutes to allow flocculation of the ferric oxide particles, then passing the mixture through a gravity separation device, such as a tube or plate settler.

It is a characteristic of nearly all heavy metal ions that they are only weakly soluble in water and tend to adsorb onto the surface of small particulates in suspension whenever they are available. Iron flocs, precipitated by bubbling air through wastewater containing dissolved ferrous ions, present just such a surface.

On the basis of the above two characteristics of metal ions, an extremely simple and inexpensive system, consisting of an aeration step, a flocculation step, and a sedimentation step, can be used to remove the ions of many metal species to concentrations of between 2 and 6 mg/L. This often a very worthwhile pretreatment prior to ion exchange or any other more expensive process; it can then be used to polish the wastewater to very low concentrations suitable for discharge. The inexpensive pretreatment process can greatly prolong the life of the more expensive process, resulting in a lower total cost.

Wastewater from a Parts Cleaning Process

Wastewater from a parts cleaning process within an automobile gear manufacturing facility was shown to contain a high concentration of oil, but the oil could not be removed in a standard API separator (plain sedimentation device). It was determined that treatment of this waste stream at the source was desirable because the FOG load contributed by this process to the total waste stream from the manufacturing facility was so great as to be a cause for failure of the main wastewater treatment facility. Investigation of the parts washing process revealed the presence of a detergent of the sulfonate type, giving rise to the thought that addition of calcium chloride, an inexpensive substance, would result in the formation of insoluble calcium sulfonate at the water-soluble sulfonate site on the detergent molecule. This caused the oil emulsion, which was stabilized by the detergent, to "break," allowing the droplets of oil to coalesce and then rise to the surface in the American Petroleum Institute (API) separator, enabling easy removal.

Air Pollution from a Trash Incinerator

A trash incinerator burning a mixture of industrial and municipal solid wastes was equipped with a full complement of scrubbers, electrostatic precipitators, bag house filters, and so on, but was discharging heavy metals in excess of permit limitations. Several different chemicals had been added to the scrubber solutions, including chelating agents, but none with success. On the basis that metal phosphates, in general, are extremely insoluble in water, it was theorized that if an inexpensive source of soluble phos-phates were added to the scrubber solution, the phosphates would react with the metals to form insoluble precipitates. The scrubber water could then be recycled with only a relatively small blow-down, and the metals would not pass out of solution and back into the air being discharged. Ordinary triple super phosphate fertilizer, as used by farmers in the area, was determined to be ideal for this purpose.

Bibliography

Seinfeld, J. H. *Air Pollution: Physical and Chemical Fundamentals.* New York: McGraw-Hill, 1975.

Streitwieser, A., Jr., and C. H. Heathcock. *Introduction to Organic Chemistry.* New York: Macmillan, 1985.

3 Laws and Regulations

Introduction

The plethora of laws, rules, regulations, ordinances, and restrictions that regulate the discharge of industrial liquid wastes, the management and disposal of industrial solid wastes, and the emission of airborne pollutants present a truly formidable challenge to anyone attempting to become knowledgeable in the area of environmental compliance. These environmental laws and regulations are published in hundreds of documents; and new laws and regulations are written, passed, and promulgated every year by dozens of local, county, state, and federal legislative bodies and executive rule-making agencies. The great majority of the presently enforceable laws and their implementing regulations have been promulgated during the past 35 years.

History of Permitting and Reporting Requirements

Most states have had regulations forbidding the "pollution" of surface water and groundwater since well before the 1950s. For instance, the Pennsylvania State Legislature passed the Clean Streams Act in 1937, which created the Sanitary Water Board and empowered it to administer the law as interpreted by the Board and implemented by a "bureau." The Clean Streams Act expressly prohibited the discharge of industrial wastes to "waters of the Commonwealth," which included groundwater as well as surface water. "Industrial waste" was broadly defined as any liquid, gaseous, or solid substance, not sewage, resulting from any manufacturing or industry. As one of its first items of business,

the Board issued a number of rules, including a requirement that all facilities for industrial waste treatment apply for and be granted a permit before commencing construction. The Pennsylvania Department of Health served as administrative and enforcement agent for the Sanitary Water Board.

Other states had similar laws, which, essentially, made it illegal for an industry to discharge wastes in such a way as to cause the receiving water to be unhealthful or unusable. The State of Maine promulgated a set of laws in 1941 that created a Sanitary Water Board and made it illegal to pollute waters used for recreation, among other restrictions. Maine's law, in fact, was similar to Pennsylvania's law regarding the intent to protect waters of the state. As further examples, the State of Illinois enacted legislation in 1929 that "establish[ed] a sanitary water board to control, prevent, and abate pollution of the streams, lakes, ponds, and other surface and underground waters in the state." South Carolina enacted a statute in 1925 that made it, "unlawful for any person, firm, or corporation to throw, run, drain, or deposit any dye-stuffs, coal tar, sawdust, poison, or other deleterious substance or substances in any of the waters, either fresh or salt, which are frequented by game fish, within the territorial jurisdiction of this State, in quantities sufficient to injure, stupefy, or kill any fish or shellfish, or be destructive to their spawn, which may inhabit said waters."

In fact, almost all the states had laws similar to those of these four states prior to 1950. However, because each of the states' environmental regulatory agencies had limited resources, and these limited resources were focused, for the most part, on wastes from municipalities, little (but nonetheless some) enforcement against industries took place.

Involvement of the federal government in prosecuting industries for pollution has a remarkably limited history prior to the 1970s and the creation of the U.S. Environmental Protection Agency (the EPA) in 1970. Prior to 1948, when Congress passed the Federal Water Pollution Control Act (FWPCA), the only legislation under which a discharger of pollutants could be prosecuted was the Refuse Act of 1899 (Section 13 of the Rivers and Harbors Act). In 1956, Congress enacted the FWPCA, thus establishing the first federal water quality law. This law was amended several times, including the Water Quality Act of 1965, the Clean Water Restoration Act of 1966, and the Water Quality Improvement Act of 1970. Here, again, however, despite the reality of federal laws, actual prosecution of industries for polluting was very limited before the 1970s.

Development of the intense regulatory climate that industrialists must operate within during the 2000s began with passage of the 1972 amendments to the Clean Water Act, Public Law 92-500 (PL 92-500), which replaced the entire language of the original 1956 Water Quality Act, including all amendments. This comprehensive legal milestone had the broad objective of returning all bodies of water in the United States to a condition in which they could be safely and enjoyably fished and swum within a few years of passage of the law itself. Specifically, a national goal was established to eliminate the discharge of pollutants into the navigable waters by 1985. A national interim goal was established that, wherever attainable, water quality would provide for the protection and propagation of fish and provide for recreation in and on the water by July 1, 1983. Many rivers, streams, and lakes in 1972 were virtually open sewers, and it was rare that any significant water body within the bounds of civilized development was not polluted, owing to the lack of enforcement of existing laws at the federal and state level.

As an example of the prevailing attitude toward clean water during the 1940s, a footbridge was built across a major river so workers could walk from a town to a large integrated pulp and paper mill in the Northeast. A garbage chute was built into the side of the footbridge as a convenience to these workers so they could carry bags of household trash and garbage on their way to work and simply drop them into the river.

There is, however, a tendency to overextrapolate the apparent indifference toward pollution of the environment that prevailed prior to the 1970s. Synthetic organic chemicals with a high degree of toxicity were not widely available and not widely recognized as a threat to environmental quality before the 1960s. The prevailing thought was that all garbage and other household wastes were biodegradable and would ultimately "disappear" by, eventually, simply being incorporated into the environment "from whence it came." The Superfund sites and other problems of gross environmental degradation of the 1980s and 1990s for the most part involved the historical release and disposal of synthetic chemicals and other hazardous substances such as PCBs, pesticides, herbicides, and other chlorinated hydrocarbons.

In addition to the unacceptable state of the nation's waterways, a significant portion of the nation's groundwater became contaminated by the unregulated disposal of solid and liquid wastes in open, unlined dumps. Much of this material was toxic. Beginning in the 1980s, a number of lawsuits and federal and state environmental enforcement actions were brought against industries whose waste disposal practices resulted in contamination of groundwater supplies that served entire cities.

The uncontrolled emission of air pollutants has also degraded the quality of air in the United States to the point where significant numbers of people with respiratory illnesses have been pushed to the point of death, and many thousands of others have suffered health impairment. Visibility has become substantially reduced in many regions of the country. It has even reached the point where nations have threatened legal action against other nations because massive quantities of air pollutants have crossed international boundaries. As with the quality of the nation's water resources, a low point was reached during the 1970s and 1980s. Due to the installation and operation of air pollution control technologies at industrial facilities and in motor vehicles, the overall quality of air in most regions of the United States has improved. There are still many regions of the United States, however, that are considered to be in "nonattainment" of specified national air-quality standards, and thus the business of air-quality protection is a robust one.

Water Pollution Control Laws

The foundation of all federal, state, and local water pollution control laws and regulations continues to be PL 92-500, the 1972 amendments to the FWPCA, commonly referred to as the Clean Water Act. Prior to passage of this watershed legislation, water pollution control laws were based on an approach that focused on water-quality standards and effluent limitations tailored to those standards. The degree of treatment required of a given industrial discharger depended upon the assimilative capacity of the receiving water body. The foundation of this approach was that the "solution to pollution is dilution," and the amount of dilution available was the basis of the degree of treatment required. One doctoral thesis, written during the late 1960s, used dissolved oxygen monitors in a river to control a valve to regulate the rate of discharge from an industrial facility. The flow of treated industrial wastewater to the river was from a reservoir into which the effluent from the treatment plant flowed. Treated wastewater would build up in volume in the reservoir when the allowable discharge to the river was low, and the reservoir would be discharged into the river during

periods of times when the allowable discharge rate (as indicated by a relatively high concentration of dissolved oxygen in the river) was high.

The approach used in developing modern environmental policy, which is found in the Clean Water Act and its implementing regulations, was to require equal treatment by all dischargers, regardless of the assimilative capacity of the receiving water. The assimilative capacity of the receiving water took on a different role, which was that of an indicator of minimum degree of treatment required, as opposed to its former role as an indicator of maximum degree of treatment required. The modern use of mathematical water-quality models is to determine whether the "categorical pretreatment standard" for a specified industrial waste stream will ensure that the water body never violates legislated water-quality standards. If the model indicates a potential water violation, additional treatment is typically required. Accordingly, under the Clean Water Act, no pollutant may be discharged that would interfere with the attainment or maintenance of water-quality standards or cause the water body to fail to attain its designated uses.

PL 92-500 has been amended several times since 1972; it is a comprehensive law that is difficult to summarize in a few pages. Some of the more important requirements are identified in the following statements:

- All dischargers must have a permit under the National Pollutant Discharge Elimination System (NPDES). The permit has three major parts: (1) effluent limitations, (2) compliance schedules, and (3) monitoring and reporting requirements.
- All dischargers must meet applicable requirements for: effluent limitations; standards of performance; toxic and pretreatment effluent standards; and monitoring and self-reporting (discharge monitoring reports or DMRs).
- Each industry has been classified under the Standard Industrial Classification

(SIC). One of the principal purposes of the CWA is to maintain a baseline set of minimum requirements among industries of the same type, so as to prevent a competitive disadvantage on the part of any individual industrial plant. The EPA has developed national standards of performance for most industrial categories; these standards are designed to reflect the application of the best-available demonstrated control technology.

- Individual states are authorized to adopt and enforce their own laws to regulate discharge of pollutants, which may be more restrictive than the corresponding federal effluent limitation, standard of performance, or toxic pretreatment standard in effect.
- Within most every SIC category, a study, leading to production of a guidance document called a "development document," has been carried out. Each development document identifies typical characteristics of wastewater from that industry, discusses methods for minimizing waste production, and defines achievable as well as acceptable levels of wastewater treatment.
- Penalties for noncompliance are set forth.
- Industries that discharge to publicly owned treatment works (POTWs) are regulated by the so-called *Pretreatment Regulations*. The EPA's General Pretreatment Regulations are promulgated under *Title 40 of the Code of Federal Regulations (C.F.R.), Part 403*, and the *Categorical Pretreatment Regulations* are found at *40 C.F.R., Parts 405* through *471*.
- Receiving waters for which the provisions of PL 92-500 are applicable include all rivers, streams, brooks, creeks, lakes, ponds, bogs, swamps, and territorial seas within the three-mile limit, as well as wetlands, drainage ditches, and intermittent streams. In short, if a receiving body of water has a free surface that is open to

the atmosphere, it is wise to assume that the provisions of PL-92-500 apply.

- Discharge of toxic pollutants is regulated under separate provisions from discharge of conventional pollutants.
- In 1983 an amendment was enacted that required each state to adopt an antidegradation policy, known as "antibacksliding," which prohibits the relaxation of any permit limitation unless one of the specified exceptions is applicable.
- Nonpoint sources of pollution, including stormwater runoff from industrial sites, municipal stormwater systems, and construction projects over one acre in size, are now regulated under so-called Phase I and Phase II stormwater regulations.
- Combined sewer overflows (CSOs) are regulated.
- Development and implementation of area-wide waste treatment management plans are authorized.
- Water-quality standards are established for water bodies and for water bodies that are considered "impaired," and implementation plans for achieving those standards are required.
- If the desired quality in the receiving water (water-quality standard) cannot be achieved by the so-called "categorical effluent limitations" for a given discharger, additional treatment is required.
- New sources are held to more restrictive discharge limitations than existing sources—that is, sources that were in existence and in operation at or prior to the promulgation of the applicable effluent limitation, standard of performance, or applicable pretreatment standard.
- Before a federal NPDES permit can be granted, the industry must obtain a state certification to ensure compliance with applicable water-quality requirements.
- States can be delegated the authority to issue and administer NPDES permits by the EPA.

Each state has its own body of water pollution control laws. Any individual state's laws and/or regulations can be more restrictive than the comparable federal law or regulation.

Groundwater Pollution Control Laws

by Adam Steinman

Whereas PL 92-500 is primarily addressed to protection of the quality of waterways, other laws have been enacted that have protection of the ground, the groundwater, and the air as primary objectives. The primary federal law that is addressed to protection of both the groundwater and the ground itself is the Resource Conservation and Recovery Act (RCRA) (PL 94-580), passed by the U.S. Congress in 1976. RCRA Subtitle C establishes a federal program to manage hazardous waste from "cradle to grave" through an extensive set of rules that regulates the identification, storage, treatment, manifesting/transportation, and disposal of hazardous wastes. RCRA thereby restricts activities that would lead to pollution of the ground, directly, and the groundwater, via the formation of leachate and subsequent percolation down through the soil to the groundwater. RCRA also protects the groundwater by strictly regulating and licensing hazardous waste Treatment, Storage, and Disposal (TSD) facilities.

RCRA, as amended, along with its implementing regulations, also defines the term *hazardous waste* and identifies substances to which this term is to be applied (*40 C.F.R. Part 261*). These definitions are discussed in Chapter 6 of this text. Disposal of all "solid wastes" (the term includes solid wastes, liquid wastes, and contained gases) that are determined to be hazardous wastes (by characteristic or listed) can be done only as allowed in the provisions of RCRA.

RCRA completely replaced the Solid Waste Disposal Act of 1965 and supplemented the Resource Recovery Act of 1970.

RCRA was amended extensively in 1980 and again in 1984 by the Hazardous and Solid Waste Amendments (HSWA).

The principal objectives of RCRA are to:

- Promote the protection of human health and the environment from potential adverse effects of improper solid and hazardous waste management
- Conserve material and energy resources through waste recycling and recovery
- Reduce or eliminate the generation of hazardous waste as expeditiously as possible

RCRA authorizes the EPA to regulate the generation, management, treatment, storage, transportation, and disposal of hazardous wastes, solid wastes, and underground storage tanks. This authorization is put forth in the form of nine subtitles (A–I), three of which contain the specific laws and regulations that industrial facilities must operate in compliance with. Subtitles C and D contain the programs for hazardous wastes and non-hazardous wastes, respectively. Subtitle I contains the program for underground storage tanks.

Subtitle C authorizes the EPA to:

- Promulgate standards governing hazardous waste generation and management
- Promulgate standards for permitting hazardous waste treatment, storage, and disposal facilities (TSDs)
- Inspect hazardous waste management facilities
- Enforce RCRA standards
- Authorize states to manage the RCRA Subtitle C program, in whole or in part, within their respective borders, subject to EPA oversight

Through RCRA, as amended, each state is both required and authorized to set up and administer regulations that govern the generation, management, and disposal of all substances that fall within the definition of "hazardous waste."

The 1984 Hazardous and Solid Waste Amendments (HSWA) to RCRA added requirements for the handling and disposal of present wastes and regulated the cleanup (site investigation and corrective action) resulting from past disposal of solid (including hazardous) wastes.

The Comprehensive Environmental Response, Compensation, and Liability Act (CERCLA) and the 1986 Superfund Amendments and Reauthorization Act (SARA) are laws in addition to RCRA and HSWA upon which solid (including hazardous) wastes are regulated. CERCLA and SARA provide for and regulate the cleanup and restoration of abandoned hazardous waste disposal sites.

Among the more prominent provisions of RCRA is the requirement to document the transportation and ultimate disposal point of all hazardous wastes through a comprehensive manifest system. Known as "cradle to grave" documentation, this section of RCRA requires that a manifest, or written record, accompany all hazardous wastes from their generation to disposal. An example manifest is shown in Chapter 6. RCRA's core regulations establish the "cradle to grave" program through the following major sets of rules:

- Identification and listing of regulated hazardous wastes (*40 C.F.R. Part 261*)
- Standards for generators of hazardous wastes (*40 C.F.R. Part 262*)
- Standards for transporters of hazardous wastes (*40 C.F.R. Part 263*)
- Standards for owners/operators of hazardous waste treatment, storage, and disposal (TSD) facilities (*40 C.F.R. Parts 264, 265*, and *267*)
- Standards for the management of specific hazardous wastes and for specific types of hazardous waste management facilities (*40 C.F.R. Part 266*)
- Land disposal restrictions (*40 C.F.R. Part 268*)
- Requirements for the issuance of permits to hazardous waste facilities (*40 C.F.R. Part 270*)
- Standards and procedures for authorizing state hazardous waste programs to be operated in lieu of the federal program (*40 C.F.R. Part 271*)

The EPA may authorize a state to administer and enforce its hazardous waste program in lieu of the federal Subtitle C program (pursuant to Section 3006 of RCRA). For a state to receive authorization for its program, its rules must:

- Be consistent with and no less stringent than the federal program
- Provide adequate enforcement to ensure compliance with Subtitle C requirements

In practical terms, state programs must follow the same general approach and be at least as stringent as federal hazardous waste rules. State law must include penalties that are at least equal to federal penalties (although the penalties do not have to be identical); and state enforcement activities must be equivalent to those performed by the EPA. Because state rules can be more stringent than federal rules, and each state's rules are somewhat different, large corporations that have manufacturing plants or other facilities in more than one state must make accommodations for the differences.

States have generally received authorization incrementally, consistent with the gradual implementation of the federal RCRA program. This is due to the unavoidable lag between federal promulgation of Subtitle C standards and adoption of similar standards by the states. Consequently, a state may be authorized to administer and enforce its programs regulating certain types of waste management units and practices within the state but may not be authorized for other types of units. Typically, states are more likely to be authorized to enforce rules adopted pursuant to RCRA than rules adopted pursuant to HSWA. As a result, some facilities in a state may be subject to state enforcement, while others are subject to federal enforcement through EPA regional offices. Some facilities may be subject to joint federal/state enforcement.

Some federal rules do not apply in a state with an authorized program until the state adopts those federal rules. Federal rules that are adopted under HSWA apply as a matter of federal law when they are promulgated, even if a state's rules have not incorporated them. However, only the EPA may enforce these rules.

In general, where a facility is subject to joint federal/state authority, compliance inspections may be conducted by the EPA, the state, or both.

Any industrial establishment that generates hazardous wastes and intends to dispose of some or all of them on the land is subject to certain requirements referred to collectively as "land disposal restrictions" (LDRs). All waste generators are required to determine the concentrations of certain constituents in their wastes. Depending on the constituents present, and their concentrations, the generators may be required to treat their wastes, or the residues from treatment of these wastes, using certain specified technologies. This requirement, referred to as the "Universal Treatment Standards," is contained in *40 C.F.R. §268.42.* In order to determine which treatment standard is applicable, the generator must determine whether or not a listed waste exhibits any characteristic (*40 C.F.R. §262.11[c]*) and, if it does, whether the listed waste treatment standard specifically addresses the characteristic. For example, F005 wastes are listed for both toxicity and ignitability. For treatment standard purposes under *40 C.F.R. Part 268*, the waste must be considered as F005/D001, because the F005 treatment standard does not specifically address ignitability.

If a waste is restricted, the generator must certify, on a land disposal restriction notification form that accompanies the hazardous waste manifest, whether it:

- Meets applicable treatment standards or exceeds applicable prohibition levels at the "point of generation"

- Can be land-disposed without further treatment
- Is subject to a "national capacity variance" or a "case-by-case extension"

A generator of a hazardous waste may not rely on transporters or TSD facilities to make determinations regarding land disposal restrictions on the wastes they generate. The regulations are clear that the responsibility for these determinations rests solely with the generators themselves. Although this does not appear to be a widely known or appreciated reality, it is very important. If a transporter or TSD facility neglects to make a determination, or makes an incorrect determination, it is the generator who is subject to enforcement for violating LDR rules. If a facility treats hazardous waste to meet applicable treatment standards, it must develop and make available a written waste analysis plan. The plan must describe how the procedures it uses results in compliance with the LDRs (*40 C.F.R. §268.7 [a][4]*).

The LDRs prohibit dilution from being used in any way to achieve compliance with any of the requirements or restrictions. The LDR dilution prohibition states that:

1. No one shall in any way dilute a restricted waste, or residual from treatment of a restricted waste, as a substitute for adequate treatment to achieve compliance with Subpart D, to circumvent effective dates, or to circumvent a statutory prohibition under RCRA §3004.
2. Dilution of wastes that are hazardous only because they exhibit a characteristic in a treatment system that subsequently discharges pursuant to a permit issued under §402 of the CWA or pretreatment of waste discharged under §307 of the CWA, or dilution of D003 reactive cyanide waste, is permissible unless a specific treatment method is specified as the treatment standard at *40 C.F.R. §268.42.*

For instance, if a waste is hazardous only because it is an acid and therefore exhibits the characteristic of corrosivity, and if that waste stream is subsequently treated in a simple pH neutralization system and then further treated in a treatment system that discharges under a CWA permit, mixing that waste with other waste streams does not constitute prohibited dilution under LDR regulations. Furthermore, the residuals from that treatment system do not, for reasons of only the acid waste stream, have to be managed under LDR regulations. However, if that same acid waste stream also contains cadmium in concentrations that exceed applicable prohibition levels, then it does fall under *40 C.F.R. §268.3* and cannot be diluted in any way to achieve compliance with Subpart D. However, one of the specific exceptions to the dilution prohibition is that it is permissible to combine waste streams for centralized treatment if appropriate treatment of the waste is occurring. Therefore, if the centralized wastewater treatment plant has one or more processes that are specifically designed and operated to remove cadmium (as well as other heavy metals), combining the waste streams is not considered inappropriate or prohibited under LDRs.

Air Pollution Control Laws

General

This section presents, first, a brief history of the development of air pollution control laws and regulations, as well as a synopsis of the provisions of each of the major laws. Then a synopsis of the laws and regulations that were in effect as of the year 2005, including a description of the major requirements of the Clean Air Act (CAA) that are pertinent to the regulations of air emissions from industrial facilities, as affected by those regulations, is presented.

Prior to 1963, the only federal law under which an industrial facility could be penalized or otherwise required under law to control (manage) discharges to the air was under either general nuisance statutes or public health laws. General nuisance laws had their roots in the 600-year-old rule of common law—"*sic utere tuo, ut alienum non laedas*" (use your own property in such a manner as not to injure that of another) (*Black's Law Dictionary*, 1551 [4th ed., 1951]).

Federal involvement in air pollution control had humble beginnings in 1955, with passage of the Air Pollution Control Act of 1955, Public Law 84-159. This act was very narrow in scope, and, because of the reluctance of Congress to encroach on states' rights, considered prevention and control of air pollution to be primarily the responsibility of state and local governments. In 1955, the federal government considered itself a resource, as opposed to an enforcer, and this perception was reflected in the provisions of the Air Pollution Control Act, which were as follows:

- The Public Health Service was mandated to initiate research on the effects of air pollution
- There were provisions for:
  - Technical assistance to the states
  - Training of individuals in the area of air pollution
  - Research on air pollution control

Although modest in its impact at the time, the 1955 law served as a wake-up call to states that air pollution was to be taken seriously and that enforceable laws regulating emission of pollutants would be forthcoming.

The 1955 law was amended in 1960 to the extent that it directed the Surgeon General to conduct research into the health effects of automobile exhaust. A report was submitted in 1962, and, as a result, the 1955 law was further amended to require the Surgeon General to conduct still more research. The result of the further research was the Clean Air Act of 1963, Public Law 88-206 (CAA), which has been amended several times, the most dramatic (in fact, earth-shaking) being the 1970 amendments, the 1977 amendments, and the present, prevailing law, the 1990 amendments.

The Clean Air Act of 1963 provided for:

- A stepped-up research and training program
- A matching grants program, whereby states and local agencies would receive federal assistance in promulgating air pollution regulation
- The development of air-quality criteria
- Federal authority to require abatement of interstate flow of air pollutants

The 1963 Clean Air Act designated six pollutants as "criteria pollutants," thought to be the most important substances affecting the public's health and welfare. These criteria pollutants, still regarded as such in the year 2005, are as follows:

- Sulfur dioxide (SO)
- Nitrogen oxides (NO_x)
- Carbon monoxide (CO)
- Lead (Pb)
- Ozone (O_3)
- Particulate matter (Pm)

The 1970 amendments established the basic framework for regulation of air emissions and air quality from the time of its enactment through 2005. The 1970 amendments authorized the EPA to establish National Ambient Air Quality Standards (NAAQS) to "protect the public health" and "public welfare." Following promulgation of an NAAQS, states are required to develop plans (State Implementation Plans) to attain and maintain compliance with the NAAQS. If a state fails to submit a sufficient plan, the EPA promulgates a plan for the state. The 1970 amendments also authorize the EPA to establish emission standards for categories of sources that, in the EPA's judgment, cause or contribute significantly to air pollution. Under this authority, the EPA has promulgated standards (New Source Performance Standards) for dozens of categories of sources (e.g., electric utility steam-generating units; industrial, commercial, institutional

steam-generating units; sewage treatment plants; and kraft pulp mills).

The 1977 amendments provide the EPA with the additional authority to develop emission standards for categories of existing sources of hazardous air pollutants. Under this authority, the EPA promulgated relatively few National Emission Standards for Hazardous Air Pollutants (NESHAPs). The slow rate of progress for developing NESHAPs was one of the primary drivers for enactment of Title III of the 1990 amendments (discussed in the sections that follow).

The 1977 amendments established new air permitting requirements for the construction of new major sources or major modifications to existing sources. The applicable permitting requirements depend on whether or not an area experiences air quality not meeting NAAQS, in which case it is considered a "nonattainment area." New major sources and major modifications in attainment areas are subject to Prevention of Significant Deterioration (PSD) requirements. In short, under PSD, a facility is required to obtain a preconstruction permit from the EPA or a state with an EPA-approved PSD program. To obtain a permit, the proposed source must demonstrate that the new emissions will receive Best Available Control Technology (BACT) and that the emissions, in conjunction with other nearby sources, will not cause or contribute to an exceedance of the NAAQS. BACT is considered the best level of control that is technically and economically feasible. In nonattainment areas, new major sources and major modifications must demonstrate that emissions will receive the Lowest Achievable Emission Rate (LAER), which is essentially the lowest level of emission attained in practice by a similar source. Unlike BACT, LAER does not consider economic feasibility. In addition to LAER, the new source or modification must obtain offsets, which are emission reductions from other facilities in an amount equal to or greater than the proposed emissions increases.

The 1990 amendments to the Clean Air Act made additional sweeping changes, especially with regard to industrial sources. Title V established a new operating permitting system that had the effect of permanently changing the way environmental managers in industry must do their jobs. Among the major provisions of Title V are monitoring and reporting requirements that are greatly expanded, compared with previous requirements. It is now required to identify all "regulated pollutants" emitted by a facility, to monitor emissions (continually or periodically), to operate the equipment in compliance with standards written into the permit, and to certify compliance with all standards in the permit on an annual basis.

In addition to the new permitting, monitoring, and reporting requirements, the number of designated hazardous air pollutants or air toxics (Title III) was increased to 189 and subsequently (in 1999) reduced by one to 188. Also, significant changes were made regarding nonattainment areas, emissions from automobiles (Title II), acid rain provisions (Title IV), and stratospheric ozone production provisions (Title VI), all of which are pertinent to the job of the industrial environmental manager.

Air Pollution Control Law, as of the Year 2005

Major sources of criteria pollutants and hazardous air pollutants (HAPs) are regulated by the Clean Air Act as amended in 1990 and as administered by state or local air-quality management agencies in which the industry operates. Some industrial establishments may be regulated by one or more state or agency requirements that are more restrictive than the CAA. The federal government, through the EPA, issues regulations that must be followed by the administrating authorities and oversees that administration.

As prescribed by the CAA, as amended, the EPA is charged with setting National Ambient Air Quality Standards (NAAQSs), a process that began in the 1970s and undergoes revision periodically. Each state has developed a state implementation plan (SIP) to attain those standards.

Industrial establishments, referred to as "sources," are categorized as a "major source" if their air emissions exceed certain specified amounts. The specified amounts are different, depending on whether or not the source is in an attainment area or a nonattainment area, and on whether that nonattainment area has been designated as either "marginal," "moderate," "serious," "severe," or "extreme," or if the source is located in a designated "ozone transport region." For purposes of PSD preconstruction permitting requirements under Title I, a source is categorized as "major" in an attainment area if it is in one of 27 listed categories and emits, or has the potential to emit, 100 tons per year (tpy) or more of any criteria pollutant. If it is not in one of the 27 listed categories, the source is major if it emits, or has the potential to emit, 250 tpy or more of any criteria pollutant. Sources in nonattainment areas have lower major source thresholds depending on the nonattainment designation. For purposes of operating permit requirements under Title V, a source is major if it emits, or has the potential to emit, 100 tpy of any regulated pollutant, or lesser amounts in certain nonattainment areas. Further, a source is considered major for both Title V permitting and for purposes of hazardous air pollutant (HAP) regulation under Title III if it emits, or has the potential to emit, 10 tpy or more of any single HAP or 25 tpy or more of all HAPs combined. Major sources have significantly more requirements regarding applying for and operating under a CAA permit than do nonmajor sources. If a source is subject to new source performance standards (NSPS), is subject to national emissions standards for HAPs (NESHAPs), or is an "affected source" subject to the acid rain program under Title IV, it also has significantly greater requirements.

For industries in one of the 27 listed categories, "fugitive emissions" are to be included in the emission totals. "Fugitive emissions" are those that issue from open windows and doors; cracks in buildings or ductwork; or, in general, via any outlet other than a stack, vent, or other device specifically designed and built to discharge substances to the air.

It is possible for industrial establishments to avoid major source classification and thus become a "synthetic minor source" by agreeing to limit emissions to below the designated maximums (even though they have the potential). Some states have actively encouraged such agreements in order to reduce the considerable cost of administrating and enforcing the air permitting program.

The CAA is organized into 11 titles, 6 of which (listed below) are of direct concern to the industrial environmental manager.

- Title I, Attainment and Maintenance of National Ambient Air Quality Standards (NAAQS)
- Title III, Air Toxics Control
- Title IV, Acid Rain Control
- Title V, Permits and Reporting
- Title VI, Stratospheric Ozone Protection
- Title VII, Enforcement

The following paragraphs present a synopsis of the provisions of each of these six titles that are of most concern to the industrial environmental manager.

Title I, Attainment and Maintenance of National Ambient Air Quality Standards (NAAQS)

If an industrial facility is located in a nonattainment area, the classification of that area has a significant influence on the financial burden that facility must bear to maintain compliance with requirements of the CAA. For instance, if an industrial plant is located in an ozone nonattainment area classified as "moderate," it will be subject to "reasonably available control technology" (RACT) requirements if it emits more than 100 tons per year of volatile organic compounds (VOCs). If the same plant is located in a non-attainment area classified as "extreme," it will be subject to RACT requirements if it emits more than 10 tons per year of VOC emissions. The basic structure for designations on nonattainment areas resides in Title I.

Simply put, an area is designated as "nonattainment" if the ambient air is not of a specified quality. Nonattainment, then, refers to a deficiency in quality regarding one or more specific substances. If an area is designated as nonattainment in particulate matter, that situation may be of little consequence to an industry having no significant particulate matter emissions. The opposite is equally important; an industry having significant particulate emissions would be well advised not to locate in this area.

Title III, Air Toxics Control

The requirements of Title III, control of hazardous air pollutants, include limiting the release of 188 substances referred to as "air toxics." The approach that Congress adopted in writing and promulgating the requirements of Title III was to mandate the publishing by the EPA of emission standards for each of the 188 hazardous substances, based on what could reasonably be expected to be achievable by the best technology available. It was intended that the EPA would issue the national standards for significant sources for 40% of the source categories by November 15, 1992; another 25% of source categories by November 15, 1994; another 25% by November 15, 1997; and the remaining by November 15, 2000. Compliance with these national standards was to be complete within no more than three years of issuance of the standards (facilities are allowed to delay compliance for up to six years if they reduce emissions before standards are issued).

Maximum Achievable Control Technology (MACT)

Under Title III, MACT can include process changes, materials substitutions, enclosures, and other containment strategies, as well as active treatment for pollutant removal.

Record keeping

Industrial facilities permitted under Title V and subject to the requirements of Title III must keep records of processing and monitoring for at least five years and retain them on site for at least two years.

Title IV, Acid Rain Control

Strategies that are being pursued for the purpose of acid rain control are based on the realization that the mobility of the acidic compounds in the atmosphere, which result from emissions of sulfur and nitrogen oxides, make acid rain impossible to control on a local basis. The essence of these strategies is to achieve reduction of SO_2 emissions by use of a market-based approach, and NO_x emissions through emission limits. For instance, sources that generate heat and/or power by burning coal are allocated certain "allowances" regarding SO_2 emissions. The allowances that have been allocated were done so with the goal of reducing annual emissions of SO_2 by 10 million pounds from 1980 levels nationwide. What is unique about the acid rain reduction approach is that the allowances can be bartered (bought and sold on a market). If a given source emits less than the allowance, it can sell the excess or "bank" it for future use.

A given source has the options of selecting low-sulfur fuel, making use of emission treatment technologies, or buying allowances from another source to meet its own allowance. A given source can also sell excess allowance that it has obtained by selecting low-sulfur fuel, for instance, to reduce the net cost of the selected option.

NO_x emission limits are levied on the coal-burning, electricity-generating units, with the nationwide goal of reducing NO_x emissions by 2 million tons a year from 1980 levels. The strategy is to achieve attainment of acid rain control goals through the emission limits, then to institute an allocation bartering system similar to that in use for SO_2 emissions to achieve regional ozone level issues if necessary.

Title V, Permits and Reporting

The requirements that pertain to applying for, obtaining, and operating within compliance of air discharge permits is a major concern of the industrial environmental manager. Either the states or the EPA can enforce the permits. The cost of administration is recovered through a fee system, through which each permitee pays not less than $25 per ton of regulated pollutant (excluding carbon monoxide). This fee can be adjusted each year, based on the consumer price index.

Application

Each state has developed standard forms for applying for air discharge permits. The forms differ somewhat from state to state, but, in accordance with Title V, certain "key elements" must be included. One key element is identification and description of each and every emission point. Also required are a complete list of regulated substances to be discharged; compliance and monitoring plans; an assessment of past compliance, alternative operating scenarios; and identification and description, including location, of any and all air pollution control equipment. Determination of whether or not the source is "major" is one of the most important requirements of the application.

Regulated Air Pollutants

The substances that are regulated under the CAA include the six criteria pollutants that were originally regulated under the CAA of 1963 (sulfur dioxide, nitrogen oxides, carbon monoxide, lead, ozone, and particulate matter) and volatile organic compounds (VOCs), as well as those substances already regulated under new source performance standards (NSPS), which include hydrogen sulfide, reduced sulfur compounds, total reduced sulfur, sulfuric acid mist, dioxin/

furan, fluorides, and hydrogen chloride. In addition, 188 hazardous air pollutants are regulated on a "technologically achievable" basis, including substances that deplete ozone and those chemicals that are subject to the accidental release provisions.

National Emission Standards for Hazardous Air Pollutants (NESHAPs)

NESHAPs are addressed directly in Title III, and are very important regarding completion of the permit application. In the case of major sources, the NESHAPs become MACTs, "maximum achievable control technologies." In the case of small or area sources, they become GACTs, "generally available control technologies." These control technologies are intended to reduce emissions of each of the 188 designated HAPs. Affected sources are generally given three years to comply with a newly promulgated MACT standard. If the EPA has not issued a MACT standard for a particular type of major source, then the state permitting authority will determine MACT for that source on a case-by-case basis.

Monitoring and Reporting

Operating permits, issued under Title V, contain the monitoring procedures and test methods to be used for each substance regulated under the permit. Reporting is normally required at six-month intervals. Certain industrial facilities are subject to "compliance assurance monitoring" (CAM), which involves additional requirements. These facilities, generally, are those major sources that rely on pollution control equipment to comply with the terms of the permit, as opposed to restricting operations, or any other strategy. In addition to semiannual reports of deviations from air permit conditions, Title V permitted sources must submit to the state permitting authority or the EPA a certification of compliance with all Title V permit terms on an annual basis.

New Source Performance Standards (NSPS)

NSPSs have been issued for several types of industrial establishments. (A new source is one that commences construction or reconstruction after a standard that applies to that source has been proposed.) In general, NSPSs reflect cutting-edge control technology for major sources. In some cases, monitoring and reporting requirements are more comprehensive. For instance, regarding VOC emissions, requirements for recording include the quantities of solvents used in manufacturing processes and calculations of solvent usage versus the quantity emitted to the air.

In summary, each source that is subject to the operating permit program is required to prepare and submit an application for a permit. The application must describe each and every source of air pollutants, as well as pertinent air pollution control requirements and standards. Whenever a facility is in a noncompliance situation, compliance plans and remedial measures must be developed and submitted. Actual emissions must be monitored, and monitoring reports must be submitted periodically. At least once a year, each source must certify its status of compliance.

Whenever the operating status of a source changes, applications for permit modifications must be submitted. Also, it is required that permit renewals be submitted at least every five years. The sources that are subject to the operating permit program include the following:

- Major sources, as defined by the CAA, or (more restrictively) by the applicable state or local air-quality control agency
- "Affected sources," which includes any stationary source that contains one or more units subject to an acid rain emission limitation, or reduction (Title IV)
- Any source (including area sources) that is subject to new source performance standards (NSPSs)

- Any source (including area sources) that is subject to standards, limitations, or other restriction of the NESHAPs.

Title VI, Stratospheric Ozone Protection

Certain requirements under the CAA relate to the prevention of leaks and servicing of stationary and mobile air conditioning and refrigeration units containing chlorofluorocarbons (CFCs) or hydrochloroflourocarbons (HCFCs). These requirements are contained in Sections 608 and 609 of the CAA. These provisions prohibit knowingly venting CFCs and HCFCs to the atmosphere from air conditioners or refrigeration equipment; require the repair of certain leaks of CFCs and HCFCs; and require that certain practices be followed for the service, repair, and disposal of air conditioning or refrigeration units.

Title VII, Enforcement

The best strategy for environmental managers to follow regarding enforcement is to use all reasonable means to avoid enforcement actions. Enforcement actions can be initiated by the federal government, the state government, local agencies, citizen groups, or individual citizens.

The foundation of avoiding enforcement actions is in corporate policy toward compliance. "Fighting city hall" is exceedingly expensive and has a poor track record. Experience bears out that getting actively involved in the permit process, developing a good working relationship with regulators, fostering among employees a sound policy of careful handling and use of potential environmental pollutants, substituting nonpolluting substances where possible, and prudently maintaining and operating pollution control equipment are far preferable to dealing with enforcement actions.

Bibliography

Army Corps/U.S. Environmental Protection Agency, *Section 404(b) (1) Guidelines Mitigation MOA.* February 7, 1990. U.S. Environmental Protection Agency/Army Corps. *Memorandum to the Field: Appropriate Level of Analysis Required for Evaluating Compliance with the Section 404(b) (1) Guidelines Alternatives Requirements.* August 23, 1993.

Clean Water Act. 33 U.S.C. 1251 et seq. (1948–1987).

Clean Water Act. 404, 33 U.S.C. 1344.

Coastal Zone Management Act. 16 U.S.C. 1451 et seq. (1972–1986).

Comprehensive Environmental Response, Compensation and Liability Act. 42 U.S.C. 9601 et seq. (1980–1987).

Council on Environmental Quality. *Implementation of Procedural Provisions of NEPA, Final Regulations.* 43FR, No. 230, November 19, 1978.

Endangered Species Act. 16 U.S.C. 153 1-1542 et seq. (1973–1984).

Federal Insecticide, Fungicide, and Rodenticide Act. 7 U.S.C. 136 et seq. (1972–1991).

Fish and Wildlife Coordination Act. 16 U.S.C. 661 et seq. (1958–1965).

National Academy of Sciences, National Academy of Engineering. *Water Quality— 1972.* Washington, DC: U.S. Government Printing Office, 1974.

National Environmental Policy Act. 42 U.S.C. 4321 et seq. (1970–1975).

National Historic Preservation Act. 16 U.S.C. 470-470t et seq. (1966–1992).

National Technical Advisory Committee. Federal Water Pollution Control Administration. *Water Quality Criteria.* Washington, DC: 1968.

Pollution Prevention Act. PL 101-508, Title VI, subtitle F. Sections 6601–6610) et seq. (1990).

Resource Conservation and Recovery Act. 42 U.S.C. 6901 et seq. (1976–1992).

Safe Drinking Water Act. 42 U.S.C. 300f et seq. (1974–1996).

Superfund Amendments and Reauthorization Act. 42 U.S.C. 11001 et seq. (1986).

Toxic Substances Control Act. 15 U.S.C. 2601 et seq. (1976–1988).

U.S. Environmental Protection Agency. *Legislation, Programs and Organization.* Washington DC: Government Printing Office, 1979.

U.S. Environmental Protection Agency. "National Discharge Elimination System." *Code of Federal Regulations.* Title 40, Part 122. Washington, DC: U.S. Government Printing Office, 1983–1995.

U.S. Environmental Protection Agency. "National Emission Standards for Hazardous Air Pollutants." *Code of Federal Regulations.* Title 40, Part 61. Washington, DC: U.S. Government Printing Office, 1997.

U.S. Environmental Protection Agency. "National Primary and Secondary Ambient Air Quality Standards." *Code of Federal Regulations.* Title 40, Part 50. Washington, DC: U.S. Government Printing Office, 1997.

U.S. Environmental Protection Agency. "Pretreatment Standards." *Code of Federal Regulations.* Title 40, Part 403. Washington, DC: U.S. Government Printing Office, 1997.

U.S. Environmental Protection Agency. *Quality Criteria for Water.* Washington, DC: U.S. Government Printing Office, 1976.

U.S. Environmental Protection Agency. "Secondary Treatment Regulation." *Code of Federal Regulations.* Title 40, Part 133. Washington, DC: U.S. Government Printing Office, 1985.

U.S. Environmental Protection Agency. "Standards of Performance for New Stationary Sources." *Code of Federal Regulations.* Title 40, Part 60. Washington, DC: U.S. Government Printing Office, 1997.

U.S. Environmental Protection Agency. Office of Water Program Operations. *Federal Guidelines—State and Local Pretreatment Programs.* EPA-43019-76-017c. Washington, DC: U.S. Government Printing Office, January 1977.

Wild and Scenic Rivers Act. 16 U.S.C. 1271 et seq. (1968–1987).

4   Pollution Prevention

Pollution Prevention Pays

Waste minimization has been a primary objective of wastewater, hazardous waste, air, and solid waste management programs since the earliest days of industrial waste treatment. Many academic programs have instructed that a crucial responsibility for an environmental engineer is to reduce the amount of pollutants that require treatment prior to discharge.

The Clean Water Act (CWA) is the cornerstone of surface water-quality protection in the United States. It was enacted to sharply reduce direct pollutant discharges into waterways, finance municipal wastewater treatment facilities, and manage polluted runoff. The broader goal is restoring and maintaining the chemical, physical, and biological integrity of the nation's waters so that they can support "the protection and propagation of fish, shellfish, and wildlife and recreation in and on the water."[1]

What may not be included in environmental engineering curricula is the concept that pollution prevention and waste minimization can result in a significant decrease in overall operating costs and a consequent increase in profitability. During the 1980s, the U.S. Congress authorized in-depth studies to analyze the financial impacts of pollution prevention on businesses and industries. The result was the emergence of pollution prevention as a central concept within industry. According to the U.S. Environmental Protection Agency, "the nation is coming to understand pollution prevention's value—as an environmental strategy, as a sustainable business practice, as a fundamental principle for all our society" (U.S. EPA, 2004).[2].

The term *pollution prevention* includes all aspects of waste minimization and pollution reduction and includes a thorough consideration of each product throughout its life cycle, from initial product development to final disposal. For each stage in a product's life cycle, an engineer must consider the pollutants and potentially toxic wastes that could be discharged to the atmosphere, surface water bodies, and the land. Existing processes and facilities must minimize flows and loads, and nontoxic substances must be substituted for toxic substances, to the maximum extent practicable. While the product life cycle analysis is vital to a successful pollution prevention program, it is also the natural precursor to significant cost savings.

National Pollution Prevention Policy

Congress established the Pollution Prevention Act[3] in 1990, indicating that the following should be adhered to, whenever feasible:

- Pollution should be prevented or reduced at the source.
- Pollution that cannot be prevented should be recycled in an environmentally safe manner.
- Pollution that cannot be prevented or recycled should be treated in an environmentally safe manner.

[1] http://www.epa.gov/watertrain/cwa/
(accessed May 24, 2005).

[2] http://www.epa.gov/p2/
(accessed November 30, 2004).

[3] http://www.epa.gov/opptintr/p2home/p2policy/
definitions.htm (accessed November 30, 2004).

- Disposal or other releases into the environment should be employed only as a last resort and should be conducted in an environmentally safe manner.

Important definitions to grasp within the Pollution Prevention Act include "pollution prevention" and "source reduction." The Act defines "pollution prevention" as source reduction and other practices that reduce or eliminate the creation of pollutants through:

1. Increased efficiency in the use of raw materials, energy, water, or other resources
2. Protection of natural resources by conservation

"Source reduction" is defined to mean any practice that reduces:

- The amount of any hazardous substance, pollutant, or contaminant entering any waste stream or otherwise released into the environment (including fugitive emissions[4]) prior to recycling, treatment, or disposal
- The hazards to public health and the environment associated with the release of such substances, pollutants, or contaminants

Methods of source reduction include equipment or technology modifications; process or procedure modifications; reformulation or redesign of products; substitution of raw materials; and improvements in housekeeping, maintenance, training, or inventory control. This chapter focuses on some of these methods more specifically, but the possibilities for employing one or several of them within an industrial pollution prevention program are limitless.

[4] Fugitive emissions refer to pollutant emissions that cannot reasonably be collected or vented through one or more point sources (e.g., vent or stack).

One important distinction to consider with respect to the National Pollution Prevention Policy and the concept of pollution prevention itself is that recycling, energy recovery, treatment, and disposal are NOT included within the definition of pollution prevention. If materials require recycling, they have already become waste that needs to be managed. Similarly, treatment and disposal are activities restricted to waste management. Pollution prevention means much more than preventing pollution—it means preventing waste to begin with. Within the greater context of this book, pollution prevention, if successfully incorporated into the industrial process, should reduce or potentially eliminate the need for industrial waste treatment.

The positive impact that pollution prevention can have on industries and our environment is potentially so great that environmental engineers, environmental scientists, and all others involved in the environmental field should consider it a professional obligation to communicate this message to clients, regulators, colleagues, and communities.

Considerations of Cost

During the 1990s, it became well established that the conscientious application of pollution prevention principles, including a thorough consideration of industrial activities and their costs, could result in increased profitability for industries and the business community as a whole. A salient realization during the inception of a pollution prevention program is that emissions of all forms—water pollutants, air pollutants, and solid wastes—are, in fact, materials that originated from purchased raw materials and that emissions represent a quantifiable loss. It would be grossly simplistic to view the cost of environmental protection as merely the cost to treat and dispose of waste. Rather, the true cost is calculated after a complete accounting of all costs involved in an industrial process, including purchase of raw materials; equipment maintenance; waste management systems; and costs associated with recycling, treatment, transportation, and disposal. For example, one pound of hydrocarbon emitted from a smoke stack at a power-generating station was previously paid for as a pound of coal or oil. If combustion performance can be improved, more power can be produced, and fewer emissions require treatment.

Regulatory Drivers

Industries are required to comply with a variety of federal, state, and local environmental regulations. Many of the applicable regulations (e.g., Resource Conservation and Recovery Act, Toxic Substances Control Act, Clean Water Act, Clean Air Act) require regulated activities and operations to be conducted to minimize potential environmental impacts. Therefore, while meeting mandatory regulations, industries should already be devoting time, energy, and resources to ensure compliance and, where required, minimize their waste generation. For example, adopting and implementing a Hazardous Waste Minimization Plan (which is required of all large-quantity hazardous waste generators) not only will reduce the regulatory burden associated with hazardous wastes, but also will reduce the amount of waste generated, lower disposal and treatment costs, and minimize the amount of potential environmental impacts associated with hazardous wastes. If the plan includes process optimization, the same amount of product can be produced using fewer raw materials, and the quantity of waste generated can be reduced. This is just one example of how environmental compliance and pollution prevention go hand in hand. In fact, environmental sustainability is often the next step once compliance is achieved.

Pollution Prevention Leads to Environmental Sustainability

Once pollution prevention practices are implemented throughout the industrial process, a business will be well on its way to achieving environmental sustainability. Sustainability, as defined by the Brundtland Commission in 1987, is "development that meets the needs of the present without compromising the ability of future generations to meet their own needs." While cost savings and regulatory drivers are important, environmental sustainability represents a higher goal—one that should be strived for because it's the right thing to do.

The benefits of pollution prevention and environmental sustainability not only include cost savings and regulatory compliance, but also improved working conditions for employees, competitive advantages with environmental-savvy clients and consumers, and improved community and regulator relations.

Benefits of Pollution Prevention

In addition to satisfying key policy requirements of federal and state regulatory agencies, an effective pollution prevention program can have major benefits. A new look at existing and traditional production methods can lead to improvements in production efficiencies, reduced costs for treatment and disposal, and reduced risk. Other benefits can include:

- Reduced liability as a future "Potentially Responsible Party" (PRP) at off-site final disposal locations
- Improved relations with customers seeking associations with environmentally conscious or "green" suppliers
- Reduced costs for handling and storage of wastes
- Improved economic and environmental bottom lines
- Improved public image
- Increased production capacity
- Enhanced regulatory relations
- Better work environment for employees
- Increase in positive community relations—facilities are welcomed, rather than tolerated, by communities

General Approach

When a pollution prevention program is implemented, either to comply with regulatory requirements, pursue environmental sustainability, or as part of necessary facility upgrades, the primary steps include:

1. Secure unequivocal support from management.
2. Develop a company-wide philosophy of waste minimization.
3. Clearly establish realistic objectives and targets.
4. Establish a baseline and ways to measure progress.
5. Develop an accurate cost accounting system.
6. Implement a continual improvement program.

While implementation methods may vary between facilities, these five steps provide a starting point. Each step is described in greater detail below.

Secure Unequivocal Support from Management

It is vital to get as much support as possible before a pollution prevention program can get started. Having the support of people who agree that pollution prevention "is a good idea" is helpful, but without support from top management and the necessary resources to effect operational changes, material substitutions may not be available. Even if top management is in agreement that pollution prevention is necessary, they must be willing to offer genuine, active support, including budgetary help and a strong voice of support when internal obstacles arise. Internal obstacles may include: lack of understanding of the benefits of pollution prevention, not allowing employees adequate time or incentives to complete pollution prevention tasks, and lack of capital for required process improvements.

When top management assigns or delegates the "responsibility" for implementing a pollution prevention program, real progress is often difficult to achieve and sustain. There have been numerous attempts to institute waste minimization programs in the absence of overt support from top management. Progress can be made soon after program implementation, only to gradually revert to former wastefulness and inefficiency when effective management support is absent. The unequivocal and visible support of top management, who are instrumental to program success, is required to make the paradigm shifts necessary to implement an effective pollution prevention program.

The key to gaining enthusiastic support from the top management of an industrial facility is to educate managers about the potential increased profitability and concrete cost savings associated with waste reduction, pollution prevention, and sustainability. Managers can be taught that emissions of "pollutants" should be viewed as expenditures that have been lost. Once pollutants are included in the accounting ledger as lost resources, their reduction shows up as a smaller loss. Additionally, managers must be coached into realizing that the costs for treatment and disposal of pollutants decrease as the quantity, and especially the toxicity, of those pollutants decreases. Again, it is the ethical responsibility of environmental management professionals to carry this message forward and to spearhead the paradigm shift from pollution control as a bottomless pit of expense to pollution prevention as a way to increase company profits.

In addition to securing support from management, it is good practice to invite others to participate in the process. Managers whose work influences the quantity and characteristics of wastes produced, as well as employees who are responsible for purchasing raw materials, should be involved at the inception of a pollution prevention program. Each stakeholder must be involved in developing an explicit scope and clear objectives and targets for their facility's pollution prevention program. The team of stakeholders must be completely cross-functional, thus representing all of the activities at the facility that in any way affect the quantity and characteristics of wastes produced. At a typical industrial facility, the stakeholders would include:

- Plant manager
- Operations management
- Engineering
- Maintenance
- Purchasing
- Environmental health and safety
- Production employees and their representative

The following is an example of how purchasing can be directly involved in the waste minimization aspect of pollution prevention. Purchasing is directly involved in communicating with suppliers regarding packaging of items that will be delivered to the industrial facility. On one end of the scale there are returnable (and reusable) containers. On the other end of the scale are multilayered, nonrecyclable packages, characterized by excessive bulk, that must all be discarded and disposed of. It is not unusual for electronic equipment to be delivered in plastic wrapping, surrounded by foam blocks, packed inside a box, covered by paper, covered in turn by plastic, and all bound by plastic or metal strapping. In many cases purchasing agents must be guided to "think outside the box" and to demand packaging from suppliers that minimizes creation of solid wastes. One of the very first examples of pollution prevention was when Henry Ford required vehicle components to be delivered in crates made of oak slats. Instead of throwing the crates away, Ford workers disassembled the crates, and used the oak slats as automobile floorboards.

Clearly Establish Objectives and Targets

Measuring the success of a pollution prevention program is impossible without the ability to compare progress with established goals. However, if unrealistic goals are made, program participants may end up discouraged because success seems so unattainable.

Significant work is involved in developing an explicit scope for a pollution prevention program. It is also critically important that all of the stakeholders are actively involved in scope development. The worst approach is

Table 4-1   Environmental Objectives and Targets

Objectives	Targets
Prevent Chemical Releases	No chlorine gas releases
	No liquid/solids/chemical spills greater than the Reportable Quantity
	No chemical dust releases
Improve Workman Compensation Statistics	Eliminate confined space in the workplace
	No chemical exposure to staff
	Maintain pump noise at less than 70 db
	No vehicle accidents
Sustainability	Reduce energy use by 10% by December 2006
	Reduce truck miles by 10% by December 2006
	Operate alternative or hybrid vehicle by December 2006
	Eliminate the use of chlorine gas in water treatment plant by December 2005
	Reduce hazardous materials use by 50% by December 2006

for one person to take on the entire task of developing the scope of a pollution prevention program. Not only will it be regarded as that person's demands on coworkers, but worse, the person may not be sufficiently familiar with the industrial processes that create the pollutants.

The International Standard for Environmental Management Systems (ISO 14001) defines an environmental objective as an overall environmental goal that is consistent with an organization's environmental policy. An environmental target, by contrast, is a detailed performance requirement that arises from the environmental objectives and that needs to be set and met in order to achieve the objectives. In other words, the objective is general and the target is specific. ISO 14001 also specifies that objectives and targets should be measurable and consistent with environmental policies, including commitments to pollution prevention, compliance with applicable legal requirements and other facility requirements, and continual improvement. Other considerations when developing objectives and targets include: legal requirements; environmental impacts; technological options; financial, operational, and business requirements; and stakeholder interests. Table 4-1 provides examples of good and bad objectives and targets.

Establish a Baseline and Ways to Measure Progress

After a detailed scope has been developed, a clear set of objectives and targets for the pollution prevention program can be developed. Here, it is important to shoot high, but not so high as to intimidate those who will implement the program. It is imperative that the stated objectives can be understood, are achievable, and can be evaluated by readily understood measurements and comparisons.

Success cannot be measured without a baseline. Existing conditions must be measured to develop an accurate baseline against which pollution prevention achievements

can be compared. Establishing a baseline that encompasses all potential levels of pollution prevention may be too much. The baseline should be developed only for matrices directly related to the objectives and targets for the pollution prevention program.

Once a clear set of objectives and targets has been developed, and measurement methods have been determined and agreed upon, action must begin and be sustained. Measurements that are truly indicative of success or failure must be taken. "What gets measured gets done" is a good guiding rule. A meaningful reward system should be implemented to acknowledge success and to encourage continuous improvement.

Another necessity is establishment of a baseline from which to evaluate the degree of success or failure of the pollution prevention program. Whichever accounting system is agreed upon must be applied to the facility before implementation of the pollution prevention program. Also, the cost accounting system must be applied at regular intervals in order to keep track of progress or lack of progress.

Accurate Cost Accounting System

For a pollution prevention program to be successful, prevailing attitudes that waste reduction and environmental compliance cost money and cut into company profits must be replaced with an understanding that a well-designed and -executed pollution prevention program can reduce costs and add value to the product, thus bolstering company profits. An accounting system that can accurately track all the true costs of production, distribution, and final disposal of the product, as well as the costs of managing, handling, and disposal of all wastes—solid, liquid, and air—is essential to convince both management and production personnel of the true value of a pollution prevention program.

It is important here to note that the idea of the "bottom line" is increasingly being replaced with the "Triple Bottom Line." The

Triple Bottom Line is a measurement of a company's economic state, environmental stewardship, and social responsibility. Corporations that are striving for sustainability and committed to more than just profits are increasingly relying on the Triple Bottom Line.

In order to enable advantageous application of an accurate cost accounting system, an accurate materials balance, inclusive of the entire life cycle of a product, is needed. Although this may seem a daunting task, and costly in and of itself, a materials balance is a necessary ingredient in the total picture of cost effectiveness. It is another example of an investment that leads to significant savings in the long run.

It has been found to be highly advantageous to employ so-called "activity-based costing" (ABC) to evaluate the degree of success or failure of a given pollution prevention program. ABC involves meticulous identification of each of the cost items within the general ledger that are related in any way to a given activity. As an example, in identifying all costs related to emission control, the costs for chemicals used in wet scrubbers are identified. Wet scrubbers are part of the air pollution control system; they are, therefore, part of the overall pollution control system. As another example, the proportion of each person's time spent in the performance of duties related to pollution control is determined. For example, if a maintenance worker spends one hour per week adding chemicals to a wet scrubber, then the corresponding proportion of that individual's salary and benefits is allocated to emission control.

An effective cost accounting method that often produces pollution prevention benefits entails assigning the cost of waste disposal or energy or water use to individual departments, based on their individual use. When the costs are allocated directly to individual department managers, they become aware of the true cost of these items, creating an incentive to conserve and reduce.

Companywide Philosophy of Waste Minimization

In addition to the absolute requirement for unequivocal support from top management for a pollution prevention program to succeed, it is equally important that everyone involved in receiving, preparation, production, packaging, storage, and shipping believe in waste minimization as a necessary component to the company's financial success and, thus, the security of their jobs. As explained below, a pollution prevention program consists of active waste minimization at each stage of the life of a product, from initial development, through manufacturing, and on to final disposal of the product at the end of its life. Every person involved in every stage of the product has an influence on the efficiency of use of raw materials, including leaks and spills, cleanup, and damage to raw materials; intermediate stages, or final product; and containment of wastes. A pervasive, company-wide belief in the direct influence of a well-executed pollution prevention program on the security of each person's job is needed for success of the program.

Promoting a belief in the benefits of waste minimization can be achieved in a variety of ways. Specific strategies include communicating how extra work may not always be required to achieve waste minimization, and day-to-day tasks may actually become easier. Certain individuals or groups who are working toward waste minimization and the progress of the pollution prevention program should be congratulated for successes, and these successes should be communicated to the entire organization. Additionally, employee incentive programs are an excellent way to recruit new help and reward stakeholders.

Implement Environmental Policy

A significant step in establishing a successful pollution prevention program is developing a facility environmental policy. According to ISO 14001, a company's top management should develop an environmental policy, and the policy should:

- Be appropriate to the nature, scale, and environmental impacts of its activities, products, and services
- Include a commitment to continual improvement and prevention of pollution
- Include a commitment to comply with applicable legal requirements and with other requirements to which the organization subscribes that relate to its environmental aspects
- Provide the framework for setting and reviewing environmental objectives and targets
- Be documented, implemented, and maintained
- Be communicated to all persons working for or on behalf of the organization
- Be available to the public

Continual Improvement and Education

New technology is being developed and brought to light through various publications on a continuing basis. New and improved techniques and technologies for improving efficiencies in product production, materials handling, substitution of non-toxic substances for toxic substances, and waste handling and disposal can significantly aid in furthering the objectives of a pollution prevention program. It is vitally important that key participants in an industry's pollution prevention program attend seminars, short courses, and regional and national meetings at which new techniques and technologies are presented. It is equally important that these key participants then hold in-house seminars and meetings where the new information is shared with other participants in the pollution prevention program.

Pollution Prevention Assessment

An "environmental audit" (Chapter 5) has the purpose of assessing a company's compli-

ance (or risk of noncompliance) with environmental regulations. A "waste minimization audit" has the objective of assessing opportunities to improve materials utilization efficiency at an operating industrial facility or individual process. A "pollution prevention assessment" is a more comprehensive program conceived to determine each source at which wastes are generated, from product development through manufacturing, use of the product, and on through to the end of the life of the product. The general approach of a pollution prevention assessment has two phases. The first phase identifies and quantifies the types of waste generated and analyzes the manner in which those wastes are generated. The second phase involves performing a materials balance on each of those sources, determining alternatives for reducing or eliminating those wastes, and analyzing the benefits and costs of each alternative. The following step-by-step process outlines the sequence of a pollution prevention assessment.

Phase I

Determine how wastes will be identified and measured and develop baseline. Aspects of the industrial process that are important to measure include:

1. Raw materials used
2. Utility use (electricity, fossil fuels, water, steam)
3. Air emissions (direct and fugitive)
4. Wastewater (flows and loads)
5. Solid and hazardous wastes

It is best to quantify these items on a per unit product made basis, so that their cost can be allocated as such.

1. For each waste stream, identify the stage of the product's life cycle responsible for its generation. For example, wood pallets received at a facility from suppliers are generated at the procurement stage and should be the responsibility of product

developers or purchasing personnel to eliminate or minimize.

2. Identify the source of each waste constituent. For example, soluble BOD discharge from an industrial wastewater treatment plant at a food processing facility can be traced to production vessel heels containing acetic acid (vinegar).

3. Quantify each waste stream and associated waste constituent.

Phase II

1. Develop a mass balance for the manufacturing process, quantifying each waste constituent from raw material receipt to waste discharge. Quantify each waste source and perform a comprehensive materials balance around each stage of the life of the product.

2. Prioritize target waste streams based on the realistic, measurable goals developed at program inception. Issues to address in this prioritization process include:

   • Which processes are generating the most waste?
   • Which are inflicting the most environmental impact?
   • Are there any quick fixes or low-hanging fruit?
   • Which waste streams can be significantly changed?

3. Brainstorm alternatives and potential solutions to waste-generating activities, industry examples, and experience. Employees who work on the waste-generating processes on a daily basis must be involved, as they might be able to provide valuable insights.

4. Analyze the technical and economical feasibility of the alternatives for eliminating or minimizing waste sources by:

   • Elimination of the source
   • Raw materials that could result in fewer waste streams and/or less environmental impacts
   • Improved housekeeping
   • Increased efficiency through manufacturing process optimization, production scheduling, or preventive maintenance

5. Perform a life-cycle cost-benefit analysis on the complete set of feasible alternatives to identify areas where cost savings can be realized (Table 4-2). It is important to compare the costs for pollution prevention program implementation.

6. Select alternatives to implement based on the cost-benefit analysis and meeting stated environmental objectives and targets.

Table 4-2 Typical Costs and Benefits of Pollution Prevention Program Implementation

Costs	The following may be costs or benefits, so they must be quantified:
Capital Equipment	Operation and maintenance
Engineering	Labor
Installation	Chemicals
Operational Costs	Raw materials
Downtime	Energy
Benefits	
Higher Raw Material Yield	
Reduced Waste and Waste Treatment Costs	
Reduced Labor, Chemicals, and Energy	
Reduced Energy	
Reduced Waste Disposal Fees	

Hierarchy of Potential Implementation Strategies

Once an organization has committed to developing a pollution prevention program, developed and communicated an environmental policy, established a baseline with objectives and targets, and identified waste sources, it must begin implementing strategies to minimize pollution and achieve the objectives and targets. This should involve shifting the focus from the end-of-pipe treatment back into the process and process equipment within the facility—from initial phases of product development, through manufacturing, to final disposal of the product.

The strategies can be selected at the final stages of the Pollution Prevention Assessment, as described above. Feasible options should be developed in detail, including a full description of each option, the requirements for implementation, and the costs and benefits. Implementation of the option or options finally agreed upon must include a thorough and accurate performance evaluation. An accurate cost accounting system must be in place for this process to produce results of value (see cost-benefit discussion, above).

The following sections present a hierarchy of pollution prevention strategies based on the following six steps.

1. *Prevent* pollution through product and process design, housekeeping, process changes, increased raw material yield, and preventive maintenance.
2. *Reduce* the amount and toxicity of wastes generated by process optimization, material substitution, and manufacturing scheduling.
3. *Reuse* materials that would otherwise be considered waste. Develop new or alternative uses for "waste."
4. *Recycle* materials that cannot be reused.
5. Treat waste using effective and efficient methods, while minimizing energy usage

and the creation of new waste streams (e.g., wastewater sludge).

6. *Dispose* any wastes that can not be reused, recycled, or treated in the most environmentally safe manner possible and in compliance with all applicable federal, state, and local regulations.

Potential Implementation Strategies

Prevention

Prevention is the strategy with the highest potential for minimizing wastes and preventing pollution. If operations, manufacturing processes, or activities can be altered to ensure that pollution prevention is achieved and the generation of waste is prevented, often raw material yield and the cost per piece of manufacturing is lowered. For example, waste generated from product packaging can be reduced or eliminated if the amount of materials used to package a product is reduced. Similarly, operational processes can be changed and equipment upgraded to reduce the amount of waste generated in an industrial operation.

Immaculate housekeeping is considered one important way to prevent pollution. The adjective "immaculate" is used here to emphasize that extra effort, beyond the routine "good" housekeeping, is required to achieve and maintain the levels of performance in housekeeping that are expected for

---

A food processing facility added sequence manufacturing campaigns and dedicated manufacturing lines to its largest soluble BOD-generating product. As a result, the facility minimized reactor vessel cleanout and thereby reduced wastewater treatment plant energy and chemical use, extended the treatment capacity of its plant, and avoided treatment plant expansion as production levels increased.

---

an effective pollution prevention program. Constant attention must be devoted to preventing waste before it occurs through preventive maintenance. Waste must be prevented; if it is not, it must be cleaned up immediately, and then an analysis must be conducted to determine why the waste occurred, followed by action taken to prevent the same type of waste from reoccurring. Then procedures must be developed and implemented to prevent reoccurrence.

Spill containment and isolation techniques must be developed and continually improved. A key characteristic of immaculate housekeeping is that no leak, spill, or correctable inefficiency occurs twice. There must be an effective program in place whereby each undesirable occurrence is reported and analyzed, its cause is determined, and a corrective action is implemented.

A food processing facility was able to develop housekeeping procedures and provide sanitation staff with equipment capable of capturing soluble BOD-containing wastes, such as marinade and batter, that were not treated in their DAF wastewater pretreatment system. This practice lowered surcharge fees and helped the facility avoid a costly upgrade to biological treatment.

Reduce

"Reduce" is very similar to "prevent," with one notable difference. Methods of prevention may prevent the waste from being initially generated, whereas methods of reduction reduce the amount of waste that actually is generated.

Material substitution is an important method to reduce wastes. There are many instances where nontoxic substances can be substituted for toxic substances in industrial processes. Many more will be available in the future, as research and development efforts produce alternatives to former processes that use toxic materials. Keeping up with these new developments is crucial to continual improvement and education, which are discussed above.

A few highly successful examples of nontoxic materials replacing toxic materials throughout industry are:

- The use of oxygen, rather than chlorine, to bleach wood pulp
- The use of alcohol for pickling, rather than acid, in the manufacture of copper wire
- The use of ozone or ultraviolet light for disinfection rather than chlorine
- The use of water-based, rather than oil-based, paints (eliminates the need for solvents and thinners for cleanup)
- The use of nonphenolic industrial detergents, rather than those that contain phenolics
- The use of water-soluble cleaning agents or citrus degreasers, rather than organic solvents

---

Substandard biological wastewater treatment at a power plant was found to be caused by toxic zinc levels in the sanitary wastewater coming from the office building. This situation was corrected by substituting a metal-free floor wax for the zinc-laden wax used by custodians, which had caused the problem.

Wastes can also be prevented and/or reduced by changing manufacturing processes and equipment. Many products can be manufactured by use of two or more alternative processes. Often, one of the process types will involve the use of substances that are less toxic than others. In addition, within any single process type, there is usually a choice to make between several sources for the equipment, and one type may be more desirable from a pollution prevention standpoint than others. For instance, an item of equipment that is air-cooled might perform as well as an item that is water-cooled but would also preclude the need to discharge waste cooling water. Of course, it should be ascertained that the air used to cool the equipment would not become degraded in quality before a decision is made regarding replacement.

———————

A textile mill avoided possible discharge permit excursions for metals by analyzing all process chemicals, reducing overall chemical use by implementing process automation, changing to metal-free chemicals where possible, and scheduling production to average out metal-loading to its wastewater treatment plant. This product substitution and manufacturing scheduling approach avoided a costly metal removal treatment project.

———————

In cases where equipment is old, worn, and subject to leaks, spills, and inefficiencies, it might be cost effective to replace it, based on the savings in cost of materials, cost of operation, and cost of handling and disposing of wastes. However, a comprehensive and accurate materials balance around the entire life cycle of the product, from initial development to final disposal, is needed to make the correct decisions.

———————

A textile mill replaced large, manual-fed dye baths with smaller, computer-controlled chemical feed systems that prevented as much as 80% of bath waste at the end of each run.

———————

Cleaning and washing are activities within an industrial process that almost always produce wastes (e.g., solvents) that require management and disposal. Often, cleanup wastes contain underutilized raw material and

actual product, in addition to the chemical constituents of the cleaning agent.

The following is a list of some examples of changing manufacturing processes and equipment for pollution prevention within the cleaning and washing areas:

- Conical-bottom tanks to reduce the amount of heel after a production run
- Line-pigging systems to capture residual product and prevent heavy waste loads from line cleaning
- Counterflow washing and rinsing system in continuous manufacturing process to reduce water and energy use
- Heat exchangers on hot water cleaning or sanitation wash water prior to discharge for energy conservation
- Microfiltration membrane systems to capture raw material or product normally washed, to drain for reuse
- Use of high-pressure wands to reduce water use and energy

Reuse

If wastes cannot be prevented, and reduction methods have been implemented, facilities can develop ways to reuse wastes. Wastes can be returned to the industrial process to be used again or reused for a different purpose.

Water is an example of a "waste" that is often returned to the industrial process for reuse. If a plant's fresh water supply has to be treated before use in the process, it might be less expensive to treat the wastewater and reuse it, rather than treat the wastewater for discharge, in addition to treating more fresh water for "once-through" use. There is a limit to the extent to which water, or any other substance, can be reused. The reason is that water evaporates and leaves nonvolatile substances, such as salts, behind. The result is that nonvolatile substances build up, increasing one or more undesirable characteristics such as corrosivity and/or scaling.

The following example illustrates a methodology for determining the rate of blowdown required to maintain a minimum

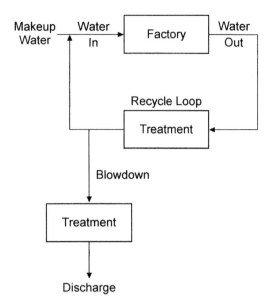

Figure 4-1   Water recycle loop with treatment and blow-down.

desired level of water quality for the water undergoing recycle for reuse.

Example 4-1

The schematic diagram presented in Figure 4-1 shows a water recycle loop in use at a factory. There is a treatment system for removal of organics and TSS. However, dissolved inorganics, such as chlorides, are not removed. These dissolved inorganics, therefore, build up in the recycle loop and can be prevented from building to above a desired maximum concentration only by use of a blow-down, also illustrated in Figure 4-1.

Problem

For a blow-down rate of 10%, calculate the equilibrium concentration of chloride ions in the water in the recycle loop, given the following information:

- Total volume of water in the entire factory recycle loop, including the in-line treatment system = 1.0 million gallons.
- Chloride ion (Cl⁻) concentration in raw water supply = 10 mg/L as Cl⁻

- Chloride ion (Cl⁻) added by factory = 90.5 lb/day
- 0 mg/L Cl⁻ removed by in-line treatment
- Blow-down = 10% = 100,000 gpd
- Evaporative loss = 5% = 50,000 gpd

Solution

Chloride in system on the first day:

$$10 \text{ mg/L} \times 8.34 \times 1.0 = 83.4 \text{ lbs}$$

Chloride added each day in make-up water:

$$10 \text{ mg/L} \times 8.34 \times (0.1 + 0.05)$$
$$= 12.5 \text{ lb/day}$$

Chloride removed each day via blow-down:

$$X \text{ mg/L} \times 8.34 \times 0.1 = 12.5 + 90.5$$
$$= 103 \text{ pounds.}$$

The concentration, X, of chloride ions in the blow-down must equal the concentration (x) in the recycle loop:

$$X = \frac{103}{8.34 \times 0.1} = 123 \text{ mg/L}$$

Recycle

Recycle is a term that is often used in conjunction with reuse, as the terms are very similar. While people may often associate recycling with being "green" or "environmentally friendly," recycling should be considered as similar to disposal. If a waste is ready to be recycled, it has already been generated. Preventing waste generation in the first place precludes the need to recycle.

However, once wastes have been prevented, reduced, and reused to the extent possible, recycling is the next step. Recycling water can be a significant part of an industrial pollution prevention program. Even if some degree of treatment is required before two or several reiterative uses, this may be much less costly than the "once-through-use" approach. The quality of water, in terms of conventional and/or priority or other pollutants, may not need to be nearly as "good" for the industrial process as for discharge in compliance with an NPDES or other discharge permit. Therefore, treatment for recycling purposes might be less costly than treatment for discharge.

A cotton-dying operation was paying a high COD surcharge to the wastewater utility. Consideration of pollution prevention alternatives led to pilot-scale studies that showed dye baths could be reused up to five times, rather than dumped after each run, by making up less than 25% of new raw materials. Implementation of this dye bath reuse not only resolved the high COD surcharge fee, but dramatically reduced chemical costs with no effect on product quality.

Recycling Smokestack Emissions

Increasingly, there are technologies available for reusing industrial waste streams. Smokestack emissions are traditionally considered a final "waste" product. However, there are technologies available for recycling these emissions. One example is an algae bioreactor system developed by GreenFuel Technologies Corporation in Cambridge, MA. "Using the sun as a free energy source, GreenFuel's proprietary algae bioreactor system produces commercial-grade biodiesel from smokestack emissions. In most cases, our biodiesel costs the same as or less than conventional petroleum diesel and recycles up to 85% of NO_x and 45% of CO_2 from the smokestack stream" (http://www.greenfuel-online.com/index.htm). A technology of this kind, which turns a waste product into a usable product, is important to research when considering potential recycling efforts within a mature pollution prevention program.

A food processor reduced its food waste disposal costs by 50% by changing its disposal outlet from a rendering facility to a composting facility.

It has become common in the textile industry to use polyvinyl chloride as a sizing agent. While it is more expensive than the traditionally used starch, which imparts a substantial organic load on wastewater treatment plants, polyvinyl chloride is readily recoverable and reusable through the implementation of membrane filtration technology.

A metal foundry had to install a sodium bisulfite dechlorination system to remove chlorine from its once-through cooling water system (using chlorinated public drinking water) to meet state discharge permit limitations. A cooling tower was installed and the cooling water loop closed, saving $500 per day in water use and eliminating the need for the dechlorination system and the discharge permit.

Treat

Treating wastes prior to disposal, provided the treatment is conducted in accordance with applicable regulations, can reduce the quantity of wastes that require disposal. If the treatment costs are reasonable, this can reduce the overall cost of waste disposal. In many cases it is advantageous to isolate one (or more) waste stream in an industrial plant and treat it separately from the other waste streams (e.g., gaseous, liquid, or solid) rather than allowing it to commingle with other waste streams prior to treatment. The segregated waste stream can then be treated and recycled, mixed with other treated effluents for discharge to the environment, or discharged separately. The following are among the many advantages of wastes segregation:

- Many substances are readily removed by specialized techniques when they are in the relatively pure and concentrated state, but are difficult to remove after being mixed with other substances and diluted by being mixed with other waste streams. For instance, certain organics, such as chlorophenols, that slowly biodegrade, are efficiently removed by activated carbon. Treating a waste stream containing these substances at the source may be more cost effective than mixing the waste with other substances that are biodegradable and on their way to a biological treatment system.
- There is more likelihood of producing a usable by-product from a segregated, relatively pure waste stream.
- There is more likelihood that the stream can be recycled if it has not been mixed with other waste streams.
- The segregated stream can be treated on a batch basis, or other campaign basis, depending of the operating schedule of the process.

A variation of the wastes segregation approach involves selective mixing of certain waste streams. This can take advantage of commonality in compatibility to certain treatment processes. Also, acid waste streams can be mixed with caustic waste streams for mutual neutralization before additional treatment.

One industry's waste might be a valuable resource for another. For instance, waste acid from one industry might be suitable for either a processing step or for neutralization of caustic wastes from another industry. Participation in waste exchanges is an attractive alternative to waste treatment and disposal. There are many networks of business that are constantly seeking uses for the wastes they generate. In addition, there may be waste exchange brokers operating in a given industry's geographical area.

A promising opportunity to reduce operating costs at a manufacturing facility is to conduct a comprehensive review of all treatment operations. This is especially important after a facility has implemented successful pollution prevention measures, especially water use reduction measures. Treatment systems designed 10 and 20 years ago might not be properly arranged for today's waste flows and loads, and minor adjustments could save considerable costs in chemical and energy use and could dramatically improve treatment efficiency.

Dispose

Regardless of the success of a pollution prevention program, some quantity of waste will require disposal. Proper disposal is a crucial part of a successful pollution prevention program, as improperly disposed wastes have historically led to significant pollution. The following guidelines should be considered when considering disposal options: perform a hazardous waste characterization for each waste stream to determine if it is regulated as hazardous waste. If it is hazardous waste, ensure that it is managed in accordance with all applicable hazardous waste regulations. If it is not hazardous waste, ensure that it is managed and disposed of in accordance with potentially applicable solid waste or other regulations.

------•------

A heavy industrial manufacturing facility was paying for 30 cubic yard dumpster removal on a cubic yard basis, and the dumpsters were picked up weekly, regardless of their content. While the facility had a long-term contract that was not easily broken or amended, it did cease using these dumpsters for wood pallet disposal. The wood pallets, in all phases of disrepair, shape, and size, took up a large volume and contributed very little weight to the dumpsters. The facility began a wood program, in which employees were allowed to salvage the wood for their own use. Eventually, the facility negotiated better terms that included dumpster rental and waste disposal by weight.

------•------

Waste disposal in total is often a significant operating cost, and facilities are often involved in long-term disposal contracts. It is often useful to undertake a careful review of how each waste stream is being disposed of, in some instances with the assistance of current or potential future waste disposal contractors. It is extremely important for both environmental consideration and corporate risk management to fully understand who is transporting and who is accepting your waste. It is facility environmental management's responsibility to properly characterize each waste stream and to ensure each waste stream is being disposed of properly, legally, and with the minimum present and future environmental risk.

Bibliography

GreenFuel Technologies Corporation. Web site: http://www.greenfuelonline.com/index.htm.

Miller, H. "In-Plant Water Recycling—Industry's Answer to Shortages and Pollution." *Design News—OEM* (November 5, 1973): 81–88.

U.S. Environmental Protection Agency. *RCRA Waste Minimization National Plan.* EPA 530-D-94-001. Washington, DC: U.S. Government Printing Office, 1994.

U.S. Environmental Protection Agency. *Summary of the RCRA Hazardous Waste Minimization National Plan—Draft.* EPA 53 0-5-94-002. Washington, DC: U.S. Government Printing Office, 1994.

U.S. Environmental Protection Agency. *Waste Minimization National Plan.* EPA 530-F-97-010. Washington, DC: U.S. Government Printing Office, 1997.

U.S. Environmental Protection Agency. Hazardous Waste Engineering Research Laboratory. *Waste Minimization Opportunity Assessment Manual.* EPA/625n-881003. Cincinnati, OH: 1988.

U.S. Environmental Protection Agency. *Waste Minimization—Environmental Quality with Economic Benefits.* EPA 530-SW-87-026. Washington, DC: U.S. Government Printing Office, 1987.

U.S. Environmental Protection Agency. *WasteWise.* Web site: http://www.epa.gov/wastewise/.

U.S. Environmental Protection Agency. *Pollution Prevention.* Web site: http://www.epa.gov/p2/.

U.S. Environmental Protection Agency. *The National Waste Minimization Program.* Web site: http://www.epa.gov/wastemin/.

U.S. Environmental Protection Agency. *Center for Sustainability.* Web site: http://www.epa.gov/region03/chesapeake/center.htm.

U.S. Environmental Protection Agency. *Pollution Prevention Information Clearinghouse.* Web site: http://www.epa.gov/opptintr/library/ppicindex.htm.

U.S. Environmental Protection Agency. *Greening the Supply Chain.* Web site: http://www.epa.gov/opptintr//dfe/tools/greening.htm.

5 Waste Characterization

Waste Characterization Study

Waste characterization is the term used for the process of determining the chemical, biological, and physical characteristics, as well as the quantity, mass flow rates, strengths (in terms of concentration), and discharge schedule, of a wastewater stream, air discharge, or solid waste stream. A waste characterization program must be carefully thought out and properly executed. The foundation of the study is a sampling and analysis program, which must be performed on representative samples. The equipment used to measure rates of flow and to physically obtain samples must be appropriate to the application and accurately calibrated.

There are three general categories of waste characterization study in common use: the Wastewater (or Air Discharge, or Solid Wastes Stream) Characterization Study, the Environmental Audit, and the Waste Audit. The appropriate choice among these three categories, for a given application, depends upon the principal purpose of the study. A Waste Characterization Study (Wastewater, Air Discharge, or Solid Wastes Stream) is usually carried out for the purpose of obtaining design criteria for a waste treatment facility, with a concurrent pollution prevention program. An Environmental Audit is performed for the purpose of assessing a plant's state of compliance with various environmental regulations, and a Waste Audit is carried out for the purpose of assessing opportunities to minimize the amount of waste generated through improved efficiency or through substitution of nonhazardous substance(s) for one or more of those that are classified as hazardous. In each case, choices

are made as to the location of sampling points, equipment to be used, the sampling schedule, and the laboratory and field analyses to be performed. There is always a balance to be struck between the cost of the program and the ultimate value of the data obtained.

Choice of Sampling Location

Since pollution prevention is always a primary objective of any waste management program, waste sampling programs should always be designed to determine at which locations in an industrial processing plant significant amounts of waste are generated. Otherwise, it would be necessary to sample only the final composite effluent from the entire plant. The following example illustrates some of the choices to be made when designing a wastes sampling program.

Figure 5-1 is a schematic of an electroplating shop with four different plating processes, designated Process 1, Process 2, Process 3, and Process 4. At the present time, all four processes discharge to a common drain that leads to the municipal sewer system. The task at hand is to develop a waste sampling and analysis program to provide data for a waste reduction program, as well as to enable calculation of design criteria for one or more treatment devices to pretreat the wastewater prior to its discharge into the municipal sewer system, within compliance with all applicable regulations. If the sole objective were to treat the wastewater to within compliance with the regulations, it would make sense to locate one composite sampler at the end of the building to sample the mixed effluent from all four plating processes. The

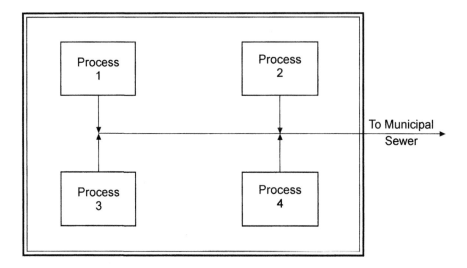

Figure 5-1 Schematic of an electroplating shop with four different processes.

questions then, would be, "How many days should the sampling period cover?" and "Over how long a time should each compositing period take place?"

The answer to the first question depends on the processing schedule and whether or not different processes are run on a campaign basis in one or more of the four processing units. It is more or less standard practice to sample the wastes from a given process (or set of processes) over a three-consecutive-day period. Five would be better than three, but a decision has to be made between the greater costs for the longer sampling period and the greater risk associated with the shorter sampling period. A prudent engineer will develop more conservative design criteria if the risk of not having accurate waste characteristics is higher. The higher cost for the more conservatively designed treatment system may well be more than the higher cost for the longer sampling period.

The second question addresses the length in time of each compositing period. Four six-hour composites per day produce four discrete samples to be analyzed, whereas two 12-hour composite samples taken each day will cost only half as much to have analyzed. Using any statistical approach available, the more discrete samples taken during the 24-hour operating day (that is, the shorter the compositing periods), the more accurate the results of the wastes characterization study will be. Here, again, a prudent engineer will recognize that more conservatism, and, therefore, higher cost, will have to be designed into a system. When the compositing periods are long, the number of discrete samples each day is low, and the risk of not having accurate, detailed characterization information is higher.

If the four plating processes are quite different from each other, a less expensive overall treatment system might result if one or more are treated separately. If such is the case, it would be appropriate to locate composite samplers at the discharge point of each of the four processes. Now, the number of samples to be analyzed for a given number of sampling days and a given number of composites each day is multiplied by four. Still, the considerations of risk, conservatism in design, and total cost apply, and it is often cost effective to invest in a more expensive wastes characterization study to obtain a lower total project cost.

It is seldom prudent to consider that the sole reason for carrying out a wastes characterization study is to obtain data from which to develop design criteria for a wastes treatment system. Rather, pollution prevention should almost always be a major objective, as it should be with any wastes management initiative. As discussed in Chapter 4, the many benefits of pollution prevention include lower waste treatment costs, as well as lower costs for disposing of treatment residuals.

When taken in the context of a pollution prevention program, a wastes characterization study takes on considerations in addition to those discussed above. Using the same example illustrated in Figure 5-1, it is seen that locating only one composite sampler to sample the combined wastewater from all four plating processes would yield little information useful for pollution prevention purposes. For pollution prevention purposes, it is necessary to locate at least one composite sampler at the wastes discharge from each of the four plating processes. Furthermore, there is an important consideration of timing regarding execution of the sampling program. In order to enable measurement of the effectiveness and therefore the value, in terms of cost savings, of the pollution prevention program, a complete wastes characterization study should be carried out before wastes minimization or other aspects of pollution prevention take place. These data, however, will not be useful for developing design criteria for wastes treatment, since implementation of the pollution prevention program will, hopefully, significantly change the characteristics of the waste stream to be treated.

A second wastes characterization study, then, should be conducted after the implementation and stabilization of the pollution prevention program. Stabilization is emphasized here, because improved housekeeping—in the form of spill control, containment, and immediate in-place cleanup; water conservation, containment, and recycling of "out of spec. product or intermediate" (rather than dumping these "bad batches" to the sewer); and other process efficiency improvement measures—is implemented (as part of the pollution prevention program). If

some of the former poor housekeeping and materials control inefficiency creeps back into the industry's routine operations, treatment processes designed using data obtained during full implementation of the pollution prevention program will be overloaded and will fail.

The principal objectives of a waste management program, which include pollution prevention along with wastes characterization, are to ensure: (1) that truly representative samples are taken, (2) that the appropriate samples are taken and the appropriate analyses performed, as dictated by the *Clean Water Act* and *RCRA*, (3) that the information obtained is appropriate and sufficient to produce an optimal waste-minimization result, and (4) that the optimum balance is struck between the cost of the waste characterization study and the cost for the treatment facilities ultimately designed and constructed.

Sampling Equipment

In general, there are two types of automatic samplers: (1) discrete and (2) integrated, or totalizing. Discrete samplers place each individual sample into its own container. These samplers are used when it is deemed to be worth the extra expense to determine the variability of the waste stream over the sampling period. Integrated samplers place each individual sample into a common container. Figure 5-2 shows a picture of a typical automatic sampler.

Sample Preservation

It is always desirable to perform laboratory analyses of samples as soon after the sample is taken as possible. However, appropriate measures must be taken to ensure that the characteristics of the sample that are to be measured in the laboratory will not change, no matter how soon (or late) after the samples are taken the analyses are performed. One of the most common problems that results in changes in sample characteristics is

Figure 5-2   Illustration of a typical automatic wastewater sampling device. (Courtesy of ISCO, Inc.)

bacterial action. If conditions within the sample allow bacterial metabolism, the sample will not be representative of the waste stream it was taken from by the time it arrives at the laboratory.

Two common methods of preventing or minimizing bacterial growth are: (1) lowering the temperature of the sample, and (2) reducing the pH of the sample.

The rate of bacterial metabolism or, indeed, the metabolism of most life forms, decreases by half for every 10-degree centigrade decrease in temperature. Metabolism essentially stops at 0°C. For this reason, refrigeration, mechanically or with ice, is accepted as a good method for preserving samples. It must be determined prior to sampling, however, whether or not the refrigeration itself will change the characteristics of the sample. A possible reason for such change is decreased solubility of one or more substances in the sample, which might then result in precipitation of that substance.

A second method of preserving samples is to lower the pH to between 1 and 2. Here

again, care must be taken not to change the character of the sample. The best way to determine the effect of sample preservation is to perform a small number of laboratory analyses immediately after taking a sample and compare the results to samples subjected to exactly the same preservation protocol, including elapsed time between sample-taking and preservation and laboratory analyses, as is anticipated for the actual waste characterization program.

Sampling for Oil and Grease

Special sampling techniques must be used when assaying a waste stream for certain substances. Nonmiscible substances in wastewater, such as oils, greases, and waxes, are examples of substances where special sampling is required. In general, these substances must be sampled by taking grab samples, using a dipping action that ensures taking some liquid from throughout the depth of wastewater flow, including the surface. The current edition of *Standard Methods*, as well as any special instructions issued by the EPA or other appropriate authority, must always be consulted to determine the currently accepted techniques.

Sampling for Volatile Substances

Substances such as trichloroethylene (TCE) that have vapor pressures significantly higher than water and that will volatilize at ambient temperatures are additional examples of substances that require special sampling techniques. Again, the current edition of *Standard Methods*, in addition to any special instructions (current) issued by the EPA or other appropriate authority, must be consulted before embarking on a sampling program.

Waste Audit

A Waste Audit has the primary purpose of assessing opportunities to improve efficiencies, decrease waste, or substitute nonhazardous materials for hazardous materials, and

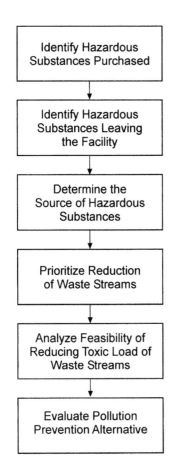

Figure 5-3   Waste Audit procedure.

thereby minimize waste generation. An accounting procedure is to be used, along with a materials balance approach to account for the fates of as many important substances as possible. Figure 5-3 presents a suggested work plan for a Waste Audit whose focus is on hazardous substances. The sections that follow address each step in the Waste Audit.

Identify Purchased Hazardous Substances

The records of the purchasing agent or the purchasing department, along with inventory records, can be examined to determine all basic and proprietary chemicals and products that are brought into the facility. In some cases, records are maintained at corporate headquarters. Materials purchased for research and development use and for laboratory use must also be scrutinized. It is most important, especially in the case of proprietary products, to determine each individual chemical component.

As an example, at a very large aluminum die cast plant belonging to one of the giants of the automobile industry, there was a problem with excessive quantities of phenolic compounds in the effluent from the wastewater treatment facility. Noncompliance problems forced extensive work to try to improve removal of phenolics by the wastewater treatment equipment. At the same time, a thorough, and unsuccessful, search for the source of the phenolics was carried out. Finally, the facility management decided to purchase and install a large activated carbon system to use as tertiary treatment to remove the phenolic substances. It was also determined that the activated carbon system would have to be preceded by a large sand filter system. While the preliminary engineering for the new tertiary system was in progress, the search for the source of the phenolic substances was greatly intensified. It was finally determined that a detergent that was used every night by an outside cleaning service was the source. The service changed to a different detergent and the entire problem disappeared. Fortunately, the sand filter–activated carbon tertiary treatment system had not yet reached the construction stage.

Identify Hazardous Substances Leaving the Facility

All discharges of wastewater, solid wastes, and air must be thoroughly characterized to determine which hazardous substances leave the facility and in what form. This may include raw materials that either were not consumed in the process, or became a component of the final product, a portion of which might be discharged as waste.

For example, chlorine is used in the manufacture of paper from wood pulp. One of the uses of chlorine is to solubilize lignin fragments that cause the pulp and, consequently, the paper to have a brown color. Another function that chlorine performs is to add across the double bonds in the organic structure of the lignin molecule, which changes the molecule (or molecular fragment) from having a brown color to being colorless. In summary, chlorine is used to "bleach" wood pulp in order to make white paper by allowing the removal of one portion of the lignin (which becomes solubilized, then rinsed away with water) and making the other portion, which stays with the pulp (and ultimately the paper), colorless. The rinse water is then wastewater, which must be treated and discharged. The chlorine that enters into the process of adding across double bonds goes out of the plant with the paper (product).

For each pound of chlorine purchased, a portion leaves the plant as a component of the wastewater, and a portion leaves the plant as a component of the salable product of the plant. Closer scrutiny will reveal that yet another portion leaves the plant as a component of the airborne emissions, some with the fugitive emissions and some with the stack emissions. This is because chlorine is a gas at ambient temperatures, and it volatilizes at every opportunity. It requires a very thorough and detailed study to determine the fate of chlorine that is purchased by an integrated pulp and paper mill.

Determine the Source of Hazardous Substances

Once each waste stream leaving the facility (solid, liquid, and airborne) has been thoroughly characterized, and all hazardous substances are identified and/or accounted for, the next step is to determine the source of each one. This can be done by use of an accounting procedure, together with principles of chemistry and knowledge of the individual industrial processes. The accounting procedure is used as a format to ensure that all chemical reactants and products are accounted for.

For some large industrial facilities, such a detailed accounting is a very significant undertaking. In the long term, however, the entire process is almost guaranteed to produce a financial gain, rather than a long-term loss, for the industrial facility. The example of the floor washing detergent given above is a graphic demonstration of the benefit of being tenacious in determining the source of each hazardous substance.

Prioritize Reduction of the Waste Streams

The objective of this step is to identify, in order of degree of adverse effect on the environment, which waste streams should be addressed first, to achieve maximum benefit to the environment. Although in some cases such prioritization will present itself as obvious, in other cases an in-depth study of the affected environment will be required to determine proper prioritization. For instance, if chlorinated (halogenated) organics are being discharged to an extremely large river, and, at the same time, a much smaller quantity per day of chlorine gas is being discharged to the atmosphere, but the plant is in a nonattainment area for ozone (another strong oxidizing substance), the waste stream that should receive the higher priority is not obvious. On the other hand, if the facility has a solid waste stream that is very concentrated in a chlorinated solvent that could be reduced dramatically by installation of a new, highly efficient vapor degreaser plus a new still for recovery and reuse of the solvent, the choice of top priority for action is more easily identified.

Analyze the Feasibility of Toxic Load Reduction

To continue with the example of the paper mill, an alternative process for accomplishing the bleaching of wood pulp to obtain white paper is to use oxygen, rather than chlorine, as the bleaching agent to remove about 50% of the lignin. However, to make the change, new equipment would have to be purchased and installed, at significant capital expense, and the value of the paper (product) might well be reduced because of a lesser degree of whiteness, though not necessarily. In order to make an informed decision, then, a detailed analysis of both technical feasibility and economic feasibility must be carried out.

Regarding technical feasibility, there are many questions, including:

- Can the oxygen process, known as "oxygen delignification," produce pulp that can ultimately be "bleached" to the desired degree of whiteness?
- Is the necessary equipment available?
- Can the existing water pollution, air emission, and solid waste control facilities function properly if the change is made? Will they perform better? Will any components be redundant?
- Are there other facilities and/or equipment in the plant that will have to be changed?
- Will the environmental permit requirements be met using the new process?

Regarding financial feasibility, the questions include:

- Does a cost-benefit analysis support the financial viability of making the change?
- Can the facility actually obtain sufficient funds to make the change?
- Can the facility survive the period between the time of initial investment and the time when overall net savings become a reality?

Evaluate the Economics of Pollution Prevention

It is one thing to ascertain the financial feasibility of a project; it is quite another thing to determine the short-term and long-term effects on the overall profit and loss position of the company. Evaluation of the economics of pollution prevention versus other approaches to waste management means

evaluating the financial effects on all phases of the product, from initial production, throughout the useful life, to final disposal. This evaluation is a very extensive undertaking, and interim findings must be continuously updated as new information becomes available. Continuing with the paper mill example, the decision to be made is whether to continue with the (ever-increasing) expense of managing the wastes from the chlorine bleaching process or to invest a large capital expenditure in an oxygen delignification system and thereby reduce waste management costs. In this case, the costs of management throughout the useful life and the costs for final disposal will not be affected, because the product is, for practical purposes, identical whether chlorine bleaching or oxygen delignification is used. However, there are large differences in:

- Costs for the "bleaching agent" itself (chlorine compounds versus oxygen compounds)
- Costs for processing equipment
- Costs for air pollution control
- Costs for wastewater treatment
- Costs for management of the solid wastes (certain chlorine wastes must be managed as "hazardous waste")
- Costs for obtaining discharge permits
- Costs for monitoring air discharges, wastewater discharges, and solid wastes disposal facilities

There are many more categories of costs that are affected by a major change in industrial process, especially one that changes from a process that generates hazardous wastes to a process that does not. In the final analysis, it is the size of the long-term net financial gain or loss on which the decision of whether or not to change a process should be based.

A properly executed Waste Audit should result in an overall savings in the cost of operation of an industrial facility, especially as regards the impact of the cost of wastes handling and disposal on the annual cost of plant operations.

Environmental Audit

An Environmental Audit is normally conducted to assess the state of compliance of an industrial facility with laws and regulations. As such, there is no need to sample and analyze waste streams, since typically the audit would include analyzing previously collected sample data. What is important, however, is to ensure that the sampling and analysis work that has been done is appropriate to the law or regulation under consideration.

The two federal laws that waste (solid and liquid) discharged from industrial facilities must comply with are the Clean Water Act (CWA) and the Resource Conservation and Recovery Act (RCRA). These two laws not only place restrictions on different substances, but require different protocols for sampling and analysis work in the case of some of the substances. A third law, the Clean Air Act, which sets compliance standards for air emissions, will be discussed later in this chapter.

Table 5-1 presents an example of a list of substances for which the CWA and RCRA, respectively, may specify restrictions. Notice that the CWA requires more compounds to be analyzed than RCRA. For example, the CWA restricts the discharge of 23 metals, whereas RCRA prohibits the discharge of 8 metals above certain concentrations. Minimum testing requirements for the CWA are defined by Tables I through III of Appendix D to *40 C.F.R. Part 122*. The facility may also be expected to report on compounds listed in Tables IV and V, if there is reasonable expectation that the pollutant may be present. Table 5-1 is not intended to be used as guidance for what a facility would be expected to test for. Rather, it is intended to illustrate that different regulations require testing for different pollutants.

Table 5-1   Requirements for Chemical, Physical, and Bacterial Analyses

Parameters	CWA NPDES List (1)	RCRA Hazardous Waste (2)	SDWA SDWA List (3)
Volatile Organics			
Acrolein	X		
Acrylonitrile	X		
Benzene	X	X	P-R
Bromobenzene			U
Bromodichloromethane	X		
Bromoform	X		
Bromomethane	X		U
Carbon tetrachloride	X	X	P-R
Chlorobenzene	X	X	P-R
Chloroethane	X		
2-Chloroethylvinyl ether	X		
Chloroform	X	X	
Chloromethane	X		
Dibromochloromethane	X		U
1,2-Dibromoethane	X		P-R
Dibromomethane	X		

Table 5-1   Requirements for Chemical, Physical, and Bacterial Analyses *(continued)*

Parameters	CWA NPDES List (1)	RCRA Hazardous Waste (2)	SDWA SDWA List (3)
Dichlorodifluoromethane	X		
1,1-Dichloroethane	X		U
1,2-Dichloroethane	X	X	P-R
1,1-Dichloroethene	X	X	P-R
cis-1,2-Dichlorethene	X		P-R
trans-1,2-Dichloroethene	X		P-R
1,2-Dichloropropane	X		P-R
1,3-Dichloropropylene	X		U
2,2-Dichloropropane	X		U
1,1-Dichloropropene	X		U
cis-1,3-Dichloropropene	X		U
trans-1,3-Dichloropropene	X		U
Epichlorohydrine			P-R
Ethylbenzene	X		P-R
Ethylene dibromide			P-R
4-Isopropyltoluene	X		U
Methylene chloride	X		P-R
Methyl ethyl ketone		X	
n-Propylbenzene	X		
Styrene	X^5		P-R
1,1,1,2-Tetrachloroethane	X		
1,1,2,2-Tetrachloroethane	X		U
Tetrachloroethene	X		P-R
Toluene	X		P-R
1,2,3-Trichlorobenzene	X		
1,2,3-Trichlorobenzene			P-R
1,1,1-Trichloroethane	X		P-R
1,1,2-Trichloroethane	X		U
Trichloroethene	X	X	P-R
1,2,4-Trimethylbenzene			U
Vinyl chloride	X	X	P-R
m-Xylene	X^5		P-R
o-Xylene	X^5		P-R
p-Xylene	X^5		P-R
Total Trihalomethanes			P-R
Semivolatile Organics			
Acenaphthene	X		
Acenaphthylene	X		

Table 5-1   Requirements for Chemical, Physical, and Bacterial Analyses *(continued)*

Parameters	CWA NPDES List *(1)*	RCRA Hazardous Waste *(2)*	SDWA SDWA List *(3)*
Anthracene	X		
Benzidine	X		
Benzo(a)anthracene	X		
Benzo(a)pyrene	X		P-R
Benzo(b)fluoranthene	X		
Benzo(g,h,i)perylene	X		
Benzo(k)flouranthene	X		
bis(2-Chloroethoxy)methane	X		
bis(2-Chloroethyl)ether	X		
bis(2-Ethylhexyl)phthalate	X		P-R
4-Bromophenyl-phenylether	X		
Butylbenzylphthalate	X		
Carbofuran	X^5		P-R
4-Chloro-3-methlyphenol	X		
2-Chloronaphthalene	X		
2-Chlorophenol	X		
4-Chlorophenyl-phenylether	X		
bis 2-chloropropylether	X		
Chrysene	X		
Di-n-butylphthalate	X		
Di-n-octylphthalate	X		
Dibenz(a,h)anthracene	X		
1,2-Dibromo-3-chloropropane			P-R
1,2-Dichlorobenzene	X		P-R
1,3-Dichlorobenzene	X		
1,4-Dichlorobenzene	X	X	P-R
3,3-Dichlorobenzidine	X		
2,4-Dichlorophenol	X		U
Di(2-ethylhexyl)adipate			P-R
Diethylphthalate	X		
2,4-Dimethylphenol	X		
Dimethylphthalate	X		
4,6-Dinitro-2-methylphenol	X		
2,4-Dinitrophenol	X		U
2,4-Dinitrotoluene	X	X	U
2,6-Dinitrotoluene	X		U
1,2-Diphenylhydrazine	X		U
Fluoranthene	X		

Table 5-1 Requirements for Chemical, Physical, and Bacterial Analyses *(continued)*

Parameters	CWA NPDES List (1)	RCRA Hazardous Waste (2)	SDWA SDWA List (3)
Fluorene	X		
Hexachlorobenzene	X	X	P-R
Hexachlorobutadiene	X	X	
Hexachlorocyclopentadiene	X		P-R
Hexachloroethane	X	X	
Indeno(1,2,3-cd)pyrene	X		
Isophorone	X		
2-Methylphenol		X	U
3,4-Methylphenol		X	
Naphthalene	X		
Nitrobenzene	X	X	U
2-Nitrophenol	X		
4-Nitrophenol	X		
N-Nitroso-di-n-propylamine	X		
N-Nitrosodimethylamine	X		
N-Nitrosodiphenylamine	X		
Pentachlorophenol	X	X	P-R
Phenanthrene	X		
Phenol	X		
Pyrene	X		
Pyridine		X	
Simazine			P-R
1,2,4-Trichlorobenzene	X		P-R
2,4,5-Trichlorophenol		X	
2,4,6-Trichlorophenol	X	X	U
Pesticides/PCBs			
Arochlor-1016	X		
Arochlor-1221	X		
Arochlor-1232	X		
Arochlor-1242	X		
Arochlor-1248	X		
Arochlor-1254	X		
Arochlor-1260	X		
Polychlorinated biphenyls (as Arochlors) (as Decachlorobiphenyl)			P-R
Aldrin	X		
alpha-BHC	X		
beta-BHC	X		

Table 5-1   Requirements for Chemical, Physical, and Bacterial Analyses *(continued)*

Parameters	CWA NPDES List (1)	RCRA Hazardous Waste (2)	SDWA SDWA List (3)
delta-BHC	X		
gamma-BHC (Lindane)	X	X	P-R
Chlordane	X	X	P-R
4,4'-DDD	X		
4,4'-DDE	X		U
4,4'-DDT	X		
Dieldrin	X		
Endosulfan I	X		
Endosulfan II	X		
Endosulfan sulfate	X		
Endothall			P-R
Endrin	X	X	P-R
Endrin aldehyde	X		
Heptachlor	X	X	P-R
Heptachlor epoxide	X	X	P-R
Methoxychlor		X	P-R
Oxamyl			P-R
Toxaphene	X	X	P-R
Herbicides			
Alachlor			P-R
Atrazine			P-R
2,4-D		X	P-R
Dalapon			P-R
Dinoseb			P-R
Diquat (as Diquat dibromide monohydrate)			P-R
Glyphosate			P-R
Picloram			P-R
2,4,5-TP (Silvex)		X	P-R
Metals			
Aluminum, total	X[5]		S-R
Antimony, total	X		P-R
Arsenic, total	X	X	P-R
Barium, total	X[5]	X	P-R
Beryllium, total	X		P-R
Boron, total	X[5]		U
Cadmium, total*	X	X	P-R
Chromium, total*	X	X	P-R
Cobalt, total	X[5]		

Table 5-1   Requirements for Chemical, Physical, and Bacterial Analyses *(continued)*

Parameters	CWA NPDES List (1)	RCRA Hazardous Waste (2)	SDWA SDWA List (3)
Copper, total	X		P-R
Iron, total	X[5]		S-R
Lead, total*	X	X	P-R
Magnesium, total	X[5]		R
Manganese, total	X[5]		S-R
Mercury, total*	X	X	P-R
Molybdenum, total	X[5]		
Nickel, total	X		
Selenium, total*	X	X	P-R
Silver, total*	X	X	S-R
Sodium, total			P-R
Thallium, total	X		P-R
Tin, total	X[5]		
Titanium, total	X[5]		
Zinc, total	X		S-R
General Chemistry			
Ammonia (as N)	X		
Asbestos	X[5]		P-R
Biochemical oxygen demand (BOD)	X		
Bromide	X[5]		
Chemical oxygen demand (COD)	X		
Chloride			S-R
Chlorine, total residual	X[5]		
Coliform, fecal	X[5]		P-R[†]
Coliform, total			P-R
Color	X[5]		S-R
Corrosivity		X	S-R
Cyanide	X		P-R
Flow	X		
Flouride, soluble	X[5]		P-R
Foaming Agents			S-R
Ignitability		X	
Nitrate	X		P-R
Nitrite	X		P-R
Nitrogen, total Kjeldahl	X		
Nitrogen, total organic	X		
Odor			S-R
Oil and Grease	X[5]		

Table 5-1   Requirements for Chemical, Physical, and Bacterial Analyses *(continued)*

Parameters	CWA NPDES List (1)	RCRA Hazardous Waste (2)	SDWA SDWA List (3)
pH	X		S-R
Phenols	X		
Phosphorus, total (as P)	X⁵		
Radioactivity, total alpha	X		P-R
Radioactivity, total beta	X		P-R
Radium 226, total	X		P-R
Radium 228	X		P-R
Radium, total	X		
Radon, 222	X		P-R
Reactivity		X	
Silica			P-R
Sulfate (as SO4)	X⁵		S-R
Sulfide (as S)	X⁵		
Sulfite (as SO3)	X⁵		
Surfactants	X⁵		S-R
Temperature	X		P-R
Total dissolved solids (TDS)	X		S-R
Total organic carbon (TOC)	X		
Total suspended solids (TSS)	X		
Turbidity	X		P-R
Uranium			P-R
Other			
2,3,7,8 –Tetrachloro-dibenzo-p-dioxin	X		P-R

Notes:

P = Primary drinking water analyte.

R = Regulated parameter.

S = Secondary drinking water analyte.

U = Unregulated parameter.

X = Parameter regulated by this list.

(1) This table is an example of what could be required for an NPDES permit; refer to *40 C.F.R. Part 122* for specific details. NPDES = National Pollutant Discharge Elimination System.

(2) Current reference list: 55FR 11862, March 29, 1990, as amended at 55 FR 22684, June 1, 1990; 55 FR 26987, June 29, 1990; 58 FR46049, August 31, 1993; 67 FR 11254, March 13, 2002.

(3) Current reference list: *40 C.F.R. Part 141* Subpart F (56 FR 3593, January 30, 1991, as amended).

(4) Information provided by Katahdin Analytical Services, Westbrook, ME.

*RCRA regulates hazardous waste based on toxic characteristic leaching procedure (TCLP) and the characteristics of ignitability, corrosivity, and reactivity.
†Only required if total Coliform analysis is positive.

To further illustrate the point that it is important to understand which testing is required under the governing permit, the Safe Drinking Water Act (SDWA) contains a list of restricted substances that is different in some respects from either CWA or RCRA and requires different protocols to analyze for some substances. It is very important when engaging the services of an analytical laboratory to inform the analysts of the appropriate law (CWA, RCR, SDWA, etc.) to which the analyses should conform.

RCRA requires waste generators to determine whether each segregable waste stream is a hazardous waste. The determination can be made one of two ways: either by sampling and specified analyses, or based on the generator's knowledge of the process generating the waste and the chemistry of the waste. For purposes of determining whether a waste stream is hazardous waste, the determination must be made at the point of generation, as opposed to after it has been commingled with other (separate) waste streams. In some cases, it is necessary to store certain waste streams and manage them separately, rather than discharge them to an industrial waste treatment facility or to a POTW.

Characteristics of Industrial Waste

Industrial wastes are classified as wastewater, solid wastes, or air discharges. There is some overlap of physical characteristics of the substances contained in each of these three classifications of wastes, since wastewater can contain dissolved gases and suspended and settleable solids, solid waste streams can include compressed gases, liquids, and certain solids, and air discharges can contain vaporized liquids, liquids in the aerosol state, and solid particles known as particulate emissions. The basis of the classification is the environmental medium to which the waste is discharged, and the characterization of the waste is normally based on the body of laws and regulations that govern that medium.

Characteristics of Industrial Wastewater

Priority Pollutants

Whereas the term *hazardous waste* is normally associated with solid waste and is regulated under *40 C.F.R.* Subpart C, the term *priority pollutant* is used in association with wastewater constituents that are of a hazardous nature. At the writing of this text, there were 126 substances, in the five categories of (1) Metals, (2) Pesticides/PCBs, (3) Semivolatile Organics, (4) Volatile Organics, and (5) General Chemistry, designated by the EPA as priority pollutants, as shown in Table 5-2.

Atomic adsorption (AA) analysis and inductively coupled plasma (ICP) analysis are typical methods used to determine concentrations of metals. A technique that combines gas chromatography with the mass spectrometer, referred to as "GC-mass spec," is used to analyze for the volatile and semivolatile organics listed in Table 5-2. In this technique, the gas chromatograph is used to separate the organics from each other, and the mass spec is used to determine the concentration of each specific organic compound.

The characteristics of industrial wastewaters are determined in accordance with several sets of laws and regulations that govern the quality of water bodies, both surface water and ground water. These laws and regulations include the Clean Water Act (CWA), the Resource Conservation and Recovery Act (RCRA), and the Safe Drinking Water Act (SDWA), as explained above. Each of these laws defines a number of physical, chemical, and biological characteristics and specifies a protocol to be used in determining each of the characteristics. The following text presents general descriptions of many of the characteristics regulated by one or more of the CWA, RCRA, or SDWA.

BOD

The standard five-day BOD test is the most commonly used method to estimate the total quantity of biodegradable organic material

Table 5-2   Priority Pollutants

Parameter	U.S. EPA Method #	Typical Laboratory PQL
Volatile		Ug/L
Acrolein	624	5
Acrylonitrile	624	50
Benzene	624	5
Carbon tetrachloride (tetrachloromethane)	624	5
Chlorobenzene	624	5
1,2-dichloroethane	624	5
1,1,1-trichloroethane	624	5
1,1-dichloroethane	624	5
1,1,2-trichloroethane	624	5
1,1,2,2-tetrachloroethane	624	5
Chloroethane	624	10
2-chloroethyl vinyl ether (mixed)	624	10
Chloroform (trichloromethane)	624	5
1,1-dichloroethylene (dichloroethene)	624	5
1,2-trans-dichloroethene	624	5
1,2-dichloropropane	624	5
1,2-dichloropropylene (1,3-dichloropropene)	624	5
Ethylbenzene	624	5
Methylene chloride (dichloromethane)	624	10
Methyl chloride (chloromethane)	624	5
Methyl bromide (bromomethane)	624	10
Bromoform (tribromomethane)	624	5
Dichlorobromomethane (dibromochloromethane)	624	5
Chlorodibromomethane (bromodichloromethane)	624	5
Tetrachloroethylene	624	5
Toluene	624	5
Trichloroethylene	624	5
Vinyl chloride (chloroethylene)	624	10
Semivolatile		Ug/L
Acenaphthene	625	5
Benzidine	625	25
1,2,4-trichlorobenzene	625	5
Hexachlorobenzene	625	5
Hexachloroethane	625	5
Bis(2-chloroethyl) ether	625	6
2-chloronaphthalene	625	5
2,4,6-trichlorophenol	625	5
Parachlorometa cresol	625	5

Table 5-2   Priority Pollutants *(continued)*

Parameter	U.S. EPA Method #	Typical Laboratory PQL
2-chlorophenol	625	5
1,2-dichlorobenzene	625	5
1,3-dichlorobenzene	625	5
1,4-dichlorobenzene	625	5
3,3-dichlorobenzidine	625	10
2,4-dichlorophenol	625	5
2,4-dimethylphenol	625	5
2,4-dinitrotoluene	625	5
2,6-dinitrotoluene	625	5
1,2-diphenylhydrazine	625	5
Fluoranthene	625	5
4-chlorophenyl phenyl ether	625	5
4-bromophenyl phenyl ether	625	5
Bis(2-chloroisopropyl) ether	625	5
Bis(2-chloroethoxy) methane	625	5
Hexachlorobutadiene	625	5
Hexachloromyclopentadiene	625	5
Isophorone	625	5
Naphthalene	625	5
Nitrobenzene	625	5
2-nitrophenol	625	5
4-nitrophenol	625	5
2,4-dinitrophenol	625	25
4,6-dinitro-o-cresol (4,6-dinitro-2-methylphenol)	625	25
N-nitrosodimethylamine	625	5
N-nitrosodiphenylamine	625	5
N-nitrosodi-n-propylamine	625	5
Pentachlorophenol	625	5
Phenol	625	5
Bis(2-ethylhexyl) phthalate	625	5
Butyl benzyl phthalate	625	5
Di-N-butyl phthalate	625	5
Di-n-octyl phthalate	625	5
Diethyl phthalate	625	5
Dimethyl phthalate	625	5
1,2-benzanthracene (benzo(a) anthracene)	625	5
Benzo(a)pyrene (3,4-benzo-pyrene)	625	5
3,4-Benzofluoranthene (benzo(b) fluoranthene)	625	5
11,12-benzofluoranthene (benzo(b) fluoranthene)	625	5

Table 5-2   Priority Pollutants *(continued)*

Parameter	U.S. EPA Method #	Typical Laboratory PQL
Chrysene	625	5
Acenaphthylene	625	5
Anthracene	625	5
1,1,2-benzoperylene (benzo(ghi) perylene)	625	5
Fluorene	625	5
Phenanthrene	625	5
1,2,5,6-dibenzanthracene (dibenzo(a,h) anthracene)	625	5
Indeno(1,2,3-cd) pyrene (2,3-o-pheynylene pyrene)	625	5
Pyrene	625	5
Pesticides/PCBs		Ug/L
Aldrin	608	0.05
Dieldrin	608	0.1
Chlordane (technical mixture and metabolites)	608	0.05
4,4-DDT	608	0.1
4,4-DDE (p,p-DDX)	608	0.1
4,4-DDD (p,p-TDE)	608	0.1
Alpha-endosulfan	608	0.05
Beta-endosulfan	608	0.05
Endosulfan sulfate	608	0.1
Endrin	608	0.1
Endrin aldehyde	608	0.1
Heptachlor	608	0.05
Heptachlor epoxide (BHC-hexachlorocyclohexane)	608	0.05
Alpha-BHC	608	0.05
Beta-BHC	608	0.05
Gamma-BHC (lindane)	608	0.05
Delta-BHC (PCB-polychlorinated biphenyls)	608	0.05
PCB-1242 (arochlor 1242)	608	0.5
PCB-1254 (arochlor 1254)	608	0.5
PCB-1221 (arochlor 1221)	608	1.0
PCB-1232 (arochlor 1232)	608	0.5
PCB-1248 (arochlor 1248)	608	0.5
PCB-1260 (arochlor 1260)	608	0.5
PCB-1016 (arochlor 1016)	608	0.5
Toxaphene	608	1.0
Metals		mg/L
Antimony	200.7	0.008
Arsenic	200.7	0.005
Beryllium	200.7	0.005

Table 5-2   Priority Pollutants *(continued)*

Parameter	U.S. EPA Method #	Typical Laboratory PQL
Cadmium	200.7	0.010
Chromium	200.7	0.015
Copper	200.7	0.025
Lead	200.7	0.005
Mercury	245.1	0.2 ug/L
Nickel	200.7	0.040
Selenium	200.7	0.010
Silver	200.7	0.015
Thallium	200.7	0.015
Zinc	200.7	0.025
Other		
2,3,7,8-tetrachloro-dibenzo-p-dioxin (TCDD)	613	10 ppt
Cyanide, total	335.4	0.01 mg/L
Asbestos	100.2	7MF/L >10um

Notes:

40 C.F.R. Part 423-126 Priority Pollutants, Appendix A.

PQL = Practical Quantitation Limit.

Laboratory limits are from a commercial laboratory, Katahdin Analytical Services, Inc., Westbrook, ME.

in wastewater. The results of the five-day BOD test (abbreviated BOD_5) are considered to be estimates of the amount of oxygen that would be consumed by microorganisms in the process of using the organic materials contained in a wastewater for food for growth and energy. Some of the organic material will thus be converted to additional microorganisms, some will be converted to carbon dioxide, and some to water. Oxygen is needed for all three purposes, as seen in equation 5-1.

$$\text{Organic matter} + \text{microorganisms} + O_2 \rightarrow \text{more microorganisms} + CO_2 + H_2O \qquad (5\text{-}1)$$

COD

Chemical Oxygen Demand (COD) is a second method of estimating how much oxygen would be depleted from a body of receiving water as a result of bacterial action. While the BOD test is performed by using a population of bacteria and other microorganisms to attempt to duplicate what would happen in a natural stream over a period of five days, the COD test uses a strong chemical oxidizing agent (potassium dichromate or potassium permanganate) to chemically oxidize the organic material in the sample of wastewater under conditions of heat and strong acid. The COD test has the advantage of not being subject to interference from toxic materials, as well as requiring only two or three hours for test completion, as opposed to five days for the BOD test. It has the disadvantage of being completely artificial, but is nevertheless considered to yield a result that may be used as the basis upon which to calculate a reasonably accurate and reproducible estimate of the oxygen-demanding properties of a wastewater. The COD test is often used in conjunction with the BOD test to estimate the amount of nonbiodegradable organic material in a wastewater. In the case of biodegradable organics, the COD is normally in the range of 1.3 to 1.5 times the BOD. When the result of a COD test is more than twice that of the BOD test, there is good reason to suspect that a significant portion of the organic material in the sample is not biodegradable by ordinary microorganisms. As a side note, it is important to be aware that the sample vial resulting from a COD test can contain leachable mercury above regulatory limits. If such is the case, the sample must be managed as a toxic hazardous waste.

Ultimate BOD

The term "ultimate BOD," designated by BOD_u, refers to the quantity of oxygen that would be used by microorganisms in converting the entire amount of organic material in a given volume of wastewater to carbon dioxide and water, given unlimited time as opposed to a time limit of five days. The ultimate BOD of a known substance can be estimated as shown in the following example.

Example 5-1

Estimate the BOD_u of a waste stream containing 100 mg/L of pure ethanol.

Solution

The first step is to write out and balance an equation depicting the complete oxidation, with oxygen, of a molecule of the appropriate substance (ethanol, in this case):

$$C_2H_5OH + 2\,O_2 \rightarrow 2CO_2 + 3H_2O \tag{5-2}$$

Next, using the mole ratio of oxygen to ethanol as calculated by balancing the equation, as well as the respective molecular weights of each substance, multiply the concentration in mg/L of ethanol in the wastewater by this mole ratio, and by the molecular weight of oxygen divided by the molecular weight of ethanol, to obtain the BOD_u of that wastewater:

Mole ratio, oxygen/ethanol = 2/1 = 2.0
mol. wt. of ethanol = 38
mol. wt. of oxygen = 16

Therefore, given the ratio of molecular weights, oxygen/ethanol = 0.42, and the ultimate BOD of the solution of ethanol is calculated by:

$$100 \text{ mg/L} \times 2 \text{ mole O}_2/\text{mole ethanol} \\ \times 0.42 = 84 \text{ mg/L} \tag{5-3}$$

pH, Acidity, and Alkalinity

The term *pH* refers to the concentration of hydrogen ions in an aqueous solution, where "aqueous solution" means either pure water or water with small (in terms of molar amounts) quantities of substances dissolved in it. Strong solutions of chemicals, such as one-molar sulfuric acid or a saturated solution of sodium chloride, do not qualify as aqueous solutions. In those solutions, the normal pH range of 0 to 14, which equals the negative logarithm of the hydrogen ion concentration in moles per liter, has no meaning. Since the pH of an aqueous solution is numerically equal to the negative log of the hydrogen ion concentration (in moles per liter), it can be readily calculated using the following equation:

$$pH = \log (1/[H^+] \tag{5-4}$$

It is therefore indicative of the acidic or basic condition of a wastewater (pH values between 0 and 7.0 indicate acidic conditions, and pH values between 7.0 and 14 indicate alkaline conditions). However, pH is not equivalent to acidity or alkalinity. A wastewater may have a pH of 2.0 but have lower acidity than another wastewater having a pH of 4.0. Likewise, a wastewater having a pH of 9.0 may, or may not, have more alkalinity than a wastewater having a pH of 10.6.

Alkalinity and acidity are defined as the ability of an aqueous solution to resist a change in pH. Alkalinity and acidity are measured by determining the quantity of a solution of acid or base, as appropriate, of known concentration that is required to completely neutralize the acidity or alkalinity of the aqueous solution.

Aqueous solutions (wastewaters, for example) that are high in either acidity or alkalinity are said to be highly buffered and will not readily change in pH value as a result of influences such as bacterial action or chemical reaction. Depending on the anion species involved in the alkalinity or acidity, an aqueous solution can be buffered in low pH ranges, high pH ranges, or neutral (near pH 7) pH ranges.

D.O.

Dissolved Oxygen (D.O.) concentration is usually determined by either a wet chemistry method known as "the azide modification of the Winkler method," or by use of a probe and meter. The wet chemistry method is considered to be the standard for comparison, when the color and other properties of the aqueous solution (industrial waste and so on) do not preclude its use. However, the probe and meter method is the most commonly used.

The standard D.O. probe consists of two solid electrodes emersed in an electrolyte and separated from the solution being tested by a plastic membrane that is permeable to oxygen. The oxygen can be in the dissolved state or the gaseous state. When the probe is emersed in the test solution, oxygen passes from the solution through the membrane and into the electrolyte in direct proportion to its concentration in the solution. As the oxygen enters the electrolyte, it changes the conductivity. This change is detected by the two electrodes and is registered on the meter.

Oxygen is only sparingly soluble in water, and its solubility decreases with increasing temperature. Increasing dissolved solids concentrations also decrease the solubility of oxygen. The saturation concentration of oxygen in deionized water at 1.0°C and standard air pressure is 14.2 parts per million (ppm); this is equivalent to about one drop from an eye-dropper in one gallon. This is in contrast to the concentration of oxygen in air, which is about 230,000 ppm, or 23%. Still, the species that live in water are sensitive to relatively small changes in the already extremely low quantity of oxygen available to them. Trout, for instance, do very well when the D.O. concentration is 7 ppm, but die when the D.O. concentration falls below 5 ppm. The microorganisms in activated sludge thrive when the D.O. concentration is 1.5 ppm, a seemingly negligible amount (or, to use the previous example, about one drop in ten gallons). Moreover, small differences in D.O. concentration in aerobic wastewater treatment systems, such as activated sludge, result in significant changes in the populations of different species of microorganisms. These changes, in turn, change the settleability of the sludge, the clarity of the treated effluent, and the overall performance of the treatment system.

Metals

While the term *heavy metals* can mean different things depending on the context in which it is used, for the purpose of this discussion, heavy metals are those elements located in or near the middle of the periodic table. Figure 5-4 shows the periodic chart with the elements considered heavy metals highlighted. Many of these heavy metals are toxic to living entities, including humans and bacteria. For this reason, heavy metals are regulated through the various environmental laws.

Table 5-3 shows the heavy metals that are regulated by the Clean Water Act (CWA), RCRA, and the SDWA. Table 5-4 lists all the metals that are regulated by the CWA, as well as the applicable U.S. EPA Method Number and the corresponding "Practical Quantitation Limit," which is essentially the limit of detection for that method. The standard method for measuring the concentration of all of the heavy metals listed in Table 5-3 is by atomic adsorption analysis (AA).

Color

Color is measured using visual comparison to an arbitrary standard. The standard is made, normally, by dissolving potassium chloroplatinate (K_2PtCl_6) and cobalt chlo-

Mokeur's Periodic table of the elements

Legend box:
- Symbol
- Atomic number
- Relative atomic mass
- Electronegativity
- Most frequent oxidation number
- Name

```
H1
1.00794
2.1      1+
Hydrogen
```

	1 IA	2 II A	3 III B	4 IV B	5 V B	6 VI B	7 VII B	8	9 VIII	10	11 I B	12 II B	13 III A	14 IV A	15 V A	16 VI A	17 VII A	18 VIII A
1	H1 1.00794 2.1 1+ Hydrogen																	He 2 4.002602 - - Helium
2	Li 3 6.941 1.0 1+ Lithium	Be 4 9.012182 1.5 2+ Beryllium											B 5 10.811 2.0 3+ Boron	C 6 12.011 2.5 4+,4- Carbon	N 7 14.00674 3.0 3+,3- Nitrogen	O 8 15.9994 3.5 2- Oxygen	F 9 18.9984032 4.0 1- Fluorine	Ne 10 20.1797 - - Neon
3	Na 11 22.989768 0.9 1+ Sodium	Mg 12 24.3050 1.2 2+ Magnesium											Al 13 26.981539 1.5 3+ Aluminum	Si 14 28.0855 1.8 4+ Silicon	P 15 30.973762 2.1 5+ Phosphorus	S 16 32.066 1.8 4+ Sulfur	Cl 17 35.4527 3.0 1- Chlorine	Ar 18 39.948 - - Argon
4	K 19 39.0983 0.8 1+ Potassium	Ca 20 40.078 1.0 2+ Calcium	Sc 21 44.955910 1.3 3+ Scandium	Ti 22 47.88 1.5 4+ Titanium	V 23 50.9415 1.6 3+ Vanadium	Cr 24 51.9961 1.6 3+ Chromium	Mn 25 54.93805 1.5 3+ Manganese	Fe 26 55.847 1.8 3+ Iron	Co 27 58.9332 1.8 2+ Cobalt	Ni 28 58.6934 1.8 2+ Nickel	Cu 29 63.546 1.9 2+ Copper	Zn 30 65.39 1.6 2+ Zinc	Ga 31 69.723 1.6 3+ Gallium	Ge 32 72.61 1.8 4+ Germanium	As 33 74.92159 2.0 3+,3- Arsenic	Se 34 78.96 2.4 4+ Selanium	Br 35 79.904 2.8 1- Bromine	Kr 36 83.80 - - Krypton
5	Rb 37 85.4678 0.8 1+ Rubidium	Sr 38 87.62 1.0 2+ Strontium	Y 39 88.90585 1.3 3+ Yttrium	Zr 40 91.224 1.4 4+ Zirconium	Nb 41 92.90638 1.6 5+ Niobium	Mo 42 95.94 1.8 6+ Molybdenum	Tc 43 98.9063 1.9 7+ Technetium	Ru 44 101.57 2.2 3+,4+ Ruthenium	Rh 45 102.9055 2.2 3+ Rhodium	Pd 46 106.42 2.2 2+ Palladium	Ag 47 107.8682 1.9 1+ Silver	Cd 48 112.411 1.7 2+ Cadmium	In 49 114.82 1.7 3+ Indium	Sn 50 118.71 1.8 4+ Tin	Sb 51 121.757 1.9 3+,3- Antimony	Te 52 127.60 2.1 4+ Tellurium	I 53 126.90447 2.5 1- Iodine	Xe 54 131.29 - - Xenon
6	Cs 55 132.90543 0.7 1+ Cesium	Ba 56 137.327 0.9 2+ Barium	La 57 138.9055 1.1 3+ Lanthanum	Hf 72 178.49 1.3 4+ Hafnium	Ta 73 180.9479 1.5 5+ Tantalum	W 74 183.85 1.9 6+ Tungsten	Re 75 186.207 1.9 7+ Rhenium	Os 76 190.2 2.2 4+ Osmium	Ir 77 192.22 2.2 4+ Iridium	Pt 78 195.08 2.4 4+ Platinum	Au 79 196.96654 2.4 3+ Gold	Hg 80 200.59 1.9 2+ Mercury	Tl 81 204.3833 1.8 1+ Thallium	Pb 82 207.2 1.9 2+ Lead	Bi 83 208.98037 1.9 3+ Bismuth	Po 84 208.9824 2.0 2+ Polonium	At 85 209.9871 2.2 1- Astatine	Rn 86 222.0176 - - Radon
7	Fr 87 223.0197 0.7 1+ Francium	Ra 88 226.0254 0.9 2+ Radium	Ac 89 227.0278 1.1 3+ Actinium	Rf 104 261.11 Rutherfordium	Db 105 262.11 Tantalum	Sg 106 263.12 Seaborgium	Bh 107 262.12 Bohrium	Hs 108 264 Hassium	Mt 109 266.1378 Maitnerium	Uun 110 269 Ununiilium	Uuu 111 272 Unununium	Uu 112 277 Unubium		Uuq 114 289 Ununquadium		Uuh 116 289 Ununhexium		Uuo 118 293 Ununoctium

6	Ce 58 140.115 1.1 3+ Cerium	Pr 59 140.90765 1.1 3+ Praseodymium	Nd 60 144.24 1.1 3+ Neodymium	Pm 61 144.9127 1.1 3+ Promethium	Sm 62 150.36 1.2 3+ Samarium	Eu 63 151.965 1.2 3+ Europium	Gd 64 157.25 1.2 3+ Gadolinium	Tb 65 168.92534 1.2 3+ Terbium	Dy 66 162.50 1.2 3+ Dysprosium	Ho 67 164.93032 1.2 3+ Holmium	Er 68 167.26 1.2 3+ Erbium	Tm 69 168.93421 1.2 3+ Thulium	Yb 70 173.04 1.1 3+ Ytterbium
													Lu 71 174.967 1.2 3+ Lutetium
7	Th 90 232.0381 1.3 4+ Thorium	Pa 91 231.03588 1.5 5+ Protactinium	U 92 238.0289 1.4 6+ Uranium	Np 93 237.082 1.3 5+ Neptunium	Pu 94 244.0642 1.3 4+ Plutonium	Am 95 243.0614 1.3 3+ Americium	Cm 96 247 1.3 3+ Curium	Bk 97 247.0703 1.3 3+ Berkelium	Cf 98 251.0796 1.3 3+ Califomium	Es 99 252.03 1.3 - Einsteinium	Fm 100 257.0951 1.3 - Fermium	Md 101 258.01 1.3 - Mendelevium	No 102 259.1009 1.3 - Nobelium
													Lr 103 260.1053 1.3 - Lawrencium

Figure 5-4 The periodic table.

Table 5-3   Heavy Metals Regulated by the Clean Water Act (CWA)

Parameter	U.S. EPA Method No. (4)	Priority Pollutants List (1)	Typical Laboratory PQL (2)
Metals			mg/L
Aluminum, total	200.7		0.10
Antimony, total	200.7	X	0.008
Arsenic, total	200.7	X	0.005
Barium, total	200.7		0.005
Beryllium, total	200.7	X	0.005
Boron, total	200.7		0.10
Cadmium, total	200.7	X	0.010
Chromium, total	200.7	X	0.015
Cobalt, total	200.7		0.030
Copper, total	200.7	X	0.025
Iron, total	200.7		0.05
Lead, total	200.7	X	0.005
Magnesium, total	200.7		0.050
Manganese, total	200.7		0.015
Mercury, total	245.1	X	0.20 ug/L
Molybdenum, total	200.7		0.10
Nickel, total	200.7	X	0.040
Sodium, total	200.7		1.0
Selenium, total	200.7		0.010
Silver, total	200.7		0.015
Thallium, total	200.7	X	0.015
Tin, total	200.7		0.10
Titanium, total	200.7		0.015
Vanadium, total	200.7		0.025
Zinc, total	200.7	X	0.025

Current reference list: *40 C.F.R. Part 136*, Inorganic Test Procedures.

PQL = Practical Quantitation Limit.

PQLs are from Katahdin Analytical Services, Inc., Westbrook, ME.

Other Approved U.S. EPA Methods are available; please refer to the regulation for optional methods.

ride in deionized water. The color produced by 1 mg/L of potassium chloroplatinate is taken as one standard unit of color. Cobalt chloride is used in very small amounts to tint the color standard to match the test solution. Cobalt chloride does not affect the intensity of the color of the standard.

In some cases, an agreement can be made with regulatory authorities to make the arbitrary standard by simply diluting the receiving water serially with deionized water, or by serially diluting the effluent in question with deionized water.

Turbidity

Turbidity refers to the light scattering properties of a sample. Turbidity can be described as "haziness" or "milkyness," and is caused by fine particles scattering light at more or less 90 degrees to the direction from which the light enters the sample. Turbidity is not to be confused with color, nor color with turbidity.

Turbidity is normally measured using an electronic device in which a beam of ordinary white light is directed through a certain path length of the sample. Photometers placed at right angles to the direction of travel of the light beam detect the amount of light diverted, which is directly proportional to the turbidity, expressed in NTUs (Nephelometric Turbidity Units).

Compounds Containing Phosphorus

When treated wastewaters are discharged to natural water systems, various effects can result, depending on which substances remain in the treated effluent and in which form. One of the more significant effects is enhancement of eutrophication within the receiving water system if significant quantities of phosphorus, in forms available to algae, are discharged.

Eutrophication, or the growth and proliferation of plant life, including algae, is always under way in any natural fresh water system: rivers, streams, ponds, or lakes. However, it is limited by the scarcity of nutrients such as phosphorus and nitrogen. At a minimum, plants, including algae, need carbon, nitrogen, and phosphorus in relatively abundant amounts, and potassium, sodium, calcium, magnesium, sulfur, and a host of additional elements in smaller amounts. Any one of these required substances can be the limiting substance that holds growth to a given level, when the substance becomes depleted as a result of that growth. In a majority of cases in the natural environment, that limiting substance is phosphorus.

If treated wastewater discharged to a natural water system significantly increases the inventory of phosphorus, the natural control on the quantity of plant growth that can be supported by that natural water system will be removed, and a disastrous level of

Table 5-4 Phosphorus Compounds Commonly Encountered in Environmental Engineering Practice

Name	Formula
Orthophosphates	
Trisodium phosphate	Na_3PO_4
Disodium phosphate	Na_2HPO_4
Monosodium phosphate	NaH_2PO_4
Diammonium phosphate	$(NH_4)_2HPO_4$
Polyphosphates	
Sodium hexametaphosphate	$Na_3(PO_3)_6$
Sodium tripolyphosphate	$Na_5P_3O_{10}$
Tetrasodium pyrophosphate	$Na_4P_2O_7$

eutrophication can result. For this reason, many waste discharge permits include a restriction to limit the discharge of phosphorus.

Phosphorus exists in wastewater in many forms, the most basic of which is orthophosphate, $PO_4^=$. Several forms of orthophosphate are encountered in industrial wastewater. A list of the most common is presented in Table 5-4. Table 5-4 also lists several polyphosphates encountered in industrial wastewater treatment. Polyphosphates are simply chains of orthophosphate $(PO_4^=)$ units formed by dehydration. All polyphosphates in aqueous solution gradually hydrolyze and revert back to the orthophosphate form in the natural environment.

Compounds Containing Nitrogen

As explained in the previous section on Compounds Containing Phosphorus, disastrous eutrophication can result in a natural water body if the substances that constitute the growth limiting nutrients for algae and other plants are augmented in quantity by the discharge of treated wastewaters. Since nitrogen, needed for production of proteins and other components of living cells, can be the substance that limits the level of algae growth, it is often one of the substances restricted in wastewater discharge permits.

Nitrogen can exist in as many as seven valence states, as shown:

$$3^- \quad 0 \quad 1^+ \quad 2^+ \quad 3^+ \quad 4^+ \quad 5^+$$
$$NH_3 \quad N_2 \quad N_2O \quad NO \quad N_2O_3 \quad NO_2 \quad N_2O_5$$

$$(5\text{-}5)$$

All but the valence states of 1^+ and 4^+ are often encountered in industrial waste discharge management (the 2^+ valence state is very often encountered in discharges to the air), and each valence state can be converted to any other valence state by natural biological processes. Accordingly, the total amount of nitrogen, or the sum of all the different forms, is commonly limited by wastewater discharge permits.

Total Solids

Substances can exist in aqueous solution in either the dissolved or the nondissolved state. The residue that is left after evaporating a sample of water at 103°C is referred to as the total solids value of that sample. It is generally regarded as everything that was in the sample that was not water. However, any of the substances originally present, organic or inorganic, that volatilized at 103°C, or less, will not be in the residue. Figure 5-5 presents a schematic breakdown of substances that can be measured as total solids in a sample of wastewater.

Suspended Solids

Solids that will not pass through a 0.45-micron filter are referred to as total suspended solids (TSS). Since the standard method for measuring TSS involves shaking the sample thoroughly before filtering, the TSS actually includes all nondissolved solids, as opposed to simply the dissolved solids that will not settle under the influence of gravity. Results of the test for settleable solids cannot be subtracted from the results of the test for total suspended solids to obtain the value of the quantity of nondissolved solids in a sample that will not settle under the influence of

gravity, because the standard test for settleable solids yields a value in terms of ml/L, while the standard test for TSS yields a value in terms of mg/L. If it is desired to estimate the quantity of nonsettleable TSS, a special procedure must be devised and then described thoroughly in the presentation of results.

Settleable Solids

Settleable solids are nondissolved solids that will settle to the bottom of a container under the influence of gravity. The standard device for measuring settleable solids is the Imhoff cone, pictured in Figure 5-6. The standard test for settleable solids involves shaking a sample thoroughly, filling the 1-liter Imhoff cone to the full mark, and then allowing the sample to settle for one hour under quiescent conditions. Settleable solids are reported as milliliters of settleable materials per liter of sample (ml/L). In order to obtain a value of the weight of settleable material, it would be necessary to remove the supernatant and to evaporate to dryness that volume of sample that contained the settleable solids. The result, of course, will include the dissolved solids contained in the liquid remaining with the settled solids. If this volume is small compared to the total (one liter) sample, the "error" will be small. This procedure is never done in the context of a wastewater discharge permit but may be done for the engineering or planning purposes of the discharger.

Total Volatile Suspended Solids (TVSS)

The standard method for estimating the quantity of nondissolved organic material in a wastewater sample is to perform a TSS test, as explained above, and then to place the 0.45-micron filter with the solids deposited thereon in a furnace at 600°C for a sufficient time to "burn" all of the material that will oxidize to carbon dioxide at that temperature. The material that remains on the filter after this procedure is the material that will not combine with oxygen and volatilize as

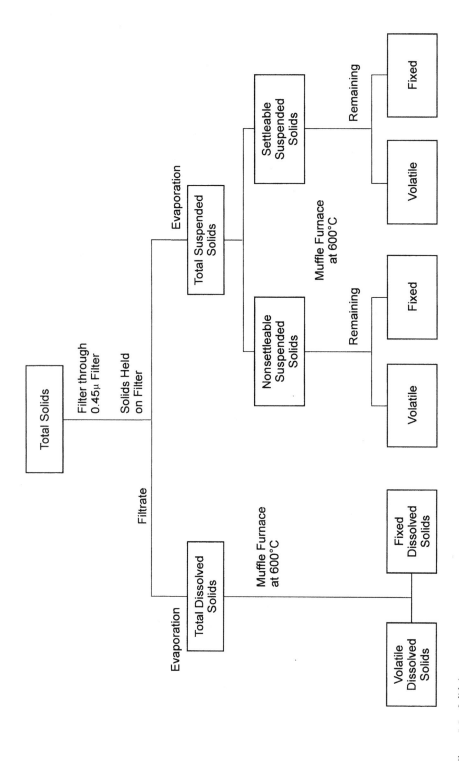

Figure 5-5   Solids in wastewater.

Figure 5-6 Imhoff cones used to measure settleable solids.

carbon dioxide at 600°C. These solids are referred to as "fixed" solids and are considered to be the inorganic portion of the suspended solids in the wastewater sample. This value, subtracted from the TSS value, produces the total volatile suspended solids portion of the wastewater sample and is thus an estimate of the quantity of undissolved organic material in the original wastewater sample.

Oil and Grease

There are three methods of estimating the quantity of oil and grease in a wastewater sample, each of which is considered a standard method. When reporting results, however, the method used is actually part of the result. In other words, as stated in *Standard Methods*, ". . . oils and greases are defined by the method used for their determination." The three methods are:

1. The partition-gravimetric method
2. The partition-infrared method
3. The Soxlet extraction method

All three involve extracting the oil and grease substances from the raw sample by use of a solvent. Also, all three require that the sample be acidified to pH 2.0 or lower by adding hydrochloric or sulfuric acid before the extraction process takes place.

The Partition-Gravimetric Method

The partition-gravimetric method involves extracting dissolved or emulsified oil and grease by intimate contact with an extracting solvent, which can be one of the following:

- n-Hexane
- Methyl-tert-butyl ether (MTBE)
- Solvent mixture, 80% n-hexane/20% MTBE, v/v

The sample and the solvent are placed together in a separatory funnel and shaken. Then the solvent (with dissolved oil and grease from the sample) is drawn off and passed through a funnel containing filter paper and 10 g sodium sulfate. The solvent is evaporated at 85°C, and the quantity of oil and grease is determined gravimetrically.

The Partition-Infrared Method

The partition-infrared method makes use of trichlorotrifluoroethane as the extraction solvent. Use of this solvent allows use of infrared light, which is absorbed by the carbon-carbon double bond that is characteristic of oil and grease substances. Thus, substances that are dissolved by trichlorotrifluoroethane and are volatilized at 85°C are detected by this method. The extraction procedure is the same as for the partition-gravimetric method. Following the extraction, the quantity of oil and grease is determined using photometric techniques with light in the infrared zone (3,200 cm^{-1} to 2,700 cm^{-1}).

The Soxlet Extraction Method

The Soxlet extraction method uses the same three choices of solvent as for the partition-gravimetric method: n-hexane, MTBE, or an 80/20 v/v mixture of the two. The extraction procedure, however, is carried out in a reflux apparatus as opposed to a separatory funnel. After extraction, the quantity of oil and grease is determined gravimetrically.

The oil and grease tests are subject to large error, since anything that will dissolve in the extraction solvent, whether it is truly oil or grease or not (in the sense that it will cause problems by coating sewerage, forming greaseballs, or will exhibit resistance to biodegradation), will be measured as oil and grease. In addition to this problem, the difficulty of obtaining a representative sample of the wastewater for measurement of oil and grease content must be kept in mind when interpreting or evaluating laboratory results. The reason for this is that oil and grease substances are only sparingly soluble in water and tend to either float on the surface, as is the case with animal fat, or sink to the bottom, as is the case with trichloroethylene. Accordingly, an automatic sampler that takes samples from the bulk flow is not useful. To take samples for laboratory determination of fat, oil, and grease (FOG) content, it is necessary to hand dip to obtain grab samples, attempting, in doing so, to obtain portions of the flow from throughout the depth of wastewater flow.

"Other Characteristics" (Pollutants)

Chlorine Demand and Chlorine Residual

Chlorine, in the form of either chlorine gas or one of the hypochlorites, is routinely added to wastewaters for the purpose of disinfection. However, disinfection will not take place if there are substances in the wastewaters that will quickly react with chlorine or hypochlorite, becoming oxidized themselves and reducing the chlorine to chloride ion (Cl^-), which is nonreactive. These substances are "reducing agents" relative to chlorine. Therefore, enough chlorine must be added to react with all of the reducing agents before any chlorine becomes available for disinfection. "Contact time" is then required for the chlorine to kill the microorganisms and thus accomplish disinfection. The amount of contact time required is in inverse proportion to the concentration of chlorine residual, which is the concentration of chlorine that remains at any point in time after reactions with other chemicals or substances (reducing agents) have taken place. The mathematical relationship between chlorine dose, chlorine residual, and chlorine demand is as follows:

$$\text{chlorine residual} = \text{chlorine dose} - \text{chlorine demand} \qquad (5\text{-}6)$$

The quantity of chlorine "used up" as a result of the disinfection process is considered part of the chlorine demand. Therefore:

$$\text{chlorine demand} = \text{chlorine reacted with reducing agents} + \text{chlorine used for disinfection} \qquad (5\text{-}7)$$

Chlorides

The importance of the presence of chlorides in wastewaters relates to their ability to participate in the conduction of electric current, and, therefore, their active role in enhancing corrosion, particularly galvanic corrosion. Chlorides are otherwise benign, being unreactive and unlikely to participate in any precipitation activity except the salting out process. Chlorides do affect the osmotic pressure of an aqueous solution and, as such, can have a deleterious effect on a biological wastewater treatment system. It has been determined through both laboratory experimentation and full-scale treatment plant operation that biological treatment systems are able to operate under conditions of chloride concentrations up to 2,000 mg/L, but the chloride concentration must be reasonably stable. Once the population of microorganisms acclimates to a given range of chloride concentration, a significant increase or decrease in chloride concentration will adversely affect the treatment process, probably because of the change in osmotic pressure and its effect on the microbial cell membranes.

Hardness

The test for hardness in water was originally developed to determine the soap-consuming properties of water used for washing. The original test involved placing a measured amount of a standard soap in a bottle containing the water being tested, shaking the mixture in a standard manner, then observing the nature of the suds, if any, on the surface of the water. Then, more soap was added, using a step-by-step procedure, until a permanent layer of suds remained on the surface of the water. The total quantity of soap required to produce the stable suds layer was proportional to the hardness of the water. It was eventually realized that the hardness was due almost entirely to the presence of calcium and magnesium ions in the water, and as soon as analytical procedures were developed that would analyze for the concentrations of these ions directly, rather than indirectly, as with the shaking bottle test, the newer analytical procedures became the standard for the hardness test. In the context of use of water in industrial processes, however, and especially the reuse of treated wastewaters in a recycle and reuse system, the significance of hardness is in the scale-forming properties of calcium and magnesium salts.

Iron and Manganese

Iron and manganese are normally considered together as problems in raw water supplies for industrial process use, as well as in process effluents. The reason is that these metals are very often found together in groundwater supplies and in some surface water supplies, under certain conditions. Iron and manganese are objectionable for two principal reasons. First, iron can be oxidized by oxygen in water to insoluble ferric oxide, which precipitates to form fine granules that can foul distribution systems and cooling devices. Second, both iron and manganese react with chlorine to form highly colored chlorides that stain many of the objects they come in contact with.

Characteristics of Discharges to the Air

There are three categories of air pollutant characterization: (1) stack discharge characterization, (2) fugitive emissions characterization, and (3) ambient air quality characterization. The three involve quite different sampling procedures but similar, and in most cases identical, analysis methods. Stack discharge and fugitive emission characterization are done primarily to determine the state of compliance with one or more discharge permits. Ambient air quality characterization is done primarily to determine the quality of air in a given area.

Ambient air quality data are used for many purposes, including:

- Issuance of construction permits for industrial projects
- Determination of state of compliance with National Ambient Air Quality Standards (NAAQS)
- Determination of the effectiveness of ameliorating activities
- Establishment of baseline information prior to construction of a significant contributor of substances to the atmosphere

Stack Sampling

The most common reason for conducting a stack sampling program is to determine the state of compliance with regulatory requirements. As such, the substances sampled for are usually dictated by the list of substances included in the air discharge permits issued to the facility.

The purpose of stack sampling is to determine, with as much accuracy as is practicable, the quantity (magnitude) of the total gas source flow rate and the quality (types and amounts of air contaminants) of the total source gas discharge. The equipment included in a typical stack sampling station includes devices to measure characteristics from which gas flow rate can be calculated, devices to measure certain characteristics directly, and equipment to collect and store samples for subsequent analyses in the laboratory.

In general, the equipment used to obtain data from which to calculate gas flow rate includes pitot tubes to measure gas velocity; a device to measure the static pressure of the stack gas; and devices to measure barometric pressure, moisture content, and temperature. Equipment used to characterize stack discharges in terms of specific substances is classified in two broad categories: particulate and gaseous. The objective of this equipment in both categories is to quantitatively remove air contaminants in the same condition as they occur when they are discharged to the air. Many individual devices as well as integrated systems are commercially available to accomplish this objective.

Sample Collection

Samples of ambient air or gas streams (including stack emissions and fugitive emissions) are sampled to determine the presence of, and concentrations of, particulate and gaseous pollutants by use of the following equipment and mechanisms:

- Use of a vacuum pump, hand operated or automatic
- Vacuum release of an evacuated collection container
- Tedlar bags
- Adsorption on a solid
- Condensation (freeze-out) in a trap

Vacuum Pumps

Vacuum pumps are the standard type of equipment used to draw samples of ambient air of stack gas or other gas stream through or into collection devices. Hand-operated vacuum pumps are used extensively to obtain grab samples. Motor-driven vacuum pumps are the standard for continuous or intermittent monitoring. Vacuum pumps can be fitted with, or connected in series with, gas flow meters to obtain data for calculation of concentrations.

Vacuum Release of an Evacuated Collection Container

Containers (having appropriate linings) can be evacuated by use of a vacuum pump before traveling to sampling locations. The collection apparatus (soil gas sampling well, for instance) can be connected to the evacuated container, the valve on the container opened, and a sample of known volume will be collected.

Tedlar Bags

Tedlar bags are made of a nonreactive, non-adsorbing (relatively) material, and are standard equipment for sampling gaseous substances in the air of a stack or other emission. They are purchased as a completely empty bag and are inflated with the collected sample of air or other gas. Samples should be analyzed as soon as possible after being collected in the Tedlar bag. Tedlar bags can be purchased in different volume capacities, including 1-liter and 10-liter bags.

Adsorption on a Solid

Solid adsorbents, such as activated carbon, can be used to collect certain airborne substances, after which the substances can be desorbed for further analysis or other work. Normally, a vacuum pump is used to draw a volume of air or other gas through a container of the adsorbent. In the case of certain substances (e.g., carbon monoxide), an indicator chemical can be incorporated with the adsorbent to directly indicate the presence of the substance.

Condensation (Freeze-Out) in a Trap

Certain substances present in air or other gas stream as a result of volatilization can be collected by drawing a stream of the carrier air or other gas through a trap held at low temperature.

Equipment used to collect particulate matter and to determine the concentration, in ambient air as well as stack emission and other gas streams, uses one or more of the following mechanisms:

- Filtration
- Electrostatic impingement
- Centrifugal force
- Dry impingement
- Wet impingement
- Impaction

Filtration

The standard method for determining the concentration of particulate matter in emissions from stationary sources, as published in the Federal Register (*40 C.F.R. Part 60*, July 1, 1998), is to withdraw, isokinetically, particulate-laden air (or other gas) from the source and collect it on a glass fiber filter maintained at a temperature in the range of $120 + 14°C$ ($248 + 25°F$), or such other temperature as specified by an applicable subpart of the published standards, or approved by the Administrator, U.S. Environmental Protection Agency for a particular application. The total quantity of particulates sampled is then determined by weighing the dried filter. The standard sample train for this determination, illustrated in Figure 5-7, includes equipment to measure gas flow rate as well as total gas volume sampled. These data can then be used to calculate particulate matter concentration.

Electrostatic Impingement

Figure 5-8 shows a schematic of a typical electrostatic impingement device for collecting particulate matter from ambient air or other gas source. In order to determine the concentration of particulate matter in the gas source sampled, additional equipment, discussed above, must be used to determine the rate of flow of the gas sampling system, or the total volume of source gas from which the particulates were extracted. Then, the total weight of particulate matter collected must be determined by weighing.

The principle of operation of the electrostatic impingement apparatus is that of electrostatic attraction. The electrostatic impingement surface is given an electrostatic charge of polarity opposite to the polarity (positive or negative) of the particles in question.

Centrifugal Force

A typical particulate matter collector that operates on the principle of centrifugal force is shown in Figure 5-9. This device accepts the sample of particulate-laden air, which can be ambient air, stack gas, or another gas stream, and directs it through a circular path on its way to the outlet. In traveling the circular path, particulates are forced, by centrifugal action, to the collection device shown. As with other particulate collection devices, equipment to measure the rate of flow of sample taken, or total volume from which the particulates were extracted, must be used to enable calculation of particulate matter concentration.

Dry Impingement

Figure 5-7 shows a "sample train" containing four impingers. These impingers can be used either wet or dry and can be used to collect particulates from ambient air of stack gases or particulates from other gas streams.

Wet Impingement

Wet impingement methods for collecting particulate samples from ambient air, a stack emission, or other gas train operate by trapping the particulates in a liquid solution. The total weight of particulate matter is then determined by filtering and weighing the entire volume of the liquid or by evaporating the liquid and weighing. A volume of fresh liquid is also evaporated, and the residual (if any) is weighed and compared with the weight of the dried sample to determine the actual weight of particulate matter. Again, equipment to measure the sample flow rate or the total volume sampled is necessary in order to determine particulate concentration.

Impaction

Figure 5-10 shows a schematic of a typical impaction particulate collector. The principle upon which this collection device operates is that of entrapment on a plate with a film of sticky material on its surface. Collection of pollen for pollen assay purposes makes use of this type of equipment.

Particulates can be physically extracted from glass fiber filters or from the other collection devices for analysis as to individual constituents, such as metals, radioactive ele-

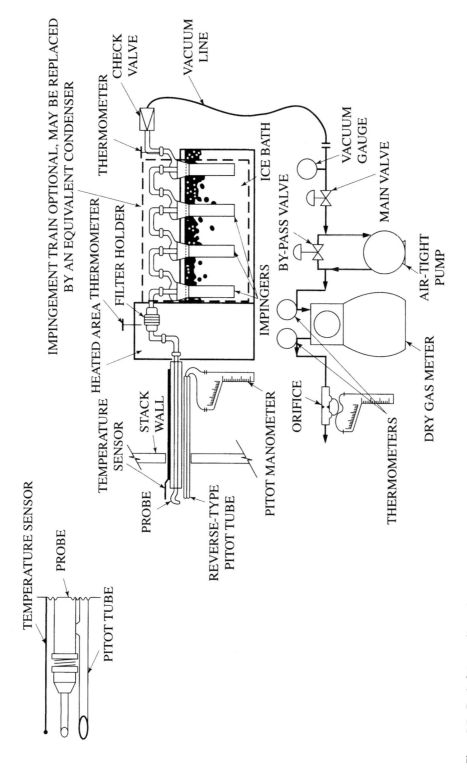

Figure 5-7   Particulate sampling train.

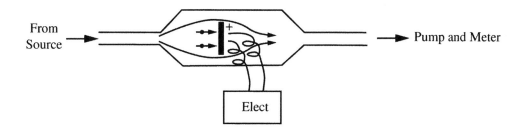

Figure 5-8   Schematic of a typical electrostatic impingement collector.

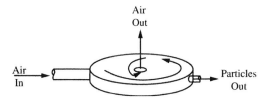

Figure 5-9   Schematic diagram of a centrifugal particle collector.

ments or other materials. The analyst should consult the appropriate Code of Federal Regulations to determine acceptable procedures.

Figure 5-11 presents a schematic of an example of a stack sampling system and illustrates several of the methods employed for data management.

Sample Analysis

Although it is realized that once a sample is collected it is subject to change as a result of chemical reaction, chemical degradation, absorption or adsorption onto the walls of the container or to other substances in the container, or other phenomena. Not much is presently known with certainty about how much change will take place in a sample once it is collected. For this reason, it is extremely important to perform the chemical or other

analyses as soon as possible after the samples are collected.

Ambient Air Sampling

Determination of the quality of ambient air as it relates to the presence and concentration of substances regarded as pollutants is the objective of ambient air sampling. The specialized devices and techniques for carrying out this task have been developed over half a century. Here, again, obtaining representative samples is a major objective of the work plan. Decisions about air sampling must strike a balance between the cost of the characterization program and the value of the data; these decisions include the duration of the sampling period, number of discrete samples taken, the size of each sample, and the number of substances sampled for.

Particulate matter in ambient air is measured by use of a "high-volume sampler," which is an integrated filter holder/vacuum pump (high volume). A glass fiber filter is held in the filter holder, and a high flow rate of ambient air is drawn through it over a measured period of time. Calculations of particulate matter concentration in the ambient air are carried out using the weight of particulate matter collected on the filter

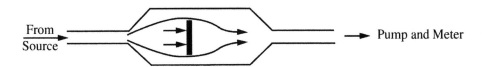

Figure 5-10   Schematic of a typical impaction particle collector.

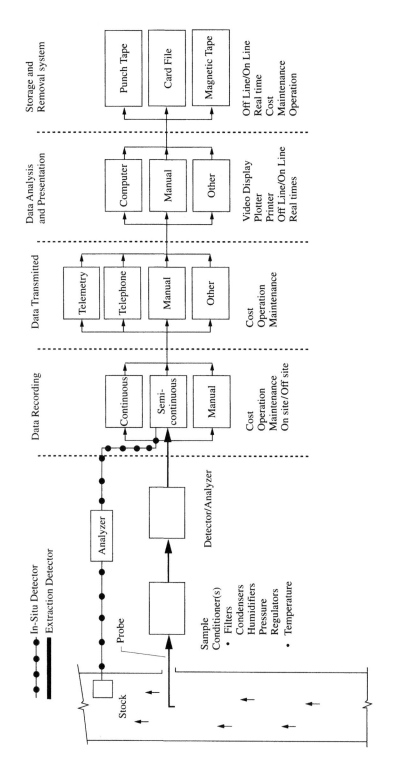

Figure 5-11   Continuous monitoring system.

and the flow rate (or total volume) of air drawn through the filter.

Air Pollutants

The following paragraphs present a brief discussion of the major air pollutants that are regulated by the Clean Air Act, as amended.

Air Toxics. Control of the release of so-called "air toxics" took on a new importance—in fact, a dominant importance—with promulgation of Title III of the 1990 Clean Air Act Amendments (CAA). Previous to these amendments, the EPA had issued standards for only eight hazardous substances over a 20-year period. The 1990 CAA lists 188 substances that must be controlled.

The designated air toxics are required to be managed by use of "Maximum Achievable Control Technology" (MACT). In addition, residual risk after implementation of MACT must be assessed. Industries are also subject to a time schedule, by source category, for implementation of MACT. All sources must be in compliance with the applicable standard within a published time schedule. The EPA published the initial list of source categories in 1992 and since that time has issued several revisions/updates to the list and promulgation schedule.

There are significantly different requirements for sources that qualify as a "major source" compared with those that do not. A major source is any stationary source that emits in excess of 10 tons per year of any of the listed 188 hazardous substances or 25 tons per year or more of any combination of those substances.

The list of 188 air toxics includes pesticides, metals, organic chemicals, coke oven emissions, fine mineral fibers, and radionuclides. The EPA is required to add to this list pollutants that may be shown to present, through inhalation or other routes of exposure, a threat of adverse effect on human health or the environment. The EPA may also remove substances from the list if it can be shown that the reasons for placing them on the list were in error.

There are also significantly elevated control requirements for the sources that represent 90% of the area sources that emit the 30 hazardous air pollutants presenting the greatest threat to public health in the largest number of urban areas.

Ozone. Ozone has the chemical formula O_3 and is a relatively strong oxidizing agent. Ozone is emitted directly by some sources, and is also a product of chemical and/or photochemical reaction between other air pollutants in the atmosphere. Ozone is an irritant and can damage sensitive tissues in animals (including humans) as well as plant tissues.

Oxides of Sulfur (SO_x). Sulfur dioxide is the major sulfur oxide of concern in the atmosphere, although sulfur trioxide (SO_3) and sulfate (SO_4) are important. The primary source of sulfur oxides in the nation's atmosphere is the burning of fossil fuels (oil and coal) to generate electrical power and heat. Since sulfur is an important component of protein, and protein is an important component of virtually all plants and animals, sulfur is consequently a component of the fossil remains of these once-living entities. When oil or coal is burned, oxygen combines with the sulfur and the resulting sulfur oxides are emitted with the gaseous releases from the combustion process. The principal harmful effects of sulfur oxides are their eventual reaction with atmospheric moisture and return to the earth as acid rain, as well as their participation in photochemical reactions in the atmosphere to produce smog.

Oxides of Nitrogen (NO_x). Nitrogen oxides, principally nitrogen dioxide, NO_2, and nitric oxide, NO, are produced during the combustion of all types of fuels. These gases are referred to together as NO_x. Since air consists of nitrogen in the N_2 gas form, and about 21% oxygen in the O_2 gas form, there is an unlimited supply of both nitrogen and oxygen available to react with each other, which they will do at elevated temperatures.

The other source of nitrogen that results in the formation of significant amounts of NO_x is the nitrogen content of the fuel itself. For instance, Number 6 fuel oil typically contains from 3% to 7% by weight of nitrogen. It has been observed that the NO_x content of stack gases increases linearly with a linear increase in the nitrogen content of the fuel oil. It is therefore to be concluded that: (1) the nitrogen content of fuel oil is converted to NO_x during the combustion process, and (2) a reasonable strategy for helping to comply with limits on NO_x emissions is to obtain and burn fuel oil having a relatively low nitrogen content.

The source of nitrogen in fossil fuels is similar to that of sulfur—namely, the living tissue from which the fossil remains were derived. All living entities, since life began, have had deoxyribonucleic acid (DNA), ribonucleic acid (RNA), and protein materials as key components. Nitrogen has always been a component of these three substances, and the fossil remains contain nitrogen as well.

The principal harmful effect of the nitrogen oxides is their reaction with atmospheric hydrocarbons and other substances to form smog.

Carbon Monoxide (CO). Carbon monoxide is the product of incomplete combustion of organic matter, including fossil fuels. While complete combustion of material composed of organic carbon produces carbon dioxide, CO_2, incomplete combustion, due to inadequate supply of oxygen, results in the production of some carbon monoxide, CO. Carbon monoxide is hazardous to the health of humans and other animals.

Carbon Dioxide (CO_2). Carbon dioxide is a product of complete combustion. The principal sources of carbon dioxide in the atmosphere are the respiration of plants, animals, and microorganisms, which use oxygen as an electron acceptor, and the burning of fossil and other fuels for generation of heat and power.

The principal harmful effect of carbon dioxide in the atmosphere is its contribution, along with methane, to the promotion of global warming due to the so-called "greenhouse effect."

Particulates. Particulates, known as "fine particulate matter" or "PM10," are very small particles of any substance. The harmful effects of particulates are their tendency to lodge in the lungs and their objectionable effect on visibility.

The sources of particulate matter in the atmosphere are many and varied. Every time something is burned, there is generation and release of particulate matter. Wind generates particulate matter by blowing dust and other particulates from places of deposition into the ambient air. Automobiles and moving heavy equipment disturb land and generate airborne particulate matter.

Volatile Organic Carbon Compounds (VOCs). Volatile organic carbon compounds are a class of chemicals that is emitted directly to the air as a result of evaporation or other type of volatilization. Sources include stored gasoline, stored solvents and other industrial chemicals, and certain industrial processes. Incomplete combustion of fuels of many types is also an important source of VOC discharge to the ambient air.

The principal harmful effects of VOCs are toxicity, possible contribution to smog via photochemical reactions in the atmosphere, and possible contribution to the greenhouse effect and consequent global warming.

Malodorous Substances. There are many substances, including compounds of sulfur in the reduced state (such as hydrogen sulfide, methyl and ethyl mercapatans, and dimethyl sulfide) and degradation products of proteins (such as amines, amides, putricine, and cadaverine) that have extremely objectionable odors. Many of these compounds can be detected by the human olfactory apparatus in atmospheric concentration ranges of only a few parts per billion by volume. Although their common objectionable property is their bad smell, some of them are toxic.

Hydrogen Sulfide (H_2S). Hydrogen sulfide is toxic as well as malodorous. Worse yet, it has the characteristic of being able to desensitize the olfactory apparatus in a few minutes' time if the concentration is higher than a few parts per billion by volume. Persons who have entered confined spaces containing hydrogen sulfide gas have died because, soon after entering the contaminated space, they were unable to smell the H_2S, did not realize they were breathing H_2S, and were overcome by a fatal dose.

Other Reduced Sulfur Compounds. A reduced sulfur compound is a chemical substance of low molecular weight that contains one or more sulfur atoms in the minus-two (sulfide) valence state. These substances, along with hydrogen sulfide, are referred to in the aggregate as TRS. Many air discharge permits contain restrictions on TRS. Some TRS compounds are toxic. All are objectionable because of their extremely strong, foul odors.

Common sources of hydrogen sulfide and other reduced sulfur compounds include tanneries, rendering plants, kraft pulp mills, and malfunctioning POTWs.

Organics. Several categories of organic material are regulated through restrictions contained in air discharge permits.

Hydrocarbons. Low molecular weight organic compounds consisting of carbon and hydrogen are discharged to the air via evaporation and as a result of incomplete combustion of fossil fuels, including gasoline. These substances undergo photochemical reactions with other substances in the atmosphere to form smog.

Methane. Methane, the lowest molecular weight hydrocarbon, enters the atmosphere as a result of natural gas extraction, coal extraction, management of solid waste, anaerobic degradation of organic material in the natural environment, and the gaseous expulsions of cattle and other animals associated with agriculture.

The principal harmful effect of methane released to the atmosphere is its contribution, along with carbon dioxide, to the cause of global warming, or the greenhouse effect.

Sampling Methods for Air Toxics. Compliance with the requirements of *Title III* of the CAA is based on the implementation of specified (MACT) control technologies and/or achieving specific HAP limits. Ultimately, sampling and analysis work are required to

determine compliance with one or more of the 188 listed substances present in the discharge. This information is needed, in some cases, to determine whether or not a given control technology must be installed.

Characteristics of Solid Waste Streams from Industries

The Resource Conservation and Recovery Act (RCRA), as amended, is the primary law governing the handling, transportation, and disposal of solid and hazardous wastes. The law is contained in its entirety, including the regulations that specifically regulate the sampling and analyses of hazardous and nonhazardous solid waste streams, in volume 40 of the Code of Federal Regulations. Subtitles C and D are addressed to the management of hazardous and nonhazardous wastes, respectively.

Hazardous Wastes

In *40 C.F.R. 260*, the EPA defines a generator as "any person, by site, whose act or process produces hazardous waste identified or listed in Part 261 or whose act first causes a hazardous waste to become subject to regulation." Further, the EPA has established three categories of generator, depending on the quantity of hazardous waste generated per month. Table 5-5 provides a description of generator categories. As the quantity of waste that is generated per month increases, the regulatory requirements also increase. A very important first order of business, therefore, is to determine whether or not a given industrial facility qualifies as a generator, and, if so, which category of generator the facility is. This situation could change at any time, due to changes in raw materials used, production chemicals and other materials used, manufacturing processes, and even changes in state or federal laws and/or regulations.

The generator categories described in Table 5-5 are based on EPA definitions. Most states have been delegated the authority to

Table 5-5   Quantity Determines Which Regulations Apply

Generator	Quantity	Regulation
Large Quantity (LQG)	1,000 kg/month (approximately 2,200 lb)	All Part 262 Requirements
	> 1 kg/month acute (approximately 2.2 lb)	
	> 100-kg residue or contaminated soil from cleanup of acute hazardous waste spill	
Small Quantity (SQG)	Between 100 and 1,000 kg/month (approximately 220 to 2,200 lb)	Part 262, Subparts A, B, C (§262.34[d] is specific to SQGs); Subparts E, F, G, H if applicable; and portions of Subpart D as specified in §262.44
Conditionally Exempt Small Quantity (CESQG)	100 kg/month	§261.5
	1 kg acute	
	100-kg residue or contaminated soil from cleanup of acute hazardous waste spill	

implement RCRA regulations and have the authority to make more stringent regulations. It is important to check the regulations for the individual state where the facility is located.

RCRA defines hazardous wastes in terms of specific properties. According to RCRA, a solid waste is hazardous if it meets one of three conditions and is not excluded from regulation as a hazardous waste:

1. Exhibits, on analysis, any of the characteristics of a hazardous waste
2. Has been named a hazardous waste and appears on an appropriate list
3. Is a mixture containing a listed hazardous waste and any other solid waste material

RCRA has identified four characteristics, which, if exhibited by a solid waste, designate it as a hazardous waste:

- Ignitable
- Corrosive
- Reactive
- Toxic

These four characteristics are described as follows:

Ignitable

- A liquid, except aqueous solutions containing less than 24% alcohol, that has a flash point of less than 60°C (140°F)
- A substance that is not a liquid and is capable, under standard temperature and pressure, of causing fire through either friction, contact with moisture, or spontaneous chemical change
- A substance that is an ignitable compressed gas per Department of Transportation (DOT) regulations
- An oxidizer per DOT regulations
- Examples of ignitable substances are waste solvents, paints, and some waste oils

Corrosive

- An aqueous material having a pH less than or equal to 2.0 or greater than 12.5
- A liquid that corrodes steel at a rate greater than 1/4 inch per year at a temperature of 55°C (130°F)
- Examples of corrosive wastes are automobile battery acid and waste pickle liquor from the manufacturing of steel

Reactive

- Reacts violently with water
- Normally unstable and reacts violently without detonating

- Generates toxic gases, vapors, or fumes on mixing with water
- Forms a potentially explosive mixture with water
- Contains cyanide or sulfide and generates toxic gases, vapors, or fumes at a pH of between 2.0 and 12.5
- Capable of detonation if heated under confinement or subjected to a strong detonating force
- Capable of detonation at standard temperature and pressure
- Listed as a Class A or B explosive by the Department of Transportation

Toxic

A solid waste exhibits the characteristic of toxicity if the extract from a sample obtained from conducting the Toxicity Characteristic Leaching Procedure (TCLP) contains any of the contaminants listed in Table 1 of *40 C.F.R. 261.24* at concentrations equal to or greater than the respective value given in that table.

The Toxic Characteristic Leaching Procedure (TCLP) test was developed as a method for determining whether or not a waste material, after being placed in a landfill, would leach metals or other substances at rates greater than what was considered acceptable. The TCLP test attempts to simulate worst-case landfill leaching conditions, where low pH precipitation (acid rain) would percolate down through the landfilled wastes and cause metals to dissolve into solution.

In performing the TCLP test, the material to be landfilled is first pulverized to particle sizes no larger than one millimeter in diameter. Then, 5.0 grams of the solid phase of the waste are placed in a 500-ml beaker or Erlenmeyer flask. Next, 96.5 ml of "reagent water" (deionized water) are added; then the beaker or flask is covered with a watchglass and stirred vigorously for five minutes. This procedure is normally carried out in a beaker, with a magnetic stirrer keeping the granules of waste material suspended and well mixed with the reagent water.

It is very important that the instruction "cover with a watchglass" be followed. After the five minutes of stirring, the pH of the mixture is determined. If the pH is below 5.0, the mixture is subjected to the specified extraction procedure using extraction fluid #1, which is prepared by mixing 5.7 ml glacial acetic acid, 500 ml of reagent water, and 64.3 ml 1N sodium hydroxide. The pH of this mixture is 4.93 + 0.05. If the pH of the mixture of pulverized waste and reagent water (after mixing for five minutes) is higher than 5.0, the mixture is subjected to the specified extraction procedure using extraction fluid #2, which is prepared by diluting 5.7 ml glacial acetic acid to a volume of 1 liter with reagent water. Extraction fluid #2 has a pH of 2.88 + 0.05, is thus a stronger acid solution than extraction fluid #1 (no sodium hydroxide is added during preparation of extraction fluid #2), and will more strongly dissolve metals from the waste material.

The reason it is important to keep the beaker covered with a watchglass during the stirring process is that carbon dioxide gas will either escape from, or dissolve into, the mixture, causing the pH to change and possibly resulting in the requirement to use the stronger extraction solution. Analysts sometimes do not use the watchglass in order to enable leaving pH probes in the mixture during the stirring process.

After the extraction process has been carried out, the extract is analyzed for suspected or possible substances. If the results of this analysis show that any of the substances are present in the extract in an amount exceeding the published limit, the waste is deemed to be "hazardous" and must be handled and disposed of in accordance with all the applicable hazardous waste laws and regulations. See Table 5-5.

Exemptions

Certain materials under certain circumstances have been specifically exempted from having to be handled and disposed of as hazardous waste. When applicable conditions are met, these exempted materials can be handled and disposed of as ordinary solid wastes.

Delisting

Some solid wastes can be removed from the "hazardous waste" category by going through a process that enables the waste to be delisted. Any generator or waste handler can petition the EPA to exclude a listed waste from regulation under Subtitle C. The petitioner must prove to the EPA that, due to facility-specific differences in raw materials, processing, or other factors, the waste is not hazardous and will therefore pose no risk to persons, animals, or the environment. If the EPA, upon examining all other factors in addition to those cited by the petitioner, finds no reason for not delisting the waste, the waste may be handled and disposed of as ordinary solid waste as regulated under *40 C.F.R., Subtitle D*. It is noted here that the delisting process can be time consuming and expensive. Delisting is typically sought for high-volume waste streams.

Cradle to Grave Manifesting

A significant requirement for all shipments of hazardous wastes subject to the regulations of Subtitle C is that they be accompanied by a written document called a manifest. Figure 5-12 shows an example of the standard form, titled "Uniform Hazardous Waste Manifest." At the time of this printing, the manifest was undergoing changes to make it federally standardized. The manifest provided in Figure 5-12 is intended to be illustrative of the type of information that would be required when filling out a manifest. It is the responsibility of the generator of any hazardous waste to initiate the manifest by writing in the appropriate information. If the waste is hauled away from the generator's facility by a second party, the hauler assumes possession of and responsibility for the manifest, although the generator retains a copy. Then, when the waste reaches the place of ultimate treatment, storage, or disposal (the

grave), the manifest again changes hands, with the hauler retaining a copy.

As shown on the example manifest presented in Figure 5-12, information contained on the completed manifest may include, but is not limited to, the following:

- Name and EPA identification number of the generator, the hauler(s), and the facility where the waste is to be treated, stored, or disposed of
- U.S. EPA and U.S. DOT descriptions of the waste
- Quantities
- Complete address of the treatment, storage, or disposal facility

In addition, the manifest must certify that the generator has in place a program to reduce the volume and toxicity of hazardous wastes to the degree that is economically practicable, as determined by the generator, and that the treatment, storage, or disposal method chosen by the generator is a practicable method currently available that minimizes the risk to human health and the environment.

Once the waste is delivered to the final place of disposition (treatment, storage, or disposal facility), the owner or operator of that facility must sign the manifest, retain a copy, and return a copy to the generator of that waste. If 35 days pass from the date on which the waste was signed for by the initial transporter, and the generator has not received a copy of the manifest from the final site of disposition, the generator must contact the initial transporter to determine the fate of the waste. If 45 days pass and the generator still has not received a copy of the manifest, the generator must submit an exception report.

Nonhazardous Solid Wastes from Industries

Some industrial solid wastes can be disposed of in municipal solid wastes landfill facilities (MSWLF). The federal law that governs such disposal is contained in its entirety in *40 C.F.R. Part 258*. (Part 257 contains the provi-

Please print or type (Form designed for use on elite (12 - pitch) typewriter) Form Approved OMB No 2050 - 0039 Expires 9 - 30 - 91

UNIFORM HAZARDOUS WASTE MANIFEST

1 Generator's US EPA ID No. Manifest Document No.

2. Page 1 of

Information in the shaded areas is not required by Federal law

3. Generator's Name and Mailing Address

A. State Manifest Document Number

B. State Generator's ID

4. Generator's Phone ( )

5. Transporter 1 Company Name

6. US EPA ID Number

C. State Transporter's ID

D. Transporter's Phone

7. Transporter 2 Company Name

8. US EPA ID Number

E. State Transporter's ID

F. Transporter's Phone

9. Designated Facility Name and Site Address

10. US EPA ID Number

G. State Facility's ID

H. Facility's Phone

GENERATOR

11. US DOT Description *(Including Proper Shipping Name, Hazard Class, and ID Number)*	12. Containers No.	Type	13. Total Quantity	14. Unit Wt/Vol	I. Waste No.
a.					
b.					
c.					
d.					

J. Additional Descriptions for Materials Listed Above

K. Handling Codes for Wastes Listed Above

15. Special Handling Instructions and Additional Information

16. **GENERATOR'S CERTIFICATION**: I hereby declare that the contents of this consignment are fully and accurately described above by proper shipping name and are classified, packed, marked, and labeled, and are in all respects in proper condition for transport by highway according to applicable international and national government regulations.

If I am a large quantity generator, I certify that I have a program in place to reduce the volume and toxicity of waste generated to the degree I have determined to be economically practicable and that I have selected the practicable method of treatment, storage, or disposal currently available to me which minimizes the present and future threat to human health and the environment; **OR**, if I am a small quantity generator, I have made a good faith effort to minimize my waste generation and select the best waste management method that is available to me and that I can afford.

Printed/Typed Name | Signature | *Month Day Year*

TRANSPORTER

17. Transporter 1 Acknowledgement of Receipt of Materials

Printed/Typed Name | Signature | *Month Day Year*

18. Transporter 2 Acknowledgement of Receipt of Materials

Printed/Typed Name | Signature | *Month Day Year*

FACILITY

19. Discrepancy Indication Space

20. Facility Owner or Operator: Certification of receipt of hazardous materials covered by this manifest except as noted in item 19.

Printed/Typed Name | Signature | *Month Day Year*

EPA Form 8700 - 22 (Rev. 9 - 88) Previous editions are obsolete.

Figure 5-12 Uniform Hazardous Waste Manifest.

sions for hazardous wastes). Within Part 257, "industrial solid waste" means any solid waste generated by manufacturing or industrial processes that is not a hazardous waste, regulated under Subtitle C of RCRA. Such waste may include, but is not limited to, waste resulting from electric power generation; fertilizer/agricultural chemicals production; food and related products or by-products production; inorganic chemicals manufacturing; iron and steel manufacturing; leather and leather products production; nonferrous metals manufacturing; organic chemicals production; plastics and resins manufacturing; pulp and paper manufacturing; rubber and miscellaneous plastic products manufacturing; the manufacture of stone, glass, and concrete products; textile manufacturing; transportation equipment manufacturing; and water treatment. Not included are wastes resulting from mining or oil and gas production.

MSWLF facilities are required by Subtitle D regulations to have, as a minimum, a liner system composed of a single composite liner that is part of a leachate collection and removal system. A low-permeability cover must be installed when the landfill reaches maximum capacity. In addition, a ground-water monitoring system must be used to detect liner failure during the 30-year mandated postclosure care period.

States, of course, may promulgate laws and regulations that are more restrictive than the federal laws and regulations. Again, it is the responsibility of the owner of the facility to determine which set of laws and regulations is the most restrictive, and this determination should be used as the basis for design of discharge control equipment.

Bibliography

American Society for Testing and Materials. *A Standard Guide for Examining the Incompatibility of Selected Hazardous Waste Based on Binary Chemical Mixtures.* Philadelphia, PA: 1983.

De Renzo, D. J. *Solvent Safety Handbook.* Park Ridge, NJ: Noyes Publications, 1986.

Dyer, J. C., A. S. Vernick, and H. D. Feiler. *Handbook of Industrial Wastes Pretreatment.* New York: Garland STPM Press, 1981.

Flick, Ernest W. *Industrial Solvents Handbook.* 3rd ed. Park Ridge, NJ: Noyes Publications, 1985.

Grant, D. M. *Open Channel Flow Measurement Handbook.* Lincoln, NE: Instrumentation Specialties Co., 1979.

Harris, J. P., and S. A. Kacmar. "Flow Monitoring Techniques in Sanitary Sewers." In *Deeds and Data,* Water Pollution Control Federation: July 1974.

Higgins, I. J., and R. G. Burns. *The Chemistry and Microbiology of Pollution.* New York: Academic Press, 1975.

Jain, R. K., L. V. Urban, G. S. Stacey, and H. E. Balbach. *Environmental Assessment.* New York: McGraw Hill, 1993.

Leupold and Stevens, Inc. *Stevens Water Resources Handbook.* Beaverton, OR: 1975.

Merrill, W. H., Jr. "Conducting a Water Pollution Survey." *Plant Engineering* October 15, 1970.

Merrill, Wm. H., Jr. "How to Conduct an Inplant Wastewater Survey." *Plant Eng.* January 7, 1971.

Sax, N. I. *Dangerous Properties of Hazardous Materials.* 6th ed. New York: Van Nostrand Reinhold, 1984.

Shelley, P. E. "Sampling of Wastewater." U.S. Environmental Protection Agency *Technology Transfer,* June 1974; *The Merck Index.* 9th ed. Rahway, NJ: Merck and Co., Inc., 1976.

Shelley, P. E. *An Assessment of Automatic Sewer Flow Samplers.* U.S. Environmental Protection Agency. Office of Research and Monitoring. EPA-R2-73-261. June 1973.

Shelley, P. E., and G. A. Kirkpatrick. "Sewer Flow Measurement—A State of the Art Assessment." EPA-60012-75-027. 1975.

Sittig, M. *Handbook of Toxic and Hazardous Chemicals and Carcinogens.* 2nd ed. Park Ridge, NJ: Noyes Publications, 1985.

Sittig, M. *Priority Toxic Pollutants—Health Impacts and Allowable Limits.* Park Ridge, NJ: Noyes Publications, 1980.

Standard Methods for the Examination of Water and Wastewater. 20th ed. New York: American Public Health Association, 1998.

Thompson, R. G. "Water-Pollution Instrumentation." *Chemical Engineering* June 21, 1976: 151–154.

U.S. Department of the Interior. "Methods for Collection and Analysis of Water Samples for Dissolved Minerals and Gases." In *Techniques of Water Resources, Investigations of the United States Geological Survey*. Book 5. Chapter Al. 1970.

U.S. Department of the Interior. Bureau of Reclamation. *Water Measurement Manual*. Denver, CO: 1979.

U.S. Environmental Protection Agency. *Manual of Methods for Chemical Analysis of Water and Wastes*. Methods Development and Quality Assurance Research Laboratory, National Environmental Research Center. Cincinnati, OH: 1974.

U.S. Environmental Protection Agency. *Method for Chemical Analysis of Water and Waste*. EPA 600/4-79/020. Washington, DC: U.S. Government Printing Office, 1979.

U.S. Environmental Protection Agency. "Announcement of the Drinking Water Contaminant Candidate List." 63 FR 10273. Washington, DC: U.S. Government Printing Office, March 1998.

U.S. Environmental Protection Agency. "Designation of Hazardous Substances." *Code of Federal Regulations*. Title 40, Part 116. Washington, DC: U.S. Government Printing Office, 1980.

U.S. Environmental Protection Agency. Office of Water. "Drinking Water Program Redirection Strategy." EPA 810-R-96-003. Washington, DC: U.S. Government Printing Office, June 1996.

U.S. Environmental Protection Agency. "General Provisions for Effluent Guidelines and Standards." *Code of Federal Regulations*. Title 40, Part 401. Washington, DC: U.S. Government Printing Office, 1997.

U.S. Environmental Protection Agency. "Guidelines Establishing Procedures for Analysis of Pollutants." *Code of Federal Regulations*. Title 40, Part 136. Washington, DC: U.S. Government Printing Office, 1979.

U.S. Environmental Protection Agency. "Hazardous Waste Management System: General." *Code of Federal Regulations*. Title 40, Part 260. Washington, DC: U.S. Government Printing Office, 1980.

U.S. Environmental Protection Agency. "Identification and Listing of Hazardous Waste." *Code of Federal Regulations*. Title 40, Part 261. Washington, DC: U.S. Government Printing Office, 1980.

U.S. Environmental Protection Agency. "Interim Status Standards for Owners and Operators of Hazardous Waste Treatment, Storage, and Disposal Facilities." *Code of Federal Regulations*. Title 40, Part 265. Washington, DC: U.S. Government Printing Office, 1987.

U.S. Environmental Protection Agency. "National Primary Drinking Water Regulations." *Code of Federal Regulations*. Title 40, Part 141. Washington, DC: U.S. Government Printing Office, 1975.

U.S. Environmental Protection Agency. "National Secondary Drinking Water Regulations." *Code of Federal Regulations*. Title 40, Part 143. Washington, DC: U.S. Government Printing Office, 1988.

U.S. Environmental Protection Agency. "Standards Applicable to Transporters of Hazardous Waste." *Code of Federal Regulations*. Title 40, Part 263. Washington, DC: U.S. Government Printing Office, 1986.

U.S. Environmental Protection Agency. "Toxic Pollutant Effluent Standards." *Code of Federal Regulations*. Title 40, Part 129. Washington, DC: U.S. Government Printing Office, 1977.

U.S. Environmental Protection Agency. "Standards Applicable to Generators of Hazardous Waste." *Code of Federal Regulations*. Title 40, Part 262. Washington, DC: U.S. Government Printing Office, 1980–1987.

U.S. Environmental Protection Agency. Office of Research and Development. *A Method for Determining the Compatibility of Hazardous Wastes.* EPA 60012-80/076. Cincinnati, OH: 1980.

U.S. Environmental Protection Agency. *Analytical Methods for Determination of Asbestos Fiber in Water.* Washington, DC: U.S. Government Printing Office, 1983.

U.S. Environmental Protection Agency. *Handbook for Monitoring Industrial Wastewater.* Technology Transfer. Washington, DC: U.S. Government Printing Office, August 1973.

U.S. Environmental Protection Agency. Office of Research and Development. Environmental Monitoring and Support Laboratory. *Handbook for Sampling and Sample Preservation of Water and Wastewater,* EPA-600/4-82-029. Cincinnati, OH: 1982.

U.S. Environmental Protection Agency. *Microbiological Methods for Monitoring the Environment.* EPA 600/8-78/017. Washington, DC: U.S. Government Printing Office, 1978.

U.S. Environmental Protection Agency. Office of Public Awareness. *Hazardous Waste Information.* SW-737. Washington, DC: U.S. Government Printing Office, 1980.

U.S. Environmental Protection Agency. Office of Water and Waste Management. *Hazardous Waste Information—Identification and Listing.* SW-850. Washington. DC: U.S. Government Printing Office, 1980.

U.S. Environmental Protection Agency. Office of Water and Waste Management, *Hazardous Wastes Information—Transporters.* SW-830. Washington, DC: U.S. Government Printing Office, 1980.

U.S. Environmental Protection Agency. Office of Water and Waste Management. *Hazardous Waste Information—Facility Permits.* SW-852. Washington, DC: U.S. Government Printing Office, 1980.

U.S. Environmental Protection Agency. Office of Water and Waste Management. *Hazardous Wastes Information—Generators.* SW-839. Washington, DC: U.S. Government Printing Office, 1980.

U.S. Environmental Protection Agency. *Samplers and Sampling Procedure for Hazardous Waste Streams.* EPA 600/2-80/018. Washington, DC: U.S. Government Printing Office, 1980.

U.S. Environmental Protection Agency. Effluent Guidelines Division. *Sampling and Analysis Procedure for Screening of Industrial Effluents for Priority Pollutants.* Washington, DC: U.S. Government Printing Office, 1977.

U.S. Environmental Protection Agency. *Test Methods for Evaluating Solid Waste—Physical/Chemical Methods.* 2nd ed. SW-846. Washington, DC: U.S. Government Printing Office, 1985.

U.S. Environmental Protection Agency. *Water Quality Standards Handbook.* 2nd ed. EPA-823-B-94-005a. Washington, DC: U.S. Government Printing Office, 1994.

U.S. Environmental Protection Agency. "Standards for Owners and Operators of Hazardous Waste Treatment, Storage, and Disposal Facilities." *Code of Federal Regulations.* Title 40, Part 264. Washington, DC: U.S. Government Printing Office, 1980–1987.

Vernick, A. S. "How to Conduct a Wastewater Survey." *Plant Engineering* July 7, 1977.

Vernick, A. S. "How to Conduct a Wastewater Survey." *Plant Engineering* August 4, 1977.

Weiss, G. *Hazardous Chemical Data Book.* 2nd ed. Park Ridge, NJ: Noyes Publication, 1986.

6   Industrial Stormwater Management

General

Precipitation of all types falls on industrial facilities and, in so doing, transports chemicals and inert solids from wherever it physically contacts them to other parts of the environment. Any industrial site will have some contamination on all surfaces, including roofs, parking lots, storage facilities, roads, sidewalks, and grassy areas.

Since all substances are soluble to some extent in water (the universal solvent), any chemical substances in either liquid or solid form will become dissolved (to an extent equal to or less than the solubility limit for that substance) and will either percolate into the ground or be carried with the stormwater runoff. Particles that are not dissolved will be transported with the runoff, which will also contain some amount of gases that are dissolved in the runoff. Most, if not all, of the precipitation that percolates into the ground will eventually reach groundwater.

The types and amounts of materials that become incorporated into stormwater runoff depend upon the state of cleanliness of the industrial facility (i.e., the roof areas, parking lots, roadways, etc.), as well as the characteristics of the precipitation itself, in terms of intensity, duration, pH, temperature, and chemical constituents. As well, factors such as the topography of the plant site, the characteristics of the surfaces over which the runoff flows, and the stormwater management practices that have been employed will have a major influence on the quality of the stormwater as it flows off the site or percolates into the ground.

It is the quality and, in some cases, the quantity of the stormwater that leaves the industrial site, whether via overland flow to surface waters or percolation into the groundwater, that are of importance from the standpoint of compliance with environmental laws, regulations, and permits. As such, stormwater runoff quality and quantity are the focus of an overall stormwater management strategy, which incorporates, as appropriate, comprehensive pollution prevention practices; collection, diversion, and treatment facilities; and regulatory compliance monitoring.

Prevention and control of contamination of groundwater and surface water by stormwater percolation or runoff are a matter of preventing or managing the following activities:

- Contamination of surfaces with which stormwater can come into contact, such as roofs, parking areas, plant roadways, outdoor storage areas, industrial yard areas, tanks, piping systems, and outdoor equipment
- Spills, leaks, or releases of pollutants as a result of accidents or as a consequence of inadequate preventive maintenance

Prevention of spills, leaks, and releases is best accomplished through proper design and operation of industrial equipment and storage facilities, an exemplary preventive maintenance program, and thorough worker training programs.

Federal Stormwater Program

Federal regulation of stormwater originated with the *1987 Clean Water Act* amendments, which established the authority for the EPA to develop a phased approach to stormwater discharge permitting and management. Two stormwater rules followed in 1990 and 1992: the stormwater application rule and the stormwater implementation rule. The stormwater application rule of November 1990 identified the types of facilities subject to permitting under the National Pollutant Discharge Elimination System (NPDES) program (found at *40 C.F.R. Part 122*), and the stormwater implementation rule of April 1992 described the requirements for NPDES stormwater permits.

Under the Phase I regulations, promulgated in 1990, a stormwater discharge permit is required for:

- "Stormwater discharges associated with industrial activity" as defined by *40 C.F.R. §122.26(b)(14)* (primarily relating to specific Standard Industrial Classification [SIC] codes)
- Stormwater discharges from large and medium municipal separate storm sewer systems (MS4s). A large MS4 is defined as a separate storm sewer with a service population of more than 250,000 people, and a medium MS4 is a system that serves more than 100,000 and less than 250,000 people
- Stormwater discharges from large construction sites (more than five acres of disturbed land)
- Facilities already covered by an NPDES permit for a combined wastewater/ stormwater discharge
- Facilities that the EPA or a NPDES state administrator determines to have stormwater discharges contributing to a violation of water quality or that are "significant contributors" of pollutants to waters of the United States

Phase II of the federal stormwater permitting program was adopted on December 8, 1999, and includes permit requirements for the following types of discharges:

- Stormwater discharges from "regulated" small MS4s in urbanized areas (UAs)[1]
- Stormwater discharges from small construction sites (between one and five acres of disturbed land)
- Stormwater discharges associated with industrial activities from small municipal facilities temporarily exempted from the Phase I requirements by provisions within the Intermodal Surface Transportation Efficiency Act (ISTEA)

The Phase II rules also contain a conditional no-exposure exclusion that is available to Phase I dischargers who can certify that all industrial activities and materials at their site are protected from exposure to precipitation. This "conditional no exposure certification" allows such facilities to "opt out" of the federal stormwater program.

It is important to note that the federal stormwater rules are not applicable in the following situations (though there may be state or local rules that are applicable):

- Nonpoint source discharges of stormwater
- Discharges of stormwater to municipal sewer systems that are combined stormwater and sanitary sewers
- Discharges of stormwater to groundwater

State Stormwater Permitting Programs

Provisions within the Clean Water Act allow states to request authorization to implement the federal NPDES permitting program within their states. As of 2006, only five states

---

[1] An "urbanized area" is defined as a land area comprising one or more places—central place(s) and the adjacent densely settled surrounding area—that together have a residential population of at least 50,000 and an overall population density of at least 1,000 people per square mile.

(Alaska, Idaho, Massachusetts, New Hampshire, and New Mexico) and the District of Columbia do not have NPDES permitting authority for stormwater. The EPA remains the NPDES permitting authority for stormwater, as well as wastewater, discharges in these jurisdictions.

Most states with NPDES permitting authority have followed the EPA's lead and developed general permits for stormwater discharges to surface waters based on the EPA's original baseline general permit. The state permitting programs must be at least as stringent as the EPA's program and may be more stringent. States that have been delegated NPDES permitting authority often incorporate elements of their preexisting stormwater management programs into a general permitting process. Industrial dischargers located in NPDES-delegated states must check with their state agencies to determine their stormwater permit options and requirements. Stormwater that discharges to groundwater, as opposed to surface water, will likely be regulated under the state's groundwater discharge laws and regulations.

Types of Stormwater Permits

Industrial facilities are required to comply with the stormwater rules if they meet the following three criteria:

- Does the facility discharge stormwater via one or more point sources (e.g., a pipe, swale, or ditch) into waters of the United States (either directly or indirectly through a municipal separate storm sewer)?
- If the facility falls within one of the following categories:

  — Engaged in industrial activity (defined by SIC code)
  — Already covered under an NPDES permit
  — Identified by the EPA as contributing to a water-quality violation

There are currently two types of federal stormwater discharge permits issued to industrial dischargers by the EPA: the Multi-Sector General Permit and the Individual Permit. NPDES-delegated states typically offer both general and individual permits, as well. Most permits are issued for a maximum term of five years, after which they may be renewed.

Multi-Sector General Permit

The NPDES Stormwater Multi-Sector General Permit (MSGP) is the simplest form of NPDES permit coverage an industrial facility can obtain, although there are circumstances that would cause a facility to be ineligible for MSGP coverage. The original MSGP was issued in the Federal Register on September 29, 1995 and incorporates requirements for 29 industrial sectors. On September 30, 1998, the modified MSGP was issued to cover additional industrial dischargers not previously included in the MSGP. In 2000, the MSGP was reissued for another five-year term, which ends on September 30, 2005. The EPA is currently in the process of renewing the permit prior to its expiration.

Industrial facilities that have activities covered under one or more of the industrial sectors in the MSGP are eligible for coverage. To obtain MSGP coverage, the facility must submit a Notice of Intent (NOI) for coverage and prepare and implement a Stormwater Pollution Prevention Plan (SWPPP). The MSGP contains some general requirements specific to all permittees, as well as industry-specific requirements for stormwater monitoring, reporting, and, in some cases, specific best management practices (BMPs) to minimize contamination of runoff.

Individual Permit

The NPDES Individual Permit requires the preparation and submittal of NPDES Forms 1 and 2F, which request specific information about the facility; the industrial operations; and the results of stormwater sampling, analysis, and flow measurement. A facility-specific Individual Permit is issued by the NPDES permitting authority and typically contains discharge limits, monitoring and reporting requirements, and may require implementation of BMPs or pollution prevention measures. Industrial facilities that are required to apply for an Individual Permit include:

- Facilities that are not covered under the MSGP
- Facilities that have previously had an NPDES permit for stormwater
- Facilities that are determined by the permitting authority to be contributing to a violation of a water-quality standard

Because of the backlog of applications and the lengthy application review and permit-writing process for Individual Permits, permitting authorities typically recommend that dischargers seek coverage under the MSGP, if eligible. In some EPA regions, applicants seeking Individual Permits have waited three to five years or more to receive their permits.

Construction-Related Permits

One additional stormwater permitting issue that an industrial or environmental manager must be aware of is the NPDES Stormwater Construction General Permit. This permit is applicable to construction projects that disturb one or more acres of land area. The permitting process is the same as for the MSGP: submittal of an NOI for coverage and implementation of a Stormwater Pollution Prevention Plan (SWPPP) that focuses on BMPs relating to sediment and erosion control during construction. NPDES-delegated states offer similar permits for stormwater discharges from construction sites. In some states, local governments also have site plan approval requirements relating to stormwater runoff that overlap with NPDES stormwater requirements. When planning for construction projects at an industrial facility, an environmental manager must consider the

potential applicability of stormwater permitting at all levels of government—federal, state, and local. Also, construction projects that result in changes in the site drainage characteristics, exposure of industrial activities, or potential pollutant sources could trigger the requirement to update the facility's stormwater permit and/or the SWPPP.

Stormwater Pollution Prevention Plan (SWPPP)

Among the important requirements of the federal MSGP is the development and implementation of a Stormwater Pollution Prevention Plan (SWPPP). The goal of the SWPPP is to reduce or eliminate the amount of pollutants in stormwater discharges from an industrial site. The SWPPP should be developed with the involvement of and input from a designated stormwater pollution prevention team comprised of facility staff representing key manufacturing, operations, and environmental management areas. The SWPPP must identify all potential pollutant sources and include descriptions of control measures to eliminate or minimize contamination of stormwater. The SWPPP must contain:

- A map of the industrial facility, identifying the contributing areas that drain to each stormwater discharge point
- Identification of the manufacturing or other activities that take place within each area
- Identification of the potential sources of pollutants within each area
- An inventory of materials that can be exposed to stormwater
- An estimate of the quantity and type of pollutants likely to be contained in the stormwater runoff
- A history of spills or leaks of toxic or otherwise hazardous materials from the past three years

Best Management Practices (BMPs) must be identified. BMPs should include good housekeeping practices, structural control measures where needed, a preventive maintenance program for stormwater control measures, and procedures for spill prevention and response. Traditional stormwater management controls, such as oil/water separators and retention/equalization devices, must also be addressed as appropriate.

- For facilities that are subject to Section 313 of the Emergency Planning and Community Right-to-Know Act (EPCRA 313) reporting, the SWPPP must address those areas where the listed Section 313 "water priority chemicals" are stored, processed, or handled. These areas typically require stricter BMPs, in the form of structural control measures, and additional stormwater monitoring.

- The facility must have a Certification of Non-Stormwater Discharges, certifying that their stormwater discharge does not contain actual or potential sources of nonstormwater discharges such as wash waters, contact cooling waters, and floor drain connections. It should be noted that the MSGP contains a list of permissible nonstormwater discharges, including fire hydrant flushings and building wash water that does not contain detergents. However, no other types of materials are permissible under this permit. In order to make this certification, a comprehensive facility survey should also be conducted to identify activities that may result in unpermitted discharges. As-built piping diagrams should be reviewed to confirm there are no non-stormwater connections to the storm sewer. Otherwise, all outfalls must be tested during dry weather conditions, using a dye or other tracer, to ensure that there are no discharges.

- A recordkeeping system must be developed and maintained, and there must be an effective program for training employees in matters of controls and procedures for pollution prevention.

The SWPPP should also describe the stormwater monitoring program and reporting procedures.

Prevention of Groundwater Contamination

Measures and facilities to prevent contamination of groundwater should be developed concurrently with those that have the purpose of protecting against surface water contamination. The most important of these measures include:

- Construction of impermeable barriers, such as concrete pads, to prevent percolation of stormwater after it has become contaminated
- Installation of foolproof automatic shut-off devices to prevent spills from overflowing tanks
- Alarms
- An aggressive preventive maintenance program to prevent the occurrence of leaks
- Control of particulate and aerosol emissions and routine cleaning of all surfaces on the industrial site

Stormwater Management Concepts

Development of a successful industrial stormwater management program requires consideration of three important concepts:

- Pollution prevention
- Source segregation
- Collection and treatment

Pollution prevention applies to both routine and accidental exposure of stormwater to pollutants. Pollution prevention measures may be functional (requiring people to implement them) or structural (not dependent upon people); they must be practiced and maintained on an ongoing basis.

Source segregation is the separation of "clean" and "dirty" areas and activities, so as to result in stormwater runoff from those areas that is clean or dirty. It can be an element of both pollution prevention and treatment. Source segregation has the effect of maximizing the concentration of contaminants in runoff since, presumably, the quantity of contaminants at a given industrial site at a given time is fixed by the circumstances and events that have occurred prior to a storm event. Therefore, the smaller the amount of stormwater in which the contaminants are dissolved or suspended, the more concentrated they will be. Source segregation is an important precursor to stormwater treatment, since for treatment devices such as oil skimmers and sedimentation basins, a higher pollutant concentration will lead to more effective treatment. Furthermore, a reduction in the volume of runoff to be collected and treated will result in more cost-effective treatment.

It is the goal of pollution prevention and source segregation to eliminate and/or reduce contamination of stormwater runoff at the site to the point where compliance with applicable regulations and permits occurs without the necessity of expensive collection, retention, treatment, and discharge facilities. Stormwater treatment should be the final step in a stormwater management program, after pollution prevention and source segregation measures are already in effect. However, for facilities that are regulated under permits with strict effluent limits (or that choose to treat and recycle stormwater in lieu of discharging it), treatment may be required to attain the desired water quality.

Stormwater Treatment System Design Considerations

Quantity and Quality

Designing for stormwater treatment involves determining the quantity and quality of the stormwater. The quantity is determined by studying the hydrology at the site, and the quality is based upon the water-quality limits specified in the discharge permit, or the desired water quality criteria required for facilities that recycle stormwater. Control of the quantity and quality of stormwater discharges is possible only if effective segregation, collection, retention, and treatment are in use.

Basic Hydrology

The quantity of stormwater runoff from an industrial site must be determined in order to design a collection and treatment system. To do this, the rainfall intensity, duration, and frequency must be determined for the given geographical location. The runoff from the site must then be determined; it is dependent upon the topographic features at the location and the time it takes the rainfall to travel to the outlet.

Precipitation

Rainfall at a given location can be quantified using several different methods. Some of the more common methods and sources of data are as follows:

- *Gauged Data*—Available from the National Weather Service or the U.S. Army Corps of Engineers.
- *Synthetic Distributions*—The Soil Conservation Service's (SCS) 24-hour rainfall distributions (Types I, IA, II, and III), available dimensionless, can be applied to different rainfall depths.
- *I-D-F Curves*—Statistical methods used to create intensity-duration-frequency curves for several design storms.
- *Design Storm*—Reoccurrence frequency of a storm event. Typically 2, 5, 10, 25, 50, or 100-year storms. By definition, there is a 4% probability that the 25-year storm will be exceeded in any given year.

It is also important to consider snowfall and snowmelt as a significant contribution to runoff.

Runoff

The rate of stormwater runoff is of concern to prevent erosion of downstream receiving waters. The volume of runoff from a developed site will be greater and reach the outlet faster due to impervious surfaces that prevent infiltration. Because the increased runoff volume reaches the outlet faster, it leads to a faster time of concentration. Time of concentration is defined as the time for a wave to propagate from the most distant point in the watershed to the outlet.

In the sections that follow, two common methods of determining runoff are presented: the rational method and the hydrograph method. However, there are several other methods that can be used, each with their own applications and limitations.

The peak flow from a site is typically estimated using the rational method. The rational method assumes that equilibrium is reached within the watershed (i.e., inflow equals outflow). Therefore, the storm duration must be as long as the time of concentration to achieve steady-state conditions. For this reason, this method should not be used for large areas (generally, more than 200 acres). The rational method is typically used for sizing storm sewer systems because of its simplicity. However, the method does not calculate flow rate versus time or the volume of runoff. Therefore, to design a downstream treatment system, the development of a hydrograph is necessary.

The stormwater discharge from a watershed is represented by a hydrograph, which is a continuous plot of instantaneous discharge versus time. Hydrographs are a representation of the physical geography of and the meteorological conditions in a watershed, and they include the combined effects of climate, hydrologic losses (e.g., evaporation, infiltration, and so on), surface runoff, subsurface stormwater flow, and groundwater flow. The peak of the hydrograph indicates the peak flow of a storm event, which is used to determine whether the flow is too high for downstream receiving waters. The area under the curve of the hydrograph represents the volume of water discharged during the storm event. The hydrograph method provides the information needed for sizing downstream retention/detention ponds and other treatment systems.

Selecting a Design Storm

In general, the size of the collection, retention, and treatment facilities is derived from precipitation records and selection of a design storm. Since the concept of a design storm event implies periodic "failure," it is necessary to include, within the design of stormwater facilities, provisions to prevent damage to those facilities or violations of permit when the design storm capacity is exceeded.

The 25-year, 24-hour storm event is considered an appropriate design basis for most stormwater management facilities at an industrial plant when "conventional" pollutants are the only substances of concern. In situations where PCBs or other toxic substances are potential pollutants, a 50- or 100-year storm would be more appropriate. A risk assessment should be carried out to determine if the benefits of designing for a storm event that would yield a greater volume and/or peak flow rate, such as the 50-year, 24-hour storm, or even the 100-year storm, would outweigh the risk of an overflow and, potentially, a permit exceedance. Factors to be considered in such a risk assessment include the water quality standards of the receiving water, the discharge permit limits, and the potential enforcement consequences should a permit violation occur.

Collection System Design

The typical stormwater collection system consists of roof drains, catch basins, storm drains, pumping stations, and open channels. The design storm peak runoff rate is used to size the collection, conveyance, and pumping systems. Collection systems must be well maintained and kept clean and free of leaks. Also, easy access to the collection system for sampling and flow measurement, when required, should be designed and built into the system.

Stormwater Retention/Detention

Stormwater retention or detention can be accomplished in lined, earthen basins, or in above- or below-ground concrete or steel tanks. Retention basins are also referred to as wet ponds or wetlands, because they retain a permanent pool of water. Detention basins are referred to as dry ponds, because they remain dry once they have drained after a storm event. An industrial plant in a nonurban location with large areas of unused land would consider the cost effectiveness of a lined earthen basin first, while a plant in an urban location with limited available unused land might first consider an above- or below-ground concrete basin. The size of the stormwater retention facility is typically based on the total volume from a design storm event. Additional factors that must be included are:

- Precipitation that falls directly into the retention device
- The rate at which water is taken out of the basin, as it relates to the probability of another storm event taking place very soon after the design storm has occurred

The procedures presented in Chapter 7 for designing the size (volume) of a flow equalization facility are appropriate to be used for the design of a stormwater retention basin, once the total volume and characteristics of the design storm have been determined.

Table 6-1 outlines the advantages and disadvantages of various options for retention and detention, as well as other methods of reducing the volume and rate of stormwater runoff.

System Failure Protection

Since, by definition, the stormwater management system will "fail" periodically, in terms of capacity of the retention device and/or the conveyance facilities, it is extremely important to design and construct overflow devices and other excess water management facilities. The overflow and excess water conveyance devices must protect against erosion and contamination of surface water or groundwater. These devices must also be maintained and kept clean, and should, therefore, be designed to allow easy access. Overflow facilities should be designed with sufficient capacity to accommodate any conceivable storm event.

Technologies to Achieve Treatment Goals

Design of treatment systems for stormwater runoff is always based upon the degree and type of treatment required, which depend upon:

- The water-quality limits specified in the discharge permit
- The water-quality criteria required for use in the plant

Table 6-2 presents typical stormwater pollutant parameters, some treatment options, and considerations for design.

Design of treatment facilities for stormwater runoff must always include consideration of future requirements, such as more stringent limits in future NPDES permits. Such consideration might result in simply providing room to add additional treatment devices. However, it is possible that the most cost-effective procedure is to provide a high level of treatment and reuse stormwater in one or more processes within the industrial plant. This solution would require the periodic discharge, or "blow-down," of water to control TDS and/or temperature. The blowdown would have to be discharged as treated process water in compliance with the applicable permit(s).

Table 6-1   Stormwater Retention/Detention Options

Option	Description	Advantages	Disadvantages
Wet Retention Pond	Wet basin or pond maintains a permanent pool of water, in addition to temporarily detaining stormwater	• Applicable to all site sizes, depending on available space • Can achieve high removal rate of pollutants • Most cost effective for larger, intensively developed sites.	• May have to be pumped out if insufficient topographic relief available • No exfiltration with high groundwater or underlying bedrock • Expensive to construct in bedrock • Wildlife habitat may become problematic at certain sites
Detention Basin	Basin designed to detain stormwater runoff from a storm for some minimum duration (e.g., 24 hours)	• Applicable to all site sizes depending on available space • Easy to clean and maintain	• May have to be pumped out if insufficient topographic relief available • No exfiltration with high groundwater or underlying bedrock • Expensive to construct in bedrock
Inline Stormwater Detention Systems	Underground chamber used to provide temporary storage of excess stormwater runoff	• Entire structure is underground • Easy to construct	• Confined access • Difficult to monitor and clean
Grassed Swale	Broad, shallow earthen channels (usually with a side slope of a minimum of 3:1)	• Inexpensive to construct • Easy to construct and maintain	• Can be subject to severe damage during heavy storms • Can affect downstream drainage patterns
Permanent Seeding/ Planting	Grass areas and shrubs/ plantings that decrease runoff by slowing velocity and permitting greater infiltration	• Relatively inexpensive • Easy to construct and maintain • Improves aesthetics of a site • Provides stabilization and filtering of sediments	• May require significant amount of space • Not effective for intense rainfall events • May require irrigation to sustain cover during dry weather • Dependent on climate and weather
Infiltration Basin/ Trench	Stone-filled excavation used to temporarily store runoff so it can infiltrate into the ground	• Easy to construct • Suitable to most sites	• Cannot be used with a high groundwater table • Not applicable in bedrock and clay subsoil • Usually used for small areas

Stormwater as a Source of Process Water Makeup

If collected stormwater is to be used as makeup water for one or more manufacturing, cleaning, cooling, or other processes in the facility, treatment and considerations can become much different from those that are appropriate to simple discharge to comply with a permit. For example, a storage facility

Table 6-2   Stormwater Treatment Design Considerations

Parameter	Treatment Options	Design Considerations
pH	• pH adjustment (measurement, chemical feed, and mixing)	• Includes flow measurement and control
Fats, Oils, and Grease (FOG)	• Oil skimming • Sand or other filtration • Dissolved air flotation (DAF)	• Does not remove emulsified FOG • Chemical augmentation might be necessary • Solids loadings
Total Suspended Solids (TSS)	• Retain stormwater runoff for a considerable period of time to allow sedimentation in the retention basin • Chemical coagulation possibly followed by sand or other filtration • Dissolved Air Flotation (DAF)	• Include provision for removal of the solids and maintenance of sufficient freeboard to contain the next one or more storms while sedimentation is taking place • Refer to Chapter 7 • TSS and FOG removal would take place simultaneously
Total Dissolved Solids (TDS)	• If the stormwater is to be used as source of process water makeup, it is likely that dilution is the only feasible solution.	• Other methods for removal of TDS can be seen in Figure 7-1, presented in Chapter 7

for the treated stormwater will likely be needed. In addition, if a facility elects to recycle treated stormwater, provisions must be included to prevent the inevitable buildup of TDS, BOD, and heat to undesirable levels, due to cycles of concentration. Whether or not a facility intends to ever discharge any stormwater, an NPDES permit should be obtained, because the possibility—in fact, the probability—of discharging stormwater runoff as a result of an unexpectedly large storm, or a problem with the stormwater recycle system, will always be present.

Dissolved Solids (TDS)

As explained in Chapter 7, recycling water in an industrial plant results in increasing TDS concentration, because of evaporation. High levels of TDS can lead to scaling and/or corrosion of piping and equipment. Often, the most cost-effective solution is to discharge a certain quantity of the recycle water each day as blow-down, and make it up with water of low TDS concentration. However, in locations where water of low TDS concentration is not available, or is available at high cost, the best solution may be to remove dissolved solids by either side-stream treatment or full-

flow treatment by one of the methods presented in Chapter 7. A mass balance analysis must be performed, as described in the case study provided at the end of this chapter, to properly design the blow-down, makeup, and/or treatment facilities.

Biological Oxygen Demand (BOD)

The buildup of BOD in a water recycle system will occur in exactly the same manner as TDS. Rather than a scaling or corrosion problem, however, the buildup of organic solids associated with BOD will result in the growth of microbes in the water conveyance and use system. This condition, referred to as "biological fouling," can result in disaster and is to be avoided, to say the least. In general, there are three approaches to avoiding the problem of biological fouling:

• Employing sufficient blow-down and makeup that biological fouling can be controlled with disinfection
• Removing dissolved organics by activated carbon, ultrafiltration, or other physical treatment process, probably in conjunction with blow-down, makeup, and disinfection

- Removing BOD with biological treatment, using one of the methods presented in Chapter 7, probably in conjunction with blow-down, makeup, and disinfection

Again, a mass balance must be performed, as was the case for TDS, to properly design BOD control facilities for water recycle and reuse systems.

Heat

In certain situations, heat can build up to undesirable levels in a water recycle/reuse system. There are, in general, three approaches to solving this problem:

- Passing the recycle water through a heat exchanger to transfer some of the heat to a process stream or to plant utility waters
- Making use of a cooling tower to waste some of the heat
- Diluting the recycle stream with cold water from another source

In all three of the above approaches, the quantity of water in the recycle system must be maintained within an acceptable range by adding makeup water or discharging excess water.

Case Study: Metals Forging Manufacturing Facility

A metals forging manufacturing facility is located on a 189-acre parcel of land alongside a river having a 7Q10 flow of 240 cfs. The plant performs various metalworking operations, such as forging, heat treating, chemical etching, and grinding. The manufacturing complex is comprised of two primary manufacturing buildings, a forge shop, and a process and maintenance (P&M) building. Other support buildings include an office and administration building, several air pollution control (APC) buildings, a rinsewater pretreatment plant (RPP), and an oily wastewa-

ter treatment plant (OWTP), as shown in Figure 6-1.

Stormwater runoff from the building roof and yard areas in the northern, eastern, and northwestern portions of the site flows by gravity to an existing impoundment (Impoundment 001) and is discharged at an NPDES-permitted Outfall 001. Remaining roof and yard areas are collected and directly discharged at four other NPDES-permitted Outfalls onsite. In addition to stormwater runoff, Impoundment 001 receives some pretreated process wastewaters from the forge shop and P&M building, fire protection water from the forge shop, and noncontact cooling water. The plant uses lubricating and cutting oils, greases, coolants, and acids in its forging and associated processes.

Forge Shop Area Stormwater Project

The Forge Shop Area Stormwater Project addressed the stormwater and process wastewater that drained to NPDES Impoundment 001. The objectives of this project were threefold:

- Segregate uncontaminated and contaminated stormwaters
- Construct a new Runoff Management Facility to replace Impoundment 001
- Identify Best Management Practices to manage stormwater quality

Preliminary Engineering Studies

As a result of studies to identify and characterize the sources of flows to Impoundment 001, it was determined that the storm drainage system had been used for the discharge of some process wastewaters, that there was a potential for spills to enter the system, and that a limited number of source areas were responsible for a large portion of the total constituent loading to Impoundment 001.

Next, an evaluation was conducted of available options for eliminating or segregating process wastewaters and contaminated stormwater from uncontaminated stormwater. The results of this work indicated that:

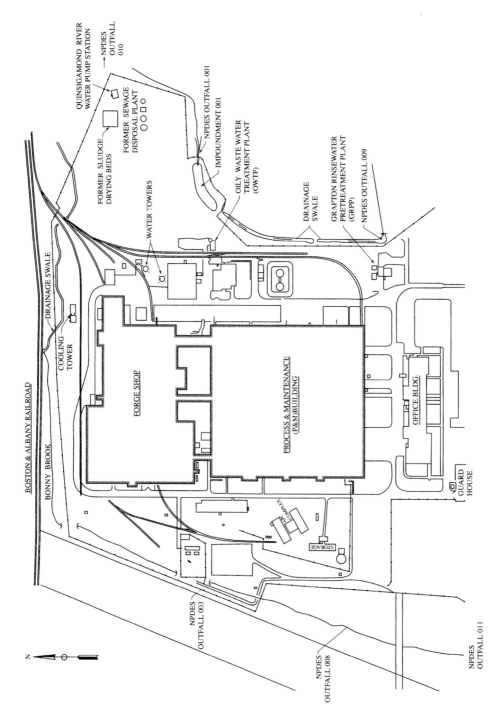

Figure 6-1 Rinsewater pretreatment and wastewater treatment facilities plan.

- Many sources could be eliminated at low capital cost.
- The eastern half of the P&M building roof was not contaminated and could be segregated, thereby eliminating 7.2 acres of runoff area from the total 29-acre drainage area to Impoundment 001.

Treatability studies were conducted on the expected influent stream to Impoundment 001, based on elimination of the identified process sources and the clean portion of the P&M building roof. A stormwater management concept of containment of the design storm, treatment and controlled release of the effluent to meet mass-based permit lim-

its, was developed as the initial basis of design for the Runoff Management Facility (RMF).

Initial Design of RMF

Piping system modifications were designed and constructed that would eliminate process wastewater and segregate the clean portion of the P&M building roof. The uncontaminated roof runoff was discharged without treatment at NPDES Outfall 009. The areas that were subject to impact from industrial activities remained tributary to NPDES Outfall 001.

The RMF design included runoff collection, treatment, and direct discharge at a new

Table 6-3   NPDES Outfall 001 Permit Limits[*]

Constituent	Monthly Average	Daily Maximum
Fats, Oil, and Grease	—	15
pH (range)	6.5–8.0	
Temperature, °C	20	28.3
Total Suspended Solids (lb/day)	70	210
Total Aluminum	0.087	0.75
Total Arsenic	<0.01	<0.01
Ammonia	0.096	0.122
Total Copper	0.006	0.0084
Total Iron	0.3	0.3
Total Lead	<0.010	<0.010
Total Mercury	<0.0002	0.0002
Total Molybdenum	0.004	0.004
Total Selenium	0.010	0.05
Total Thallium	0.0136	0.0136
Total Zinc	0.054	0.059
Total Cyanide	0.005	0.022
Trichloroethylene	0.0027	0.005
2,4,6-Trichlorophenol	0.0012	0.0036
Methylene Chloride	0.05	0.05
Tetrachloroethylene	0.0008	0.0008
Biological Toxicity:		
C-NOEC	>100% Effluent	
LC50	>100% Effluent	

[*]Limit stated in mg/L except as indicated.

NPDES Outfall 001. The treatment concept consisted of grit removal, sedimentation, oil skimming, and pumped discharge of the effluent, at a rate that would meet current mass-based NPDES permits, as follows:

Total Suspended Solids (TSS)	70 lb/day
Fats, Oil, and Grease (FOG)	50 lb/day

Provisions were included for containment of spills within the drainage area of the RMF, and consideration was given to future effluent polishing (by means of polymer addition and adsorption clarification) in the event that the RMF was not able to meet the limits of the NPDES permit. Subsequent to the development of this concept, the EPA issued a renewed NPDES permit containing the water quality–based limits presented in Table 6-3, which could not be met with the proposed treatment scheme.

Revised Design of RMF

Because of the high cost for an advanced treatment system, and the lack of guarantee of meeting the new discharge limits, it was decided to institute a stormwater recycle program to eliminate the regular discharge at Outfall 001, and reduce the demand for water from process water supply sources. The design concept was thus revised to incorporate sand filtration and recycle of the treated water for use in the manufacturing process. The original design features of grit and oil removal, sedimentation, and equalization were retained.

The following design criteria for the RMF were agreed upon:

Design storm:	25-year, 24-hour storm event
Mode of operation:	Incorporate process water effluent into stormwater retention and treatment system. Recycle and reuse continuously. Blow-down as necessary to maintain desired water quality.

Recycle water quality:	TSS: < 1–10 mg/L
	FOG: < 1 mg/L
	pH: 5.0–8.0

In order to determine the buildup of TDS, TSS, BOD, and heat due to the "cycles of concentration" effect, mass balances of water, constituent substances, and heat were performed.

Water Balance

Since the RMF would receive both stormwater runoff and process wastewaters, dry and wet weather flows had to be considered. A water balance was performed to represent the flow of waters before and after the recycle system was put into effect. Incoming water to the site originated from several sources:

- On-site wells for process use and fire protection
- Municipal water for process and sanitary use
- River water for process use and fire protection
- Stormwater

Flow measurement data were collected, and average annual rainfall data were used to estimate average wet weather flow. The water balance representing the "before recycle" scenario is presented in Figure 6-2.

Discharges from the site originated from the following sources:

- NPDES permitted outfalls
- RPP effluent
- Sanitary wastewater
- Miscellaneous losses

As with the incoming flows, flow data were collected and water use records were used to estimate sanitary and process wastewater flows. An internal cooling water tower was responsible for most of the losses, due to evaporation. Also, some water was lost from the system as dilution water for acid and caustic solutions. The "after recycle" water

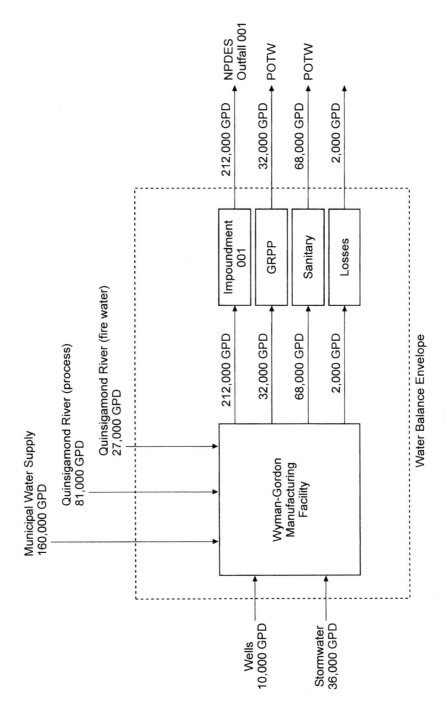

Figure 6-2   Water balance before recycle.

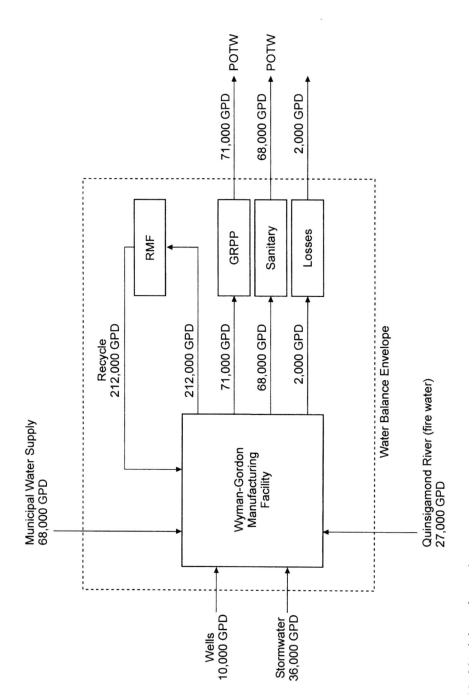

Figure 6-3   Water balance after recycle.

balance, presented in Figure 6-3, shows that all stormwater runoff could be used as process water makeup. In addition to reducing the water withdrawn from the river, less municipal water would be used (68,000 vs. 160,000 GPD).

Mass Balance

Although the existing levels of TDS, TSS, and BOD were very low, the mass balance showed that levels of BOD in the recycle loop could reach 10 to 15 mg/L, indicating a need for disinfection to prevent biological fouling of the water distribution piping and equipment. It was determined that TDS would not reach problematic levels, and because of the sand filter in the treatment train, TSS was not expected to be a problem. A heat balance showed that the temperature within the recycle loop could reach 102°F during the summer, indicating the possible need for a heat exchanger or cooling tower. A cooling tower, of course, would have the effect of increasing the concentration of TDS, TSS, and BOD and would require revision of the mass balances if considered further.

Stormwater Collection and Retention

The RMF was designed to serve three storm drain lines and the effluent from the proposed Oily Wastewater Pretreatment System (OWPS), a pretreatment system that was being designed to remove free oils and emulsified oils from presswaters. All of these flows were routed to the eastern portion of the site to Outfall 001, but required interception prior to the Outfall and rerouting to the northeast to the proposed RMF location. This was accomplished via the design of a cast-in-place concrete junction chamber at the head end of the Impoundment. The junction chamber was designed to intercept and reroute the flows without "throttling" the storm drainage system or allowing solids to settle out. The structure was designed so that it could be constructed in a phased approach that would maintain the discharges

to Impoundment 001 until the RMF was ready to receive flow.

The sedimentation basin was sized to retain the design storm volume, provide the surface area and retention time for solids settling, and serve as an equalization tank. Design of the basin (configuration and depth) was controlled by site hydraulic limitations imposed by the invert elevations of the existing storm drain lines relative to the water surface elevation of the receiving water for the overflow discharge.

Treatment and Recycle Storage

The layout of the RMF site is shown in Figure 6-4. The unit processes of the RMF are represented in Figure 6-5. The unique design aspects of each unit process are described in the following sections.

PLC Control and Operator Interface

Control of the RMF was semiautomatic with the use of a programmable logic controller (PLC) for pump control and alarms. An Operator Interface Terminal (OIT) was placed in the RMF for operator control. A second OIT was placed in the plant engineering office for monitoring purposes.

Grit Chamber

The grit chamber was designed to lower the influent flow velocity, thereby allowing for the removal of particles with a specific gravity greater than 2.65 and of a size that would be retained on a 100-mesh screen. Initial oil skimming was also performed by means of a slotted pipe skimmer that drained the skimmed oil and water mixture to the sump within the RMF Treatment Building, where the oil was removed. The grit chamber also provided added storage capacity (188,000 gals.).

Sedimentation Basin

The sedimentation tank was a below-grade concrete tank, open to the air and divided by a center wall. The tank was sized to contain

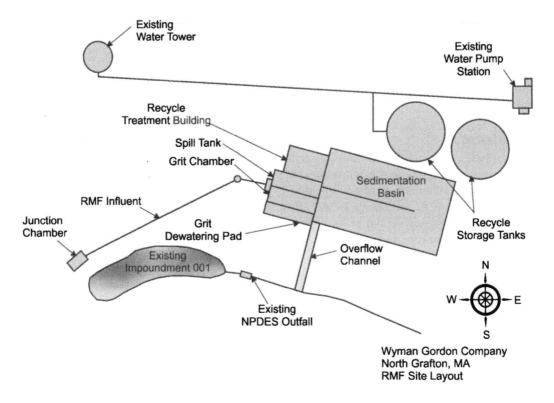

Figure 6-4 RMF site layout.

the design storm runoff volume of 2.06 million gallons and allow for two feet of freeboard, for a total capacity of 2.34 million gallons. The tank size (100 ft × 180 ft × 17 ft) and configuration provide the surface area and retention time for additional settling of solids. Due to a high groundwater table elevation, the structural design of the tank incorporated a combination of antiflotation techniques that added mass to offset buoyant forces.

Oil Skimming
Oil skimming is accomplished in two locations: (1) in the grit chamber (as previously described), and (2) in the baffled effluent sump located inside the RMF Treatment Building. A floating tube–type oil skimmer will remove floating oils, which will be temporarily stored for use as a fuel source in a

waste oil burner in the RMF Treatment Building. In the state where the facility was located, waste oil was considered a hazardous waste; therefore, secondary containment of 100% of the storage tank volume was required.

pH Adjustment
Due to the acidic nature of rainfall in the Northeast (pH levels as low as 5.8 had been measured in the existing Impoundment 001), provisions for pH adjustment were included in the RMF design. Sodium carbonate was fed into the line as the wastewater was pumped from the sedimentation tank to the sand filters. In addition to meeting the process water pH needs, a more neutral pH would meet the NPDES permit limit range of 6.5 to 8.0 in the event of an overflow, intended or not, through the Outfall 001.

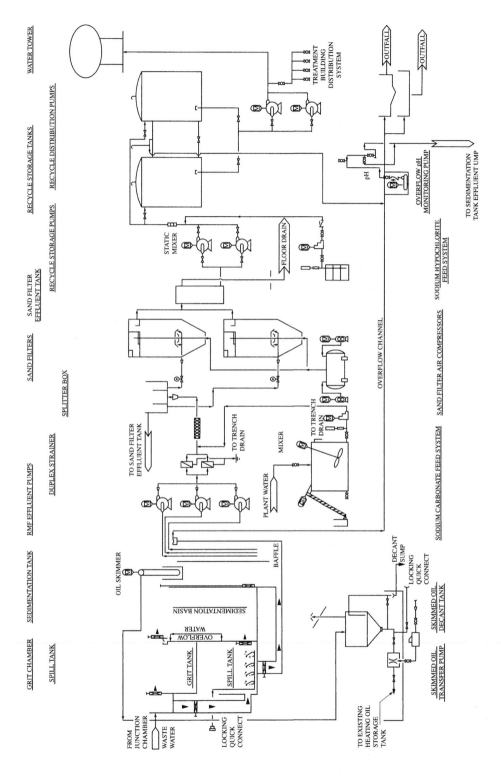

Figure 6-5 Process flow diagram.

Sand Filtration

Filtration was achieved by two Parkson upflow sand filters. This type was selected because the continuous backwash system eliminated the need for large volumes of backwash water (backwash flow was approximately 5% to 7% of the forward flow for the upflow filter).

Recycle Water Disinfection

It was recognized that recycle of the treated runoff could result in increased BOD concentrations in the recycle water and create a problem with biofouling in the process water distribution system. A sodium hypochlorite feed system was therefore included.

Recycle Water Storage and Distribution

Recycle water was stored in two A. O. Smith glass-fused-to-steel aboveground tanks (ASTs) with aluminum domes, each with a capacity of 1.5 million gallons. This type of tank requires no sandblasting, welding or painting and, therefore, little maintenance. Recycle water was pumped from the ASTs to the existing process water distribution system as needed. In the event that there is not enough recycle water in storage, the existing river water pumps receive a signal to provide supplemental water.

Hydraulic Overflow

The RMF was designed to handle the 25-year, 24-hour design storm without a discharge to NPDES Outfall 001. In the event that there is a storm in excess of the design storm, or a series of small storms occurring in rapid succession so as to exceed the capacity of the RMF, two options have been included for hydraulic relief. Minor overflows (up to 1.4 cfs) can be routed to NPDES Outfall 010 via an existing outfall pipe that discharges directly to the river. Hydraulic overflows in excess of 1.4 cfs spill over a kneewall in the overflow channel and discharge at NPDES Outfall 001, in a location tributary to the river. Flow measurement at Outfall 010 is achieved using an orifice plate located in the floor of the overflow channel. An H-flume can be used to handle the larger range of flows that may be experienced at Outfall 001.

Spill Containment

A spill containment tank has been located at the influent end of the sedimentation tank, adjacent to the grit chamber, to provide plant personnel with the opportunity to divert a spill from the sedimentation basin. Spill-contaminated runoff can be contained separately, rather than be allowed to further contaminate a potentially large volume of runoff contained in the sedimentation basin.

Treatment Residuals

Wastewater treatment residuals generated by the RMF consist of skimmed oil, grit, and sludge from the sedimentation tank. Skimmed oil is burned in the RMF waste oil burner to recover the heat value. The grit chamber and sedimentation tank are cleaned on an annual basis. The method of disposal of grit and sludge depend on their characteristics.

Operations Strategy

In order to minimize the size of the sedimentation basin, it was designed for an operating depth between 1.5 and 3.0 feet during dry weather conditions. The 3.0-foot level gives the treatment system eight hours of continuous operation, thereby avoiding frequent equipment starts and stops. With this type of operations strategy, the sedimentation basin has been capable of containing the 25-year, 24-hour design storm runoff volume without an overflow. However, it was also necessary to consider what would happen during a storm event in excess of the design storm. The approach was to fully utilize both the in-ground and aboveground storage capabilities by stopping the pumps that fill the ASTs if they are at the overflow level and the sedimentation basin is not full. At the point where the sedimentation basin reaches its capacity, treatment will resume and overflow should occur from the ASTs to Outfall 010 (depending on the overflow rate), therefore eliminating the discharge of untreated stormwater runoff.

Summary

The design of the RMF, the subject of this case study, was driven by a need to comply with NPDES permit limits at Outfall 001. However, the benefits that have been realized from this case study and design include the following:

- Reduced rates of water withdrawal from the river and potable water supply
- Reduced annual costs for potable water supply usage
- Increased ability to capture spills within the drainage area of the RMF
- Reduced potential for discharge of pollutants to the environment and NPDES permit excursions
- Reduced capital and operating costs for compliance with the NPDES permit

Bibliography

Aron, G., and C. E. Egborge. "A Practical Feasibility Study of Flood Peak Abatement in Urban Areas." In Report to U.S. Army Corps of Engineers, Sacramento District, Sacramento, CA; 1973.

Bendiet, P. B., and W. C. Huber. *Hydrology and Floodplain Analysis.* 2nd ed. Reading, PA: Addison-Wesley, 1992.

Chow, V. T. *Open Channel Hydraulics.* New York: McGraw-Hill, 1959.

Chow, V. T. (ed.). *Handbook of Applied Hydrology.* New York: McGraw-Hill, 1964.

Federal Aviation Agency. "*Department of Transportation Advisory Circular on Airport Drainage.*" Report A/C 150-5320-SB. Washington, DC: 1970.

Henderson, F. M. *Open Channel Flow.* Toronto: Macmillan, 1966.

Kibler, D. F., et al. *Recommended Hydrologic Procedures for Computing Urban Runoff from Small Developing Watersheds in Pennsylvania.* University Park, PA: Institute for Research on Land & Water Resources. 1982.

Morgali, J. R., and R. K. Linsley. "Computer Simulation of Overland Flow." *Journal of the Hydraulics Division ASCE,* 1965.

Rantz, S. E. "Suggested Criteria for Hydrologic Design of Storm-Drainage Facilities in the San Francisco Bay Region, California." U.S. Geological Survey Professional Paper 422-M, 1971.

Schulz, E. F., and Lopez, O. G. "Determination of Urban Watershed Response Time." Colorado State University Hydrology Paper No. 71. Fort Collins, CO: 1974.

U.S. Army Corps of Engineers. Hydrologic Engineering Center. "Storage Treatment, Overflow, Runoff Model: STORM." *User's Manual.* Davis, CA: 1976.

U.S. Department of Agriculture. Soil Conservation Service. "A Method for Estimating Volume and Rate of Runoff in Small Watersheds." TP-149. n.d.

U.S. Soil Conservation Service. "Urban Hydrology for Small Watersheds." *SCS Technical Release,* No. 55. Washington, DC: 1975.

Viessman, W., Jr., T. E. Harbaugh, and I. W. Knapp. *Introduction to Hydrology.* New York: Intext Educational Publishers, 1972.

Wanielista, M. P. *Storm Water Management: Quantity and Quality.* Ann Arbor, MI: Ann Arbor Science Publishing, 1978.

7   Methods for Treating Wastewaters from Industry

General

Technologies for treating industrial wastewaters can be divided into three categories: chemical methods, physical methods, and biological methods. Chemical methods include chemical precipitation, chemical oxidation or reduction, formation of an insoluble gas followed by stripping, and other chemical reactions that involve exchanging or sharing electrons between atoms. Physical treatment methods include sedimentation, flotation, filtering, stripping, ion exchange, adsorption, and other processes that accomplish removal of dissolved and nondissolved substances without necessarily changing their chemical structures. Biological methods are those that involve living organisms using organic or, in some instances, inorganic substances for food. In so doing, the chemical and physical characteristics of the organic and/or inorganic substance are changed.

Most substances found as pollutants in industrial wastewaters can be categorized as to whether chemical, physical, or biological treatment should be the most appropriate. For instance, dairy wastewater should most appropriately be treated by biological means, because the bulk of the pollution load from a typical dairy is organic material from whole milk, which is readily biodegradable. As a general rule, biological treatment is more economical than any other type of treatment where reasonably complete treatment is required and wherever it can be made to work successfully.

It is very often possible to make preliminary selections of candidate treatment technologies based on fundamental properties of the pollutants and experience. For instance, when candidate treatment technologies to treat wastewaters from a metal plating operation are being considered, none of the biological treatment technologies would be appropriate, since metal ions are not biodegradable. However, both chemical precipitation (a chemical treatment technology) and ion exchange (a physical treatment technology) should work well, based on the fundamental properties of the substances to be removed (dissolved inorganic cations and anions). The question then reduces to a comparison between the advantages and disadvantages of these two technologies, and experience provides much of the information appropriate to this evaluation.

For example, experience has shown that, for most metal plating wastewaters, chemical precipitation is far less costly than ion exchange; however, chemical precipitation is not reliably capable of reducing metal concentrations to less than approximately 5 mg/L, principally because the process of removing precipitated metals by settling in a clarifier typically does not remove the very small particles of precipitate. Sand (or other) filtration effectively removes most of the particles of metal precipitate that will not settle. The concentrations of dissolved metals even after chemical precipitation and sand filtration are still no lower than 1 to 2 mg/L, at best. Furthermore, ion exchange can "polish" the effluent from chemical precipitation and sand filtration to very low concentrations (20 to 50 ppb). Ion exchange could do the entire job of removing metals from industrial wastewater to very low concentrations without being preceded by chemical precipitation and sand filtration, but usually the cost of doing so is much higher than the cost of the three processes in combination.

It follows, then, that it would not be prudent to spend effort, time, and money to conduct a large-scale investigation into technologies for treating wastewaters from metal plating beyond the line of thinking outlined above. The pollutants in these wastewaters are not organic and therefore not biodegradable; extensive experience has shown that:

- Chemical precipitation is the most cost-effective method for removing the bulk of the dissolved metals.
- Sand, diatomaceous earth, or other media filtration is the most cost-effective "next step" to follow the chemical precipitation process.
- If still further reduction in metals concentration is required, ion exchange is the best candidate.

Having said that, it must now be said that in many cases there will be substances in certain metal plating wastewaters that require more than straightforward alkaline precipitation, filtration, and ion exchange. For instance, if chelating agents are present, it may be necessary to destroy or otherwise inactivate them, in order to expose the metal ions to the full effect of the precipitating anions. In other cases, if the concentration of organic matter is high, it may interfere with the precipitation process and have to be removed by biological or other treatment, prior to the metals removal steps.

Figure 7-1 presents a categorization of the components of industrial wastewater and preliminary selections of treatment technologies based on the appropriateness of the mechanism of each technology compared with the fundamental properties of the pollutants. Different versions of Figure 7-1 could be generated by beginning with a characterization other than dissolved or undissolved—for instance, organic or inorganic—but all would ultimately result in the same list of appropriate treatment technologies.

In Figure 7-1, the first level of categorization of pollutant characteristics is that of dissolved or undissolved state. For instance, trichloroethylene as a pollutant in wastewater would be dissolved (albeit to only very low concentrations), organic and volatile. Figure 7-1 can be used to develop a list of candidate technologies. The list can be narrowed further by considering characteristics such as biodegradability, technical feasibility, and economic feasibility. Candidate technologies, then, would be stripping, activated carbon adsorption, and chemical oxidation.

Figure 7-2 presents a schematic of "the industrial waste system," showing that raw materials, water, and air enter the system, and, as a result of the industrial process(es), products and by-products exit the system, along with airborne wastes, waterborne wastes, and solid wastes. Since discharge permits are required for each of the waste-bearing discharges, treatment systems are required. Each of the treatment systems has an input, the waste stream, and one or more outputs. The output from any of the treatment systems could be an air discharge, a waterborne discharge, and/or a solid waste stream.

Principle and Nonprinciple Treatment Mechanisms

Most treatment technologies will remove substances other than the target substances. For instance, "biological treatment" can effectively remove a certain amount of metal ions from wastewater. Because metal ions do not particularly like to be dissolved in water (they are hydrophobic), they are driven by the second law of thermodynamics to be adsorbed on the surface of solids—just about any solids, including activated sludge solids.

This mechanism for removing metals from wastewater is often undesirable, since the presence of the metal ions in the waste sludge may render the sludge unsuitable for a desired disposal method. Composting with wood chips to produce a horticultural soil

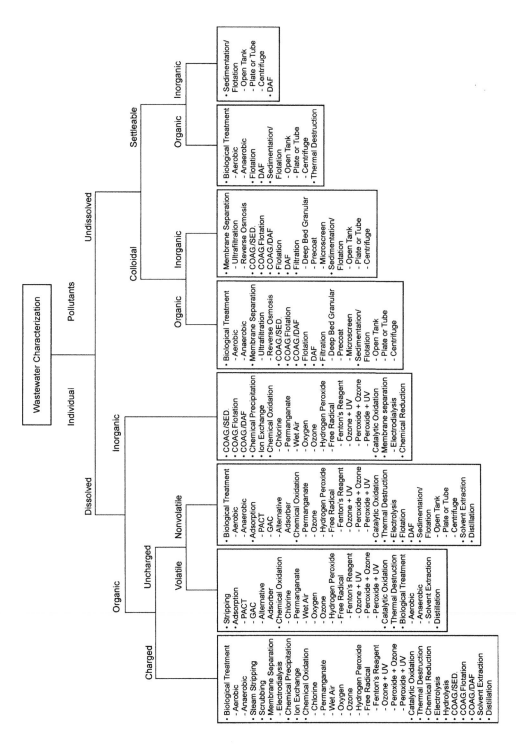

Figure 7-1   Candidate treatment technologies, based on fundamental characteristics of pollutants.

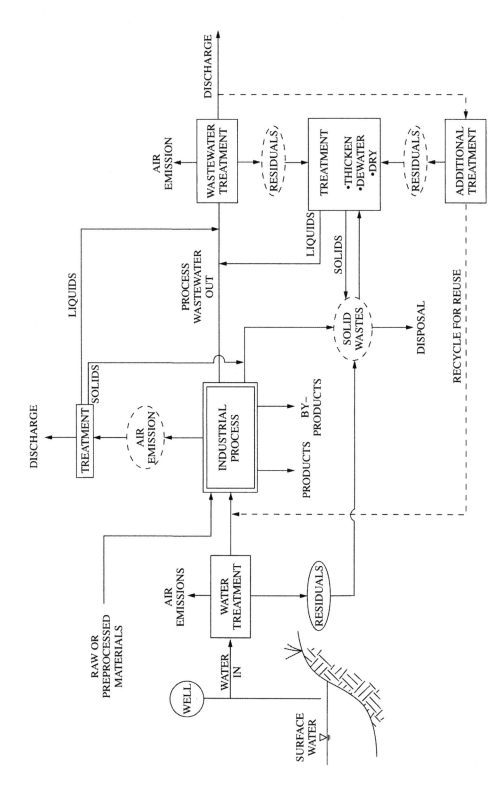

Figure 7-2   The industrial waste system.

conditioner is an example. Another example is simple disposal in an ordinary municipal landfill. In other instances, the removal of metal ions from a wastewater that has very low concentrations of them may serve as a fortuitous polishing step. In any case, biological treatment is a "principle" technology for removal of organics from wastewaters, and its unintended removal of metal ions is a "nonprinciple" mechanism.

Waste Equalization

Among the most effective waste management procedures is equalization of the waste stream. Equalization can be of two types: flow equalization and constituent equalization. Flow equalization refers to changing the variations in rate of flow throughout the processing and cleanup cycles to a more steady flow rate that is more nearly equal to the average flow rate for that period of time. Constituent equalization refers to the concentration of the target pollutants in the waste stream. Throughout the 24-hour day, the concentrations of individual constituents in a given industrial waste stream typi-

cally vary over wide ranges, as processes are started up, operated, shut down, and cleaned. Waste treatment systems that are designed for given ranges of concentration of target pollutants often do not perform well when those constituents are in concentrations significantly different from the design values.

Equalization can be either online, as diagrammed in Figure 7-3(a), or off-line, as diagrammed in Figure 7-3(b). On-line flow equalization is accomplished by allowing the waste stream to flow into a basin. The waste is then transferred from the basin to the treatment system at a constant (or more nearly constant) rate. The fundamental requirement of the basin is that it be sufficiently large that it never overflows and that it always contains enough waste that it never becomes empty, causing the flow to the treatment system to stop.

As shown in Figure 7-3(b), offline equalization is accomplished by restricting the flow into the treatment system by means of either a flow-regulating valve or a constant-speed positive displacement pump. When there is excess waste flow, it is directed to the

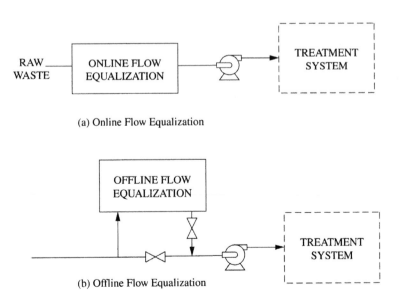

(a) Online Flow Equalization

(b) Offline Flow Equalization

Figure 7-3   Flow equalization configurations.

equalization tank. When there is insufficient flow, it is made up from the equalization tank.

With respect to constituent equalization, offline equalization can be used advantageously when the generation of wastes at night is significantly less than during the day. A portion of the strong daytime wastes can be stored in the equalization facility and then directed to the treatment system at night. The treatment system can be significantly smaller, since it is not required to treat wastes at the high rate that they are generated during the daytime.

A certain amount of constituent equalization will take place as a consequence of flow equalization, but this amount will not be optimal. Flow equalization is best done within a cycle in which the equalization basin is close to overflowing during highest flows and close to empty during lowest flow rates. Constituent equalization, in contrast, is best done by dilution. There are several different approaches to constituent equalization by dilution, ranging from dilution to a constant concentration of the target substance with clean water to simply retaining the waste stream in a completely mixed basin of constant volume.

The following are alternative approaches to constituent equalization:

1. *Batch treatment.* This is the ultimate in constituent equalization. The processing and cleanup wastewaters are collected in a well-mixed basin that is large enough to contain the entire processing and cleanup flows. Treatment can take place subsequently, either in the same tank or by being pumped at a constant rate through a continuous flow treatment system.
2. *Offline equalization tank.* As described previously, a portion of the flows containing high concentrations of pollutants are diverted to an offline equalization basin and are mixed with less concentrated flows at a later time.
3. *Completely mixed, inline equalization tank.* A tank that is equipped with sufficient mixing capability to maintain completely mixed conditions and that has sufficient volume to hold the flow between peak high and low flow rates, is the most common type of constituent equalization device. The larger the volume, the better the constituent equalization, but the higher the cost to both construct, maintain, and mix. The tank is maintained full; therefore, this device does not achieve flow equalization.
4. *Dilution with clean water or treated effluent.* Stormwater runoff, cooling water, or other previously used but clean water relative to the wastewater being treated may be used. Target substances that can readily be measured with a probe and meter, such as a specific ion probe and meter, are the best candidates for this type of substance equalization. This method of constituent equalization, of course, increases the total flow through the treatment system.

As mentioned earlier, the principal value of waste equalization is that in most cases the treatment system can be made smaller, since the maximum values for both flow rate and constituent concentration will be reduced (method number 4 is an exception). The treatment system will, therefore, have a lower capital cost as well as lower operating and maintenance costs; hence, by definition, it will have a lower life-cycle cost.

Flow Equalization

Figure 7-3(a) illustrates flow equalization by means of an online flow equalization basin. Flow into the basin is by gravity and varies as the waste generation rate varies. Flow out of the basin to the treatment system is made constant by either an appropriate valve or a positive displacement pump. An aerator and/

or a mixer are provided to prevent undesirable occurrences, such as settling out of solids; biological activity that would result in anaerobic conditions with consequent odor problems; or chemical or biological reactions that would change the nature of the wastes.

The principal design parameter of a flow equalization basin is size. Since the cost of the basin will be a direct function of its size, and since operation of the basin, in terms of maintaining complete mixing and maintaining aerobic conditions, is also a direct function of basin size, the flow equalization basin should be no larger than necessary to accomplish the required degree of equalization.

Development of a "mass diagram" is an excellent method for determining the required size of a flow equalization basin. A mass diagram is developed by plotting accumulated quantity of flow of the waste stream versus time.

A truly representative period of time may be one day, one week, or even longer. Then, as illustrated in Figure 7-4, the line that is representative of a constant rate of flow for the accumulated volume flowing at a steady rate over the total time period is drawn. Then, lines are drawn that are (1) parallel to the line that represents the average rate of flow, and (2) tangent to the mass diagram at points that are the maximum distances above and below the average-rate-of-flow line. The storage volume required for the equalization basin is that volume represented by the vertical distance between the two tangent lines. In some cases, the mass diagram may be at all

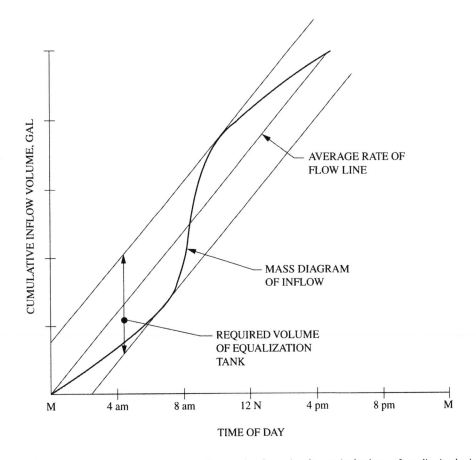

Figure 7-4   Mass diagram with illustration of the method used to determine the required volume of equalization basin.

points below, or in other cases at all points above, the average rate of flow line. In these cases, there will be only one tangent line, which will be drawn at the point on the mass diagram that is furthest distant from the average-rate–of-flow line. The volume required for the equalization basin in these cases will be the volume represented by the vertical distance between the single tangent line and the average rate of flow line.

As with many other unit processes, it is usually prudent to provide an equalization basin that is 10% to 20% larger than the size determined by the mass diagram method. This will account for changes in flow rate greater than experienced when the data were taken for the mass diagram, or other uncertainties. In fact, the percent increase in size should be inversely proportional to the degree of confidence the design engineer has in how accurately the data obtained are representative of the full range of conditions to be experienced by the treatment system.

Example 7-1 illustrates the procedure for determining the size of a flow equalization basin. Example 7-2 illustrates how a flow equalization basin can significantly reduce the size of treatment facility required for a given industrial facility.

Example 7-1

The flow of wastewater from a poultry processing plant is shown in Table 7-1. If it is desired to pump this flow at a constant rate over a 24-hour period, what is the minimum size required for an online flow equalization tank?

Example 7-2

Determine the size, in terms of design flow rate, of a dissolved air flotation (DAF) system to treat the wastewater described in Example 7-1: (a) without an equalization tank, and (b) with an equalization tank of 310,000 gallons, plus 10% to 20%.

1. The data presented in Example 7-1 and diagrammed in Figure 7-5 show that the

Table 7-1 Cumulative Flow for Equalization Tank

Hour Ending	Cumulative Flow (Thousands of Gallons)
0100	5
0200	11
0300	14
0400	20
0500	25
0600	31
0700	62
0800	120
0900	200
1000	240
1100	309
1200	320

maximum hourly rate of flow occurs between 9:00 and 10:00 P.M., when cleanup operations and probably dumping baths are taking place. This equates to 154,000 gallons per hour, or 2,600 gpm. Thus, a DAF facility would have to be sized for at least 2,600 gpm if no flow equalization is provided.

2. If an inline tank of at least 310,000 gallons is provided, the design flow for the DAF system should be that of the average rate of flow for the 24-hour period, or 1.2 MG/24 hrs. = 834 gpm. Thus, the inclusion of a flow equalization tank allows the treatment facility to be 68% smaller.

Constituent Equalization

As stated previously, some amount of constituent equalization always takes place during flow equalization. In fact, it is standard practice to design for flow equalization and then operate to attain the degree of constituent equalization needed to achieve treatment objectives. One way this is done is by manually or automatically decreasing rates of flow during periods when constituent concentrations are high. The simple dilution equation

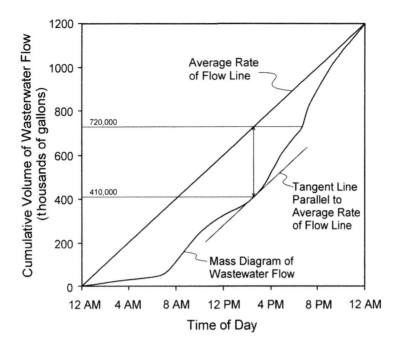

Figure 7-5   Mass diagram: Cumulative volume of wastewater flow over time.

shown as equation 7-1 can be used to calculate concentrations of constituents that result when volumes of a given concentration of that constituent are mixed with known volumes of the same constituent at a different concentration.

$$\frac{C_1(V_1)+C_2(V_2)}{V_1+V_2}=C_3 \qquad (7\text{-}1)$$

where

C_1 = Concentration of constituent in question in the first considered volume

V_1 = First considered volume

C_2 = Concentration of constituent in question in the second considered volume

V_2 = Second considered volume

C_3 = Concentration of the constituent in question when the volumes V_1 and V_2 are mixed together

pH Control

Although pH control is a form of chemical treatment, it is presented here in the general section because, like flow equalization, it is used with biological and physical treatment systems as well as with chemical treatment systems.

Water, the basic substance of all industrial wastewaters, is composed of hydrogen and oxygen. Water dissociates into hydrogen ions and hydroxide ions, as shown in equation 7-2.

$$H_2O \leftrightarrow H^+ + OH^- \qquad (7\text{-}2)$$

In accordance with the law of mass action, discussed in the section entitled "Reaction to Produce an Insoluble Solid," the quantities of hydrogen and hydroxide ions are such that the mathematical product of the hydrogen ion concentration, expressed as moles per liter ($[H^+]$), and the hydroxide ion concentration ($[OH^-]$) is always 10^{-14}, as shown in equation 7-3.

$$\left[H^+ \right]\left[OH^- \right]=10^{-14} \qquad (7\text{-}3)$$

The hydrogen ion concentration is thus a fundamental property of any aqueous solution. Any liquid for which equation 7-3 does not hold is not an "aqueous solution." A strong solution of sulfuric acid is an example. It is a strong acid solution, not an aqueous solution, and its pH value really has no meaning.

The hydrogen ion concentration of an aqueous solution, such as an industrial wastewater, has a major influence on its characteristics. Which substances will dissolve in a given wastewater, as well as how much of a given substance can be dissolved, are two important characteristics. Another is that pH strongly influences the corrosivity of wastewater. The value of the pH of an aqueous solution must be within a certain range for bacteria and other microorganisms to live and thrive and for fish and plants to live and thrive. A host of other characteristics that influence the success of wastewater treatment methods, such as chemical coagulation, activated carbon adsorption, ion exchange, chemical oxidation, and the release of gases such as hydrogen sulfide and ammonia are absolutely dependent for success on the proper range of pH.

Because the hydrogen ion concentration of wastewaters is so commonly dealt with, and since it is cumbersome to express hydrogen ion concentration in terms of molar concentrations, the concept of pH was developed. Very simply, the term p preceding any item means "the negative logarithm of that item." Thus, pKa means "the negative log of the numerical value of Ka," and pH means "the negative log of the numerical value of the hydrogen ion concentration, in moles per liter." Equation 7-4 illustrates the pH concept.

$$pH = -\log\left[H^+ \right]=\log\frac{1}{\left[H^+ \right]} \qquad (7\text{-}4)$$

Example 7-3

Calculate the pH of an aqueous solution that has a concentration of hydrogen ions equal to 2.3×10^{-3} moles per liter.

Solution:

$$pH = -\log\left[H^+ \right]=\log\frac{1}{\left[2.3x10^{-3} \right]}= 2.64$$

or

$$pH = -\log\left[H^+ \right]=-\log(2.3x10^{-3})= 2.64$$

Example 7-4

Calculate the hydroxide ion concentration of an industrial waste with a pH of 5.4.

Solution:

$$pH = \log\frac{1}{\left[H^+ \right]}=-\log\left[H^+ \right]= 5.4$$

The concentration of hydrogen ion in this wastewater, then, is $10^{-5.4}$ moles/liter, but:

$$(10^{-5.4})x\left[OH^- \right]=10^{-14}$$

or

$$OH^- =\frac{10^{-14}}{10^{-5.4}}=10^{-8.6}\ \text{moles/liter}$$

Thus, it is seen that the pH scale, which, in accordance with equation 7-3, is a logarithmic scale, ranges between the values of 0 and 14; and a change in pH value of 1.0 equates to a change in concentration of hydrogen ion of a factor of 10. If the pH of a wastewater changes from 5.1 to 7.1, the molar concentration of hydrogen ions will have decreased by two orders of magnitude.

Typically, industrial waste discharge permits require that the pH be within the values of 6.5 and 8.5, and many industrial waste treatment processes require that the pH be held within a range of plus or minus 0.5 pH units. Some treatment processes require an even smaller pH range for successful operation. For these reasons, pH control is one of the most important aspects of industrial wastewater treatment.

A common procedure used to control pH in industrial waste treatment is illustrated in Figure 7-6. Basically, a mixing chamber is used to mix acidic and/or basic reagents with the wastewater. pH electrodes are placed either in the discharge from the mixing chamber or, in some cases, in the chamber itself. The electrical signal from the pH electrodes is amplified and relayed to a controller, which activates valves or pumps to regulate the flow of acidic or basic reagent into the mixing tank.

Many pH control applications are simple, straightforward, and require only that the electrodes be kept clean and well calibrated, and the control system well-maintained, for success. If the system is "well behaved" (the pH of the wastewater entering the mixing tank does not change very often, or very much), a simple control system such as that shown in Figure 7-6 is adequate. However, if the pH, as well as the acidity and/or the alkalinity of the wastewater entering the mixing tank, changes appreciably throughout the processing and cleanup day, the simple control system depicted in Figure 7-6 may be inadequate. In fact, pH control can be extremely difficult and can require a much more extensive control system than that shown in Figure 7-6.

A basic reason for the potential difficulty with satisfactory pH control relates to the extremely large range of values of hydrogen

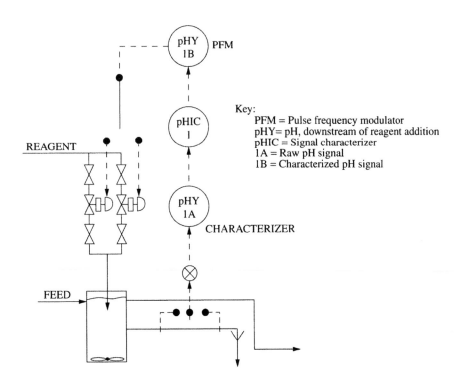

Figure 7-6 pH control system.

ion concentration within the 0 to 14 pH scale range. Fourteen orders of magnitude is a range that few, if any, other detection devices must contend with. Another reason relates to the high sensitivity of commonly available pH detection electrodes. These electrodes typically can respond to changes in pH as small as 0.001.

The extremely large range of the substance detected (hydrogen ion), coupled with the high sensitivity of detection capability, must be relayed electronically and, ultimately, mechanically to a control system that regulates the flow of reagent to the mixing tank. If the system is not well behaved, two complete control systems in series (and, in some cases, three) are required to attain satisfactory results. The multiple systems should have successively smaller control valves or pumps, in order to overcome the difficulty in translating, on the one hand, a change of several orders of magnitude in hydrogen ion concentration and, on the other hand, a very small change in a valve setting.

Chemical Methods of Wastewater Treatment

Chemical methods of wastewater treatment take advantage of two types of properties: (1) the chemical characteristics of the pollutants, regarding their tendency to react with, or interact with, treatment chemicals; and (2) the chemical characteristics of the products of reaction between pollutants and treatment chemicals, regarding their solubilities, volatilities, or other properties that relate to the inability of the product to remain in water solution or suspension.

In general, six chemical processes can be taken advantage of to remove substances from wastewater:

1. Reaction to produce an insoluble solid
2. Reaction to produce an insoluble gas
3. Reduction of surface charge to produce coagulation of a colloidal suspension

4. Reaction to produce a biologically degradable substance from a nonbiodegradable substance
5. Reaction to destroy or otherwise deactivate a chelating agent
6. Oxidation or reduction to produce a nonobjectionable substance or a substance that can be removed more easily by one of the methods listed above

Table 7-2 presents an enumeration of chemical treatment technologies and classifies them in these six categories.

Reaction to Produce an Insoluble Solid

The industry standard procedure for removing metals from wastewaters is alkaline precipitation. Alternative methods include precipitation of the metal as the sulfide; precipitation as the phosphate; precipitation as the carbonate; or coprecipitation with another metal hydroxide, sulfide, phosphate, or carbonate. All of these technologies take advantage of the law of mass action, illustrated as follows:

When a chemical system is in equilibrium, expressed as:

$$A + B \leftrightarrow C + D \qquad (7\text{-}5)$$

it must obey the equation:

$$\frac{[C][D]}{[A][B]} = K \qquad (7\text{-}6)$$

where K is a constant. In words, equation 7-6 states that for a given system of chemical substances that have reacted to the point of equilibrium, the mathematical result of multiplying the products, each expressed as moles per liter, divided by the mathematical product of the reactants, is always the same number. Consequently, for K to remain constant, an increase in either A or B shifts the equilibrium to the right, causing a corresponding decrease in B or A, respectively. Likewise, removal of some or all of either C or D will

Table 7-2 Chemical Treatment Technologies and Appropriate Technology Category

Chemical Treatment Technology	Technology Category
Alkaline precipitation of metals	1
Alkaline chlorination of cyanide	2, 6
Breakpoint chlorination removal of ammonia	2
Precipitation of metals as the sulfide	1
Precipitation of metals as the phosphate	1
Precipitation of metals as the carbonate	1
Chemical coagulation	3
Chemical oxidation of nitrite	2
Hydroxyl free radical oxidation of organics	4, 5, 6
• Ozone + hydrogen peroxide	
• Ozone + ultraviolet light	
• Hydrogen peroxide + ultraviolet light	
• Fenton's reagent (H_2O_2 + Fe ++)	
Precipitation of phosphorus as metal phosphate	1
Removal of arsenic, first by oxidation of arsenite to arsenate, followed by coprecipitation of arsenate with ferric salt	1, 6
Reduction of ionic mercury to metallic form using hydrazine, sodium borohydride, or other reducing agent	1, 6
Precipitation of barium as the sulfate	1
Reduction of hexavalent chromium to insoluble trivalent chromium, pH adjustment to 8.2–8.6	1, 6
Precipitation of fluoride as calcium fluoride, insoluble at high pH	1
Oxidation of ferrous to ferric, precipitation of ferric oxide, insoluble in the neutral pH range	1, 6

1. Identify one or more insoluble compounds of which the target pollutant is an ingredient.
2. Identify one or more soluble compounds that are reasonably inexpensive sources of the remaining substances in the insoluble compound(s), making certain that the soluble compounds are not a pollutant as well, since it is undesirable to trade one pollutant for another.
3. Perform experiments in the laboratory to confirm the technical and financial feasibility of each promising treatment method.

As an example of the foregoing procedure, consider that lead is the target pollutant. Review of Langels, or another appropriate handbook, shows that the compounds of lead shown in Table 7-3 are highly insoluble.

Of these three compounds, lead carbonate and lead sulfide are seen to be essentially insoluble in water. Executing the second step, it is determined that a relatively inexpensive source of carbonate ions is common soda ash, Na_2CO_3, while a somewhat more expensive material, sodium sulfide, is a source of sulfide ions. A promising treatment method for removing lead from an industrial wastewater, then, would be to add soda ash and precipitate the lead as the insoluble carbonate.

Table 7-4 presents a list of heavy metals with theoretical solubilities of their hydroxides, carbonates, and sulfides.

Another method of removing certain metals is illustrated in Table 7-5, which presents a summary of pH range and other conditions that have been found to effectively remove iron, aluminum, arsenic, and cadmium,

shift the equilibrium to the right, causing decreases in both A and B.

It follows, then, that if substance A is a pollutant, and substance B will react with A to produce an insoluble precipitate (C or D), which would constitute removal from solution, substance B can be added until substance A has essentially disappeared.

A step-by-step procedure that can be used to develop an effective, efficient, cost-effective treatment technology is as follows:

Table 7-3 Insoluble Compounds of Lead

Compound	Solubility in Water (mg/L)
Lead hydroxide ($Pb(OH)_2$)	2.1
Lead carbonate ($Pb(CO_3)$)	7.0×10^{-3}
Lead phosphate ($Pb_3(PO_4)_2$)	20×10^{-3}
Lead sulfide (PbS)	3.8×10^{-9}

Table 7-4 Theoretical Solubilities of Hydroxides, Sulfides, and Carbonates for Selected Heavy Metals (Palmer et al., 1988)

Metal	Solubility of metal ion; mg/L		
	As hydroxide	As carbonate	As sulfide
Cadmium (Cd^{++})	2.3×10^{-5}	1.0×10^{-4}	6.7×10^{-10}
Chromium (Cr^{+++})	8.4×10^{-4}	—	No precipitate
Cobalt (Co^{++})	2.2×10^{-1}	—	1.0×10^{-8}
Copper (Cu^{++})	2.2×10^{-2}	—	5.8×10^{-18}
Iron (Fe^{++})	8.9×10^{-1}	—	3.4×10^{-5}
Lead (Pb^{++})	2.1	7.0×10^{-3}	3.8×10^{-9}
Manganese (Mn^{++})	1.2	—	2.1×10^{-3}
Mercury (Hg^{++})	3.9×10^{-4}	3.9×10^{-2}	9.0×10^{-20}
Nickel (Ni^{++})	6.9×10^{-3}	1.9×10^{-1}	6.9×10^{-8}
Silver (Ag^{+})	13.3	2.1×10^{-1}	7.4×10^{-12}
Tin (Sn^{++})	1.1×10^{-4}	—	3.8×10^{-8}
Zinc (Zn^{++})	1.1	7.0×10^{-4}	2.3×10^{-7}

Table 7-5 Summary of pH Ranges and Conditions Found to Produce Good Removals of Indicated Metals

	Substance to be Removed			
	Iron	Aluminum	Arsenic	Cadmium
pH and other condition	7–8 all Fe oxidized to Fe+++	6–7, but up to 8.5 may be okay	Co-ppt with iron; therefore, pH 7–8	10 or so, but pH 8 with carbonate is best

Table 7-6 Common Methods and pH Values for Removal of Heavy Metals

Chromium	Reduction to trivalent state by bisulfite or metabisulfite, followed by precipitation at pH 8–9.5.
Copper	pH 10–12, or as the sulfide (by adding sodium sulfide). Evaporative recovery or ion exchange for recovery.
Lead	pH 10–11, or precipitation as the carbonate (by adding soda ash) or as the phosphate (by adding phosphoric acid or a soluble phosphate).
Manganese	Oxidation to insoluble manganous dioxide by chemical oxidants (free chlorine residual, ozone, potassium permanganate), ion exchange.
Mercury	Precipitation as the sulfide, at pH values between 5 and 8. Also, ion exchange, coagulation, and activated carbon.
Nickel	Generally, pH 11–12. In some cases, pH values ranging from 5–10 have produced good results. Precipitation as the carbonate or sulfide has worked well at pH values close to neutral.
Selenium	Dissolved selenium is removed by precipitation at pH 11–12, or by coprecipitation with iron at pH 5.5–8, or with alum at pH 6 or so. Undissolved selenium is removed by sedimentation and/or filtration.
Silver	Because of the value of silver, ion exchange removal followed by recovery of the silver is very common.
Zinc	Wide range of pH values, depending on other substances in the wastewater. Phosphate precipitation at pH 8–9 has worked well.

respectively, from typical industrial wastewaters. For some metals, the solubility is dependent on pH, such that a simple pH adjustment is all that is required to make the metal insoluble and therefore more readily removed.

Table 7-6 presents treatment methods that are most commonly used for removal of nine metals in addition to those shown in Table 7-5. Figure 7-7 presents relationships between pH value and solubility for six common heavy metals and the importance of correct pH. Notice that Figure 7-7 shows that chromium and zinc exhibit optimum pH values of 7.5 and 10, respectively, while iron,

copper, nickel, and cadmium show ever-decreasing solubility with increasing pH.

When an insoluble precipitate is formed, individual atoms share electrons to build a crystal lattice structure that results in particles that "grow" to a size that will settle in a clarifier under the influence of gravity. Often, two distinct processes take place. The first is the formation of small crystals of the substance being precipitated. The second is coagulation of the small crystalline particles, which is the clumping together of many small particles to form larger particles that settle well. Sometimes a coagulant or a coagulant aid can assist this process.

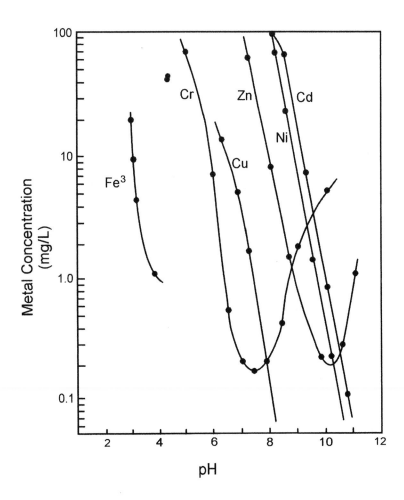

Figure 7-7   Precipitation of heavy metals as hydroxides (Lanouette, 1977).

It is common practice to include a deep bed granular filter or, in some cases, ultrafiltration as the final step in the removal of substances by the formation of an insoluble solid. Typically, chemical precipitation followed by sedimentation results in 10 mg/L or less of the target substance remaining in the treated effluent. Polishing this effluent normally reduces the residual to 1 to 3 mg/L.

It is always necessary to confirm a chemical treatment process by laboratory experimentation. Even though there are voluminous books and research papers that describe how various substances, such as heavy metals, are "removed," the presence of interfering substances such as chelating agents, complexing substances, and substances that will compete for the proposed treatment chemical can sometimes render a method ineffective in any given application.

A suggested procedure for determining, in the laboratory, optimum doses of reagent chemicals and ranges of pH is as follows:

Using a "jar test apparatus," shown in Figure 7-8:

1. Place wastewater, pretreated by sedimentation or otherwise, as anticipated for full-scale treatment plant, in "jars" of about 1,200-ml volume; 1,000-ml beakers can be used, but jars of square cross-section produce better mixing action.
2. Determine reagent doses to bracket the anticipated optima, regarding:

   a. pH
   b. Precipitant
   c. Coagulant (if any)
   d. Flocculant aid (if any)

(These determinations have to be made on the bases of (1) experience, (2) the literature, or (3) educated guesses.

Figure 7-8   Jar test apparatus.

3. With all stirrers in the "rapid mix" mode, add the selected doses to each jar as nearly simultaneously as possible. One method is to place each dose in a small beaker and have several people, with a beaker in each hand, dose each jar at the same time.

4. "Rapid mix" for 30 seconds.

5. "Slow mix," so as to achieve good coagulation. Observe the coagulation process (appearance of particles in the test solutions) and the flocculation process (building of particles into large, flocculent solids with attendant decrease in the turbidity of the test solution). Record observations. Carry on for up to 30 minutes.

6. Allow to settle. Observe flocculation and settling characteristics. Record observations.

7. Measure results of the following:

    a. Target substance concentration
    b. pH
    c. Turbidity
    d. TSS
    e. Residual reagents, if desired

Table 7-7 presents a listing of chemicals that have been used successfully to produce insoluble precipitates and thus effect removal of pollutants such as metals, phosphorus, sulfide, and fluoride.

The Use of Carbamates

Carbamates, a class of organic compounds that was developed for use as pesticides during the 1940s, have been found to function extremely well as precipitants for certain metals. In some cases, one or more carbamates have been found to be capable of precipitating metals in the presence of chelating agents and other substances that interfered with removal of the metals using more conventional methods, such as pH adjustment. Several chemical manufacturers now market

Table 7-7 Chemical Substances Commonly Used for Industrial Wastewater Treatment by Removal of Target Pollutants as a Precipitate

Chemical	Application
Lime	Heavy metals, fluoride, phosphorus
Soda ash	Heavy metals
Sodium sulfide	Heavy metals
Hydrogen sulfide	Heavy metals
Phosphoric acid	Heavy metals
Fertilizer-grade phosphate	Heavy metals
Ferric sulfate	Arsenic, sulfide
Ferric chloride	Arsenic, sulfide
Alum	Arsenic, fluoride
Sodium sulfate	Barium
Carbamates	Heavy metals

proprietary substances, which are carbamates, as effective precipitants.

Figure 7-9 presents a suggested sequential procedure for developing a treatment system, from initial concept to the point of initiating final design documents. Although the process depicted in Figure 7-9 is addressed to development of a treatment system for removal of target substances by formation of an insoluble substance, it can be easily adapted to development of many other types of treatment systems.

Disadvantages of Treatment by Production of an Insoluble Compound

Certain problems are common to treatment processes wherein the mechanisms of removal are formation of an insoluble solid, followed by separation of the solids from the liquid by sedimentation. One is the occasional inability of the precipitated solid to build into particles that are sufficiently large to settle, under the influence of gravity, in the clarifier. Another is the often-voluminous sludge, which is difficult to dewater.

Often, sludge produced by precipitation of heavy metals must be handled and dis-

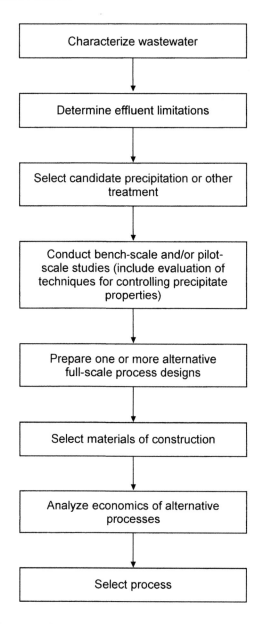

Figure 7-9   Suggested procedure for developing a treatment system.

posed of as "hazardous waste," because of its inability to "pass" the TCLP test (see Chapter 5), although there are methods of treating or conditioning these sludges to render them capable of passing. Some of these methods are proprietary; some are not. For instance, under certain circumstances, the phosphate precipitate of zinc will pass the TCLP test. Each individual sludge must be experi-

mented with in the laboratory, using principles discussed in this text, to develop a precipitation process that yields a sludge that can be dewatered economically to produce a residual that can be disposed of as nonhazardous.

One method of avoiding the expense and future liability of disposing of sludge as hazardous waste is to dewater, dry, then store the dried material for later recycle and reuse. Even though this would require licensing as a hazardous material storage facility, it might prove to be the most cost-effective option.

Suggested Approach for Treatment of Industrial Wastewater by Formation of an Insoluble Substance

Figure 7-10 presents a suggested approach, or model, for treatment of industrial wastewaters to remove substances by formation of an insoluble substance. First, as in all industrial wastewater treatment applications, the cost effectiveness of primary treatment, including plain sedimentation, should be evaluated. The second step should be pH adjustment to the range required for optimum effectiveness of the third step, which should include whichever conditioning steps are to take place prior to the principal treatment process. Examples are reduction of hexavalent chromium to the trivalent state, oxidation of arsenite to arsenate, and destruction of chelating agents by oxidation with Fenton's reagent.

Next, a second pH adjustment step may be required for optimum performance of the principal treatment process, shown in Figure 7-10 to be chemical precipitant addition, mixing, flocculation, sedimentation, and, finally, filtration. Figure 7-10 shows that there is a return of a small amount of the sludge from the sedimentation device to the influent to the flocculation device. This is labeled "return for seed." In this context,

"seed" refers to the particles of sludge, actually chemical precipitates that were formed previously, acting as nucleation sites for new precipitates. In certain applications, this greatly enhances the precipitation process, enabling a significantly higher degree of removal.

Reactions to Produce an Insoluble Gas

Considering, again, equations 7-5 and 7-6, it can be seen that if product C or D is a gas that is very poorly soluble in the wastewater being treated, it will remove itself from solution as it is formed, thus forcing the equilibrium to the right until the target substance (A or B) has essentially disappeared. An example of this treatment technology is the removal of nitrite ion by chlorination, as shown by equation 7-7:

$$2NO_2^- + Cl_2 + 8H^+ \rightarrow N_2 \uparrow 4H_2O + 2Cl^- \quad (7\text{-}7)$$

As chlorine is added in the form of chlorine gas or hypochlorite, or other chlorine compound that dissolves in water to yield free available chlorine, the nitrite ions are oxidized to nitrogen gas and water. Nitrogen gas, being only sparingly soluble in water, automatically removes itself from the chemical system, driving the equilibrium to the right until all of the nitrite ion has been removed.

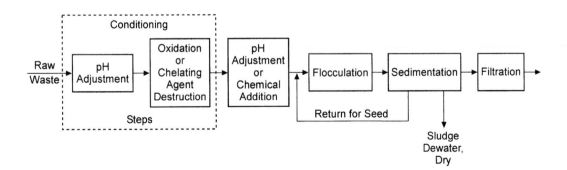

Figure 7-10   Schematic of suggested model for design of facility for removal by forming an insoluble substance.

Table 7-8   Chemical Substances that Can Be Applied to Remove Appropriate Target Substances as an Insoluble Gas

Chemical Substance	Treatment Technology
Chlorine gas or other chlorine compound (hypochlorite, or chlorine dioxide, for instance)	Breakpoint chlorination to remove ammonia
	Alkaline chlorination of cyanide
Sulfuric or hydrochloric acids (technical grade or waste acids)	Removal of sulfide as hydrogen sulfide gas

Table 7-8 presents a listing of chemical substances that can be added to remove the appropriate pollutants as an insoluble gas.

Breakpoint Chlorination to Remove Ammonia

For many years, breakpoint chlorination has been used to produce a free chlorination residual in drinking water. The basic process of breakpoint chlorination is that chlorine reacts with ammonia in four different stages to ultimately produce nitrogen gas, hydrogen ions, chloride ions, and possibly some nitrous oxide, and some nitrate, as shown in equations 7-8 through 7-12.

First, chlorine reacts with water to yield hypochlorous and hydrochloric acids:

$$Cl_2 + H_2O \leftrightarrow HOCl + Cl^- + H^+ \qquad (7\text{-}8)$$

Then, hypochlorous acid reacts with ammonia:

$$HOCl + NH \leftrightarrow NH_2Cl + H_2O \qquad (7\text{-}9)$$

$$HOCl + NH_2Cl \leftrightarrow NHCl_2 + H_2O \qquad (7\text{-}10)$$

$$HOCl + NHCl_2 \leftrightarrow NCl_3 + H_2O \qquad (7\text{-}11)$$

$$HOCl + NCl_3 \leftrightarrow N_2 + N_2O + NO_3^- \qquad (7\text{-}12)$$

Breakpoint chlorination can be used to convert ammonia to nitrogen gas in wastewater treatment. However, since many substances that are stronger reducing agents than ammonia will exert their demand, this method is suitable only if such substances are not present in significant amounts.

Alkaline chlorination of cyanide is included in Table 7-8 as a process that uses chlorine in a reaction to produce an insoluble gas. Since this reaction, taken to completion, is also an oxidation process, it is addressed later in this chapter.

Reduction of Surface Charge to Produce Coagulation of a Colloidal Suspension

A very high percentage of industrial wastewaters consist of colloidal suspensions. In fact, it is very often possible to destabilize industrial wastewaters by chemical coagulation; allow separation of the destabilized colloidal material from the water; further treat the water to achieve discharge quality by a polishing step, if necessary; and then recover the coagulant from the separated waste substances. The coagulant can be reused, and the waste substances can be further treated, if necessary. The advantage is that the polishing step can be significantly more economical than if it were used to treat the raw wastewater, and, in some cases, the separated colloidal material can be recovered as a by-product.

A colloidal suspension consists of one substance in a fine state of aggregation evenly dispersed throughout a second. The first phase, which may consist of single polymers or aggregates of smaller molecules, is called the dispersed or discontinuous phase, and the second phase is called the dispersing medium or continuous phase. The distinguishing characteristics of a colloid system are the size of the dispersed particles and the behavior of the system, which is governed by surface phenomena rather than the chemical properties of the components. The size of the dispersed particles ranges between 1 and 100 microns, placing them between molecules and true particles. Colloidal systems can be further classified into emulsions, gels, or sols.

Emulsions consist of two immiscible liquids, one being finely dispersed throughout the other. They must be stabilized by a surface-active agent, called an emulsifier, which reduces the surface tension at the interface between the two phases. Free energy from the increased surface area otherwise would tend to destabilize the emulsion. When an emulsifier is present, the repulsive forces, due to like electrostatic charges on each of the dispersed aggregates, exert the principal stabilizing force. These charges arise by several different means depending on the nature of the emulsion. The principal destabilizing forces are the Brownian movement, which causes the aggregates to come into contact with each other, and the London–van der Waal forces of attraction, which tend to cause the emulsified aggregates to coalesce after moving to within a critical distance from each other.

A gel results when organic colloids of long, thin dimensions are dispersed in a liquid medium, resulting in the formation of a nonuniform lattice structure when suitable groups on the colloidal particles come into contact. The dispersed phase is like a "brush heap" in this respect, and the gel assumes a semisolid texture. Gelatin is a familiar example of a substance that forms a gel.

The most common colloidal system encountered in industrial wastes consists of organic particles or polymers, and/or inorganic particles, dispersed in a liquid to yield a fluid system known as a sol. This differs from an emulsion in that an emulsifying agent is not required. The particles or polymers belong to one of two classes, depending upon whether or not they have an attraction for the dispersing medium: lyophobic (solvent-hating) for those substances that do not have an attraction, and lyophilic (solvent-loving), for those substances that do. In the case of industrial wastes, where the dispersion medium is water, the classes are called hydrophobic and hydrophilic.

Lyophobic Sols

Giant molecules or polymers that have no attraction for a particular liquid, and thus possess no tendency to form a true solution with the liquid, can be induced to form a lyophobic sol by the application of sufficient energy to uniformly disperse the particles throughout the liquid medium. If the dispersed particles contain groups that are ionizable in the dispersing medium, or if certain electrolytes are present in the dispersing medium, the anion or cation of which is preferentially adsorbed by the dispersed particles, the sol will be stabilized by mutual repulsion of like electrostatic charges on each of the particles of the dispersed phase.

Figure 7-11 illustrates a colloidal suspension wherein particles (which can be organic or inorganic, macromolecules or aggregates of smaller molecules, or finely divided solids) are dispersed in water. The dispersed particles have a surface charge, negative in all cases, which can have resulted from ionization of certain groups, breaking of covalent bonds, or adsorption of previously dissolved ions from the water medium.

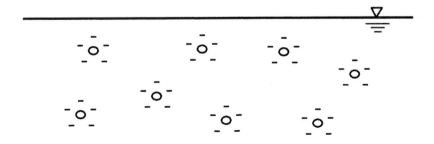

Figure 7-11 Illustration of a colloidal suspension.

An example of a hydrophobic sol, which owes its surface charge to selective adsorption of ions from solution, is the silver bromide sol, which results from mixing potassium bromide and silver chloride in water. Both potassium bromide and silver chloride are soluble in water. Silver bromide is not. Therefore, when potassium bromide and silver chloride are mixed together in water, silver bromide will be formed as an insoluble product and will precipitate from solution in the form of many tiny silver bromide crystals, in accordance with equation 7-13:

$$K^+ + Br^- + Ag^+ + Cl^-$$
$$\rightarrow AgBr + K^+ + Cl^-$$

(7-13)

When potassium bromide is in excess, the silver ions will all be tied up as the silver bromide precipitate. There will therefore be unprecipitated bromide ions in solution, and these will be adsorbed to the precipitated silver bromide crystalline particles. The sol, consequently, will possess a negative surface charge. At the exact equivalence point (equivalent silver and bromide), the silver bromide will precipitate from solution, because there will be neither bromide ions nor silver ions in excess. When silver nitrate is in excess, the sol will possess a positive charge. Mutual electrostatic repulsion results when either silver or bromide is present in excess.

Soap micelles are examples of colloid systems whose stabilizing surface charge arises from ionization of certain groups on the micelles. Long hydrocarbon "tails" of the soap molecules clump together to escape water, as illustrated in Figure 7-12 (the absence of the hydrocarbon chain in the water allows the water to form more hydrogen bonds, thus lowering the free energy of the entire system). Clumping together of the hydrocarbon chains of many soap molecules creates a micelle with many ionizable groups at its periphery. Since the charge due to the dissociated ions is the same for each micelle, electrostatic repulsion prevents the micelles from agglomerating.

Lyophobic sols are thus characterized by two phenomena—the absence of an attraction of the dispersed particles for the dispersing medium, and stabilization by mutual repulsion of like electrostatic charges on each of the suspended particles.

Lyophilic Sols

If, in the formation of a sol, the dispersed particles have an attraction for the molecules of the dispersing medium, each dispersed particle will adsorb a continuous layer of solvent molecules on its surface. The usual driving force involved in the formation of a hydrophilic sol is the reduction of free energy in the system as a result of stronger bonding between the solvent molecules and the dispersed particles than between the solvent

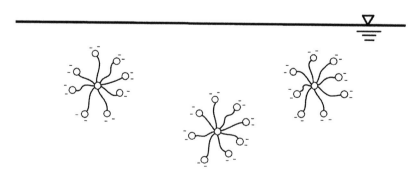

Figure 7-12   Colloidal suspension of soap micelles.

molecules themselves. This adsorbed layer serves as a protective shell for each of the dispersed particles, thus constituting the principal stabilizing factor for the sol. Additional stability can result from repulsion of like electrostatic charges on each particle, if chemical groups on these particles ionize or if ions are adsorbed onto the surfaces of the particles from the solution. These charges originate in a manner similar to those in lyophobic sols but are far less important to the stability of lyophilic sols than of lyophobic sols.

Electrokinetics of Lyophobic Sols

A graphical representation of the charge distribution on a lyophobic colloidal particle is presented in Figure 7-13. The model colloidal particle used for this figure has a negatively charged surface. This negative charge could have arisen from one or more of several electrochemical activities undergone by the particle when placed in the suspending medium, including those explained above, or simply from the adsorption of OH^- or other anions because of the greater affinity of the suspending medium for cations.

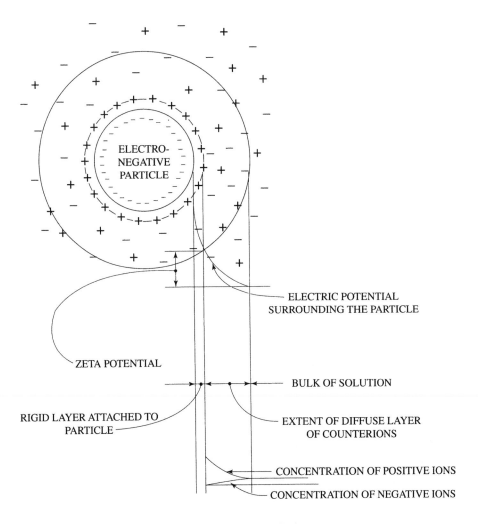

Figure 7-13   "Layers" or "shells" surrounding a negatively charged colloid.

The negatively charged particle attracts a layer of positive ions, which may originate either from dissociated electrolytes present in the suspending medium or from dissociated groups on the particle itself. These ions, oppositely charged to the inherent charge of the colloidal particle, are called "counterions." They are drawn to the particle by electrostatic attraction, while thermal agitation or Brownian motion tends to distribute them uniformly throughout the solution. The layer of counterions, referred to as the "Stern layer," is relatively rigid. The Stern layer does not entirely neutralize the charge on the colloid; the excess charge gives rise to a diffuse layer of co- and counterions, intermingled, but having a higher density of counterions close to the colloid and a higher density of coions at the outer reaches of the layer. This diffuse layer is often referred to as the "Gouy-Chapman layer." The Stern layer–Gouy-Chapman layer combination is called the "Helmholtz double layer" or the "diffuse double layer."

When the suspended colloid shown in Figure 7-13 is placed in an electric field, it will migrate toward the positive pole. As it moves through the suspending medium, the ions in the Stern layer move as a fixed part of the colloid, while those in the diffuse Gouy-Chapman layer tend to slough off, or stay behind. A plane of shear is developed at a certain distance from the surface of the colloid, and this plane of shear, or "slipping surface," defines the boundary between the Stern and Gouy-Chapman layers. Immediately below the colloid particle in Figure 7-13 is a graphical representation of the electrical potential at increasing distance from the surface of the colloid. The potential decreases linearly between the surface of the particle and the inner periphery of the Stern layer of counterions. From the inner periphery of the Stern layer outward, the potential drops at a decreasing rate. The potential at the outer periphery of the Stern layer (that is, at the surface of shear) is termed the "zeta potential." The value of the zeta potential is directly dependent upon the same factors that deter-

mine the thickness of the Stern layer, namely, the strength of the charge at the surface of the colloid; the nature—especially the value of the ionic charge—and degree of solvation of the ions in the Stern layer; and the frictional drag exerted by the suspending medium on the double layer as the particle migrates under the influence of an applied electric field.

Each of the ions in the Stern layer is solvated; thus, a layer of tightly bound solvent molecules surrounds each lyophobic colloid. In contrast to the water layer associated with hydrophilic colloids (a detailed discussion of which follows) the water layer in the case of a hydrophobic colloid is bound only by attraction to the ions in the Stern layer and not by attraction to the colloid surface itself. The zeta potential in the case of hydrophobic colloids gives a direct indication of the distance over which the colloidal particles can repel each other and thus of the stability of the colloid system.

Electrokinetics of Lyophilic Sols

When an organic macromolecular solid is placed in a given liquid, one of three possible states of solute-solvent interaction will result (solubility in this instance is defined as a limited parameter indicating compatibility with the solvent but not true solubility in the strict sense of the definition):

1. The macromolecular solid is insoluble in the liquid.
2. The solid swells, but has a limited solubility.
3. The substance is soluble in the liquid.

In cases 1 and 2 the systems are always lyophobic, whereas in case 3, the only possibility is that of a lyophilic sol. As an example of case 3, when amylose is placed in water, hydrogen bonds are formed between molecules of water and the hydroxyl groups on amylose that are at least as strong or stronger (that is, involve at least as much or more bond energy) than hydrogen bonds between the molecules of water itself. If a charge exists on the macromolecule because of factors similar to those accounting for the charge on lyophobic colloids, and if an electric field is applied across a portion of the sol, the particle will migrate toward one of the poles. As the particle moves, water molecules that are bound to the macromolecules by hydrogen bonds, plus those interlaced, again by hydrogen bonding, migrate as an integral part of the particle. This layer of water, usually monomolecular, defines what is known as the "solvated solvent layer" around the macromolecules and serves as a protective shell against influences that could be exerted by the chemical and physical properties of the suspending environment.

Secondary stabilizing forces possessed by a charged hydrophilic colloid arise principally from ionic dissociation of constituent groups on the macromolecule rather than by adsorption of ions. It is convenient to picture a charged lyophilic colloid as having a diffuse double layer of ions collected around it. The zeta potential is interpreted in the same manner as for lyophobic colloids, being the potential at the surface of shear in the diffuse double layer, and is measured in the same way, that is, by use of a zeta potential meter or by electrophoresis techniques.

Coagulation of Colloidal Waste Systems

Lyophobic Colloids

Coagulation, or agglomeration of the particles in a lyophobic colloidal system, can be brought about by neutralization of the surface charge on each particle to the point where the repulsive forces will be less than the London–van der Waal forces of attraction. Since the zeta potential directly indicates the strength of the net charge on each particle, a reduction of the zeta potential is synonymous with a reduction of the stabilizing forces of the sol. A zeta potential of zero (the isoelectric point of the system) corresponds to minimum stability. However, the zeta potential need not be zero for coagulation to take place. It must be reduced only to within a certain minimum range, referred to as the "critical zeta potential zone."

The zeta potential can be reduced by one or a combination of several methods, including increasing the electrolyte concentration of the sol, reducing the potential on the surface of the dispersed solid by external manipulation of the pH, and/or by adding multivalent counterions. The Schulze-Hardy rule states that the sensitivity of a colloid system to destabilization by addition of counterions increases far more rapidly than the increase of the charge of the ion. That is, in increasing from a divalent charge to a trivalent charge, the effectiveness of a coagulant increases far more than a factor of 3/2. Overbeek illustrated this rule by showing that a negatively charged silver iodide colloid was coagulated by 140 millimoles per liter of $NaNO_3$, 2.6 millimoles per liter of $Mg(NO_3)_2$, 0.067 millimoles per liter of $Al(NO_3)_3$, and 0.013 milllimoles per liter of $Th(NO_3)_4$.

The distinction between primary stability, imparted by the surface charge, and secondary stability, pertaining to the effective repulsion of colloidal particles because of the magnitude of the zeta potential, is useful for interpreting the coagulation of these particles by the addition of multivalent ions counter to the surface charge of the colloid. The presence of the multivalent counterions brings about a condition whereby the charges in the double layers are so compressed that the counterions are eventually contained, for the most part, within the water layers that originally contained the solvated ions in the double layer. The primary stability is thus reduced as a consequence of neutralization of the charge on the surface of the colloid, and the secondary stability is reduced since the particles can now approach each other to within a distance corresponding to coalescence of their water sheaths (water solvating the counterions and coions in the diffuse double layer) (recall that, if an ion is dissolved, it has, by definition, many water molecules surrounding it and "bonded" to it by "hydrogen bonding") without prohibitive repulsion between their respective diffuse double layers of ions.

Lyophilic Colloids

The principal stabilizing factor in the case of lyophilic sols is the solvating force exerted on the particles (not on the ions in the diffuse double layer) by the suspending medium. Electrolytic repulsion between the particles, though of lesser importance to the total stabilizing force, is also a factor that must be dealt with for the ultimate coagulation of a lyophilic colloid system. Two methods can be employed to desolvate lyophilic colloids.

In the first method, a liquid that is both a poor solvent for the suspended particle and highly miscible with the suspending medium can be added to the system. When this is done, the colloid is no longer able to form bonds with the medium that are stronger

than the internal bonds between elements of the medium, and the principal stabilizing factor is removed. This process is called "coacervation."

The other method is to add a substance, such as a sulfate salt, which can form stronger solvation bonds with the solvent than the solvent can with the suspended colloids. When this is done, the added salt effectively pulls the hydrated water layer from the surface of the colloid, again destroying the principal stabilizing factor. This process is known as "salting out."

The actual destabilizing mechanism, as well as the overall effect, of both of these methods is the same. In each case, the dispersing medium is able to form stronger bonds (thus decrease its free energy more) with the additive than with the dispersed phase. If the previously solvated particles

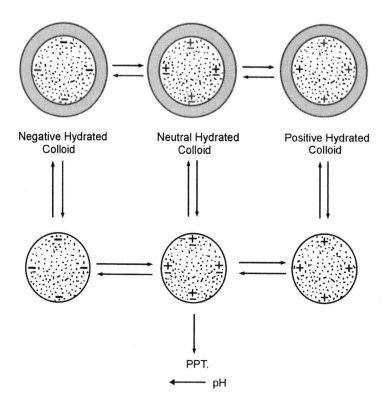

Negative Hydrated Colloid   Neutral Hydrated Colloid   Positive Hydrated Colloid

PPT.

← pH

Figure 7-14 Diagrammatic representation of the various types of colloids. The colloids in the upper row are lyophilic; those below are lyophobic. pH increases from right to left.

carry no net charge at this point, that is, they have either zero or very small zeta potential, they will flocculate and separate from the dispersing phase once coacervation has been accomplished.

If they do possess a net charge, resulting in a significant zeta potential and thus repulsive forces stronger than the London–van der Waal forces of attraction, the colloids will not coagulate but will remain in suspension as a lyophobic sol. Coagulation must then be effected according to the methods presented in the "Lyophobic Sols" section.

Figure 7-14 presents a diagrammatic representation of the various types of colloidal systems and how destabilization results in removal.

Coagulants

Coagulants such as alum, ferric sulfate, and cationic polyelectrolytes all work by suppressing the zeta potential of the colloidal system to a value sufficiently low that the colloidal particles will collide and then coalesce under the influence of slow stirring. Anionic and nonionic polyelectrolytes can greatly aid in the building of much larger flocculated particles that will both settle faster and produce a less turbid effluent. Used in this manner, the anionic and nonionic polyelectrolytes are referred to as "coagulant aids."

When alum $(Al_2(SO_4)_3.18H_2O)$ dissolves in water, some of the aluminum goes into true solution as the trivalent aluminum ion, Al^{+++}. If there are colloidal particles with a negative surface charge, the trivalent aluminum ions, plus other aluminum species such as $Al(OH)^{++}$ and $Al(OH)_2^+$, will be attracted to these negatively charged surfaces and will suppress the net negative surface charge, which is to say they will suppress the zeta potential. Other metal salts that dissolve to yield trivalent ions, such as ferric sulfate, ferric chloride, and aluminum chloride, coagulate colloidal suspensions with effectiveness similar to that of alum.

Salts that dissolve to yield divalent ions, such as calcium chloride or manganous sulfate, also effect reduction of the zeta potential

and eventual coagulation, but with an efficiency far less than the difference in ionic charge might indicate. As explained above, the Shultz-Hardy rule states this effect.

Laboratory experimentation is always required to determine the optimum doses of coagulants and coagulant aids. There is no characteristic, substance, or property of a wastewater that can be measured and then used as an indicator of the quantity of coagulant required. There is no substitute for performing "jar tests."

A suggested procedure for conducting a jar test program to determine the optimum quantities of reagents and pH range is presented in the section entitled "Reaction to Produce an Insoluble Solid." This procedure is identical to that recommended for determining the optimum doses of coagulants. The observations made during steps 5 and 6 will be oriented toward the disappearance of the turbidity originally present in the test solution, as opposed to the turbidity caused by the initial appearance of precipitated target substances (as in the case of the reactions to produce an insoluble solid).

Reaction to Produce a Biologically Degradable Substance from a Nonbiodegradable Substance

Some substances that are resistant to biodegradation can be chemically altered to yield material that is biodegradable. Examples are long chain aliphatic organics made soluble by attachment of ionizable groups, and cellulose or cellulose derivatives.

Hydrolysis, under either acidic or alkaline conditions, can be used to break up many large organic molecules into smaller segments that are amenable to biological treatment. Hydrolysis derives its name from the fact that when a carbon-carbon bond is "broken," a hydrogen atom attaches to the site on one of the carbon atoms, and an OH group attaches to the site on the other carbon atom. Heat may be required for effective hydrolytic action, and consideration of proper reaction time is very important.

Certain fats and oils are made water soluble by chemically attaching ionizable groups such as sulfonates or ammonium groups to their long-chain hydrocarbon structures for use in leather tanning, conditioning and finishing, and other industrial uses. These water soluble fats and oils are found in wastewaters from the industries that produce them, as well as in the wastes from the industries that use them. They are characteristically resistant to biodegradation, which is one of the reasons for their usefulness.

Hydrolysis of these water soluble fats and oils can break them into small segments that are still soluble in water and readily degradable by anaerobic microorganisms, aerobic microorganisms, or both. Laboratory experimentation is required to determine the technological feasibility for a given industrial waste, as well as a cost-effective process.

Substances that are made from cellulose or derivatives of cellulose are resistant to biological degradation, even though the basic building block for cellulose is glucose, a substance that is among the most readily biodegraded substances in existence. The reason for the nonbiodegradability of cellulose is that microorganisms found in aerobic biological treatment systems are not capable of producing an enzyme that can break the particular linkage structure, known as the "beta link," that joins the individual glucose molecules end to end to produce the very long chain that ultimately winds around itself to become cellulose.

Acid hydrolysis, usually requiring some heat, is capable of breaking cellulose into small segments. Just how small depends on the conditions of the hydrolysis process, as regards acidity, heat, catalysts, and reaction time. In general, anaerobic microorganisms are capable of rapidly biodegrading segments of cellulose that are in the size range of a few hundred glucose units, while aerobic microorganisms require that the cellulose be broken down into much smaller units.

A patented process that is based on the principles described above uses hydrolysis to alter the structures of refractive organics that remain in wasted biological treatment sludge after aerobic or anaerobic digestion has been carried out to essential completion. As an alternative to dewatering and disposing of this digested sludge, which, by definition, is not biodegradable, it can be broken into smaller organic units by hydrolysis, then sent back through the industry's biological treatment process from whence it came. Advantage can be taken of periods of low loading, such as weekends, holidays, periods of plant shutdown for maintenance, or during shifts when loadings to the treatment plant are low.

Reaction to Destroy or Otherwise Deactivate a Chelating Agent

Often, removal of metals from an industrial wastewater by simple pH adjustment, with or without addition of sulfide, carbonate, phosphate, or a carbamate, is ineffective because of the presence of chelating agents. Chelating agents, discussed in chapter two, are of various makeup, and include organic materials such as EDTA or inorganic materials such as polyphosphates. The following is an enumeration of methods that are candidates for solving this type of wastewater treatment problem.

Organic Chelating Agents

1. Destroy the chelating agent by acid hydrolysis.
2. Destroy the chelating agent by hydroxyl free radical oxidation, using one of the following technologies, as discussed in the section entitled "Oxidation or Reduction to Produce a Nonobjectionable Substance."

    a. Fenton's reagent (H_2O_2 + ferrous ions)

    b. Hydrogen peroxide + ultraviolet light

    c. Ozone + hydrogen peroxide

    d. Ozone + ultraviolet light

3. Destroy the chelating agent by adding potassium permanganate and heating

(the manganese ions will then have to be removed along with the target metals).

4. Pass the wastewater through granular activated carbon. In some cases, the chelating agent will adsorb to the carbon. In some of those cases, the chelated metal will remain chelated, and thus be removed on the carbon. In other cases, adsorption to the carbon effects release of the metals, which can then be precipitated without interference from the chelating agents.

Inorganic Chelating Agents

1. Add a stronger chelating agent, such as heptonic acid or EDTA, both of which are organic. Then use one of the methods given in step 1 for organic chelating agents.

2. Add a stronger chelating agent, such as heptonic acid or EDTA. Then it is sometimes possible to remove the metals by passing the wastewater through a strong cationic exchange resin generated on the hydrogen (acid) cycle.

3. If the chelating agent is polyphosphate, it may be feasible to hydrolyze it with acid and heat.

As with most cases involving industrial wastewater treatment, a proper program of laboratory experimentation followed by truly representative pilot plant work must be conducted to develop a cost-effective treatment scheme. The "throw some method," wherein one throws some of a candidate reagent into a sample of wastewater to "see if anything good happens," is perfectly proper to do during periods of break or frustration, but must never be used to eliminate forever consideration of a reagent or technology that comes to mind based on fundamental properties or chemistry.

Coprecipitation

Although technically a physical treatment method, since adsorption is the mechanism, coprecipitation is discussed here because it is often brought about as a result of chemical treatment to produce an insoluble substance of another target pollutant. For instance, both arsenic and cadmium are effectively coprecipitated with aluminum or iron. If there is dissolved iron (the soluble form is the ferric, or +3 state) and arsenic and/or cadmium in an industrial wastewater, sodium hydroxide can be added to form insoluble ferric hydroxide, which then builds a rather loose crystal lattice structure. This precipitate exists in the wastewater medium as a suspension of particulate matter. As these particles are building, arsenic and/or cadmium ions, which are relatively hydrophobic, adsorb to them, effectively removing them from solution. As the particles settle to the bottom of the reaction vessel, and the resulting "sludge" is removed, the treatment process is brought to completion. However, the removed sludge must now be dealt with. As discussed in the section entitled "Physical Treatment Methods," the usual procedure is to separate the water from the particles using vacuum filtration, sludge pressing, or other physical separation process followed by evaporation. The resulting dry mixed metals can be subjected to metal recovery processes, stored for later metals recovery, or properly disposed of.

Table 7-9 presents an enumeration of successful applications of coprecipitation for industrial wastewater treatment.

In wastewaters containing iron plus other metals, the wastewater can be aerated to oxidize soluble ferrous ions to insoluble ferric ions. As the ferric ions precipitate from solution, dissolved species of other metals will adsorb to the growing crystalline precipitates and will thus be removed from solution.

In wastewaters of low pH that contain aluminum ions, adjusting the pH to 6.3 or so will cause the aluminum to precipitate. The aluminum precipitate particles, which include oxides and hydroxides, act as effective adsorption sites for other metals.

One of the most effective methods for removing arsenic from wastewaters is to oxi-

Table 7-9   Applications of Coprecipitation of Metals as a Wastewater Treatment Method

1.	In wastewaters containing iron plus other metals, the wastewater can be aerated to oxidize soluble ferrous ions to insoluble ferric ions. As the ferric ions precipitate from solution, dissolved species of other metals will adsorb to the growing crystalline precipitates and will thus be removed from solution.
2.	In wastewaters of low pH that contain aluminum ions, adjusting the pH to 6.3 or so will cause the aluminum to precipitate. The aluminum precipitate particles, which include oxides and hydroxides, serve as effective adsorption sites for other metals.
3.	One of the most effective methods for removing arsenic from wastewaters is to oxidize all of the arsenic to arsenate, then add a source of ferric ion, such as ferric sulfate. Then, by maintaining the pH in the neutral range, and allowing oxygen to react with ferrous ions and precipitate as Fe_2O_3 crystals, the arsenate will adsorb to the growing crystals, become entrapped, and be effectively removed.

dize all of the arsenic to arsenate, then add a source of ferric ion, such as ferric chloride or ferric sulfate. Then, by maintaining the pH in the neutral range and allowing oxygen to react with ferric ions and precipitate as Fe_2O_3 crystals, the arsenate will adsorb to the growing crystals, become entrapped, and be effectively removed.

Oxidation or Reduction to Produce a Nonobjectionable Substance

Some highly objectionable substances can be chemically oxidized to produce nonobjectionable substances, such as carbon dioxide and water. An example is the destruction of the common rodenticide warfarin by potassium dichromate, acid, and heat, as represented by equation 7-14.

$$+134H^+ + 4K_2Cr_2O_7$$
$$\xrightarrow{heat} 19CO_2 + 24H_2O + 8Cr^{+++} \quad (7\text{-}14)$$

As shown in equation 7-14, highly toxic warfarin is oxidized to harmless carbon dioxide and water. Also shown in equation 7-14 is

the presence of Cr^{+++} in the effluent, which may have to be removed in a subsequent step.

The removal of chromium from industrial wastewaters by chemically reducing hexavalent chrome ions, which are soluble in water and highly toxic, to the trivalent state, which is neither soluble (in the correct pH range) nor toxic, is another example of treatment by chemical oxidation or reduction to produce nonobjectionable substances. Reducing agents that have been found to work well include sulfur dioxide, sodium or potassium bisulfite or metabisulfite, and sodium or potassium bisulfite plus hydrazine. The correct range of pH, usually in the acid range, must be maintained for each of these chemical reduction processes to proceed successfully.

Table 7-10 presents additional examples of treatment of industrial wastewater by oxidation or reduction of an objectionable substance to produce one or more nonobjectionable substances.

Alkaline Chlorination of Cyanide

Cyanide has been used for many years as the anion to associate with metals used in metal plating baths, since many metal cyanides are soluble in water to relatively high degrees. Because of the extreme toxicity of cyanide, its removal from industrial wastewaters was one of the earliest industrial wastewater treatment processes to be developed and used widely.

Table 7-10   Examples of the Use of Chemical Oxidation or Reduction to Produce Nonobjectionable Substances

1. Alkaline chlorination of cyanide to produce carbon dioxide, nitrogen gas, and chloride ion.

2. Chemical reduction of hexavalent chromium (toxic) to produce insoluble, less-toxic trivalent chromium.

3. Oxidation of (soluble) ferrous ions to the insoluble ferric state by exposure to oxygen (air).

4. Destruction of organic materials such as toxic substances (solvents and pesticides, for example) and malodorous substances (methyl mercaptan and dimethyl sulfide, for instance) by oxidation by free radicals:

    a. Hydrogen peroxide + ultraviolet light

    b. Fenton's reagent (H_2O_2 + ferrous ions)

    c. Ozone + hydrogen peroxide

    d. Ozone + ultraviolet light

5. Oxidation of organics with ozone, which may or may not involve free radicals

6. Oxidation of organics with hydrogen peroxide, which may or may not involve free radicals.

7. Destruction of toxic organics by oxidation with heat, acid, and either dichromate or permanganate. Products are carbon dioxide, water, and some nonobjectionable refractory compounds. Just about any organic compound can be destroyed by this method.

8. Wet air oxidation of various organics, such as phenols, organic sulfur, sulfide sulfur, and certain pesticides. This process takes place under pressure, with oxygen supplied as the oxidizing agent by compressed air.

9. Chlorination of hydrogen sulfide to produce elemental sulfur. Subsequent neutralization is usually required.

Chlorine will react instantaneously with cyanide, at all pH levels, to produce cyanogen chloride, as follows:

$$CN^- + Cl_2 \rightarrow CNCl + Na^+ + Cl^- \quad (7\text{-}15)$$

At pH levels other than alkaline (that is, below 8.5), the cyanogen chloride will persist as a volatile, toxic, odorous gas. In the presence of hydroxide alkalinity, however, cyanogen chloride is converted to cyanate, which is a thousand times less toxic than cyanide:

$$\begin{aligned} CNCl + 2OH^- \\ \rightarrow CNO^- + H_2O + Cl^- \end{aligned} \quad (7\text{-}16)$$

If chlorine is present in excess, that is, in an amount significantly greater than that required for stochiometric completion of the reaction indicated in equation 7-15, and the chlorine that will be consumed in reactions with reducing reagents and organics present in the wastewater (side reactions), it will oxidize the cyanate (produced as shown in equation 7-16) to carbon dioxide and nitrogen gases, both only sparingly soluble in water:

$$\begin{aligned} 2CNO^- + 4OH^- + 3Cl_2 \\ \rightarrow 2CO_2 + 6Cl^- + N_2 + 2H_2O \end{aligned} \quad (7\text{-}17)$$

which accounts for the effectiveness of the so-called "alkaline–excess chlorine process" for destruction of cyanide.

Because chlorine is subject to side reactions as indicated above, the chlorine dose required to produce satisfactory destruction of cyanide must be determined in the laboratory, and extreme caution must always be exercised to never allow the cyanide solution to attain an acidic pH, which will allow development of cyanide gas:

$$CN^- + H^+ \rightarrow HCN \quad (7\text{-}18)$$

another sparingly soluble gas, but an extremely toxic one.

Usually, the caustic–excess chlorine process for destruction of cyanide is carried out at pH 8.5. Automatic pH control is normally employed, with suitable fail-safe processes. Automatic chlorine dosing can be accomplished by use of an ORP probe and controller.

Oxidation of (Soluble) Ferric Ions to (Insoluble) Ferrous Ions by Oxygen

Iron is soluble in water in the +2 (ferrous) valence state. Ferrous ions are very easily oxidized to the +3 (ferric) valence state, soluble only in strongly acidic aqueous solutions. The oxygen content of air is a sufficiently strong oxidizing agent to bring about this oxidation.

Aeration, followed by sedimentation, has been used successfully to remove iron from industrial and other wastewaters, including landfill leachate. The aeration step accomplishes conversion of the dissolved ferrous ions to the insoluble ferric state. The ferric ions quickly and readily precipitate from solution as hydrous and anhydrous ferric oxides ($Fe(OH)_3$ and Fe_2O_3, respectively). These oxides form crystal lattices that build to particle sizes that settle well under the influence of gravity. Coagulants and/or coagulant aids are often used to assist the coagulation and flocculation process, producing a higher-quality effluent.

A benefit that often results from the removal of iron by oxidation followed by sedimentation is that other metal ions, all of which are inherently hydrophobic, "automatically" adsorb to the particles of ferric oxides as they are forming via precipitation. They are thus removed from solution along with the iron. This process, known as "coprecipitation," is sometimes augmented by actually adding more ferrous ions to the wastewater in the form of ferrous sulfate or other ferrous salt in an amount sufficient to coprecipitate other metals.

Oxidative Destruction of Organics by Free Radicals

Free radicals are powerful oxidizers that can convert many organics all the way to carbon dioxide, water, and fully oxidized states of other atoms that were part of the original organic pollutants, including sulfates and nitrates. Free radicals can be generated in a controlled manner to destroy a host of objectionable organic substances, including pesticides, herbicides, solvents, and chelating agents. As discussed previously, chelating agents interfere with the mechanisms employed to remove metals (by precipitation, for instance).

A free radical is an atom, or group of atoms, possessing an odd (unpaired) electron (one that has no partner of opposite spin). A free radical has such a powerful tendency to obtain an electron of opposite spin, and thus complete a thermodynamically stable pair, that it will easily extract one from an organic molecule. When this happens, the organic molecule, or a portion of it, becomes a free radical itself, and goes after another molecule, organic or otherwise, forming another free radical, and so on. A chain reaction is thus set up and, if managed properly, can be induced to continue until nearly all of the target substance has been removed.

Management of the chain reaction consists of maintaining the pH of the system within a range favorable for the reaction, and supplying enough of the free radical–generating substance (discussed as follows) to keep the process going. Side reactions (with reducing agents) "kill" free radicals. If the rate of "killing" free radicals exceeds the rate of generation of free radicals for a long enough time, the treatment process terminates.

A free radical, then, has one unpaired electron and has the same number of electrons as protons. A negatively charged ion, in contrast, has an even number of electrons, each paired with another electron of opposite spin, and has more electrons than protons.

Free radicals can be generated by the following methods:

- Addition of hydrogen peroxide
- Addition of hydrogen peroxide to a solution that contains ferrous ions, either

$$-CH_2C = CCH_2- + H_2O_2 \rightarrow -CH_2 --- CCH_2- + H_2O$$

$$\diagdown_O\diagup$$

$$(7\text{-}19)$$

present in the wastewater or added along with the hydrogen peroxide (Fenton's reagent)

- Addition of hydrogen peroxide, followed by irradiation with ultraviolet light
- Addition of both ozone and hydrogen peroxide
- Adding ozone and irradiating with ultraviolet light

Oxidation with Hydrogen Peroxide

Hydrogen peroxide has the chemical formula H_2O_2 and is an oxidizing agent that is similar to oxygen in effect but is significantly stronger. The oxidizing activity of hydrogen peroxide results from the presence of the extra oxygen atom compared with the structure of water. This extra oxygen atom is described as a "peroxidic oxygen" and is otherwise known as "active oxygen."

Hydrogen peroxide has the ability to oxidize some compounds directly; for instance, alkenes, as shown below. The peroxidic oxygen adds to the double bond, producing a hydroperoxide, as shown in equation 7-19.

The hydroperoxide then autooxidizes to alcohols and ketones.

Hydrogen peroxide consists of two hydrogen atoms and two oxygen atoms bonded by shared electron pairs, as shown in Figure 7-15.

The oxygen-oxygen single bond is relatively weak and is subject to breakup to yield •OH free radicals:

$$H - O - O - H \rightarrow •OH + •OH$$

or

$$peroxide \rightarrow Rad •$$

The two •OH free radicals sometimes simply react with each other to produce an undesirable result; however, the radical can attack a molecule of organic matter, and in so doing, produce another free radical. This is called a chain-initiating step:

$$•OH + RH \rightarrow H_2O + R •$$

chain-initiating step

As discussed previously, in free radical oxidation of organics, this process continues in such a way that the organics are broken

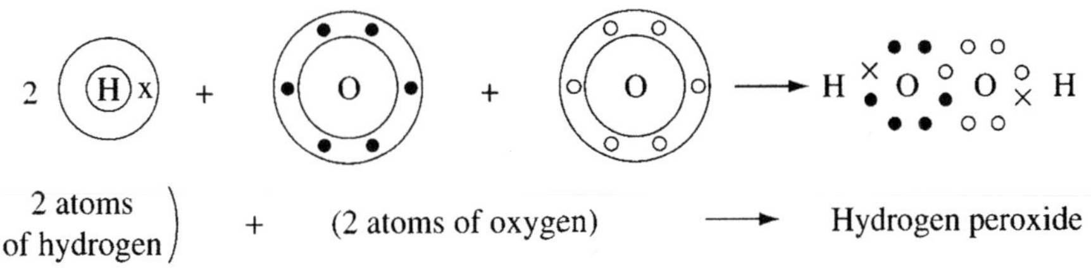

$$2H + 2O \longrightarrow H-O-O-H$$

Figure 7-15 Formation of hydroxyl free radicals.

down, all the way to carbon dioxide and water.

$$R\bullet + R - C \rightarrow R + R\bullet + C\bullet$$
chain propagation steps

$$RH + R\bullet + O \rightarrow CO_2 + H_2O$$

$$RH + C\bullet + O \rightarrow CO_2 + H_2O$$

Hydrogen Peroxide Plus Ferrous Ion (Fenton's Reagent)

Hydrogen peroxide will react with ferrous ions to produce ferric ions, hydroxide ions, and hydroxyl free radicals, as shown in equation 7-20:

$$Fe^{++} + H_2O_2 \qquad (7\text{-}20)$$
$$\rightarrow Fe^{+++} + OH^- + \bullet OH$$

The hydrogen peroxide thus dissociates into one hydroxide ion (nine protons and ten electrons [OH^-]) and one hydroxyl free radical (nine protons and nine electrons [$\bullet OH$]), as shown in equation 7-21:

$$H_2O_2 + e^- \xrightarrow{\ uv\ } \bullet OH + OH^- \qquad (7\text{-}21)$$

In this case, there is only one $\bullet OH$ free radical, as opposed to two $\bullet OH$ free radicals, when hydrogen peroxide breaks down in the absence of ferrous ions as discussed previously. The single $\bullet OH$ then attacks a molecule of organic material as discussed previously, initiating a chain reaction (chain-initiating step), with the result that the organic material is eventually oxidized all the way to carbon dioxide and water.

Hydrogen Peroxide Plus Ultraviolet Light

When hydrogen peroxide is added to an aqueous solution that is simultaneously irradiated with ultraviolet light (UV), the result is that the hydrogen peroxide more readily breaks down into $\bullet OH$ free radicals than when the UV is not present, as illustrated in equation 7-22:

$$H_2O_2 \xrightarrow{\ uv\ } 2\bullet OH \qquad (7\text{-}22)$$

There are, therefore, significantly more hydroxyl free radicals to enter into chain-initiating steps, as discussed previously, than is the case without UV.

Ultraviolet light thus greatly increases the oxidative power of hydrogen peroxide, in a manner similar to that of metal activation (Fenton's reagent). Although it has not been made clear how the reaction proceeds, it seems likely that the ultraviolet energy enables hydrogen peroxide to either separate into two hydroxyl free radicals, each having nine protons and nine electrons, as suggested by equation 7-23:

$$H_2O_2 \xrightarrow{\ uv\ } 2\bullet OH \qquad (7\text{-}23)$$

or to obtain an electron from some source, probably the target organic compounds, and thus dissociate into one hydroxide ion (nine protons and ten electrons [OH^-]) and one hydroxyl free radical (nine protons and nine electrons [$\bullet OH$]), as shown in equation 7-24:

$$H_2O_2 + e^- \rightarrow \bullet OH + OH^- \qquad (7\text{-}24)$$

The hydroxyl free radicals then go on to enter or perpetuate a chain reaction, as shown previously.

Oxidation with Ozone

Ozone, having the chemical formula O_3, is a gas at ambient temperatures. Ozone has physical characteristics similar to oxygen but is a far stronger oxidizing agent. Ozone reacts with organic compounds in a manner similar to that of oxygen, adding across double bonds and oxidizing alcohols, aldehydes, and ketones to acids. Ozone requires less assistance than oxygen does, as from heat, catalysts, enzymes, or direct microbial action.

Ozone Plus Hydrogen Peroxide

Addition of both ozone and hydrogen peroxide has the effect of oxidizing many organics to destruction much more strongly and effectively than by adding either ozone or hydrogen peroxide alone. When both ozone and hydrogen peroxide are present in water containing organics, •OH free radicals are formed through a complex set of reactions. The result of the complex reactions is that two •OH radicals are formed from one hydrogen peroxide and two ozone molecules:

$$H_2O_2 + 2O_3 \rightarrow 2(\bullet OH) + 3O_2 \qquad (7\text{-}25)$$

The •OH radicals then react with organics to form harmless carbon dioxide, water, and other smaller molecules. As an example, •OH radicals react with trichloroethylene and pentachlorophenol. The products in both cases are carbon dioxide, water, and hydrochloric acid:

$$C_2HCl_3 + 6 \bullet OH$$
$$\rightarrow 2CO_2 + 2H_2O + 3HCl \qquad (7\text{-}26)$$

$$C_6HCl_5O + 18 \bullet OH$$
$$\rightarrow 6CO_2 + 7H_2O + 5HCl \qquad (7\text{-}27)$$

The ozone plus hydrogen peroxide system has the advantage (compared to, say, Fenton's reagent) that ozone itself will react in a first-order reaction with organics, resulting in further reduction of pollutants.

In addition to the formation of •OH radicals, as shown by equation 7-28, there may also be formation of oxygen free radicals:

$$H_2O_2 + O_3 \rightarrow 2(\bullet O) + H_2O \qquad (7\text{-}28)$$

Oxygen free radicals may then enter into chain reactions to break up organics, as shown in equations 7-29 and 7-30 (below).

This chain reaction may continue so as to destroy many organics in addition to those destroyed by ozone and the •OH radicals.

Ozone Plus Ultraviolet Light

Ozone can be used in combination with ultraviolet light to, in some cases, produce more rapid and more complete oxidation of undesirable organic matter than with either ozone or ultraviolet light alone. Equation 7-31 illustrates this alternative process:

$$\begin{aligned}Organic\ matter &+ ozone + UV\ light \\ &\rightarrow CO_2 + H_2O + O_2\end{aligned} \qquad (7\text{-}31)$$

Here again, free radicals may or may not be involved.

Chlorination of Hydrogen Sulfide to Produce Elemental Sulfur

Hydrogen sulfide is objectionable for a number of reasons, including its contribution to crown corrosion in sewers and its malodorous character. Hydrogen sulfide can be oxidized to elemental sulfur, as shown in equation 7-32:

$$H_2S + Cl_2 \rightarrow 2HCl + S^0 \qquad (7\text{-}32)$$

It may then be necessary to neutralize the hydrochloric acid by one of the usual methods.

$$-CH_2C = CCH_2- + \bullet O \rightarrow -CH_2C - - - CCH_2- \qquad (7\text{-}29)$$

$$-CH_2C - - - CH_2- \rightarrow -CH_2 + \bullet O + \bullet CCH_2- \qquad (7\text{-}30)$$

Additional methods of oxidizing hydrogen sulfide to either sulfate ion or elemental sulfur (both odor free) include:

- Raising the level of dissolved oxygen (beyond the level of saturation at atmospheric pressure) by adding oxygen under pressure
- Adding hydrogen peroxide
- Adding potassium permanganate

In the cases of oxygen under pressure or hydrogen peroxide, oxygen oxidizes (takes electrons away from) sulfur atoms. Normally, if the pH is in the alkaline range, the sulfur atoms will be oxidized to elemental sulfur, and one gram of hydrogen sulfide will require about 2.4 grams of H_2O_2, depending on side reactions. If the pH is in the acid range, the sulfur will be oxidized all the way to sulfate ion, and each gram of hydrogen sulfide will require four times more hydrogen peroxide (about 9.6 grams) as the pure chemical. In contrast, each gram of hydrogen sulfide will require about 4.2 grams of chlorine (as Cl_2) or 11.8 grams of potassium permanganate ($KMnO_4$) to oxidize the sulfur to elemental sulfur.

Thermal Oxidation

Certain highly toxic organics, such as PCBs and dioxin, are best destroyed by thermal oxidation. Dow Chemical Co., as well as others, constructed incinerators in the 1940s to safely dispose of wastes from the manufacture of many different chemicals.

PCBs, which are resistant to biodegradation (meaning that they are biodegraded very slowly), are destroyed completely by incineration, but temperatures of 2,300°F or more are required. Other organics are destroyed at lower temperatures, and each application must be tested and proven.

Thermal oxidation has often been used to remediate contaminated soil. The mechanism and products of thermal oxidation are the same as for chemical oxidation with oxygen. The mechanism is electrophilic attack,

and the products are carbon dioxide, water, and oxidized ions and molecules such as sulfate, sulfur oxides, and nitrogen oxides. Equation 7-33 represents the thermal oxidation process.

$$Organic\ matter + O_2$$
$$\xrightarrow{heat} CO_2 + H_2O + SO_x + NO_x \qquad (7\text{-}33)$$

When thermal oxidation is required, as opposed to chemical oxidation, it is because the energy of activation required for the reactants shown in equation 7-33 is too great for the reaction to proceed without heat. It is heat that supplies the energy of activation needed.

Catalytic Oxidation

When the energy of activation required for oxygen, or another oxidizing agent, to react with a certain organic substance is too great for the reaction to proceed under normal conditions, there are two ways to overcome this deficiency: (1) add heat to supply the energy of activation; and (2) add a catalyst, which reduces the energy of activation. When heat is added, the process is called thermal oxidation. When a catalyst is added, the process is called catalytic oxidation. The principal products are the same, namely, carbon dioxide and water (when the oxidation process is allowed to go to completion), and the mechanism of oxidation is basically the same, except for the effect of the catalyst in reducing the activation energy. In many cases, the catalyst is not consumed in the process, as illustrated in equation 7-34.

$$Organic\ matter + O_2 + catalyst$$
$$\rightarrow CO_2 + H_2O + catalyst \qquad (7\text{-}34)$$

Solvent Extraction

Solvent extraction operates on the principle of differential solubilities. When a substance (solute) is dissolved in one solvent, and that combination is mixed vigorously with a solvent in which the substance is more soluble than it is in the first solvent (and the two solvents are not soluble in each other), the dissolved substance will pass from the first solvent into the second. The driving force behind this transfer is the second law of thermodynamics, as explained in Chapter 2. The stronger bond energies that result when the substance is dissolved in the second solvent result in a lower level of free energy than was the case with the first solvent.

Biological Methods of Wastewater Treatment

Biological treatment of industrial wastewater is a process whereby organic substances are used as food by bacteria and other microorganisms. Almost any organic substance can be used as food by one or more species of bacteria, fungi, ciliates, rotifers, or other microorganisms. In being so used, complex organic molecules are systematically broken down, or "disassembled," then reassembled as new cell protoplasm. In aerobic or anoxic systems, oxygen, which acts as an electron acceptor, is required in either the dissolved molecular form or in the form of anions such as nitrate and sulfate. The end result is a decrease in the quantity of organic pollutants and an increase in the quantity of microorganisms, carbon dioxide, water, and other by-products of microbial metabolism. In anaerobic systems, substances other than oxygen act as the electron acceptor. Each of these systems is described in greater detail later in this chapter.

Equation 7-35 (below) describes the aerobic or anoxic biological treatment process. The specific component of this process are explained in following sections.

Organic Matter

1. Is regarded as pollution prior to the treatment process.
2. Is used as food by the microorganisms.
3. Might have been formed by a natural process, by a living plant or animal, or might have been formed synthetically by a chemical manufacturing process.
4. Is composed of the elements carbon, hydrogen, oxygen, nitrogen, phosphorus, and many additional elements in much smaller amounts. These elements are connected by chemical bonds, each of which is characterized by a certain quantity of energy called "bond energy." As the microorganisms disassemble the organic matter, they are able to capture much of this energy and use it to make new chemical bonds, in the synthesis of new protoplasm. However, the process is less than 100% efficient. Fewer chemical bonds can be assembled in the process of cell synthesis than were disassembled during the microbial degradation process. Because of this, the microorganisms need a way to get rid of carbon, hydrogen, and other atoms that result from the process of degradation but for which there is not sufficient energy to form carbon-carbon and other high-energy bonds required in the cell synthesis process. Because relatively low energy bonds can be formed with oxygen, the microorganisms get rid of excess carbon atoms as CO_2 and excess hydrogen atoms as H_2O. Other elements, if in excess, can be combined with oxygen as well and passed off into solution as the oxide. Nitrate and sulfate are examples.

$$Organic\ matter + microorganisms + oxygen + nutrients$$
$$\rightarrow more\ microorganisms + CO_2 + H_2O + oxidized\ organic\ matter$$

$$(7\text{-}35)$$

Microorganisms

1. Include bacteria, fungi, protozoa, nematodes, and worms.
2. Exist in a hierarchical food chain within which bacteria and fungi feed directly on the organic matter (pollutants), and the higher life forms (protozoa, nematodes, etc.) feed on the bacteria.

Oxygen

Is referred to as:

1. A hydrogen acceptor
2. An electron acceptor

Nutrients

Include the following:

1. Nitrogen
2. Phosphorus
3. Sulfur
4. Micronutrients

More Microorganisms

1. Are the result of growth of the microorganisms originally present.
2. Must be handled and disposed of as waste sludge.
3. Typically amount, in mass, to one-third to one-half the amount of organic matter (pollutants) originally present in the untreated wastewater, when the organic material is measured as BOD and the microorganisms are measured as dried solids.

CO_2

1. Is a waste product, in that it is a method used by the microorganisms to get rid of carbon atoms that have resulted from the degradation of the organic pollutants, but for which there is not sufficient energy to make carbon-carbon and other higher-energy bonds needed to make new cell material in the cell growth process.
2. Microorganisms do not disassemble organic molecules for the fun of it. They do it because they have a compulsion to grow (i.e., increase in numbers).

H_2O

1. Is a waste product.
2. Is the mechanism used by microorganisms to get rid of excess hydrogen atoms derived during the process of disassembling organic matter.

With respect to oxidized organic matter:

1. The degradation of organic matter by the microorganisms is not 100% complete.
2. The organic matter in the untreated wastewater may contain some organic molecules that the microorganisms are unable to degrade.

All biological treatment processes—aerobic, anaerobic, suspended growth, and fixed growth (also referred to as "fixed film")—are accurately represented by this relationship. The differences in the various configurations of processes are in the speed of reaction; the form of oxygen used; the relative amounts of "more microorganisms" and "oxidized organic matter" produced; and the types of tankage and equipment and amount of land required.

The overall process depicted by equation 7-35 involves diffusion of the molecules of organic matter through the aqueous medium (the wastewater itself), and adsorption (or other type of attachment) of these organic molecules onto the surface of the microorganisms. Then, the microorganisms to which the molecules, or particles, of organic matter are attached, must manufacture enzymes capable of breaking the organic molecules or particles down into elementary segments that can pass through the microorganism's cell membranes. Then, the cells' metabolic machinery "metabolizes" the elementary segments of organic material, rearranging molecular structures and building more cell protoplasm in order to grow by binary fission and "wasting" a certain amount of the food as carbon dioxide, water, and some low-molecular-weight organics ("oxidized organic material").

Bacteria and fungi are the primary converters of whatever organic materials are in the wastewater to new cell protoplasm and waste materials. However, these single-celled microorganisms make up only a portion of the multitudinous diverse life forms that populate a biological treatment system. In a mature treatment system, a food chain hierarchy becomes established, ranging from the single-celled primary converters through a number of species of protozoa, rotifers, worms, and in some cases algae and many other types of microscopic life forms. The rotifers and successively higher life forms graze on one or more of the lower life forms (the primary converters). The trick to managing a well-operating biological treatment system is to manipulate the "feeding" of the microorganisms and remove certain quantities of the microorganisms on a periodic basis ("wasting") in such a way as to maintain optimum relative numbers of the various life forms. In activated sludge systems, this is best done by controlling sludge age to within a range that works best for each individual system.

Development of Design Equations for Biological Treatment of Industrial Wastes

Two equations are used to mathematically describe the fundamental kinetics of the "treatment" that takes place as a result of microorganisms converting organic material to new cell mass, carbon dioxide, water, and residual material (referred to as "oxidized organic material," "other low-molecular-weight compounds," or "refractory organics").") These equations are empirical, are applicable to the treatment of wastes in all environmental media (e.g., activated sludge treatment of wastewater containing organics; biofiltration of air streams containing hydrogen sulfide; and biodegradation of organics in landfilled sludge) and are stated as follows:

$$\frac{dX}{dt} = Y\frac{dF}{dt} - k_d X \qquad (7\text{-}36)$$

$$\frac{dF}{dt} = \frac{kXS_e}{K_s + S_e} \qquad (7\text{-}37)$$

where

X = Mass of microorganisms (grams or pounds)

F = Mass of organic matter used as food by the microorganisms (normally expressed as BOD) (milligrams or pounds)

Y = Constant, represents the proportion of organic matter that gets converted to new microorganism cell material (dimensionless)

k_d = Constant, represents the proportion of the total mass of microorganisms that self-degrade (endogenous respiration) per unit time (inverse days)

K = Maximum rate at which the microorganisms represented by the symbol X are able to degrade the organic matter, no matter how much organic matter is present

S_e = Mass of F at the conclusion of the degradative process; equivalent to mass of BOD in the treated effluent.

K_s = Mass of organic matter, F, that induces the microorganisms, X, to degrade that organic matter at a rate equal to one-half the maximum possible rate, k

Two additional parameters are defined as follows:

$$\Theta_c = \frac{X}{\Delta X\big/day} = mean\ cell\ residence\ time$$

$$= sludge\ age \qquad (7\text{-}38)$$

where

Θ_c = the average amount of time a component of the microbial population spends in the reactor before exiting as effluent solids or being removed as daily wasting, in order to maintain a constant amount of microorganisms in the reactor

$$U = \frac{dF/dt}{X} \qquad (7\text{-}39)$$

Figure 7-16 illustrates the parameters k and K_s. As shown, when the concentration of food, or substrate (designated by F, S, BOD, etc.) is very low, it is therefore limiting, and an increase in concentration of food will result in a proportionate (described by the slope of the line that extends from the origin in a straight manner) increase in utilization (eating) rate. Once the concentration of substrate reaches a certain level (Smax), from that concentration on, food is not limiting, and the microorganisms utilize the food at their maximum rate, k. At the concentration of substrate equal to one-half Smax, the rate of substrate utilization is equal to K_s. The numerical value of k, as described above, is specific for a given population of microorganisms feeding upon a given type, or mixture, of organic substances. As will be shown below, the value of k that is chosen by a design engineer will have a direct effect on the size of treatment system and, therefore, its cost, not to mention performance. The appropriate value of k for given treatment systems has been the subject of heated debate, and even lawsuits. More appropriately, however, it has been the subject of considerable research. Values of k have been reported throughout the environmental

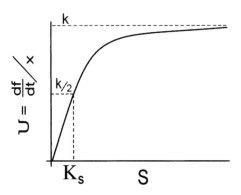

Figure 7-16   Specific utilization versus substrate concentration.

engineering literature, and have ranged from 0.01 to 5.0, a difference of a factor of 5,000! It is always necessary to understand the conditions under which a value of k was determined before using that value for any kind of calculation of size of treatment facility.

In the case of a mixture of substances such as is typical of industrial wastewater, there is a k rate that is applicable to each individual substance. Moreover, when a mixture of substances is utilized by a population of microbes, the substances that are most readily utilized (glucose, for instance) exhibit the highest value of k. As each of the more easily utilized substances in the original mixture is depleted, the apparent value of k decreases. Figure 7-17 illustrates how the apparent value of k changes with time, as a mixture of substances is utilized.

As shown in Figure 7-17, as a population of microorganisms begins utilizing a mixture of substances, the apparent value of k for the mixture is relatively high. As the most easily utilized substance becomes scarce, the microorganisms begin utilizing, successively, those substances that are more difficult to utilize, and the apparent value of k decreases. For the mixture as a whole, then, the most appropriate value for k is an approximate "average" value, or one that is observed over a lengthy period of time.

In fact, all of the rate constants in the above equations are specific for the particu-

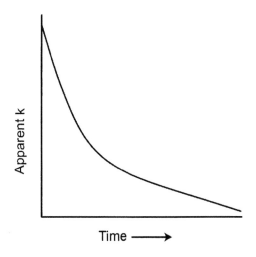

Figure 7-17 Apparent value of k versus time for a mixture of substances.

lar population or "mix" of the microbes present at any given time, as well as for the type or mixture of types of organic matter. These equations were developed using completely soluble substrate ("food") and pure cultures of bacteria. Monod, who developed equation 7-37, used pure cultures of E. coli and B. subtilis in broths of glucose and mannose.

Dividing dF/dt by X normalizes the rate of substrate removal to a unit weight of microorganisms, enabling comparison of perform-

ance between systems that have different food-to-microorganism mass-loading parameters. The result of this mathematical operation is U, the specific substrate utilization rate, as shown in equation 7-39. The parameter U_b would be equal to the food-to-microorganism ratio, F/M, if 100% treatment were taking place.

The fundamental relationships given above can be rearranged to develop equations, which can be used for design purposes. Table 7-11 presents several versions of design equations derived from one or more of the basic relationships shown previously. The parameters X, X_0, t, F, and S_e are measured directly. Θ_c can be calculated, and the parameters Y, k_d, k, and K_s can be determined from the results of either bench-scale pilot plant data or data from full-scale treatment systems, as shown in the following text.

Development of Biological Treatment Kinetics in The Laboratory

Bench scale pilot treatment systems can be used to generate values for the kinetic coefficients and constants needed in design equations as follows. Four or more treatment systems are operated for several days at steady state and are identical in all respects except

Table 7-11 Design Equations for Biological Treatment Systems

Characteristics	Complete Mix		Plug Flow with Recycle
	No Recycle	With Recycle	
Hydraulic Residence Time	$\Theta = \dfrac{V}{Q}$	$\Theta = \dfrac{V/Q}{1+R}$	$\Theta = \dfrac{V/Q}{1+R}$
Concentration of BOD_5 in Effluent	$S_e = \dfrac{K_s(1+k_d\Theta)}{\Theta(k-k_d)-1}$	$S_e = \dfrac{K_s(1+k_d\Theta_c)}{\Theta_c(k-k_d)-1}$	
Active Biomass Concentration	$X = \dfrac{[X_0+S_0-S_e)Y]}{1+k_d\Theta}$	$X = \dfrac{Y(S_0-S_e)}{1+k_d\Theta_c} \times \dfrac{\Theta_c}{\Theta}$	$X = \dfrac{Y(S_0-S_e)}{1+k_d\Theta_c} \times \dfrac{\Theta_c}{\Theta}$
Sludge Age	$\Theta_c = \Theta$	$\Theta_c = \dfrac{K_S+S_e}{S_e(k-k_d)-k_dK_S}$	$\Theta^{-1} = (k-k_d) - \left(-\dfrac{KY(1+R)}{\Theta X}\ln\dfrac{(S/S_e)+R}{1+R}\right)$

for food-to-microorganism ratio (F/M) and those parameters that F/M affects. Those parameters include mixed liquor volatile suspended solids concentration (MLVSS), which is directly related to the value of X. As the four or more laboratory bench-scale pilot units are run simultaneously, the parameters X, X_e, t, F, and S_e are determined each day for several consecutive days. Alternatively, a single larger-sized pilot unit, or a full-scale treatment plant, can be used. In this case, the system must be operated under conditions of one value of F/M at a time. After four periods of steady-state operation at four or more values of constant but different F/M values, the analysis can be carried out. Each time the F/M value is changed, the treatment unit must be operated at that F/M value for a time period equal to at least three sludge age periods before it can be ensured that "steady state" has been achieved.

The analysis consists of first plotting values of the inverse of Θ_c vs. U, as in Figure 7-18. The value of Y is the slope of the straight line of best fit for the plotted data points, as shown in Figure 7-18, and the value of k_d is that of the vertical intercept.

Next, the inverse of the values of U for each of the four or more treatment systems is

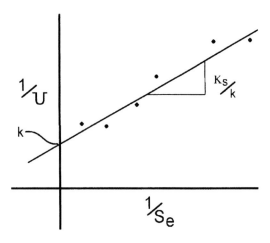

Figure 7-19   Plot of values of U vs S_e.

plotted against the inverse of the corresponding values of S_e, as shown in Figure 7-19.

The vertical intercept in Figure 7-19 is the value of k, and the slope of the line of best fit (Figure 7-19) is equal to K_s/k. Having thus determined values for all the parameters that appear in the design equations presented in Table 7-11, the design of a new or modified microbiological treatment system, or troubleshooting an existing treatment system, can proceed.

Example 7-5

Design an aeration tank to reduce the BOD_5 of an industrial wastewater to below 50 mg/L. The BOD_5 of the raw wastewater averages 1,714 mg/L. Primary treatment is expected to remove 30% of the BOD_5. The average rate of flow is 250,000 gallons per day.

Use complete mix with recycle.

Step 1: Determine design values for the appropriate kinetic constants:

$K = 0.51$ hr^{-1}
$K_s = 325$ mg/L BOD
$Y = 0.57$
$K_d = 2.4 \times 10^{-3}$ hr^{-1}
$\Theta = 12$ hrs = HRT
$\Theta_c = 10$ days = Sludge Age
$a = 1.5$ mg COD/mg MLVSS
$b = $ COD/BOD $= 1.4$

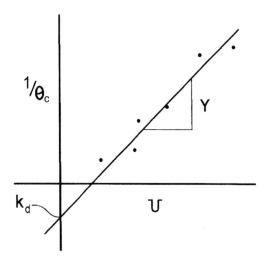

Figure 7-18   Plot of values of inverse Θ_c vs U.

Step 2: Estimate the soluble BOD of the treated effluent using the following relationship:

$$S = \frac{K_S(1 + kd\,\Theta_c)}{\Theta_c(k - k_d) - 1}$$

$$= \frac{325(1 + (2.4 \times 10^{-3} \times 24))}{10 \times 24(0.51 - 2.4 \times 10^{-3}) - 1}$$

$$= 4.24 \text{ mg/L}$$

Note that this is soluble BOD, which should always be 10 mg/L or less if biological treatment is reasonably complete. It would not be uncommon that the total five-day BOD would be in the range of 30 to 50 mg/L for a wastewater with a BOD as high as 1,714 mg/L.

Step 3: Calculate reactor volume:

$$V = Q \times \Theta$$

$$= 250{,}000 \text{ gal/day} \times 1 \text{ ft}^3 / 7.48 \text{ gal}$$
$$\times 12 / 24$$

$$= 16{,}700 \text{ ft}^3$$

Step 4: Calculate required MLVSS concentration:

$$X = \frac{Y(S_0 - S_e)}{1 + k_d\Theta_c} \times \frac{\Theta_c}{\Theta}$$

$$= \frac{0.57(1200 - 4.24)}{1 + (2.4 \times 10^{-3} \times 10 \times 24)} \times \frac{10 \times 24}{12}$$

$$= 8{,}655 \text{ mg/L MLVSS}$$

Step 5: Calculate the recycle ratio if the concentration of solids in the recycle is 20,000 mg/L:

$$R = \frac{1 - \dfrac{\Theta}{\Theta_c}}{\dfrac{X_r}{X} - 1} = \frac{1 - \dfrac{12}{10 \times 24}}{\dfrac{20{,}000}{9{,}400} - 1} = 0.84$$

Therefore, Recycle Q = 0.84 × Influent Q

Step 6: Calculate solids generation rate:

$$X_W = \frac{VX}{\Theta_c}$$

$$= \frac{16{,}700 \text{ ft}^3 \times 8{,}655 \text{ mg/L} \times \dfrac{1 \text{ lb}}{454 \times 10^3} \times \dfrac{3.78 \text{ L}}{\text{gal}} \times \dfrac{7.48 \text{ gal}}{\text{ft}^3}}{10 \text{ days}}$$

$$= 900 \text{ lb/day}$$

Step 7: Calculate the oxygen consumption rate:

$$\frac{Oxygen\ Demand}{Volume \times Time}$$

$$= \frac{S_0 - S_e}{\Theta}(1 - aY) + (0.9a \times k_d \times X)$$

$$= \frac{1.4(1200 - 4.24)}{12}(1 - 1.5(0.57) \times \frac{0.92}{1.4})$$

$$+ (0.9 \times 1.5 \times 2.4 \times 10^{-3} \times 8{,}655)$$

$$= 182 \text{ mg O}_2/\text{L/hr}$$

Because the volume of the reactor is 16,700 ft^3:

$$16{,}700 \text{ ft}^3 \times \frac{7.48 \text{ gal}}{\text{ft}^3} \times \frac{0.182 \text{ g}}{\text{L} \times \text{hr}} \times \frac{1 \text{ lb}}{454 \text{ g}}$$

$$= 190 \text{ lb Oxygen per day}$$

The Role of Oxygen in Wastewater Treatment

All forms of biological metabolism involve the disassembly of organic compounds (the food) and reassembly into new cell protoplasm (growth) and waste products. Not all of the food can be converted to new cell protoplasm, however. It takes a certain amount of energy, in the form of chemical bond energy, to assemble new protoplasm (i.e., to complete the chemical bonds that hold the carbon, hydrogen, and other elemental units together). The source of that energy is the relatively high energy bonds in the molecules of the organic matter used as food. For instance, when a molecule of glucose is disassembled to obtain energy to build protoplasm material, carbon-carbon bonds are broken with the consequent release of 60 to 110 kcal/mole of bond energy, depending on which atoms are bonded to the carbon atom being worked on.

Having obtained the energy from the carbon-carbon bond (and other bonds the carbon atoms were involved in), the organism has no further use for the carbon atoms. These "waste" carbon atoms cannot simply be discharged to the environment by themselves, however. They must be attached to other atoms and discharged to the environment as simple compounds. Since this process also requires energy, the "waste" compounds must be of lower bond energy than those of the food that was disassembled. The balance is then available for constructing new cell protoplasm.

The atoms to which the "waste" atoms are bonded are known as "ultimate electron acceptors." This name arises from the fact that the electrons associated with the waste carbon atoms are paired with electrons of the "ultimate electron acceptor" in constructing the molecule of waste substance.

The function of oxygen in cell metabolism (any cell—animal, plant, or bacterial) is that of ultimate electron acceptor. If the source of oxygen is molecular O_2 dissolved in water, the process is termed aerobic; it is depicted in equation 7-40 (below).

Bacterial growth is not shown in equation 7-40.

Notice that equation 7-40 shows that unwanted hydrogen atoms from the "food" are wasted by attaching them to oxygen and discharging them to the environment as water (H_2O).

If the source of oxygen is one or more dissolved anions, such as nitrate (NO_3) or sulfate (SO_4), and if there is no dissolved molecular oxygen present, the process is termed anoxic; it is depicted in equation 7-41 (below).

Microbiological growth is not shown in equation 7-41.

If there is no oxygen present, either in the molecular O_2 form or in the form of anions, the condition is said to be anaerobic. Under anaerobic conditions, cell metabolism takes place as a result of substances other than oxygen functioning as the ultimate electron acceptor. Equation 7-42 depicts this type of microbiological treatment.

$$R-\overset{\overset{H}{|}}{\underset{\underset{H}{|}}{C}}-\overset{\overset{H}{|}}{\underset{\underset{H}{|}}{C}}-\overset{\overset{H}{|}}{\underset{\underset{H}{|}}{C}}-\overset{\overset{H}{|}}{\underset{\underset{H}{|}}{C}}-R + O_2 \rightarrow CO_2 + H_2O \qquad (7\text{-}40)$$

organic matter + dissolved oxygen →
carbon dioxide + water

$$R-\overset{\overset{H}{|}}{\underset{\underset{H}{|}}{C}}-\overset{\overset{H}{|}}{\underset{\underset{H}{|}}{C}}-\overset{\overset{H}{|}}{\underset{\underset{H}{|}}{C}}-\overset{\overset{H}{|}}{\underset{\underset{H}{|}}{C}}-R + SO_4 \rightarrow CO_2 + H_2O + H_2S + CH_4 \qquad (7\text{-}41)$$

organic matter + sulfate anions → carbon dioxide + hydrogen sulfide + methane

$$R-\overset{\overset{H}{|}}{\underset{\underset{H}{|}}{C}}-\overset{\overset{H}{|}}{\underset{\underset{H}{|}}{C}}-\overset{\overset{H}{|}}{\underset{\underset{H}{|}}{C}}-\overset{\overset{H}{|}}{\underset{\underset{H}{|}}{C}}-R + e^- \rightarrow reduced\ organic\ compounds + CH_4 \qquad (7\text{-}42)$$

The microorganisms involved in the process depicted by equation 7-42 are extremely useful in certain industrial waste treatment applications; for instance, treatment of sugar refinery wastes and fruit processing wastes, which characteristically contain very low quantities of nitrogen and phosphorus compared with the amount of BOD. In a properly designed and operated facultative lagoon (discussed further on in this chapter), in which there is enough mixing to maintain movement of essentially all the liquid (to enhance contact between dissolved organics and to reduce short-circuiting), but not enough aeration is applied to maintain more than the upper two or three feet of the lagoon in an aerobic state, incoming dissolved BOD is rapidly "sorbed" onto microbial cells, which eventually settle to the lagoon bottom. Then, within the sludge layers at the lagoon bottom, anaerobic degradation takes place, whereby the sorbed organics are metabolized, and new (anaerobic) bacterial cells are produced as a result of bacterial "growth." The production rate of new cells, in terms of pounds of new cells produced per pound of organic matter treated (measured as BOD) is significantly less than is the normal case for aerobic treatment. The advantageous result is significantly less waste sludge requiring disposal.

As indicated in the parenthetical notes, equations 7-40, 7-41, and 7-42 show only the process of wasting excess carbon, hydrogen, and, in the case of equation 7-41, sulfur atoms, and do not include the growth process whereby some of the carbon, hydrogen, sulfur, nitrogen, and other atoms are reassembled to produce new cell protoplasm. These equations show only that the energy recovered from some of the disassembly process must be used to create bonds for waste products, such as carbon dioxide, water, and other compounds or anions.

Biological Treatment Technologies

Figure 7-20 presents an enumeration of the principal variations of biological methods of wastewater treatment. As Figure 7-20 illustrates, it is convenient to classify biological treatment processes as either aerobic or anaerobic (notice that, as used in this sense, the term *anaerobic* includes both anoxic and anaerobic). Within each of those two major categories, there are two principal types of systems: suspended growth and attached growth. The suspended growth systems all have diverse populations of microbes suspended in a mixture of liquid that includes the wastewater being treated. When the concentration of microbes is relatively high, as in the case of activated sludge, the mixture of suspended microbes, wastewater being treated, and other substances, both dissolved and suspended, is referred to as "mixed liquor suspended solids" (MLSS). The term *MLVSS* is used to designate that portion of the MLSS that is active microbes (the V in this term stands for "volatile"). That the MLVSS concentration is only an approximate indicator of the actual concentration of active microbes in a mixture of activated sludge is discussed in Chapter 5.

Attached growth systems all have masses of microbes attached to a medium. Wastewater to be treated flows in contact with this medium and, especially, the attached microorganisms. The microbes are able to access the organic matter in the wastewater as a result of the wastewater flowing over, around, and through the media to which the microbes are attached. The trickling filter and the rotating biological contactor are familiar examples of fixed-growth systems.

As explained earlier, aerobic wastewater treatment systems require that dissolved molecular oxygen, as the molecule O_2, be present and available to the microbes as they disassemble organic pollutant molecules. It is convenient to categorize aerobic wastewater treatment systems according to their relative "intensity of treatment." A treatment system of high intensity is one in which the concentrations of both pollutants and microorganisms are high. Oxygen must be added in high quantity to maintain aerobic conditions, and the system is said to be relatively highly

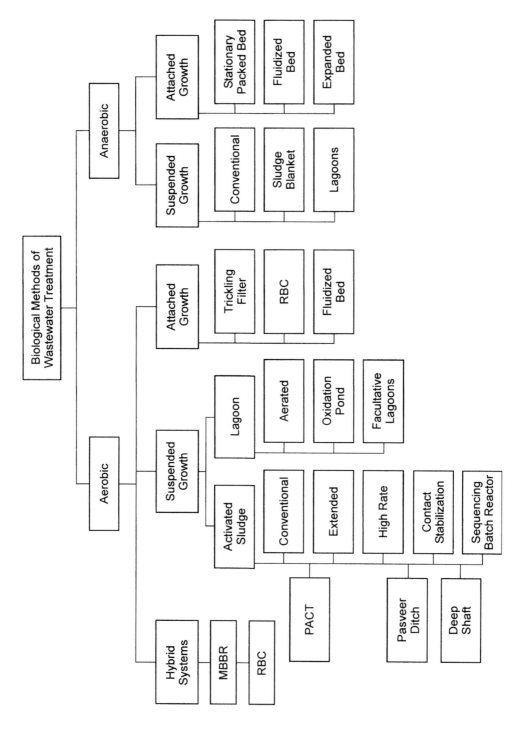

Figure 7-20   Aerobic and anaerobic methods of wastewater treatment.

stressed. Aerobic biological treatment systems range in intensity from high-rate activated sludge, which has MLVSS concentrations as high as 10,000 mg/L and hydraulic retention times as low as a few hours, to very low stressed aerobic or facultative nonaerated lagoons, which have MLVSS concentrations of less than 100 mg/L and hydraulic retention times of over 100 days. Figure 7-21 presents an ordering of aerobic suspended growth wastewater treatment systems, ranging from stabilization ponds to high-rate activated sludge. Fixed-growth systems also vary in treatment intensity, but normally over a smaller range than suspended growth systems.

Development of the most cost-effective suspended growth system is usually a matter of tradeoff between capital cost and operation and maintenance (O&M) costs. The present value of the total capital cost, spread over the useful life of the facility, and the O&M costs, also expressed as present value in order to allow direct comparison, often change in inverse proportion to each other. Low-intensity systems require larger tankage and more land area, but lower O&M costs in terms of electrical power for aeration and labor costs for operators. High-intensity systems require more skilled operators and significantly more oxygen supplied by mechanical means but smaller tankage and land area.

Treatment of Industrial Wastewaters Using Aerobic Technologies

Suspended Growth Systems: Activated Sludge

An activated sludge wastewater treatment system has at least four components, as shown in Figure 7-22: an aeration tank and a settling tank (clarifier); a return sludge

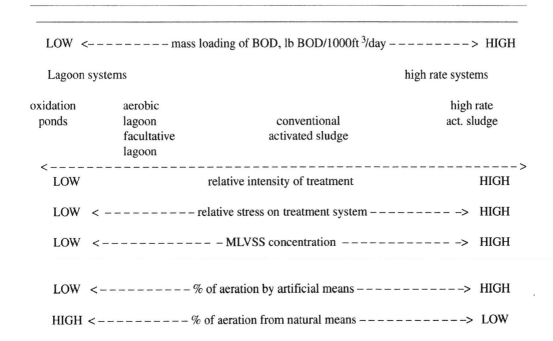

Figure 7-21   Distribution of various types of activated sludge treatment systems, based on treatment intensity, in terms of mass loading of BOD (lbs/BOD/ft³/day).

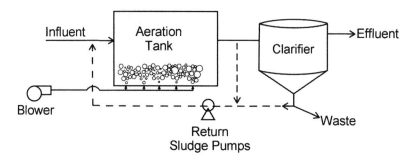

Figure 7-22   Basic components of an activated sludge system.

pump; and a means of introducing oxygen into the aeration tank. Wastewater, sometimes pretreated and sometimes not, enters the aeration tank (and is therefore the "influent"); it is mixed with a suspension of microbes in the presence of oxygen. This mixture is referred to as "mixed liquor." The microbes "metabolize" the organic pollutants in the wastewater, converting them to more microbes, carbon dioxide, water, and some low-molecular-weight organics, as depicted in equation 7-35. After spending, on the average, an amount of time equal to the hydraulic residence time (Θ) in the aeration tank, the mixed liquor flows into the clarifier, where the solids (MLSS) separate from the bulk liquid by settling to the bottom. The clarified "effluent" then exits the system. The settled solids are harvested from the clarifier bottom and are either returned to the aeration tank or are "wasted." The MLVSS solids that are returned to the aeration tank are microbes in a starved condition, having been separated from untreated wastewater for an extended period of time, and are thus referred to as "activated." It is this process of returning microbes from the clarifier to the aeration tank that enables buildup of their concentrations to high levels (1,800 to 10,000 mg/L), and that, indeed, characterizes the activated sludge process itself.

The MLSS solids that are taken out of the system and therefore referred to as "wasted" represent the main means of controlling the "mean cell residence time" or "sludge age."

Sludge age is an extremely important parameter in the successful operation of an activated sludge treatment system. Activated sludge systems that are maintained at a very low sludge age, on the order of two days or so, will contain what is known as a very young population, which is typically highly active and mobile and difficult to induce to settle well in the clarifier. Activated sludges with somewhat longer sludge ages, between 7 and 15 days, have many more microorganisms per unit of organic "food." They are, therefore, in a much more starved condition than a sludge of young sludge age and tend to predation and cannibalism. When food becomes very scarce, the microorganisms themselves become food. The live bacteria and fungi are food for higher life forms, and those that die break apart and spew their cell contents into the fluid medium, providing food for other bacteria and fungi.

To defend themselves against predation and cannibalism, some microbial species are able to exude and surround themselves with a protective mass of polysaccharide material. In addition to affording protection, this gelatinous material helps to flocculate the microbes that make up the MLVSS, enabling better settling characteristics in the clarifier.

When the sludge age increases to over 20 days or so, the microbes become so advanced in predatory behavior that they develop the ability to manufacture enzymes that can break down the polysaccharide protective material. The sludge thus loses its excellent flocculent nature and, consequently, its good settling characteristics.

The best settling activated sludge, and therefore the system that produces the clearest effluent, will be the system in which the gelatinous polysaccharide protective material is maintained in optimum amounts.

In terms of treated effluent quality, the effluent from activated sludge systems with very low sludge ages is typically high in suspended solids; those with sludge ages of around ten days have low suspended solids, and those with a very high sludge age are often very high in suspended solids.

An excellent tool for use in maintaining an optimum activated sludge culture (in terms of treatment performance, settleability, and low concentration of solids in the effluent) is the microscope. The usefulness of microscopic examination of activated sludge as an aid for process control can be explained as follows.

Consider a container of fresh, biodegradable wastewater, inoculated with a "seed" of activated sludge from a well-operating treatment system. The container is aerated, mixed well, and provided with a steady supply of biodegradable organics, but at a rate that is slower than the growth rate of the microbial population that develops. (Figure 7-22 depicts such a system.) Initially, there is a very high concentration of "food" compared with the numbers of microorganisms. Under this condition, bacteria will multiply at their maximum rate. Each individual bacterial cell will "grow," and through the process of binary fission, become two cells within a time period corresponding to the maximum attainable growth rate of that particular species, which can be as short a time as 20 minutes. A logarithmic increase in numbers of the fastest growing bacteria that can readily metabolize the organics in the wastewater takes place, and those bacteria dominate the population during the first few hours. Examination of a sample of the contents of the container, using a microscope, will show this to be the case.

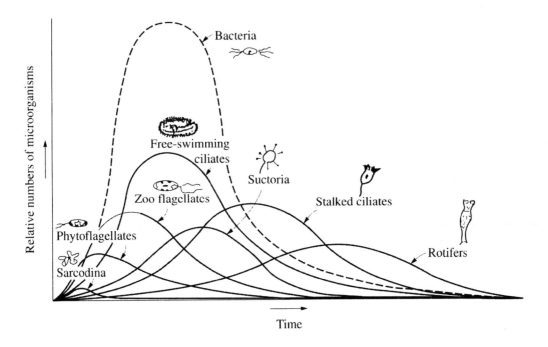

Figure 7-23   Progressive growths and deaths of microorganisms in activated sludge.

Figure 7-23 depicts the relative numbers of several of the major types of microbes as they increase and decrease, with time, under the conditions described in the preceding paragraph. As shown in Figure 7-23, during the initial hours, there will not be much growth of anything while the microbes with which the container was "seeded" become adjusted to the new environment. They need to manufacture the appropriate enzymes for the particular molecules of food available. This period of time is referred to as the "lag phase of growth."

As the first individual bacteria develop these enzymes and begin to grow, the phase of increasing growth rate occurs, and eventually, full logarithmic growth takes place and continues as long as food is unlimited and predation does not occur. Figure 7-24 shows the so-called "growth curve" that applies to

each individual species of microbe within the container. This figure shows that, sooner or later, within any biological system that can be described by the preceding paragraphs, food will become limiting and the rate of growth will decline. Some individual microbes will grow and some will die. Normally, there will be a period of time when the growth rate equals the death rate, and the population will be stable. Finally, as the food supply runs out, and/or predation exceeds growth, the population will decline.

Returning to Figure 7-23, and considering the microbial population within the container as a whole, as the bacteria that first begin growing reach high numbers, microbes that prey on them begin to grow. Then, in succession, higher forms of microorganism that can feed upon the microorganisms that grow earlier (and are thus said to be higher

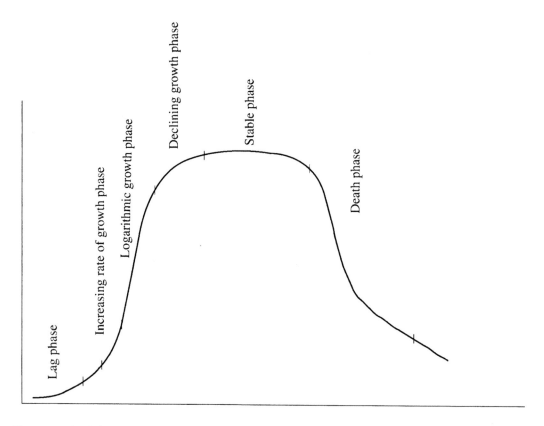

Figure 7-24   Typical growth curve.

on the food chain) go through their own growth curve process. In a biological treatment system, that succession is typically flagellated bacteria (bacteria equipped with a "tail" that propels them), free-swimming ciliates, stalked ciliates, rotifers, and finally worms. Microscopic examination of a sample of the microbial population from a given treatment system, then, can reveal the current stage of development of the system, in terms of "young sludge" or "old sludge."

Experience has shown that, usually, optimum operation of activated sludge occurs when the relative proportions of amoeboids, flagellated bacteria, free-swimming ciliates, stalked ciliates, and rotifers are as shown in the middle of the bar graph presented in Figure 7-25. Figure 7-25 shows that, when activated sludge is in a young condition, the relative numbers of flagellated bacteria and free-swimming bacteria are high, and there are almost no stalked ciliates. The effluent from the secondary clarifier will be high in suspended solids, and many of those solids will consist of long, thin bits and pieces of ill-formed activated sludge, referred to as "stragglers." The cure for this condition is to decrease the food-to-microorganism ratio by wasting less sludge and allowing the concentration of MLVSS in the aeration tank to increase.

Figure 7-25 shows further that when the relative proportions of rotifers, stalked ciliates, and nematodes become high compared with the flagellated bacteria and free-swimming ciliates, there will again be high solids in the effluent from the secondary clarifier. In this case, the solids will appear as tiny, more or less spherical bits and pieces of activated sludge, referred to as "pin floc." The cure for this condition is to increase the rate of sludge wasting, thus increasing the food-to-microorganism ratio.

Unfortunately, it is not always the case that adjusting the rate of sludge wasting will cure problems of high suspended solids in the treatment system effluent. Conditions other than sludge age that can affect effluent quality are: concentration of dissolved oxygen in the aeration tank, degree of mixing, the changing nature of the influent to the aeration tank, temperature, and the presence of toxic materials. High numbers of filamentous organisms within the activated sludge community, as explained in the section entitled "Selectors," is a common cause of poor effluent quality.

Selectors

Among the most common problems associated with activated sludge treatment systems is the poor settling of the activated sludge itself, referred to as "bulking sludge," caused by the presence of so-called "filamentous" microorganisms. Filamentous microorganisms, sometimes called "filaments," are characterized by long strands of "hair-like filaments." A small amount of these microbes is good to have, since they help to produce clarity in the effluent, but too many lead to an unacceptably slow settling rate.

There are three principal causes for the development of excess numbers of filamentous microbes within an activated sludge system: (1) low concentrations of nutrients, especially nitrogen and/or phosphorus; (2) low levels of dissolved oxygen, either throughout the aeration tank or in pockets within the aeration tank (resulting from inadequate mixing); and (3) low levels of organic loading (low food-to-microorganism ratio, F/M). So-called "selectors" have been proven to be very effective in preventing the development of dominant populations of filamentous microorganisms.

A selector is a device that has the purpose of counteracting the third of these causes. It is basically a chamber in which the activated sludge experiences high F/M conditions for a period of time that is short compared with the time spent in the aeration tank. Under the high F/M conditions, the "good" microbes out-compete the bad (filamentous) microbes, preventing their growth. In the high F/M environment, the nonfilamentous microbes adsorb most of the dissolved

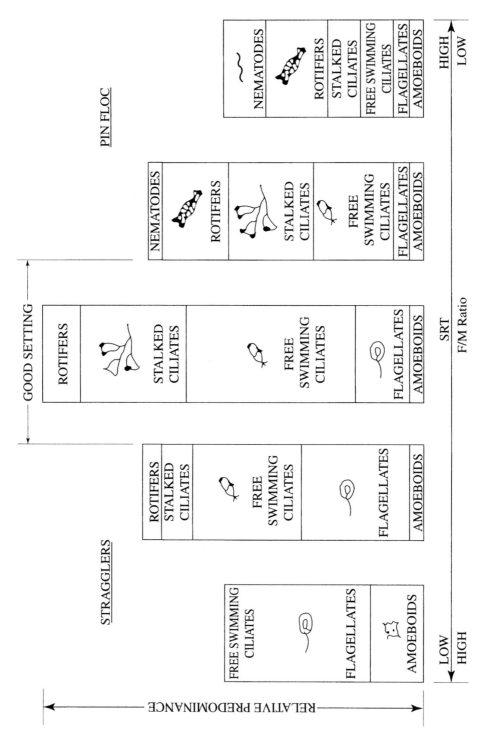

Figure 7-25   Relative predominance of microorganisms versus F/M and SRT.

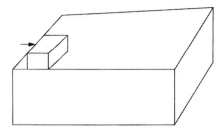

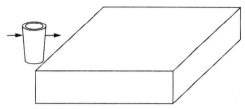

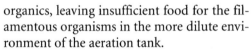

(a) Tank of 10–30 min. hydraulic retention time, located within aeration tank

(b) Tank of 10–30 min. hydraulic retention time, located outside aeration tank

Figure 7-26   Variations in selector configuration.

organics, leaving insufficient food for the filamentous organisms in the more dilute environment of the aeration tank.

The most common type of selector is a tank of 10 to 30 minutes' hydraulic retention time, completely mixed, in which return sludge and untreated wastewater are commingled before proceeding to the aeration tank. This tank can be separate from the aeration tank, or it can be a compartment within the aeration tank. Often, the latter can be done at lower construction cost. Figure 7-26 shows several variations of construction of selectors.

Selectors can be aerobic, anoxic, or anaerobic but must always be mixed well. Therefore, a selector should be equipped with both mechanical mixing and aeration, in order to provide the flexibility to operate aerobically or otherwise, as experience indicates which produces the best results.

Variations of the Activated Sludge Process

Activated sludge is essentially a biological wastewater treatment process in which microorganisms feed upon waste organic matter in an aeration tank; grow in numbers as a result; are separated from the treated wastewater in a clarifier that follows, hydraulically, the aeration basin; and are then returned to the aeration basin in a starved condition in order to greatly increase the

numbers of microorganisms in the aeration tank. The basic process is extremely flexible and capable of many different configurations. Table 7-12 lists nine alternative configurations and presents comments regarding the characteristics and appropriate use of each.

Conventional Plug Flow

If there is a standard of reference for the activated sludge wastewater treatment process, it is the conventional plug flow system. However, conventional plug flow is by no means the most common variation of biological treatment systems used for industrial wastes. Figure 7-27 presents a schematic of conventional plug flow. Table 7-12 notes several important features or characteristics that distinguish this system from other variations of the activated sludge process. As shown in Figure 7-27 and noted in Table 7-12, the most outstanding characteristic of plug flow is the long, narrow pathway the wastewater must traverse in flowing from the inlet to the outlet of the aeration tank. The reason for this is to ensure that the microbes, having entered the aeration tank as return sludge at the tank inlet, have sufficient time to be in contact with the organic matter contained in the volume of wastewater they originally mixed with, in order that as little mixed liquor as possible gets through the tank in a period of time shorter than the theoretical

Table 7-12   Variations of the Activated Sludge Process

Process Variation	Comment
Conventional Plug Flow	Aeration tanks are long and narrow to minimize short-circuiting. Operational parameters range as follows: HDT: 4–6 hrs, F/M: 0.3–0.6; MCRT: 7–14 days, F/M is high at head of tank, low at effluent end.
Conventional Complete Mix	Aeration tanks square or circular, concentration of substances everywhere equal to concentration in effluent; shock effects minimized; operational parameters as for conventional plug flow.
PACT Process	Aeration tanks either plug flow or complete mix. Powdered activated carbon added to aeration tank to remove nonbiodegradable organics. Operational parameters as for other variations.
Extended Aeration	Aeration tanks as for either plug flow or complete mix; sludge wasting rate low; MCRT: 20 days or more; MLVSS: 1,200–2,500 mg/L; F/M: 0.05–0.1; hydraulic retention time: 20 hours or more.
High Rate Aeration	Aeration tanks as for complete mix; F/M: 0.8–1.2; MLVSS: 5,000–10,000 mg/L, aeration rate very high, HRT: 3–6 hrs; MCRT: 15–25 days.
Contact Stabilization	Influent enters contact tank. HRT: 30 min, conventional clarifier removes MLVSS, return sludge to stabilization tank (similar to plug flow or complete mix); MLVSS: 3,000–6,000 mg/L; HRT: 3–8 hrs, discharge from stabilization tank mixed with influent prior to contact tank.
Sequencing Batch Reactor	Aeration tank as for complete mix, operational parameters as for conventional, aeration tank and clarifier are one and the same.
Pasveer Ditch	Aeration tank unique configuration, see description. Operational parameters similar to conventional plug flow.
Deep Shaft	Aeration tank unique configuration, see description. Operational parameters similar to conventional plug flow.

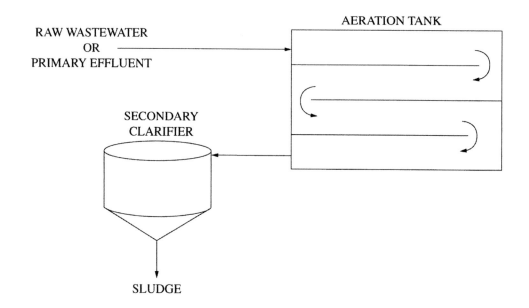

Figure 7-27   Basic flow pattern of the conventional plug flow variation of the activated sludge process.

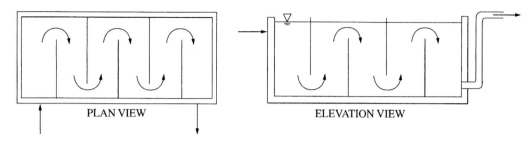

PLAN VIEW

ELEVATION VIEW

(a) Horizontal Flow Baffles

(b) Over-and-Under Baffles

Figure 7-28   Baffle arrangements in a flow-through tank.

detention time. An important design objective, then, is to minimize short-circuiting.

Several different approaches have been used to create a long, narrow flow path through the aeration tank. The most common approach has been to make use of baffles in a rectangular tank, as shown in Figure 7-28(a, b). However, research has shown that baffles create dead volumes, as depicted in Figure 7-29. Figure 7-29 shows that as the mixed liquor flows from the inlet to the out-

let, dead volumes exist in the regions of corners and wherever flow proceeds around the end of a baffle. The undesirable effects of the dead volumes are (1) reduction of the working volume of the aeration tank, with consequent reduction of actual hydraulic retention time (HRT); and (2) development of anaerobic conditions in the dead volumes, with consequent bad odors and development of filamentous bacterial populations.

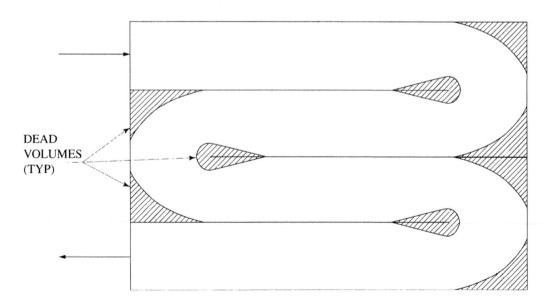

DEAD
VOLUMES
(TYP)

Figure 7-29   "Dead volume" in a baffled, flow-through tank.

Conventional Complete Mix

The "complete mix" variation of the activated sludge process is modeled after the familiar completely mixed continuous flow reactor used in many manufacturing processes. The mathematical relationships that were developed many years ago and published in the chemical engineering literature can be used for design as well as for operation. For instance, the "standard" equation that relates the change in concentration of a given substance entering a completely mixed reactor to the concentration leaving the reactor as influenced by the theoretical hydraulic retention time, and the reaction rate is given by:

$$C_{ei} = \frac{C_{oi}}{1+k_t} \tag{7-43}$$

where

C_{oi} = Concentration of substance i entering the reactor

C_{ei} = Concentration of substance i exiting the reactor

k = Rate of reaction of substance i in the reactor

t = Amount of time, on the average, molecules or other elemental quantity of substance i spends in the reactor

Stated in terms of kinetic parameters, as developed earlier in this chapter, the concentration of BOD_5 (for instance) in the effluent from a completely mixed activated sludge system is given by:

$$S = \frac{K_s(1+k_d\,\Theta_c)}{\Theta_c(k-k_d)-1} \tag{7-44}$$

where

S = Concentration of substrate (i.e., BOD_5) in the effluent

K_s = Mass of organic matter that induces the microorganisms to degrade that organic matter at a rate equal to half the maximum possible rate, k

k_d = Constant; represents the proportion of the total mass of microorganisms that self-degrade (endogenous respiration) per unit time (inverse days)

k = Maximum rate at which the microorganisms represented by the symbol X are able to degrade the organic matter, no matter how much organic matter is present

Figure 7-30 is a diagrammatical representation of a wastewater treatment plant designed to operate in the completely mixed mode.

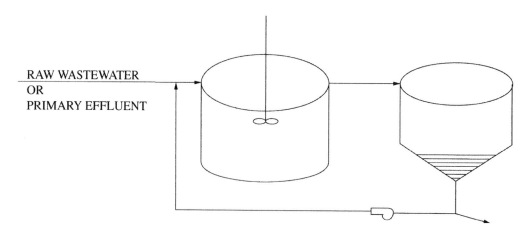

RAW WASTEWATER
OR
PRIMARY EFFLUENT

Figure 7-30   Schematic diagram of a complete mix activated sludge system.

One of the major differences between the conventional plug flow system and the complete mix system is in regards to short-circuiting. In the plug flow variation, the objective of both design and operation is to keep short-circuiting to an absolute minimum. As little as possible of the incoming pollutants should reach the effluent in amounts of time shorter than the theoretical hydraulic retention time, calculated by dividing the volume of the tank by the flow rate. In the case of the complete mix system, the objective is to manage short-circuiting, recognizing that a certain quantity of the pollutants will reach the effluent almost immediately after entering the aeration tank.

Figure 7-30 illustrates that when an incremental volume of flow enters the reactor, it is immediately dispersed throughout the reactor. The effluent, at any instant in time, is simply an instantaneous sample of the completely mixed contents of the reactor. Therefore, the concentration of any given substance in the effluent will be the same as its concentration in the reactor, and the effluent necessarily contains a small amount of unreacted substance.

The primary advantages, then, of the complete mix variation of the activated sludge process, are:

1. Slug doses of any given constituent are quickly diluted to the maximum extent afforded by the aeration tank.
2. The amount of short-circuiting, and therefore the amount of unreacted pollutant, can be controlled in a direct manner by manipulation of the flow rate (for a given volume of aeration tank) and determined by calculation.

The complete mix variation of the activated sludge process is often favored for application to industrial wastewaters, largely because of its ability to dilute slug doses of substances and to withstand periodic changes in wastewater characteristics due to changes in activity within the manufacturing plant. Either diffused or mechanical aeration can be used, and in some cases, when the concentration of substances to be treated is low (relatively, volume of aeration tank is high), a combination of mechanical mixing devices plus diffused or mechanical aeration is the least expensive system.

The PACT Modification of Activated Sludge

The PACT modification of the activated sludge process is a proprietary process that was developed during the mid-1970s to solve problems that related to the nonbiodegradability of certain synthetic organics. The PACT process is essentially one of the familiar variations of the activated sludge process, with the additional feature that powdered activated carbon is added to the aeration tank. (PACT is an acronym for "powdered activated carbon technology.") The activated carbon is incorporated into the MLSS for the purpose of adsorbing organics that are dissolved in the wastewater. The activated carbon will adsorb almost any organic, whether biodegradable or not, but the adsorbed biodegradable organics tend to be removed by microbial action, while the nonbiodegradable organics do not. The result is that the activated carbon becomes saturated with nonbiodegradable organics and eventually must be removed from the system. Figure 7-31 shows, diagrammatically, the PACT modification of activated sludge in the complete mix mode.

Since the powdered activated carbon is incorporated into the MLSS within the activated sludge system, it is removed as a consequence of the normal sludge wasting done to control the mean cell residence time (sludge age). The wasted sludge can then be incinerated to reduce the volume of excess biological solids and, at the same time, regenerate the activated carbon. Experience has shown that about 20% of the adsorptive capacity of the activated carbon is lost with each regeneration; therefore, continual makeup with new activated carbon is required.

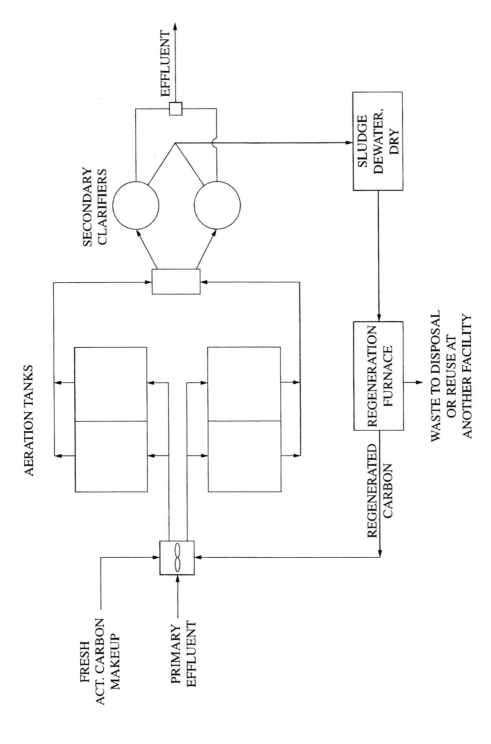

Figure 7-31   PACT modification of the activated sludge process, complete mix configuration.

The appropriate quantity of powdered activated carbon must be determined, first by laboratory experimentation and then by operating experience. To prevent discharge of excessive nonbiodegradable organics during startup and during the first few weeks of operation of a PACT system, an excessive quantity of activated carbon should be added, then the concentration reduced gradually until one or more of the target substances begins to be detected.

Predictive Mathematical Model of the PACT Process

A predictive mathematical model that can be used to analyze laboratory-scale pilot data to develop design parameters for a full-scale PACT system is presented as follows.

The parameters needed to design the biological portion of the PACT process are the same as for ordinary activated sludge—namely, the coefficients Y, k_d, k, and K_s, as explained previously. Furthermore, the laboratory procedures used to develop values for these parameters are the same as presented above—namely, the use of "block aerators" and the daily monitoring of food in, food out, sludge growth, and oxygen utilized. In the case of the PACT process, however, it is necessary to determine and then subtract the influences of organic matter removed and daily sludge increase, in order to determine those values that were due to biological activity.

Extended Aeration Activated Sludge

The extended aeration modification of activated sludge has the design objective of lower costs for waste sludge handling and disposal, at a sacrifice of higher capital costs due to larger tankage and more land area per unit of BOD loading. There may be a savings in O&M costs resulting from a lower level of intensity required from the operators, but there is an increase in the amount of oxygen required per pound of BOD in the influent, due to the increased amount of autooxidation (there is some increase in aeration via the surface of the aeration tank, due to the larger surface area). The reduction in amount of wasted sludge to be disposed of per pound of BOD removed results from the increase in autooxidation (use of the contents of the cells of dead microbes for food) because of a longer sludge age. Table 7-12 shows the range of values for F/M loading, hydraulic detention time, sludge age, and MLVSS concentrations appropriate to the extended aeration modification of the activated sludge process.

High-Rate Modification of Activated Sludge

The high-rate modification of the activated sludge process is simply one of the previously described modifications (plug flow, complete mix, or other) in which the MLVSS concentration is maintained at a significantly higher concentration than is normal for conventional processes, and hydraulic retention time is significantly shorter than "normal." The high-rate process can be used to treat readily biodegradable wastes such as fruit processing wastewaters, but effective equalization is required to prevent short-term changes in waste characteristics.

Advantages of the high-rate process are its effectiveness in overcoming the problem of the tendency toward bulking due to dominance of filamentous organisms in complete mix systems, as well as its lower construction costs due to smaller tankage. Disadvantages include relatively intense operational control requirements, as well as relatively large quantities of waste sludge due to less autooxidation.

Contact Stabilization Modification of Activated Sludge

The contact stabilization modification differs from the conventional process in that the activities that normally take place in the aeration tank (adsorption of substances onto the microorganisms, then metabolism or conversion of these substances to more micro-

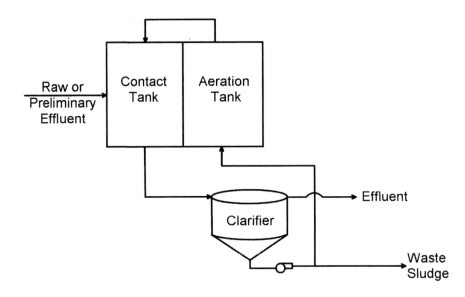

Figure 7-32   Schematic of contact stabilization modification of activated sludge.

organisms plus waste products) are separated. As shown in Figure 7-32, the untreated wastewater is mixed with return sludge in a relatively small "contact tank." It is here that the rapid process of pollutant adsorption takes place. The mixture then moves on to an aeration tank or "stabilization tank" where microbial metabolism takes place. The aeration tank is also relatively small compared with conventional activated sludge, since only the sludge, after separating from the bulk liquid in the clarifier, proceeds to the aeration tank. The bulk liquid is either discharged from the clarifier as final effluent or is subjected to further treatment, such as disinfection, sand filtration, or other process. It is because only the sludge (with its load of adsorbed organic material) proceeds to the aeration tank that the energy requirements for mixing are lower.

The principal advantage of contact stabilization is the lower capital cost due to smaller tankage. However, this modification is useful only on wastes that are rapidly adsorbed. In general, those wastes are characterized by a high proportion of dissolved, relatively simple, organics.

Sequencing Batch Reactor Modification of the Activated Sludge Process

The sequencing batch reactor process (SBR) is a modification of complete mix activated sludge. As diagrammed in Figure 7-33, the two principal sequential processes of activated sludge, aeration and settling, take place in the same tank. Settling occurs after sufficient aeration time for treatment has taken place, after turning off all aeration and mixing. Then, clarified supernatant is decanted; the reactor is refilled with fresh, untreated wastewater; and the cycle is repeated. Waste sludge is withdrawn immediately after the react phase has been completed.

An SBR is normally operated in six sequential stages, or phases, as follows:

1. *Fill phase.* The reactor is filled until the desired food-to-microorganism (F/M) ratio has been reached.
2. *React phase.* The reactor is mixed and aerated. Treatment takes place.
3. *Sludge wasting phase.* A quantity of mixed liquor that corresponds to the quantity of solids, on a dry basis, is withdrawn from the completely mixed contents of the reactor. For instance, if a ten-

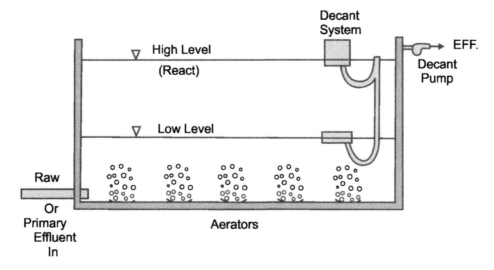

Figure 7-33   Schematic of sequencing batch reactor process.

day sludge age is desired, one-tenth of the volume of the reactor is withdrawn each day.

4. *Settle phase.* Aeration and mixing are terminated, and the reactor functions as a clarifier.

5. *Decant phase.* Clarified, treated wastewater is withdrawn from the top one-quarter to one-third of the reactor.

6. *Idle phase.* The system can be mixed and aerated at a low rate for a few days at a time needed between periods of waste generation.

There are several modifications to the basic procedure outlined previously. For instance, it has been found advantageous in some cases to either mix, aerate, or both while filling is taking place. In these cases, the first one or two stages are referred to as the "mixed fill phase" followed by the "react fill phase" (phases 1and 2) or just "the react fill phase" (phase 1). Also, some operators have found it acceptable to waste sludge from the bottom of the reactor after settling has taken place. While this practice results in handling less water with the waste sludge, it does not ensure the removal of the same fraction of the active microbial solids each day, as does

the practice of wasting a certain fraction of the mixed reactor contents each day.

Among the several important advantages of the SBR process is its capability of extending the react phase for as long as is necessary to achieve the desired degree of treatment. In this respect, it is good design practice to have at least two parallel SBR units, each of at least half the design capacity. In the absence of a parallel unit, a collection tank, designed and operated as an equalization basin, can receive and store wastewater until the SBR unit is able to receive more wastewater.

With two parallel SBR units, the following troubleshooting procedures are available to the operator:

1. If one of the units receives a large slug dose, the second unit can be used to perform double duty while the first unit is allowed to continue to react until the supernatant is suitable for discharge.

2. If one of the SBR units becomes upset due to toxicity or any other reason, the second unit can be placed in "double-duty mode." Then the contents of the upset unit can be bled into the operating unit at as slow a rate as is necessary for complete treatment as well as recovery of the upset unit to take place. During

extended periods of low waste generation, such as for yearly maintenance
shutdown of the processing plant, all the
MLSS can be placed in one of the SBR
units. It can then be "fed" a small
amount of synthetic waste to maintain
viability of the microbes.

3. As is the case for duplicate systems of any
unit process, one can be used for treatment while the other is emptied for
maintenance.

In short, the operator has more flexibility
and control with SBR technology than with
flow-through technology. Any or all of the
cycle stages can be lengthened or shortened
to achieve desired treatment. The "idle
phase" allows taking up the slack between the
total lengths of time of the other five phases
and the 24-hour day.

SBR technology has successfully, and very
cost effectively, been used for removal of
nitrogen (by the nitrification-denitrification
process). It has also, but less frequently, been
used to remove phosphorus.

The nitrification-denitrification process
can be incorporated into the operation of an
SBR system as follows:

1. Allow nitrification (conversion of
ammonia and organically bound nitrogen to nitrate ions, NO_2^-) to take place
during the react phase by:

   a. maintaining dissolved oxygen (DO)
   level above 2 mg/L
   b. maintaining the temperature within
   the reactor higher than 17°C
   c. maintaining sufficient react time

If these conditions are maintained, and if
there are no substances in the wastewater
that are toxic to the nitrosomonas species,
nitrification of the ammonia released from
the organic matter in "treating" the BOD will
take place. Then, denitrification (conversion
of the nitrate ions to nitrogen gas) can be
induced:

2. Discontinue aeration, but maintain mixing after the react phase.

During this period, the contents of the
reactor will become anoxic. Facultative bacteria (nitrobacter) will continue to oxidize
organic matter (BOD) and will use the oxygen atoms from the nitrate ions for an electron acceptor.

There are tricks to achieving successful
nitrification–denitrification, as well as
desired BOD removal, using SBR technology.
One trick is to maintain the proper amounts
of time for each phase. Another is to manage
the amount of BOD remaining after the react
cycle (by addition of methanol, if necessary)
so there is enough to support the denitrification process but not enough to cause noncompliance.

SBR technology can be used to achieve a
lower quantity of phosphorus in the industrial waste discharge, but only with the successful application of more operator skill
than for ordinary operation. Some phosphorus will always be removed with the waste
sludge, since all animal and plant cells contain phosphorus. The microbial cells of the
activated sludge can be induced to take up
more than the normal amount of phosphorus (referred to as "luxury uptake of phosphorus"). If this is done successfully, more
phosphorus than normal will be included
with each pound of waste sludge. The procedure for inducing luxury uptake of phosphorus in an SBR system is as follows:

1. Allow anaerobic conditions to develop
within the contents of the reactor by the
absence of aeration during the fill cycle.
As soon as the small amount of nitrate
that is contained in the MLSS at this
point has been used up for its oxygen
(electron acceptor) content, anaerobic
conditions result. At this point, the high-
energy phosphate bonds in the organisms are broken as an energy source,
allowing them to out-compete other
organisms. Once the air is turned back
on and conditions become aerobic,

acenitobacter (and other organisms that perform the "luxury phosphorus update") store excess phosphorus in the form of high-energy phosphate bonds. This energy source then allows them to out-compete other organisms for VFA (food) uptake when the anaerobic cycle is repeated. This is a "selection" process that favors organisms that release phosphorus in the anaerobic phase and uptake excess phosphorous in the aerobic phase.

2. Allow sufficient time during and possibly after the fill phase for the "luxury uptake" of phosphorus to take place to the maximum. Experimentation that is carefully controlled, with observations recorded, is required.

3. Conduct the sludge-wasting phase at this point, or from the settled sludge after the settling phase. An advantage to wasting the sludge at this point is sludge age is easily controlled. For example, wasting 1/20 of the volume of the reaction tank each day yields a 20-day sludge age. The advantage to wasting from the settled sludge is that less volume is handled; however, the calculations of sludge concentration and volume can be lengthy.

4. Begin aeration and, consequently, the react phase.

The contents of the reactor should be well mixed, in order to maintain contact between the microbes and the dissolved phosphate ions, throughout steps 1 through 4.

Procedure for Design of an SBR System

The following procedure can be used to design an SBR wastewater treatment system. In all instances where the word "select" appears, the design engineer must have available either extensive experience with the wastewater in question or the results of extensive laboratory and/or pilot treatability studies.

1. Select a hydraulic retention time.
2. Select an F/M loading.

3. Obtain the kinetic coefficients:
4. Select a value for the concentration of settled sludge.
5. Select a value for the specific gravity of the settled sludge.
6. Determine how much of the reactor will be decanted during the decant phase.
7. Select a depth of liquid (working depth).
8. Determine nutrient requirements.
9. Estimate, using the following calculation, the concentration of soluble BOD in the effluent:

$$S_e = \frac{K_S(1 + k_d\,\Theta_c)}{\Theta_c(k - k_d) - 1} \qquad (7\text{-}45)$$

The result of performing several trials to balance a desired value for Θ_c and the resulting value of S_e will produce the design value for Θ_c.

10. Compute the mass of MLVSS required:

$$X = \frac{Y(S_0 - S_e)}{1 + k_d\Theta_c} \times \frac{\Theta_c}{\Theta} \qquad (7\text{-}46)$$

11. Select a desired range for MLVSS concentration.
12. Calculate the size of the reactor, balancing the required volume for the desired HRT, the required volume to achieve the desired MLVSS concentration (taking into consideration the calculated MLVSS mass required), and the volume needed to decant the desired volume each day.

Pasveer Oxidation Ditch and Variations

During the 1930s, Pasveer invented the "brush aerator," which consisted of a cylindrical street brush immersed to about 20% of its diameter in the mixed liquor of an activated sludge aeration tank. Figure 7-34 illustrates how, when the brush was rotated, it would throw droplets of the mixed liquor into the air, thus accomplishing aeration as oxygen molecules from the air dissolved in the water droplets. As with other mechanical aeration devices, the very large surface-to-

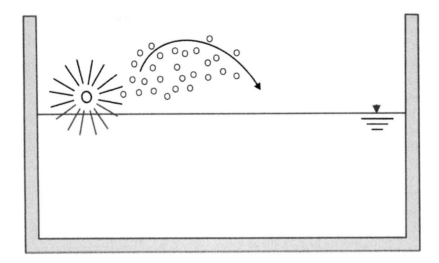

Figure 7-34   Pasveer's "brush aerator."

volume ratio of the very large number of water droplets provided for effective dissolution of oxygen into the water. This inexpensive aerator was combined with an innovative aeration basin to produce a cost-effective extended aeration activated sludge treatment system. Figure 7-35 shows that the aeration basin was built in the shape of an oval, similar in appearance to a race track. One or more brush aerators were placed so as to accomplish both aeration, by throwing droplets of mixed liquor (MLSS plus the bulk liquid) into the air, and mixing, by causing the mixed liquor to flow around and around the oval basin. In addition, a high degree of dis-

solution of oxygen into the mixed liquor through the liquid surface was accomplished. The movement of the mixed liquor, induced by the brush aerator, kept the mixed liquor mixed and continually renewed the surface with unaerated mixed liquor. Since the aeration basin was simply an oval ditch, construction cost was small, and since the surface-to-volume ratio of the mass of mixed liquor was high, surface aeration accounted for a large fraction of the total; thus, the cost for aeration was low. The system came to be known as the "Dutch Ditch," acknowledging Pasveer's nationality as well as the materials of construction of the aeration basin.

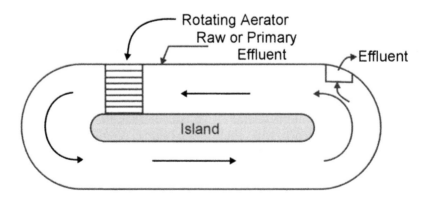

Figure 7-35   Schematic of Pasveer oxidation ditch.

Since Pasveer's time, several variations to his basic ideas have been developed. In some cases, the earthen basin has been replaced with other materials of construction, including concrete, asphalt, plastic membrane, and steel. A number of devices have been used in place of the brush aerator. All such systems, however, have in common the high surface-to-volume ratio afforded by the shallow, oval-type aeration tank configuration, enabling relatively low cost for aeration and mixing.

Deep Shaft Aeration

Two perplexing problems with activated sludge treatment systems in general, and extended aeration systems in particular, are (1) the characteristically low efficiency of transfer of oxygen from the air supplied, to actual use by the microorganisms (typically 2% to 15%), and (2) the relatively large land area required. In the case of conventional extended aeration systems, which attempt to reduce operation costs by autooxidating sludge rather than having to dispose of it by landfilling, the more extended the aeration, the more land area is required. The deep shaft aeration system was developed to solve, or significantly diminish, these two problems.

With respect to low efficiency of oxygen transfer, one of the basic causes is the low driving force for dissolution of oxygen, which, as explained earlier, is simply the difference between the saturation concentration in the mixed liquor and the concentration that actually exists. This driving force can be increased in direct proportion to the value of the saturation concentration for any given value of the actual concentration. The saturation concentration can be increased in direct proportion to the pressure of oxygen in the gas volume that is in contact with the mixed liquor, and this pressure can be increased in direct proportion to the depth beneath the surface of the aeration basin at which bubbles of gas containing oxygen exist. These facts led to the realization that if an aeration basin were configured as a very long U-tube of small diameter, and if this U-tube-shaped basin were oriented vertically so as to attain very high hydraulic pressure at the bottom, due to the very deep column of liquid above it, the saturation concentration of oxygen would be many times higher than usually experienced with activated sludge. The result would be a driving force many times greater, and transfer efficiency would increase accordingly. Thus, the deep shaft aeration system, shown diagrammatically in Figure 7-36, was developed.

Figure 7-36 shows a "deep shaft aeration system" consisting of a 20-foot diameter boring into the earth, similar to a mine shaft. A partition separates the shaft into two compartments. Raw or, in some cases, pretreated wastewater enters the compartment on the left and is drawn down toward the bottom of the shaft by a small difference in hydrostatic head between the two sides. Air bubbles are blown into the compartment containing "fresh" wastewater, only a few feet below the surface. Thus, the pressure against which the blowers work to force air into the mixed liquor is small. These air bubbles are carried down to the bottom of the shaft with the mixed liquor because the velocity of the fluid toward the bottom of the shaft is greater than the rise rate of the bubbles. As the bubbles are carried deeper into the shaft, the hydrostatic pressure becomes greater; thus, the pressure within the bubbles becomes greater. The greater pressure in the bubbles causes the saturation concentration of oxygen in the mixed liquor to increase proportionately; thus, the driving force for dissolving oxygen into water increases.

Pressure does not influence microbial activity; therefore, the bacteria and other microbes throughout the full depth of the shaft can utilize the oxygen within the air bubbles that were introduced into the mixed liquor under low-pressure conditions. The result is a profound increase in the efficiency of transfer of oxygen.

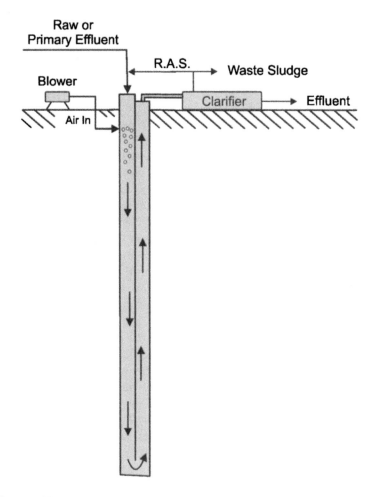

Figure 7-36   Schematic of deep shaft treatment system.

Membrane Bioreactor Modification of the Activated Sludge Process

The membrane bioreactor (MBR) modification of the activated sludge process has the objectives of (1) decreasing the tank volume necessary to accomplish treatment by providing an environment where very high concentrations of bacteria can thrive, and (2) eliminating reliance on gravity clarification. These improvements come at the expense of capital and operating cost of a membrane filtration system, which requires periodic cleaning to maintain transmembrane pressure and effective filtration. Another benefit of this system is that the concentrated mass of mixed liquor can manage fluctuating

loads, and the membrane filtration process allows all the mixed liquor to be captured and maintained in the aeration tank.

A number of manufacturers have introduced membrane systems to the marketplace. Figures 7-37, 7-38, and 7-39 show three fundamentally different configurations.

Each of these systems offers filtration surfaces with pore spaces in the range of 0.2 to 0.4 micrometers. Each needs to be placed in an agitated environment to control the buildup and plugging due to biomass accumulations, and each periodically needs to be taken out of service and cleaned using special cleaning solutions to recover the filtration capabilities. Some manufacturers offer an

Figure 7-37 A horizontal tube MBR membrane system. (Courtesy of Ionics, Inc.)

Figure 7-38 A verticle tube MBR membrane system. (Courtesy of Zenon Envronmental, Inc.)

in-place purge/cleaning system, which, it is believed, will prolong the operating life between out-of-service cleanings.

There are two popular configurations of the MBR system. The first includes a completely mixed aeration tank with membranes submerged in the tank contents. MLSS can run as high as 12,000 mg/L. The membranes are built in modules, enabling them to be removed for cleaning and maintenance while the remainder of the system stays in place for continued operation. The aeration provided to meet the oxygen needs of the MLSS is sufficient to meet the agitation/cleaning needs of the membranes during normal operation. One drawback of this system is the labor needed to remove the membrane modules

from the tank and transfer them to another tank for out-of-tank cleaning in an agitated solution (the makeup of this solution varies somewhat between membrane manufacturers).

The second configuration includes a complete mix aeration tank, again with a mixed liquor as high as 12,000 mg/L, but under this configuration there are two banks of modules placed in adjacent tanks that serve as the filtration areas. Under this configuration one of the two banks of membranes can be taken out of service for offline cleaning while the other bank provides treatment. Under this configuration the mixed liquor is removed from the offline tank, the membranes are

Figure 7-39 A flat-plate MBR membrane system. (Courtesy of Kubota, Corp.)

washed down, and the cleaning solution is added to the tank to allow cleaning to take place without removing the membranes. The drawback of this system is that aeration must be provided to both the aeration tank (to provide for the oxygen needs of the MLSS) and to the filtration tanks (to agitate and clean the membranes).

Some benefits of MBR systems include:

1. Small footprint
2. Very clean effluent, single numbers
3. Normally undesirable mixed liquor bacterial characteristics, such as filamentous or bulking sludges, are less significant in an MBR than in a conventional activated sludge process
4. Lower bacteria levels in the effluent due to microfiltration, easier disinfection
5. Modular units for ease of installation, maintenance, and replacement
6. Ability to manage fluctuating loads and flows

Some limitations of MBRs include:

1. Pretreatment needs
2. Limited ability to handle grease and oil
3. Inorganics will build to a higher level than conventional systems
4. Fine screening necessary
5. Membrane life is uncertain

Finally, things to keep in mind when considering MBR technology include:

1. Out-of-tank or out-of-process cleaning required
2. At the time of this printing, modules are more expensive than some alternatives
3. Foam and undesirable organisms get captured in the aeration tank
4. Ability to handle peak flow is limited by the number of modules online

Design and Operational Characteristics of Activated Sludge Systems

Table 7-13 presents important design and operational characteristics of five alternative configurations of the activated sludge method of wastewater treatment.

The parameters presented in Table 7-13 are "average" values, applicable to wastewaters that have concentrations of BOD, TSS, FOG, and other common parameters that are not far outside the normal ranges found in municipal wastewater. In cases of activated sludge treatment systems for industrial wastewaters that have concentrations of BOD or other characteristics that are significantly outside these ranges, laboratory and/or pilot studies must be conducted. There is no reliable way to transpose or interpolate the performance characteristics of one of these biological treatment processes (as achieved with one type of wastewater) onto another, unless the fundamental characteristics of the pollutants in the wastewater are similar.

Table 7-13   Design and Operational Characteristics of Activated Sludge Systems

Parameter System Type	HRT	Sludge Age	MLVSS mg/L
Conventional	6–8 hrs	7–12 days	2,000–3,000
Extended	20 hrs	20 days	1,200–2,500
High Rate	3–4 hrs	6–10 days	2,500–4,000
Contact Stabilization	0.5 hrs	7–12 days	2,000–3,000
Sequencing Batch Reactor	6–8 hrs	7–12 days	2,000–3,000

Aeration Systems for Activated Sludge

Air must be supplied to activated sludge systems to provide oxygen for microbial respiration. A wide range of alternative air supply systems is available, and there can be as much as a 150% difference in total annual costs from one system to another. Figure 7-40 presents an enumeration of the major types of aeration devices available. Mixing is also required in activated sludge systems, and aeration can often provide all of the mixing that is necessary. Sometimes, however, supplemental mixing is more economical.

As shown in Figure 7-40, the two principal types of aeration devices are mechanical and diffuser. The basic difference between the two is that the mechanical aerators cause small droplets of the mixed liquor to be thrown up out of the aeration tank, through the air above the tank, and back down into the tank. These mechanical devices also mix the contents of the aeration tank, with the objectives of (1) there being no "dead zones" and (2) each portion of the liquid mass in the aeration tank being thrown into the air every few minutes. Oxygen transfer takes place through the surface of each droplet. For this reason, the more efficient mechanical aerators are those that create the largest surface-to-volume ratio of the activated sludge mass per unit of energy expended per unit of time. This principle is explained in the paragraphs that follow. It should also be noted that in cold climates, mechanical aeration tends to cool the MLVSS, and diffused aeration tends to heat it. This may be a factor to consider during selection.

The driving force for oxygen transfer in the case of mechanical aerators is the gradient between the oxygen concentration in the air and the concentration of dissolved oxygen within a given droplet. As illustrated in Figure 7-41, the transfer of oxygen from the air into a droplet is a four-step process. First, oxygen diffuses through the bulk air medium to the surface of the droplet. Next, each oxygen molecule must diffuse through the double-layered "skin" of the droplet, which consists of a layer of nitrogen and oxygen molecules covering a layer of water molecules. This diffusion through the two layers can be considered one step, and is thought to be the rate-limiting step for the process as a whole. The final two steps are diffusion of oxygen into the bulk liquid of the droplet, followed by diffusion into the bulk liquid contents of the aeration tank, once the droplet returns to the tank.

The reason that diffusion through the double "membrane" at the surface of the droplet is the rate-limiting step is illustrated in Figure 7-42. Within either the bulk air or the bulk liquid, each molecule of the medium is attracted to other molecules equally in all directions (across the entire surface area of the molecule). At the interface between liquid and air, however, each molecule of the medium is attracted to other like molecules in only the directions where the like media are present (across only half of the surface area, or the bottom half in the case of the liquid molecules). Therefore, since the total attractive force is the same as in the bulk medium, but the force is distributed over only half the area, the effective attraction is essentially doubled. This causes the molecules of both gas and liquid to be more dense and, therefore, to be less permeable to the passage of other molecules.

Air diffusers introduce bubbles of air into the bulk liquid within the aeration tank, as illustrated in Figure 7-43. In this case, as opposed to the case for mechanical aerators, the oxygen transfer process is from a more or less spherical "container" of air directly to the bulk liquid. Again, the driving force for oxygen transfer is the difference in concentration between oxygen molecules in the air bubble and the concentration in the bulk liquid. There is still the process of diffusion of oxygen molecules through, first, the air, except that here, the air is contained in a small "package," which is the bubble. Next, the oxygen molecules must diffuse through the

Figure 7-40   Alternative types of devices for aeration: (a) aeration panel (courtesy of Parkson Corp.); (b) flexible membrane diffuser (courtesy of Parkson Corp.); (c) dome diffuser (courtesy of Parkson Corp.); (d) coarse bubble (courtesy of Parkson Corp.); (e) fine-bubble membrane diffuser (courtesy of Eimco Process Equipment, a Baker Hughes Company); (f) disk diffuser (courtesy of U.S. Filter/Envirex).

(g)

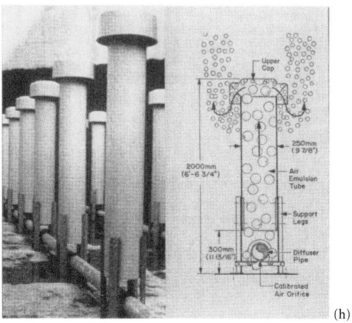

(h)

Figure 7-40   Alternative types of devices for aeration: (g) submerged static aerator (courtesy of IDI/Infilco Degremont, Inc.); (h) alternative device for aeration (courtesy of IDI/Infilco Degremont, Inc.).

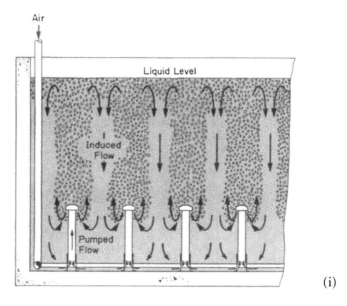

(i)

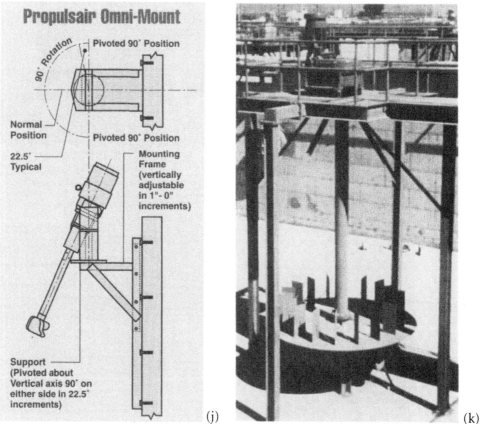

(j)

(k)

Figure 7-40   Alternative types of devices for aeration: (i) typical deep basin flow pattern (courtesy of IDI/Infilco Degremont, Inc.); (j) aeration device diagram (courtesy of Eimco Process Equipment, a Baker Hughes Company); (k) submerged turbine aerator (courtesy of IDI/Infilco Degremont, Inc.).

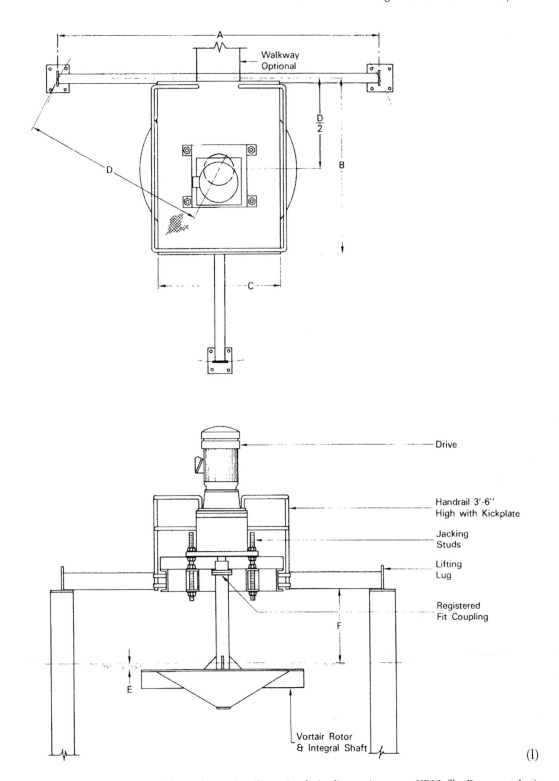

Figure 7-40   Alternative types of devices for aeration: (1) aeration device diagram (courtesy of IDI/Infilco Degremont, Inc.).

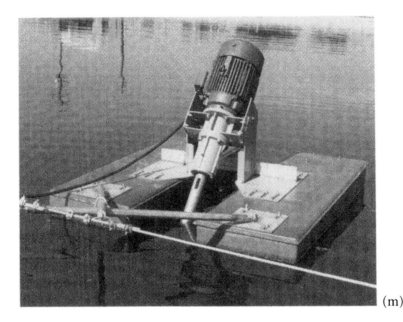

(m)

(n)

Figure 7-40   Alternative types of devices for aeration: (m) pontoon-mounted, aspirating aerator (courtesy of Eimco Process Equipment, a Baker Hughes Company); (n) fixed-mounted, aspirating aerator (courtesy of Eimco Process Equipment, a Baker Hughes Company).

double "membrane" of gas, then through liquid molecules that surround the bubble, then into the bulk liquid. Here, the process has four steps rather than five, and the rate-limiting step is still considered to be the rate of diffusion through the double-layered membrane.

Air diffusers manufactured for the purpose of supplying air to activated sludge wastewater treatment systems are divided into two categories: coarse bubble diffusers and fine bubble diffusers (also called "fine pore diffusers"). In general, coarse bubble diffusers require less maintenance than fine

(o)

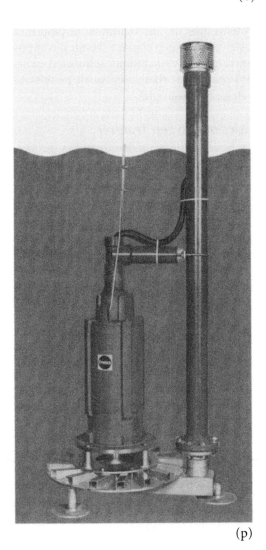

(p)

Figure 7-40   Alternative types of devices for aeration: (o) sequencing batch reactor (SBR) (courtesy of Paques ADI Systems, Inc.); (p) fixed submerged aerator (courtesy of ABS, a company in the Cardo Group).

bubble diffusers and require somewhat less air pressure to pass a given flow rate of air (therefore less power per unit of air supplied) but achieve a lower degree of oxygen transfer efficiency (OTE). Fine bubble diffusers characteristically provide higher OTE values than coarse bubble diffusers, owing to the significantly higher surface-to-volume ratio of the smaller air bubbles. Since the rate-limiting step of the oxygen transfer process is diffusion through the double layered "membrane" surrounding each air bubble, and since the flux of oxygen, in terms of pounds of oxygen per unit area of bubble surface, will be the same regardless of bubble size, increasing the bubble surface area will directly increase the transfer of oxygen.

Fine bubble diffusers have been shown to have significant disadvantages compared with coarse bubble diffusers or mechanical aerators in certain specific instances, due to a higher tendency to cause foaming and a tendency to clog or otherwise become fouled. If foaming occurs and antifoam agents are added, the antifoam agents act to cause the

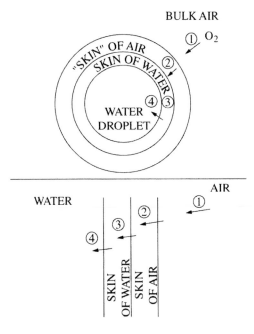

Figure 7-41   Illustration of the mass transfer of oxygen molecules from air into a droplet of water.

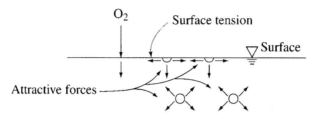

Figure 7-42   Illustration of the rate-limiting step for oxygen transfer.

fine bubbles to coalesce and become large bubbles. The tendency for fine pores to clog or become otherwise fouled results in the necessity for periodic cleaning or replacement. In addition, the lower air supply rate needed by fine bubble diffusers for the required oxygen transfer results in less air for mixing, an important component of aeration. The addition of one or more alternatives to satisfy mixing requirements, for instance, by supplying more air than is required for oxygen transfer, or making use of mechanical mixers along with the fine bubble aerators, sometimes results in the long-term economics favoring coarse bubble diffusers.

Some industrial wastes have chemical or physical characteristics that make them bad candidates for fine bubble diffusers. Sometimes, the reason is obvious. Treatment sys-

tems for potato starch processing wastewater, which foams copiously due to the types of proteins present, and treatment of pulp mill wastewaters, which contain chemical components (possibly including sulfonated remnants of lignin) that cause small bubbles to coalesce, are examples.

Basics of Oxygen Transfer

The basic relationship that describes the oxygen transfer process is as follows:

$$\frac{dC}{dt} = K_L a (C_{\text{inf}}^* - C) \qquad (7\text{-}47)$$

where

dC = Change in concentration of oxygen with time, dt

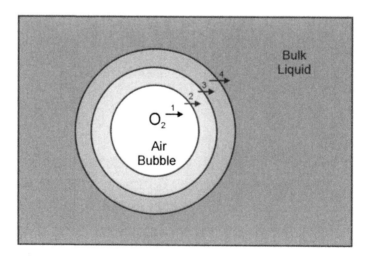

Figure 7-43   Diffusion of oxygen from inside air bubble to bulk liquid.

K_La = Overall mass transfer coefficient, made up of the coefficients K_L, the liquid film coefficient, and the unit interfacial area a.

C^*_{inf} = The saturation concentration for oxygen in water. It is the concentration to which oxygen would become dissolved in water if a gas containing oxygen was in contact with water for an infinite period of time. Its value is proportional to the mole fraction of oxygen in the gas that is in contact with water.

C = Concentration of dissolved oxygen at time t.

The value of K_La in the above equation is normally determined by experiment. C^*_{inf} and C are measured directly. K_La is deter- mined by performing a regression analysis of oxygen uptake data, obtained by conducting experiments during which oxygen concentration is measured at progressive times after aeration of a test liquid has begun, starting with a concentration of approximately zero. Figure 7-44 presents a typical graph of the concentration of dissolved oxygen in an aqueous solution after the aeration process has begun at time equal to zero, starting with a dissolved oxygen concentration of about zero. As shown in Figure 7-44, the concentration of dissolved oxygen increases at a decreasing rate (first-order kinetics) until either the saturation concentration (C^*_{inf}) is reached or an equilibrium is reached at which the rate of dissolution of oxygen is equal to the rate at which dissolved oxygen is consumed, by either microbiological respiration or chemical reaction.

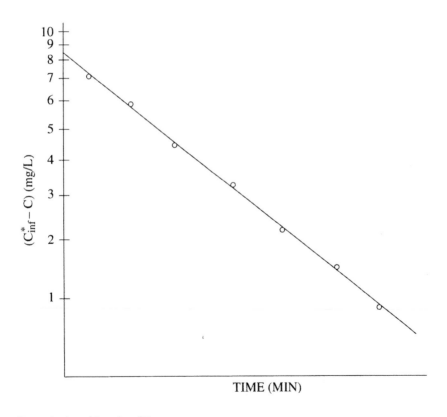

Figure 7-44   Determination of the value of K_La.

Extensive experimentation has shown that the value of K_La, the overall mass transfer coefficient, depends upon a number of characteristics and factors. In fact, within a given aeration basin in which oxygen is being dissolved into a given industrial wastewater, the value of K_La is different in different locations within that basin. Some of the characteristics and factors include:

- Temperature
- Chemical makeup of the wastewater as it changes from one location to another within the basin (the principal reason for this change is the treatment process itself)
- Liquid depth
- Barometric pressure
- Relative humidity
- Intensity of mixing or turbulence
- Variation in physical characteristics of the aeration devices at different locations within the basin (if there is, indeed, a variation; usually, there is significant variation due to different degrees of clogging and other "wear and tear" effects of service)

It is not possible, therefore, to compare the effectiveness of one aeration device with another, unless all of the influencing characteristics are equal except the aerators themselves, or unless procedures are applied to account for the differences. Such procedures have been developed and are described in *Standard Guidelines for In-Process Oxygen Transfer Testing*, American Society of Civil Engineers (ASCE) publication No. ASCE-18-96. Two approaches are described in this publication. The first is referred to as the "nonsteady-state method." The second is called the "off-gas method." A summary of these procedures is as follows.

Nonsteady-state Method

This method has the objective of determining the average oxygen transfer coefficient (K_La) under actual process conditions by measuring the change in dissolved oxygen (DO) concentration over time after producing a sudden change in the prevailing steady-state conditions. DO concentrations are then taken at successive time intervals, at one or more locations within the operating aeration basin. The DO concentration vs. time characteristics, as the contents of the basin progress toward a new equilibrium condition, are thus determined. The following assumptions accompany this procedure:

- The system is completely mixed.
- The oxygen uptake rate and K_La values remain constant during the test.
- DO probes are located so that each detects a DO concentration that is representative of equal basin volumes.

One of two methods is used to produce the sudden change in (departure from) steady-state conditions. The first is to (as suddenly as possible) change the level of power supplied to the aerators. The second is to quickly pour a volume of hydrogen peroxide (H_2O_2) into the operating basin. When the level of power is increased, the value of K_La is determined by analysis of the increase in DO concentration as it approaches the new equilibrium value. When the hydrogen peroxide method is used, the excess DO is stripped out in a manner equivalent to the manner in which it would increase. K_La is thus determined by use of a "curve," which mirrors the expected characteristics of increase to the equilibrium value. A nonlinear regression (NLR) technique is used to determine the values of K_La.

It is important to maintain constant load and oxygen uptake conditions throughout the test. This can be managed by either diverting some of the load or by conducting the test during periods when the load is expected to be constant.

Off-gas Method

This method employs a tentlike hood to capture and measure gas-phase oxygen emerging from the surface of an operating aeration basin. The mole fraction of oxygen in the "off-gas" is then compared with the mole fraction of oxygen in the air supplied to the basin. A gas-phase mass balance is then used to directly determine the oxygen transfer efficiency of the diffused air aeration devices in service at the time the test was conducted. Devices for measuring the oxygen content of air and the flow rate of air are used to determine the mass of oxygen that enters the aeration basin over a defined period of time. Equivalent devices are used to measure the mass of oxygen exiting the basin through the surface of the liquid over the same period of time. This ratio is used to calculate the mass rate of oxygen input to the aeration basin. Typically, the performance curves supplied with the air blower equipment (mass of air per horsepower-hour, for instance) are used to determine (or to check, if a flow meter is used) to determine the mass of air supplied to the aeration basin.

The tentlike device, or "hood," is equipped with pressure-sensing devices as well as oxygen concentration sensing equipment and gas flow rate measurement capability. The oxygen concentration sensing equipment is used to determine the mole fraction of oxygen in the gas that is exiting the aeration basin via the surface of the mixed liquor. The pressure-sensing devices are used to determine when the flow of gas into the hood is equal to the flow exiting the hood. This "flow equilibrium" will be shown to be the case when the pressure within the hood remains constant over a period of time. It is recommended that the hood be maintained under a slight positive pressure, with respect to the atmosphere outside the hood, i.e., + 1.27 to 2.54 mm [0.05 to 0.10 in.] water gauge.

The area of the base of the hood, expressed as a ratio of the total area of the aeration basin, is used to calculate the total mass rate of oxygen exiting the aeration basin. The difference between the quantity of oxygen entering the aeration basin in a unit of time and the quantity exiting the basin in the same unit of time is then used to determine the oxygen transfer characteristics of

the particular aeration system (blowers, air delivery system, and diffusers) used for that test.

A gas-phase mass balance of oxygen over the liquid volume is written as:

Oxygen removed from gas stream = oxygen dissolved into the liquid

or

$$\rho(q_i Y_i - q_e Y_e) = K_L a(C_{inf}^* - C) \times V \quad (7\text{-}48)$$

where

ρ = Density of oxygen at temperature and pressure at which gas flow is taking place

q_i, q_e = Volumetric rate of gas flow into and out of the volume of liquid, respectively

Y_i, Y_e = Mole fraction of oxygen in gas (air) flow into and out of the liquid, respectively

C = Concentration of dissolved oxygen in the liquid (aqueous solution)

V = Volume of basin

as illustrated in Figure 7-45.

The value of $K_L a$, then, is calculated directly by rearrangement of equation 7-48:

$$K_L a = \frac{\rho(q_i Y_i - q_e Y_e)}{(C_{inf}^* - C)V} \quad (7\text{-}49)$$

As was the case with the nonsteady-state method described above, the resulting estimates of the value of $K_L a$, oxygen transfer efficiency, and other characteristics are applicable, strictly speaking, to only those particular conditions of temperature, atmospheric pressure, chemical characteristics of the mixed liquor, and all other variables that existed when the test was conducted. The "standard oxygen transfer rate" (SOTR) equation has been developed to provide a method to enable comparison, on a best-

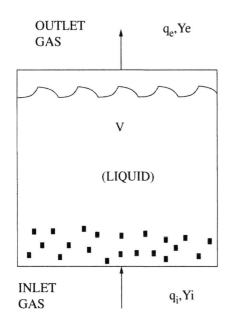

OUTLET GAS q_e, Y_e

V

(LIQUID)

INLET GAS q_i, Y_i

Figure 7-45   Gas-phase mass balance.

estimate basis, of the performances of different aeration devices by mathematically adjusting the value of K_La to account for nonstandard conditions when actual tests were conducted.

The standard oxygen transfer rate (SOTR) is defined as:

$$SOTR = K_La(C^*_{inf}) \qquad (7\text{-}50)$$

Standard conditions are considered to be as follows in the United States:

- Temperature = 2°C
- Barometric pressure = 760 mm Hg
- Tap water

In Europe, standard conditions are considered to be as follows:

- Temperature = 10°C
- Barometric pressure = 760 mm Hg
- Tap water

In the United Kingdom, a surfactant (commercial detergent) is added to the tap water.

Three factors α, β, and Θ are used to mathematically adjust the value of K_La to account for the effects of differences in wastewater characteristics (chemical content, etc.), differences in the saturation value of dissolved oxygen, and the effects of temperature, respectively. The alpha factor (α) is used to compare the oxygen transfer effectiveness in a given wastewater to its effectiveness in clean water:

$$\alpha = K_La_{ww}/K_La_{cw} \qquad (7\text{-}51)$$

where

K_La_{ww} = The value of K_La, the overall mass-transfer coefficient, in the wastewater that is under consideration

K_La_{cw} = The value of K_La in clean water, under the same conditions of temperature, barometric pressure, and relative humidity that prevailed for the determination of K_La

In other words, then, the value of alpha (α) is the ratio of the rate of oxygen transfer in the wastewater under consideration to the rate of oxygen transfer in clean water, when all other physical and environmental characteristics are equal.

Usually, the value of alpha is less than one, meaning that there are few substances normally found in wastewater that enhance the dissolution of oxygen. If one aeration device is able, somehow, to "produce" an alpha value higher than that "produced" by another (several have been reported in the literature), and if the difference in the values of K_La do not negate this effect, the aerator producing the higher alpha value is shown to be the better performer.

The beta factor, used to account for the difference in the saturation value of dissolved oxygen in a given wastewater, as opposed to the saturation value in clean water, is expressed as:

$$\beta = C^*_{inf\,ww} / C^*_{inf\,TP} \qquad (7\text{-}52)$$

The saturation concentration of the wastewater in question in contact with air can be determined by testing the wastewater in the laboratory. Published tables are used to obtain the appropriate value for the saturation concentration of oxygen in tap water that is in contact with air.

A reasonably accurate correction for differences in temperature is as follows:

$$K_L a_{(T)} = K_L a_{(20)} (\Theta^{T-20}) \qquad (7\text{-}53)$$

where:

$K_L a_{(T)}$ = Value of $K_L a$ at temperature T
  T = Temperature in °C
  Θ = 1.024

"Standard" correction factors for differences in barometric pressure and relative humidity are presented in *Guidelines for Quality Assurance of Installed Fine Pore Aeration Equipment* (ASCE, 1998).

The standard oxygen transfer rate (SOTR) is used to estimate the actual oxygen transfer rate (OTR) under "actual" (as opposed to standard) conditions by use of equation 7-54:

$$OTR = \alpha \left(\frac{\beta\, C^*_{walt} - C_L}{C^*_{inf\,20}} \right) \Theta^{T-20} \qquad (7\text{-}54)$$

where

C^*_{walt} = The saturation concentration of oxygen (from ambient air) in tap water, corrected for the increased saturation concentration at the depth of operation of the aeration devices, i.e., $C^*_{walt} = C^*_{TP} \times f_d$, where:

C^*_{TP} = The saturation concentration of oxygen (from air) in water at the prevailing temperature and atmospheric pressure
f_d = Factor accounting for increased saturation concentration at depth

$$f_d = \frac{C^*_{inf\,20}}{C_{20}}$$

C_{20} = Saturation concentration at the surface, under standard conditions
C_L = The target value of dissolved oxygen under normal operating conditions

Progressing one step further, it is convenient to determine air-flow requirements for any given wastewater treatment process by use of equation 7-55:

$$Q = \frac{SOTR}{SOTE \times \rho_a \times f_a \times 60}(100) \qquad (7\text{-}55)$$

where:

SOTR = Oxygen transfer rate under standard conditions, lb O_2/hr
SOTE = Oxygen transfer efficiency under standard conditions, %
  Q = Rate of flow of air under standard conditions, SCFM
  ρ_a = Density of air at standard conditions = 0.075 lb/ft³
  f_a = Weight fraction of oxygen in air, decimal (0.231)
  60 = Minutes per hour
  100 = Conversion from percent to decimal

Under standard conditions, then,

$$Q = \frac{96.62 \times SOTR}{SOTE} \qquad (7\text{-}56)$$

Shop Tests

An alternative method for estimating the value of K_La, as well as other parameters and characteristics of oxygen transfer effectiveness by various aeration devices, is the so-called "shop test."

A given vessel is filled with a given liquid (water or an industrial wastewater). The aerator being evaluated is activated, with the dissolved oxygen concentration equal to zero. Measurements of dissolved oxygen concentration are made at periodic time intervals, and a plot of the data (as shown in Figure 7-45) will enable determination of the value of K_La at the temperature, barometric pressure, and relative humidity that prevailed at the time of the testing. The aeration device that produces the highest value of K_La will be shown to be the most effective, as far as oxygen transfer is concerned, of those evaluated under the specific, identical conditions.

As another alternative, a slight modification of the "shop test" is to make use of a cylinder, two feet, or so, in inside diameter, mounted alongside the aeration basin, on site. The aeration device(s) to be evaluated is installed at the bottom of the cylinder, the cylinder is fitted with air-flow rate, gas phase oxygen, and dissolved oxygen sensing equipment, and the cylinder is filled with mixed liquor from the aeration basin. Tests as described above are run to determine values of both α and K_La. The primary advantages of this system are that the tests can be run on fresh MLVSS and wastewater, as opposed to having to transport these substances to the "shop," and that close control can be maintained over chemical and physical conditions, as opposed to the lack of control that attends the off-gas method described above.

Aerated Lagoons

Aerated lagoons usually consist of earthen basins equipped with mechanical or diffused aeration equipment. There is no secondary clarifier except for a quiescent zone at the outlet. There is no controlled sludge return from the bottom of this quiescent zone.

As an alternative to the quiescent zone, a separate pond is sometimes used, in which case the pond is referred to as a polishing pond. In other cases, a mechanical clarifier can be used. A considerable number of pulp and paper mills, in fact, have installed mechanical clarifiers as part of aerated lagoon systems. These alternatives are desirable if the design of the lagoon makes use of complete mix conditions as a method to avoid short-circuiting.

There are two distinct types of aerated lagoon systems: (1) aerobic and (2) partially mixed, facultative.

An aerobic lagoon must have sufficient mixing to suspend all of the solids and must have enough aeration capacity to satisfy all of the BOD removal aerobically.

A partially mixed, facultative (combined aerobic/anaerobic) lagoon requires only enough mixing to keep all of the liquid in motion. A significant portion of the biological and other solids resides at the lagoon bottom and undergoes anoxic and anaerobic degradation. Enough aeration is applied to maintain aerobic conditions in only the upper two to three feet of liquid in a partially mixed, facultative lagoon.

The advantage of a partially mixed, facultative lagoon over an aerobic lagoon is significantly lower operating cost. In a properly operated partially mixed, facultative lagoon, much of the dissolved BOD in the raw wastewater is "sorbed" (via adsorption and absorption) onto and into the microbial cells, which then settle to the bottom and undergo anoxic and anaerobic degradation. There is no requirement to supply oxygen for this degradation process. Furthermore, anoxic and anaerobic metabolisms result in the generation of far less sludge, in the form of growth of new microbial cells, per unit of BOD degraded, than is the case with aerobic metabolism.

The advantage of an aerobic lagoon over a partially mixed, facultative lagoon is in process control. There is no positive control over how much sludge settles and how much stays in suspension in a partially mixed lagoon.

Since there is no settling out of sludge solids in a properly operated aerobic lagoon, there is no lack of control. There is, however, a significant price to pay for the extra aeration capacity required for an aerobic lagoon over that required for a partially mixed, facultative lagoon.

The anoxic and anaerobic microbial activity that takes place at the bottom of a properly operated partially mixed, facultative lagoon converts organic solids within the sludge to soluble organic acids, which diffuse into the upper strata of the lagoon. These compounds are volatile and highly odorous, and constitute a potential odor problem; however, if there is sufficient aerobic microbial activity in the upper strata of the lagoon, the organic acids will be converted to carbon dioxide, water, and microbial cell protoplasm. Properly designed and operated partially mixed, facultative lagoon systems do not emit noticeable odors.

Aerated lagoons require relatively little operator attention, since there is no wasting or returning of sludge to manage sludge age. Technically, sludge age is infinite, and autooxidation is at the maximum. For this reason, the effluent TSS concentration is typically high (relatively, compared with activated sludge). Aerated lagoons eventually collect sufficient inert solids that sludge removal is required.

The frequency of sludge removal can vary from yearly to once every ten or so years. The high-cost (relatively) aspects of lagoon systems include large land area as well as the cost of construction of a liner system to protect the groundwater. In the case of partially mixed, facultative lagoons, periodic losses of solids (via the effluent) that had previously settled to the bottom, plus periodic episodes of algae blooms within the lagoon, combine to make them unreliable in complying with effluent TSS restrictions. Some facultative lagoon installations have incorporated additional facilities that are used only when needed, such as a sand filter, or a final, settling lagoon, to ensure compliance with discharge permits. Here, again, a well-developed

life-cycle cost analysis is needed to determine the wisdom of selecting any of the various lagoon alternative configurations.

Nonaerated Facultative Lagoons

Nonaerated facultative lagoons are designed and constructed similarly to aerated lagoons; but there is no aeration other than that which diffuses naturally through the surface. These systems are termed "facultative" because, properly operated, the upper third or so of the depth is aerobic, the lower third or so is anoxic to anaerobic, and the middle third phases in and out of the aerobic and anoxic states and is therefore facultative. In order to achieve aerobic conditions throughout a significant portion of the depth (one-third or so), the organic loading must be sufficiently low that the rate of diffusion of oxygen from the air above the lagoon is as high or higher than the rate of oxygen utilization by the aerobic microbial population. As with other types of lagoons, anoxic and anaerobic degradation take place within the sludge at the bottom. Organic solids are converted to dissolved organic acids, which diffuse into the aerobic region near the surface, presenting a potential odor problem. In a properly designed and operated system, however, these volatile organic acids are converted to carbon dioxide, water, and microbial cell protoplasm by the aerobic microbial population in the aerobic zone. Well-functioning facultative lagoon systems, aerated or not, do not give off significant objectionable odors.

Oxidation Ponds

Oxidation ponds are designed and constructed similarly to facultative lagoons, in that they are usually earthen basins of very long hydraulic detention time. The design objective is different from that of facultative lagoons, however. Oxidation ponds must be sufficiently shallow and have sufficiently low organic loading that aerobic conditions are maintained everywhere. For this condition to hold, the rate of production of oxygen by algae, plus the rate of diffusion of oxygen through the surface and from there to the bottom of the pond, must always be greater than the rate of utilization of oxygen by the microbes within the system as they metabolize the organic pollutants.

Because of their shallow construction, oxidation ponds are useful only in warm climates. Also, since algae are usually a principal source of oxygen, sunlight intensity is an important design consideration. The requirement for liner systems to protect the groundwater is a major cost consideration, since oxidation ponds typically have the highest area to volume ratio of the commonly used biological treatment systems.

Design of Lagoon Systems

The discussions above have shown that, in general, there is a tradeoff between detention time and aeration capacity in the design of aerobic biological treatment systems. The question, then, is, "Which combination of detention time and aeration capacity will produce the desired quality of effluent for the lowest life-cycle cost?" Factors on which the answer bears include wastewater strength,

characteristics relating to biodegradability of the organics in the wastewater, cost for land, soil type as it relates to construction costs, and temperature, as well as others.

While activated sludge systems for treatment of industrial wastes are best designed using the procedures presented previously, lagoon systems have been most successfully designed using one of the empirical approaches presented and discussed briefly in Table 7-14.

In general:

- Temperature affects both the rate of microbial respiration and the rate at which oxygen dissolves in the bulk liquid in the lagoon.
- Algae affect the quality of the effluent from a lagoon by:
  — Adding to the TSS of the lagoon effluent
  — Adding oxygen to the bulk liquid of the lagoon during the daylight hours
  — Depleting the oxygen in the bulk liquid of the lagoon during the nondaylight hours

Table 7-14   Alternate Approaches to Design of Lagoon Systems

Approach	Comment
Areal Loading	Lagoon system size is determined by simply applying an area loading rate of, for instance, 15 lb BOD/ac/d, with no more than 40 lb/ac/d in any one of several lagoons in series. Depth is a separate design parameter, which increases with colder climates. Depth is first determined to provide required detention time, then is increased to accommodate anticipated ice thickness, sludge storage, and accommodation of direct rainfall, if significant. Experience is the principal design guide.
Plug Flow	Useful when long, narrow ponds can be constructed. Construction cost is relatively high because of the high perimeter-to-volume ratio. If baffles are used to create a long, narrow flow path, considerable dead space results and must be taken into consideration regarding hydraulic detention time.
Complete Mix	Complete mix operation is useful for controlling short-circuiting. Best treatment results when three or more ponds are operated in series. Successful design depends on selecting the correct reaction rate constant, k. Value of k decreases with consequent increase in lagoon system size with decreasing operating temperature.
Marais and Shaw	A variation of the complete mix method approach. Design considerations are identical to complete mix approach.
Gloyna Method	Applicable to warm climates with high solar radiation. Assumes solar energy for photosynthesis is above saturation.
Wehner-Wilhelm	Requires knowledge of hydrodynamic equation characteristics of each pond, as well as reaction rate.

Experience has shown that the amount of algae present in the lagoon systems increases from aerobic to partially mixed, facultative to nonaerated, facultative to those highest in algae mass, oxidation ponds. In all but complete mix aerobic systems, algae are integral parts of the treatment system and must be expected. Furthermore, lightly loaded (relative to design) systems are likely to have significantly more algae. The EPA has recognized that compliance with typical secondary permit limits for total suspended solids (TSS) may be impossible for certain lagoon systems and has provided for "equivalent to secondary" alternative TSS limits on a case-by-case basis.

In the course of the normal day-in and day-out operation of a lagoon system over the full calendar year, solids, including those that were contained in the raw wastewater and the microbial solids that result from the microbial purification action of the lagoon system, and algae settle to the bottom of the lagoon to form a sludge deposit, referred to as the benthic deposit, or benthos. Anoxic/anaerobic conditions develop in the benthos, and anoxic/anaerobic bacterial action reduces the volume of the benthos and releases volatile acids to the overlying bulk liquid. This adds to the BOD load to be treated by the lagoon system. During the spring months, lagoons located in cold climates experience an overturn when the benthos rises into the bulk liquid. This is known as a "benthal release." The BOD of the bulk liquid increases significantly during these periods, typically causing nearly complete depletion of the dissolved oxygen supply. When this happens, aerobic biological treatment is limited to what can be supported by the immediate dissolution of oxygen by the aerators and by diffusion through the surface. In some locations, a similar benthal release occurs during the autumn months. Very often, under these conditions, concentrations of both BOD and TSS in the lagoon effluent will rise to exceed permit limitations. It is extremely important to include the effects of benthal release in the design of a lagoon system as well as in the plan of operation.

One solution is to design and build a three-lagoon system. Treat with two lagoons and use the third lagoon to store effluent not suitable for release (have a high- and a low-discharge pipe). The stored wastewater can be routed through the two treatment lagoons during periods of low wastewater flow.

Low-Energy Complete Mix Approach

Experience has shown that short-circuiting has been the most significant cause of failure of lagoon systems to perform as intended. Many techniques have been used to solve short-circuiting problems, including the use of baffles and the strategic placement of inlet and outlet devices. However, perhaps the most positive approach has been to employ a low-energy complete mix technique. In this approach, the shape of the individual ponds that make up the lagoon system, combined with the type of aeration device, has enabled complete mix hydrodynamics, as evaluated using chemical tracer studies.

BOD Test Considerations

When aerobic treatment systems, especially lagoons, are operated in an underloaded (relative to design) condition, there is a potential problem with use of the standard five-day biological oxygen demand (BOD_5) test for determining plant performance as well as compliance with permit limits. When the carbonaceous BOD (CBOD) has been removed, the microorganisms within the treatment system (lagoon), under certain conditions, begin oxidizing ammonia that has been liberated from proteins and other substances. This "extra treatment" leads to the development of a robust population of nitrifying bacteria (nitrosomonas and nitrobacter), which are present in the treated effluent. Consequently, when samples of the treated effluent are placed in BOD bottles, nitrification of ammonia (including that introduced with the dilution water) takes place during the five-day BOD test. The result is that the BOD_5 test will include both CBOD from the treated wastewater and an amount of nitrogenous oxygen demand (NBOD) (about 4.6 mg of oxygen are required to oxidize 1 mg of ammonia). Experience shows that a treated effluent with a CBOD of less than 10 mg/L can exhibit a BOD_5 of more than 50 mg/L using the standard five-day BOD test. For this reason, an alternative test has been developed, in which nitrification inhibitors are added to the BOD bottle in the standard five-day BOD test. This test is referred to as the CBOD test. The EPA and, essentially, all state environmental agencies can authorize this test to be used instead of the standard BOD_5 test, on a case-by-case basis. In fact, some state agencies have authorized the CBOD test to be used at the analyst's option.

Attached Growth Systems

Attached growth wastewater treatment processes are characterized by a microbiological slime mass attached to a solid surface. As the wastewater flows over the slime mass, organic substances and other nutrients diffuse into it, providing food for microbial growth. In the case of aerobic systems, oxygen also diffuses into the slime mass. Typically, the microbes grow, increasing the mass (referred to as the "slime layer"), and the slime layer develops as shown in Figure 7-46.

As depicted in Figure 7-46, organics, other nutrients, and oxygen diffuse from wastewater, which is moving over the surface of the slime layer. As the slime layer grows, it becomes so thick that oxygen becomes consumed before it can diffuse more than a few millimeters into the slime layer. An anaerobic zone thus becomes established even though the process as a whole is said to be aerobic. The anaerobic zone is usually a fortuitous development, however, because the anoxic and anaerobic activity acts to consume microbes that have grown in the aerobic zone, as well as some of the organics that

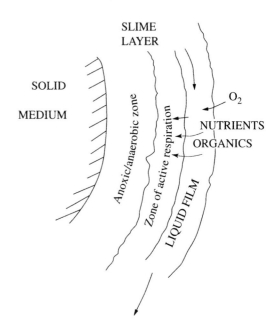

Figure 7-46 Characteristics of attached growth slime layer.

have diffused in from the wastewater. This activity reduces the quantity of waste sludge that must be handled and disposed. There is almost always some sloughing off of excess slime layer, however, and for this reason a clarifier is usually included in the attached growth treatment system.

Three aerobic attached growth wastewater treatment processes that are in common use, shown in Figure 7-47, are the trickling filter, the rotating biological contactor (RBC), and the fluidized bed.

Trickling Filters

Trickling filters make use of solid "media," usually contained in a steel or concrete tank with a perforated bottom. The tank can be round, square, rectangular, or another shape. Wastewater is applied to the top of the media by distribution devices and, under the influence of gravity, trickles down over the surfaces of the media. Air flows from the bottom up through the media to supply oxygen. Air blowers can be used for this purpose, but normally, sufficient air flows by gravity. This is because the biological activity in the filter

(a)

(b)

(c)

Figure 7-47   Aerobic fixed-film systems: (a) rotating biological contactor (courtesy of U.S. Filter/Envirex); (b) trickling filter (courtesy of NSW Corporation); (c) fluidized bed (courtesy of NSW Corporation).

gives off enough heat to cause the air to expand, thus becoming lighter and buoyant compared with the ambient air.

The earliest trickling filters (from the early 1900s through the 1950s) made use of stones of one to four inches in size for the medium over which the wastewater was caused to flow or "trickle." Stone, or in some cases slag or coal, is still used, but many trickling filter systems now use media manufactured from plastic. Also, some trickling filter systems employ horizontal flow, referred to as "cross flow." Figure 7-48 shows photographs of stone media, as well as representative types of plastic media. The plastic media are often referred to as "packing." Filters making use of stone media are normally three to eight feet

in depth. Plastic media trickling filters have been built as shallow as 6 feet and as deep as 40 feet.

The system used to distribute wastewater over the top of the media can be either moveable or stationary. A grid of stationary nozzles supplied with wastewater by pumps is common in the case of plastic media filters. So-called "rotary distributors," shown in Figure 7-49, are normally used with stone media filters. The rotary distributor is fed from the center of the circular filter. The action of the wastewater issuing from one side of the distributor arm supplies what is needed to turn the distributor, thus spreading an even application of wastewater over the top of the bed of media.

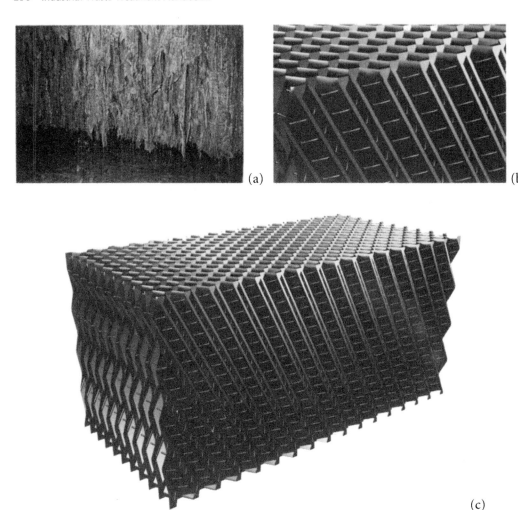

Figure 7-48 Various types of trickling filter media. (a) courtesy of NSW Corporation. (b, c) Courtesy of Marley Cooling Tower, a United Dominion Industries Company.

Recirculation

It is important to manage the operation of a trickling filter in such a way as to control the thickness of the slime layer. If the slime layer is too thin, there will be insufficient microorganisms to accomplish the desired degree of treatment. Also, a slime layer that is too thin allows the applied wastewater to flow too quickly through the filter and thus receive insufficient treatment. If the slime layer becomes too thick, the flow of air becomes impeded due to insufficient open space between adjacent surfaces of the filter media.

Successful operational performance of a trickling filter results when the organic loading rate and the hydraulic loading rate are in proper balance. The ability to recycle wastewater that has already passed through the trickling filter is the operator's best means to control this balance, as illustrated in Figure 7-50.

Figure 7-50 shows a trickling filter followed by a clarifier. A recycle pump is included so that varying effluent flow rates from the clarifier can be recycled back to the influent to the trickling filter, thus affording

Figure 7-49   Photograph of rotary distributor for trickling filter (courtesy of U.S. Filter/General Filter).

the operator a means of controlling the hydraulic loading rate to the trickling filter-clarifier system. The hydraulic loading rate will be the sum of the influent plus the recycled flow rates. As this flow rate increases, the quantity of water flushing through the filter increases. This increase has the effect of physically shearing the slime layer, reducing its thickness.

During normal operation, trickling filters are not significant sources of odors, even though, as described above, there is significant anoxic and anaerobic activity taking place within the slime layer. Normally, the aerobic microbiological activity within the few millimeters closest to the surface of the slime layer oxidizes the odorous and other

products of anoxic and anaerobic metabolism. However, if the organic loading rate is too high with respect to the hydraulic loading rate, the slime layer will grow too thick, the rate of flow of air will become insufficient to supply enough oxygen to accomplish the needed aerobic oxidization, and an odor problem will develop. Often, treatment performance becomes insufficient, as well. The problem can be corrected by increasing the recycle rate and causing hydraulic shearing of the slime layer, thus reducing its thickness.

Design

Several different approaches are in current use for designing trickling filter systems. If at

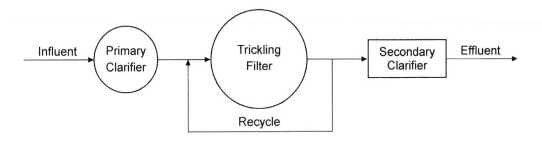

Figure 7-50   Trickling filter system with recycle.

all possible, it is best to take advantage of what has been learned by operating an existing wastewater system with as close to identical characteristics as possible. While this may be feasible for municipal wastewaters, it is often impossible in the case of an industrial wastewater. There is no substitute for a well-designed and executed laboratory study, followed by a pilot-scale program. Several manufacturers of trickling filter systems have pilot-scale treatment units that can be moved to an industrial site on a flat-bed trailer and operated for several months to obtain realistic design criteria and other valuable information. As a preliminary step in the design of a trickling filter system for an industrial application, the following methods can be used.

NRC equations, an empirical approach to designing trickling filters, were developed in the 1940s by analyzing records from a number of military installations. This approach is, therefore, more applicable to domestic wastewater, but some industrial wastes, such as certain food processing wastewaters, might be sufficiently similar in characteristics to warrant its use. The design engineer should give careful and thorough consideration to such similarities and possible differences before proceeding very far. The benefit of the NRC equation approach is the quickness and low cost of developing an initial estimate of cost for a trickling filter treatment system.

The BOD removal efficiency can be estimated by:

$$E = \frac{100}{\sqrt{1 + 0.561 \sqrt{\dfrac{W}{V \ F}}}} \qquad (7\text{-}57)$$

where

E = BOD removal efficiency (%)
W = Organic loading rate (lb. BOD/day)
V = Volume of filter media, thousands of ft^3
F = Recirculation factor (dimensionless)

The recirculation factor is determined by:

$$F = \frac{1 + R}{\left(1 + \dfrac{R}{10}\right)^2} \qquad (7\text{-}58)$$

where

R = Recirculation ratio (Q_r/Q)
Q_r = Rate of recycle flow (GPM)
Q = Rate of flow of wastewater to filter (GPM)

Usually, the organic loading rate can be determined by a wastewater characterization study, as described in Chapter 5. In the case of an industrial plant yet to be built, the organic loading rate can be estimated using information obtained from an existing facility, where it is reasonable to expect similar wastewater characteristics. Caution is warranted, however. The NRC equations have been shown to yield reasonable estimates when applied to municipal wastewater with a temperature close to 20°C. If the wastewater to be treated is expected to be of a significantly lower temperature, the expected treatment efficiency would be significantly less for a given organic loading rate.

Eckenfelder's equations, another approach to designing trickling filters, were developed by analyzing laboratory, pilot-scale, and full-scale trickling filters with various types of plastic media and treating various industrial wastes. As is the case for use of the equations presented earlier for design of activated sludge treatment systems, it is necessary to obtain a microbiological reaction rate, K, to use this approach:

$$\frac{S_e}{S_i} = e^{\left[-KS_a^m D(Q_v)^{-n}\right]} \qquad (7\text{-}59)$$

where

S_e = BOD$_5$ of effluent from filter, after clarifier (mg/L)

S_i = BOD$_5$ of wastewater applied to trickling filter (mg/L)

K = Reaction-rate constant (see below)

m = Empirical constant (see below)

n = Empirical constant (see below)

S_a = Specific surface area of trickling filter

D = Depth of trickling filter (ft)

Q_v = Hydraulic loading rate(ft^3/day/ft^2)

where

$$S_a = \frac{Surface\ area\ of\ trickling\ filter\ (ft^2)}{Volume\ of\ trickling\ filter\ (ft^3)} \quad (7\text{-}60)$$

The reaction-rate constant, K, is specific for a given depth of trickling filter. The reason for this is that as wastewater trickles down through progressive depths of the filter, the more easily assimilated organic compounds are removed first. Because of this, the "localized" rate of removal or "treatment" decreases with the increasing depth of trickling filter, and consequently the "localized" value of K decreases with increasing depth. However, the value of K for the trickling filter as a whole increases with increasing depth, since the deeper the trickling filter, the more BOD$_5$ is removed. The rate of increase in the value of K decreases with increasing depth.

This decreasing rate of increase in overall reaction rate is impossible to predict in the absence of laboratory and pilot-scale data. For this reason, values of K that are calculated from data taken at an operating, full-scale trickling filter are applicable for only the depth of that filter.

Similarly, values for the empirical constants m and n are specific for the conditions (including temperature, depth of trickling filter, wastewater, and filter media) that existed when the data from which the values were calculated were obtained. Published values of these constants, including values for the reaction-rate constant K, are useful as indicators of the relative treatability of some wastewaters, taking into account compara-ble filter depths and temperatures, but are not useful for design of new systems or retrofit of existing systems unless all of the physical and operating conditions are substantially the same. Once again, in the case of industrial wastewaters, as opposed to municipal wastewaters, there is no substitute for a well-designed and executed laboratory, followed by pilot-scale study, to generate values of design parameters for a trickling filter wastewater treatment system.

Roughing Filters

There are many instances where trickling filters have been used to reduce, or "knock down," the BOD$_5$ of industrial wasters prior to a principal, or main, treatment system. Trickling filters are useful for this purpose, because they are robust processes, are relatively resilient, and require relatively little operator attention. In short, roughing filters often present a very cost-effective alternative for significant BOD removal when reliable additional treatment facilities follow.

Roughing filters are, typically, operated at high hydraulic loading rates, in part to prevent them from becoming anaerobic and thus sources of odor nuisance, as explained above. Usually, high recycle rates are used for this purpose.

Roughing filters have been successfully used to "absorb" shock loads due to daily cleanup or occasional spills. In this role, the roughing filter protects the main treatment system and thus provides effective protection against noncompliance episodes.

Two very important effects of roughing filters on the overall wastewater treatment system are:

1. The roughing filter removes the most easily assimilated organic material. The effect on the reaction-rate constant of downstream treatment facilities must be considered in the design and operation of the main treatment system.

2. Because of the high hydraulic loading rates and the high microbiological growth rates (due to the easily assimi-

lated organics), sloughing from roughing filters is typically high. This must also be considered in the design of the main treatment system.

Usually, plastic filter media are used. Depths of roughing filters can range up to 40 feet. There is always the potential for roughing filters to become an odor nuisance. The most effective means an operator has to solve an odor problem is the recycle rate.

Rotating Biological Contactors

Rotating biological contactors (RBCs) are attached growth processes consisting of a series of parallel discs made of highly resilient plastic. The discs rotate about a metal rod, the ends of which are attached to the ends of a basin that contains the wastewater to be treated. As the discs rotate, only a portion of the disc assembly is immersed in the wastewater. While the discs are immersed, microbes that have attached themselves to the discs adsorb and absorb organic material and other nutrients from the wastewater. As each portion of each disc rotates out of the wastewater, it comes into contact with air, thus supplying oxygen for microbial assimilation of the organic substances. The emersion depth determines the relative amounts of time the discs, with their attached growth, are in contact with wastewater and air. The rotation speed determines the actual amount of time. Drive mechanisms for RBCs can be provided by mechanical systems or by air drive systems (where air drive can vary the speed of the rotation and shear forces). Usually, the depth of emersion is both a design and an operation parameter, while the speed of rotation is only a design parameter. The operator can vary the depth of emersion by adjusting overflow weirs on the basin, but seldom has the capability to vary the speed of rotation.

RBC systems can consist of a single basin containing one or more rotating disc assemblies, or can be arranged such that two or more basins, each containing one or more rotating disc assemblies, are in series. When two or more basins are in series, each is called a stage. Because the stages are separate, and because the most easily assimilated organics will be removed in the first stage and so on through the final stage, a different population of microbes will become established on each one. Also, reaction kinetics will be successively slower from the first stage to the last. This characteristic of RBC systems has been used advantageously in treatment systems where nitrification must take place. The final stage(s) can be managed so as to maintain an optimum environment for nitrosomonas organisms.

As an additional note, one or more RBC stages can be operated in the anoxic mode, following the nitrification stage, by completely immersing the rotating disc assembly in a basin of effluent from the nitrification stage. The denitrification stage, of course, should be covered. The denitrification stage can be prevented from becoming anaerobic, and thus an odor nuisance, by controlling the hydraulic residence time by use of recycle. As odors are noticed, the recycle rate can be increased, preventing anoxic respiration from going to completion.

Two problems that RBC systems have experienced, and which can be prevented by appropriate design and operation, are (1) physical failure of the shaft and/or drive assembly that turns the shaft and (2) odor nuisance.

Experience with a large number of RBC systems has resulted in knowledge of what is required, with respect to materials of construction and appropriate equipment sizing, to prevent shaft failure. The treatment process design engineer must aggressively ensure that the equipment supplier is well aware of past shaft failure problems and has specifically designed the equipment to withstand all conceivable stresses. Written guarantees are no conciliation for the devastating disruption of having to replace a rotating disc assembly.

Experience has shown that addressing the root cause of the problem can prevent odor nuisance issues, which have plagued some

RBC systems. In some cases, severe odor problems have resulted from excess growth of the microbiological slime layer. The slime layer becomes so thick that it bridges across the space between the discs, preventing air from flowing in these spaces. The consequent absence of oxygen results in anoxic and anaerobic conditions and the production of volatile acids, hydrogen sulfide, and other reduced sulfur compounds, all of which are strongly malodorous. At the lowered pH caused by the volatile acids the hydrogen sulfide becomes insoluble and passes into the atmosphere. Hydrogen sulfide can be smelled by humans in the low parts per billion ranges and can quickly cause an intolerable odor problem. This problem can be avoided by preventing the slime layer from bridging across the space between the discs. Techniques that have been successfully applied include installing a coarse bubble air diffuser (holes drilled in a pipe) at the low point of the emersion of the discs so that the bubbles travel up through the wastewater between the discs. This action causes agitation, which strips the organisms from the discs, allowing the remaining organisms access to the oxygen in the air. This technique also acts to supply additional oxygen to the microbes within the slime layer.

A second technique that has been helpful in solving odor problems associated with RBC wastewater treatment systems addresses the fact that free molecular oxygen is poisonous to the microorganisms that produce the odorous substances. Adding a small quantity of hydrogen peroxide to the influent of an RBC stage that appears to be developing an odor nuisance problem has worked well. This technique can be used while a permanent solution is being developed and implemented.

Another method to resolve odor problems has been developed by some manufacturers. This method incorporates an air drive system where cups or buckets are incorporated into the media discs such that they capture bubbles from the aeration system and cause a buoyant force that makes the shaft turn. See Figure 7-51 for a diagram of this system. The benefits of this system include:

- Shaft speed can change by varying the aeration rate.
- The agitation caused by the aeration limits the buildup of biomass on the discs.

Limitations include:

- Denitrification is not possible with an air drive system.

Design

In practice, the design of an RBC wastewater treatment facility is carried out by selecting one or more proprietary systems and working through the design procedures supplied by the vendors of those systems. Typically, the design engineer will work with a representative of the manufacturer of the proprietary system to first establish the necessary characteristics of the wastewater. In the case of an existing facility, it will probably be necessary to carry out a full-blown wastewater characterization study, as presented in Chapter 4, making sure to include the pollution prevention phases. In the case of a new industrial production facility, it will be necessary to estimate the wastewater characteristics, making maximum use of data from existing plants reasonably expected to be similar.

After design values of wastewater characteristics have been selected, the design parameters for the treatment system are determined. These parameters include the following:

1. Organic loading rate (lb BOD_5 per 1,000 ft^2 of disc surface per day)
2. Hydraulic loading rate (gal per ft^2 of disc surface per day)
3. Number of stages
4. Number of shafts per stage
5. Diameter of discs
6. Number of discs per shaft
7. Submergence depth of each rotating disc assembly

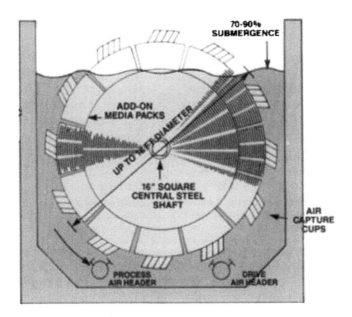

Figure 7-51   An air drive RBC. (Courtesy of US Filter Envirex Products.)

8.   Rotation rate of each shaft
9.   Hydraulic retention time for each stage

The organic and hydraulic loading rates usually follow the recommendations of the manufacturer, but the design engineer, who assumes ultimate responsibility for the successful operation of the system, must be aware of the reasonableness of the manufacturer's recommendations. The best way to accomplish this is by becoming familiar with existing installations, noting the degrees of success and problems those systems have had.

The number of stages is determined by making an estimate using all the available information, including the experience of other design engineers, of the efficiency of removal (% removal of BOD_5) in each successive stage. The first stage, of course, will remove the most easily assimilated organics and will require the most oxygen. The following stages will be characterized by successively slower treatment kinetics, and the hydraulic residence times, depths of submer-

gence, and disc rotation rates should be appropriate.

Fluidized Bed

Attached growth fluidized bed systems for treatment of industrial wastewater consist of a bed of granular material in a tank equipped with a hydraulic distribution system at the bottom and a treated effluent collection device at the top, very similar to an upflow sand filter. In fact, upflow sand filters function as fluidized bed treatment systems when their operation results in an attached growth of microorganisms [slime layer] on the sand media. Wastewater to be treated flows up through the media, usually with a pump. As the wastewater goes through the bed, its velocity and consequent drag force overcome gravity and lift the granules of media. The higher the velocity, the greater the lift.

A slime layer, consisting of microorganisms capable of assimilating the organics in the wastewater, becomes attached to the granules of media. Oxygen is made available to these microorganisms by adding air, air enriched in oxygen, pure oxygen, or another source of oxygen to the influent flow.

Because wastewater generation within a manufacturing or other industrial facility is never steady, day in and day out throughout the calendar year, it is necessary to use an effluent recycle system to maintain a constant velocity of flow. A clarifier follows the fluidized bed to separate biological material that has sloughed off from the media, as well as particles of the media itself that get carried over the effluent weir.

Design

Design of an attached growth fluidized bed wastewater treatment system is necessarily based on laboratory and pilot-scale studies.

The choice of media should take into account surface-to-volume ratio, potential reactivity with substances in the wastewater, and specific weight. Heavier media require more pump energy to attain fluidization.

Hybrid Systems

These systems are a new generation of processes that combine the strengths of suspended growth systems with those of fixed-film systems. These systems improve the capture of fines and eliminate the cloudy effluents typical for fixed-film systems, while providing the resilience of fixed-film systems and their capability to perform well in swings in flow and load and in the maintenance of system performance following shock loads. Sludge-settling characteristics are improved in comparison to some suspended growth systems. The combination allows more biomass to be incorporated in a given reactor volume, decreasing the footprint compared with conventional systems.

Moving Bed Bioreactor System (MBBR)

This system uses a complete mix activated sludge reactor with the addition of a floating inert media. The mixed liquor is removed in conventional secondary clarifiers, and return sludge is pumped to the head of the aeration tank as in conventional activated sludge systems. Control of the concentration of MLSS is accomplished through wasting sludge from the return sludge stream. A number of manufacturers offer proprietary media; see Figure 7-52. The media provide attachment sites for attached growth organisms within the MLSS.

One advantage of this system is that the volume, size, and geometry of inert media can be varied to meet the specific needs of the user in such a way that present needs may even be met without media, and growth or expansion of the industrial process can be accommodated through the addition of media. The media are retained in the aeration tank through the use of proprietary screening systems. Some of the manufacturers claim that the media cause a shearing action that enables coarse bubble diffusers to act like fine bubble systems, due to the shearing action caused by the floating inert media.

Benefits of MBBR systems include:

- More biomass per unit volume of aeration tank
- Flexibility, can add more media with industrial growth
- Improved settleability
- Higher WAS concentration
- Retrofit opportunities

Limitations include:

- Pretreatment required to eliminate plugging of the screens
- Oil and grease may coat the media

Rotating Biological Contactors (RBC) with Recycle

This system uses traditional RBC shafts in a conventional configuration and adds an aeration system and a return sludge stream to create a combination fixed-film/suspended growth system. This system is a meld of a traditional activated sludge system and an RBC system and offers similar benefits and

(a)

(b)

Figure 7-52   Typical MBBR media. (a) Courtesy of Parkson Corporation. (b) Courtesy of Anox Kaldnes.

improvements to the MBBR system described previously.

The oxygen and mixing requirements of the MLSS may be partially met by the aeration and mixing action of the mechanically driven rotating shafts. This may reduce the aeration requirements of these systems and offer a lower operation cost. The benefits and limitations are similar to the MBBR alternative, with the following additions:

Benefits of the RBC with return include:

- No screens are necessary to contain the media,

- If baffles are not installed, foam and undesirable organisms will not accumulate in the reaction tanks, and

- Low energy costs, if mechanically driven.

Limitations of the RBC:

- Traditionally, RBC basins are shallow; oxygen transfer from the diffused aeration system will be limited if new basin configurations are not provided for the aeration system, and

- If air-driven, then the aeration system necessary for the hybrid system is already in place and all that is necessary to create a hybrid and increase the treatment potential of the system is to add return sludge to the head of the RBC tank.

Treatment of Industrial Wastewaters Using Anaerobic Technologies

Anaerobic wastewater treatment, accomplished through microbiological degradation of organic substances in the absence of dissolved molecular oxygen, has undergone a complete change in role since the mid-1980s. Used for decades as a slow-rate process requiring long retention times and elevated temperatures, it was considered economically viable on only wastes of high organic strength. Its principal role in wastewater treatment was for stabilization of waste biosolids from aerobic treatment processes. It was also used as a treatment step preceding aerobic treatment, in which large, complex molecules were broken down to more readily biodegradable substances. It is now used routinely at ambient temperatures on industrial wastewaters with organic strengths as low as 2,000 to 5,000 mg/L COD. In fact, the economic attractiveness of treating wastewaters by first using anaerobic technology and then polishing with one of the aerobic technologies certainly has the potential to turn the wastewater treatment world upside down.

More recent developments have enabled use of anaerobic treatment at cold temperatures for wastewaters with COD values as low as 100 to 200 mg/L. Research conducted since the mid-1970s has shown that, by addressing the fundamental reasons for the apparently slow treatment capability of anaerobic systems, modifications could be developed to overcome them. The result has been the development of anaerobic technologies that are capable of treatment performance comparable to aerobic systems, at significantly lower overall cost. Additionally, anaerobic systems are capable of treating some substances that are not readily treated by aerobic systems, such as cellulosic materi-

als, certain aromatic compounds, and certain chlorinated solvents.

In this chapter, all microbiological mechanisms carried out in the absence of dissolved molecular oxygen (O_2), whether anoxic or truly anaerobic, are referred to as anaerobic. In this sense, the term *anaerobic* simply means "in the absence of free, molecular oxygen."

Although it can be argued that conclusive proof has yet to be produced, it is useful, in interpreting the performance of modern anaerobic treatment technologies, to assume that the fundamental reason for the apparently slow kinetics of anaerobic treatment is just that—it is a slow microbiological process. On an individual microorganism-to-organic-molecule basis, anaerobic degradation is slower than aerobic degradation. The method by which anaerobic treatment has been made capable of competing with aerobic treatment has been to greatly increase the numbers of anaerobic organisms per unit of organic matter to be treated. For instance, if aerobic metabolism is 10 times faster than anaerobic metabolism, then the time required for complete treatment by either process can be made nearly equal by increasing the number of active anaerobic organisms to ten times the number of aerobic organisms, for a given volume of wastewater.

Figure 7-53 illustrates the means by which anaerobic treatment has been successfully transformed from a slow process, compared with aerobic treatment, to one with a required hydraulic retention time that is essentially the same as what is considered normal for aerobic treatment. Figure 7-53(a) illustrates the "old" type of suspended anaerobic culture, characterized by large clumps of biological solids. The active microorganisms can be found within only a limited thickness of active biofilm on the surface. Since the surface-to-volume ratio is small, the total number of active anaerobic microorganisms is small for a given volume of reactor.

Figure 7-53(b) illustrates two of the newer types of suspended microbiological cultures.

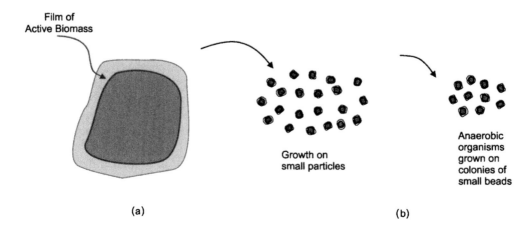

Figure 7-53   Illustration of the increase in active biomass with decreased size of film support media.

Anaerobic microorganisms are induced to grow on the surfaces of particles that are very small, compared with the clumps of biosolids shown in Figure 7-53(a), or are induced to develop small beads rather than large clumps. The surface-to-volume ratios in the (b) illustrations are orders of magnitude greater than in the (a) illustration. The result is that there are orders of magnitude more active microbes in a given volume of reactor in the newer anaerobic treatment systems.

The question arises, then, as to whether this same strategy could be used to make aerobic treatment even faster. The answer is that for aerobic treatment, the rate-limiting step becomes oxygen transfer, or getting oxygen from the outside air (or in some cases from a source of pure oxygen) to the inside of each microbe. There is no such limitation in the case of anaerobic treatment. For anaerobic treatment, the sources of oxygen are nitrate, sulfate, and other anions, already present in the wastewater and in water itself.

The principal cost-saving characteristics of the newer anaerobic treatment technologies, compared with aerobic technologies such as activated sludge, are (1) the absence of need for aeration, which represents the largest portion of O&M costs for aerobic systems, and (2) the fact that the amount of waste biosolids (sludge) that must be handled, dewatered, and disposed of is less than

that for aerobic systems by approximately a factor of ten. Added to these advantages is the cost recovery capability represented by methane. Methane recovered from anaerobic treatment processes has routinely been used as a source of energy to operate motors for pumps or for space heating, either at the treatment plant itself or in another location. As a general rule, about 5.62 ft^3 of methane can be harvested as a result of anaerobic degradation of one pound of COD.

The reason for the smaller quantity of waste biosolids is that anaerobic metabolism is much less efficient, in terms of units of cell growth per unit of organic matter metabolized, than is aerobic metabolism. Consequently, more of the organic matter being treated is used for energy, and correspondingly less is used for cell growth. For the same reason, correspondingly less nitrogen, phosphorus, and other nutrients are needed per unit of organic matter removed for treatment to take place. For most anaerobic treatment applications approximately 80% to 90% of the COD removed is converted to methane and carbon dioxide. Five percent or less becomes incorporated into new cell protoplasm, and the balance is lost as heat or refractory organic "junk."

A corollary to the characteristics of low rate of conversion of organic material to new cells and low rate of excess biosolids production is the characteristic of very long mean cell residence times, compared with aerobic treatment systems. In many anaerobic treatment systems, mean cell residence times, otherwise known as sludge age, are on the order of 100 to 200 days or more.

Two important characteristics of industrial wastewaters regarding their suitability as candidates for treatment by one of the anaerobic technologies are alkalinity and sulfur content. The anaerobic degradation of organic substances in industrial wastewaters includes conversion of complex materials to organic acids. If the alkalinity within the treatment system in insufficient, the pH will decrease to the point of toxicity to the system's microbial population. Similarly, if the sulfate content of an industrial wastewater is more than about 200 mg/L, the concentration of hydrogen sulfide, which is a by-product of the anaerobic degradation process, will increase to the range of toxicity to the system's microbial population.

Development of Anaerobic Technologies

The development of anaerobic wastewater treatment from a very slow process to a very fast treatment process is illustrated in Figure 7-54. Hydraulic retention time is used as the indication of treatment speed. The earliest anaerobic reactors were open tanks or open earthen ponds. There was no attempt to control anything. The content of the reactors was simply allowed to react at its own speed, at whichever temperature resulted, and to develop its own characteristics of biosolids, gas production, and inert solids buildup. Eventually, the organic material would become stabilized (witness the absence of massive accumulations of organics after many years).

Taking advantage of the fact that, in general, all biochemical reactions approximately double in rate for every 10°C increase in temperature, the next generation of anaerobic reactors was managed to attain faster treatment performance by heating. Usually, a degree of mixing was accomplished either as a separate objective or as a fortuitous consequence of the heating process. This also added to the increase in treatment efficiency by increasing the effectiveness of contact between the microbes and the organic material.

Mechanisms of Anaerobic Metabolism

Anaerobic treatment of organic wastes can be described as a progression of events that starts with hydrolysis, proceeds through acidogenesis, and ends with methanogenesis. These processes are symbiotic in the sense that none can proceed for very long without one or more of the others, explained as follows.

As illustrated in Figure 7-55, complex organics, such as lipids (fats), proteins, polysaccharides, polynucleotides, and aromatics, are first broken down to their elemental building blocks. Hydrolysis is the principal mechanism for this process, and there is no reduction of COD. Exoenzymes secreted by a variety of anaerobes carry out this hydrolysis. The basic building blocks include fatty acids in the case of lipids, amino acids in the case of proteins, simple sugars for polysaccharides, nucleic acids for nucleotides, and benzene derivatives for aromatic compounds. These basic building materials are further broken down, again by hydrolysis, to alcohols and then to fatty acids of relatively small molecular size. Acetic acid, plus smaller amounts of proprionic, butyric, and valeric acids, is the product of this process, which is known as acidogenesis. Molecular hydrogen is also produced during this process.

The final steps include conversion of the products of hydrolysis and acidogenesis to methane and carbon dioxide. This process is known as methanogenesis.

The portion of organic matter that becomes converted to new microbial cells is not necessarily represented in Figure 7-55. Some of the intermediate and some of the final products of hydrolysis and acidogenesis are diverted to various metabolic pathways of cell material construction. Most likely, the new cell material is made via the two-carbon

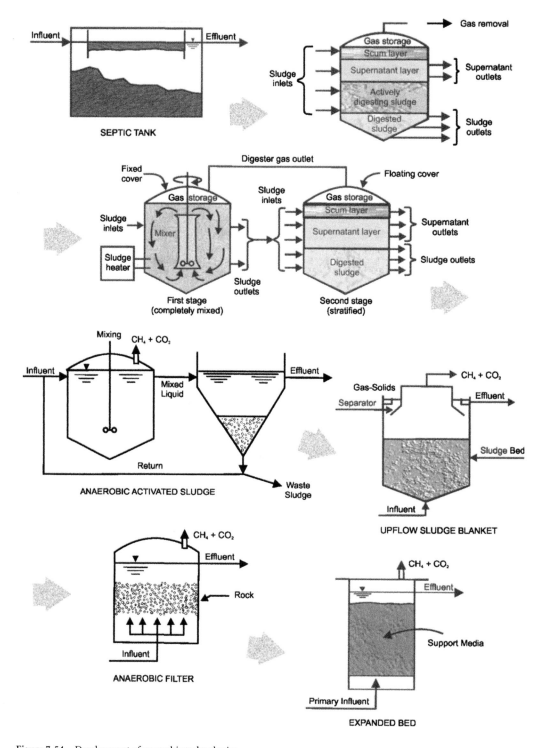

Figure 7-54   Development of anaerobic technologies.

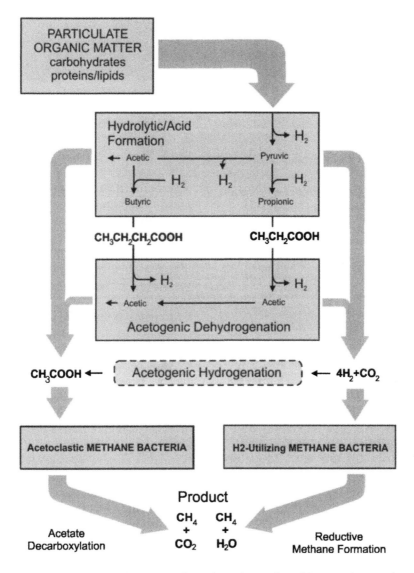

Figure 7-55   The groups of bacteria and the principal transformations performed in converting organic particulates to methane and carbon dioxide during anaerobic digestion (after Mosey and McInerny & Bryant; each rectangle represents a bacterial group).

precursor of acetate, which is ethanol, being carried into the cells. It is then acted upon by the cell's construction machinery, which includes the RNA, the DNA, and the mitochondria.

The primary product of the hydrolytic breakdown of complex organic substances is ethanol. At this point, very little, if any, COD has been removed from the wastewater, and very little, if any, energy has been captured by the anaerobes for use in reassembling some of the organic breakdown products into new cell protoplasm. The method used by most anaerobes to liberate this needed energy is to convert the ethanol to methane and carbon dioxide. This process releases almost 21 kcal per mole of ethanol converted. The anaerobes cannot convert ethanol directly to methane and carbon dioxide, however, but must first convert ethanol to acetic acid, with the consequent release of molecular hydrogen:

$$CH_3CH_2OH + H_2O$$
$$\rightarrow CH_3COO^- + H^+ + 2H_2 \qquad (7\text{-}61)$$

The acetic acid that is produced directly by hydrolysis and acidogenesis, as well as that produced as shown in equation 7-61, is converted to methane and carbon dioxide:

$$CH_3COO^- + H+$$
$$\rightarrow CH_4 + CO_2 + 6.77 \text{ kcal/mole} \qquad (7\text{-}62)$$

The energy made available by this transformation, 6.77 kcal/mole of acetate converted (minus losses due to inefficiencies), is used by the anaerobes to make new chemical bonds in the assembly of new cell protoplasm. There are many products of hydrolysis and acidogenesis other than acetic acid, however, including ethyl alcohol, propyl alcohol, propionic acid, butyl alcohol, and others. Many of these substances cannot be converted directly to methane and carbon dioxide. Current thinking is that at least three species of anaerobeic organisms are involved in a three- (or more) step process, at least one of which is an energy-consuming process. The three-step process for conversion of ethanol is shown in equations 7-63 through 7-65. Equation 7-63 shows that first ethanol is converted to acetate and molecular hydrogen, a process that consumes 1.42 kcal of energy per mole of ethanol converted:

$$CH_3CH_2OH + H_2O + 1.42 \text{ kcal/mole}$$
$$\rightarrow CH_3COO^- + H^+ + 2H_2 \qquad (7\text{-}63)$$

Then, both the acetate and the hydrogen are converted to methane and carbon dioxide, each by a different species of anaerobe:

$$CH3COO^- + H^+$$
$$\rightarrow CH_4 + CO_2 + 6.77 \text{ kcal/mole} \qquad (7\text{-}64)$$

$$2 H_2 + 1/2 CO_2$$
$$\rightarrow 1/2 CH_4 + H_2O + 15.63 \text{ kcal/mole}$$
$$\qquad (7\text{-}65)$$

As shown, 6.77 kcal of energy per mole of acetate converted is made available (minus losses due to inefficiencies) from the conversion of acetate to methane and carbon dioxide. Two and a half times more than that, 15.63 kcal/mole, is made available by the conversion of hydrogen and carbon dioxide to methane and water.

As is the case with many microbiological metabolic processes, one of the products of metabolism, as shown in equation 7-63, is highly toxic to the species that carries out the process. In the case shown by equation 7-63, the substance that is toxic to the species that carries out that reaction is molecular hydrogen. Consequently, in order for the process to continue in an anaerobic reactor, the hydrogen must be removed by the species responsible for the reaction represented by equation 7-65, almost as soon as it is formed. The two anaerobic species are thus symbiotic, since one depends on the other for food (molecular hydrogen) and the other depends upon the first to remove the hydrogen, which is toxic to it. In addition, the two species are symbiotic in that, by a means that is not fully understood at this time, some of the energy released by the reaction shown in equation 7-65 is made available and used by the species that carries out the reaction shown in equation 7-63.

In a manner similar to that shown in equations 7-64 and 7-65, propionic, butyric, and other alcohols and acids are converted to methane and carbon dioxide with the release of energy that can be used for cell synthesis. As an example, equation 7-66 illustrates the breakdown of propionate to acetate and hydrogen, which are then converted to methane and carbon dioxide, as shown in equations 7-64 and 7-65.

$$CH_3CH_2COO^- + 2H_2O$$
$$\rightarrow CH_3COO^- + CO_2 + 3H_2 \qquad (7\text{-}66)$$

Here, again, molecular hydrogen is highly toxic to the species that carries out the reaction shown in equation 7-66. The success of the overall process is dependent upon a symbiotic relationship between two anaerobic species, as described above.

Variations of Anaerobic Treatment Systems

There are two types of anaerobic wastewater treatment systems: suspended growth and attached growth, as is the case with aerobic wastewater treatment systems. Attached growth systems are commonly referred to as fixed-film (FF) systems.

Suspended growth systems are those in which anaerobic microorganisms feed upon the organic content of wastewater in a vessel or lagoon that contains no managed support medium to which the microorganisms attach. As microbial growth takes place, it is retained in the reactor by settling before the treated effluent is decanted. The microbes form particles that grow to a size dictated by the solids management characteristics of that particular system. In general, the solids management capability and characteristics differentiate between the several types of anaerobic treatment systems in common use.

Attached growth systems, otherwise known as fixed-film systems, have a support medium, often called "packing," to which the anaerobic microorganisms attach as they grow. The media can be stationary or not. Stationary media include rocks, coal, plastic or metal discs, and plastic packing. Sand is an example of media that is not stationary.

Suspended Growth Systems

Upflow Anaerobic Sludge Blanket (UASB)
Figure 7-56 presents a diagrammatic sketch of the UASB system, one of the more technologically advanced high-rate anaerobic wastewater treatment systems. These systems are capable of removing 80% to 90% of COD from wastewaters with influent COD concentrations as low as 2,000 mg/L, with

hydraulic retention times of eight to ten hours.

The distinguishing characteristic of the UASB system is the granular bead, one to two millimeters in diameter, that contains the anaerobic microbes. These active beads are developed within the anaerobic reactor by the most basic process that is characteristic of life itself, natural selection. As wastewater is induced to flow up through the anaerobic sludge blanket, hydrodynamic drag causes the blanket to be fluidized or expanded. Because the food is there, under anaerobic conditions, anaerobic microbes of all types will use it and experience growth. Those species that tend, for whatever reason, to form into solid bead-like structures will become incorporated into solids that are too heavy to be carried up and over the effluent weir. Those that do not tend to form into relatively heavy solids will be carried up, over the weir, and out of the system. Eventually, the microbes that tend to form into rather dense, bead-like solids come to predominate the biological growth and become, themselves, the sludge blanket.

If the velocity of upflow through the sludge blanket is managed by use of effluent recycle when wastewater flow rate decreases, the bead-like microbial solids that make up the sludge blanket will rub against one another and will continually roll and abrade each other's surfaces. This action keeps the sizes of the individual beads to within the desired range and maintains the desired high value of "active microorganism-to-organic substance," which accounts for the high performance of the system.

As illustrated in Figure 7-56, the principal components and operational characteristics of the UASB system are:

- An influent distribution system
- A sludge "blanket" consisting of beads of active anaerobic (and/or anoxic) microorganisms, formed as described above
- A gas collection system

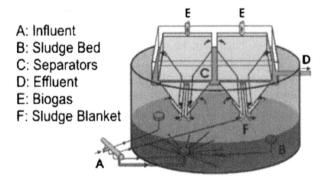

A: Influent
B: Sludge Bed
C: Separators
D: Effluent
E: Biogas
F: Sludge Blanket

Figure 7-56   Upflow anaerobic sludge blanket system (USAB) (drawing courtesy of Biothane Corporation).

- An effluent collection and discharge system that excludes air from the interior of the reactor

As influent wastewater enters the reactor via the influent distribution system, it flows up through the sludge blanket. Depending on the rate of flow, the velocity of the rising influent will cause a certain amount of expansion of the sludge. Furthermore, depending on the cross-sectional area of the sludge blanket, there will be a certain variability in the distribution of the influent wastewater. There is a choice to be made here. For a given volume of sludge blanket (i.e., quantity of active microorganisms), the smaller the ratio of the cross-sectional area of sludge blanket to its depth, the more uniform will be the distribution of influent and the greater the head against which the influent wastewater pump(s) must pump. Recently, design practice has favored deeper sludge blankets of relatively small cross-section.

Mixed, Heated Anaerobic Digester

The mixed, heated anaerobic digester, usually arranged in two stages, as illustrated in Figure 7-57, is what could be called "the typical 'high-rate' sludge digester." It represents an advanced version of the "old" anaerobic treatment technology, in which only mixing and temperature elevation were used to reduce required hydraulic retention time. The principal objective of mixing was to improve contact between active microbes and organic material, often in solid form. The objective of heating was simply to take advantage of the fact that almost all microbial metabolism doubles in rate for each 10°C rise in temperature. As an attending benefit, some organics, more soluble at the elevated temperature, are more readily metabolized because of their more direct availability to the microorganisms.

Methane harvested from the treatment process itself is normally used to heat the digester contents. This accounts for the fact that the process is simply not economically feasible if the organic content is less than that represented by 8,000 to 10,000 mg/L COD.

As shown in Figure 7-57, the first stage of a mixed, heated anaerobic digester is the reactor. Nearly all of the anaerobic degradation of organics takes place in the first stage. The second stage is not mixed. This stage normally has a floating cover, and gas produced in the first stage is piped to this vessel. The functions of the second stage are solids separated by sedimentation, sludge storage, and gas storage. Clarified supernatant is decanted and, normally, returned to the head of the treatment system, possibly representing a significant organic and TSS load. Gas and stabilized solids are periodically transferred to processing and final use or disposal.

Anaerobic Contact Reactor

The anaerobic contact reactor is a technologically advanced variation of the "old" mixed, heated anaerobic digester. It is technologi-

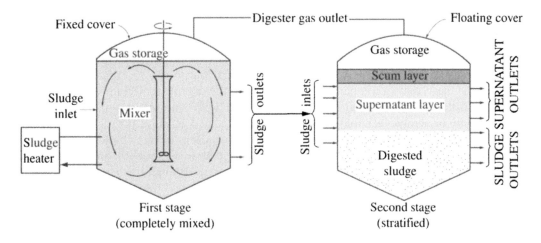

Figure 7-57   Mixed-heated anaerobic digester.

cally advanced in that it attempts to maintain a high ratio of active microbes-to-organic matter by providing sufficient mixing energy to keep the size of biological solids small by shearing action and by returning "seed" organisms to the reactor from the clarifier. Hydraulic retention times of less than a day have been reported with treatment performance approaching 90% removal of COD.

As shown in Figure 7-57, the mixed, heated anaerobic reactor is followed by a clarifier. If a plain sedimentation clarifier is used, it is normally preceded by a degasifier. Otherwise, a vacuum flotation solids separa-

tion device may be used. The raw or pretreated wastewater enters the anaerobic reactor, which is normally operated in the completely mixed mode (CMR). Active anaerobes are continually recycled back from the clarifier, maintaining high-rate treatment kinetics.

Anaerobic contact reactors have proven to be of significant value as a first stage for removing COD from high-strength wastewaters, such as those from meat packing and rendering plants. Subsequent stages have usually been aerobic. The impressive cost effectiveness of the anaerobic first stage is

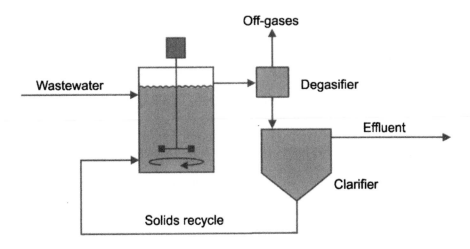

Figure 7-58   Anaerobic contact reactor.

due to (1) the rapid hydrolysis of complex organics to simple, easy-to-treat compounds with very little energy input (equation 7-61), (2) the small quantity of excess biosolids that results, and (3) the advantageous use that is made of the heat that accompanies some industrial wastewaters. This heat would be detrimental to the aerobic technologies, since the saturation concentration of oxygen, and thus the driving force to dissolve oxygen into the mixed liquor, is significantly lowered.

Fixed-Film Systems

Expanded Bed Reactor

The expanded bed system is among the most technologically advanced anaerobic wastewater treatment options. This technology, illustrated in Figure 7-59, was developed with the objectives of (1) achieving the maximum possible active microbe-to-organic matter ratio, (2) optimizing the effectiveness of contact between organic substances and microbes, and (3) minimizing the energy requirement to expand, or "fluidize," the bed, as well as to pump wastewater through the system.

As illustrated in Figure 7-59, the objective of maximizing the (F/M) ratio has been achieved through use of a "packing medium," that has a surface-to-volume ratio which is as large as possible. The objective of maximizing the efficiency of contact between microbes and organic matter was achieved by utilizing the upflow (fluidized bed) configuration. The objective of minimizing the energy needed to fluidize the bed (expand the bed, in this case) has been achieved by utilizing material of low specific weight (specific gravity only slightly greater than one) as the packing medium.

Referring again to Figure 7-59, raw or pre-treated wastewater blended with recycled effluent is pumped into the bottom of the reactor. A distribution system distributes this mixture as uniformly as possible across the full cross-section of the reactor. The wastewater then flows up through the packing medium, which is coated with a thin film of active anaerobic microbes. As mentioned above, the velocity of upflow of wastewater being treated has several purposes, and its value is critical to the success of the system. The first purpose is to expand the bed and cause the organic molecules in the wastewater to follow a tortuous path, contacting many coated grains of the medium. The second purpose is to carry biological and other solids that were knocked off the grains of medium up through the bed. The third purpose is to cause the grains of medium to abrade against one another, rubbing off excess biological and other solids and, thus, maintaining the active microbes on the medium in as thin a film as possible.

The velocity of upflow through the expanded bed system, coupled with the recycle rate, also determines the hydraulic retention time (HRT). The rate at which microbial solids are produced, removed from the grains of medium by the abrasive action of the expanded bed, and carried up through and out of the bed determines the mean cell residence time, or sludge age.

As the biological and other solids emerge from the top of the bed, the upflow velocity of flow decreases, due to the absence of solid particles of the medium from the cross-section of the flow. Here, the solids are not carried upward, but "settle." Solids thus accumulate in the settling zone above the expanded bed and must be removed at a rate that precludes them from being swept over the outlet weir at the top of the reactor.

Methane and carbon dioxide gases are continually produced within the microbial films surrounding the particles of bed medium. These gases result in bubbles that grow in size until their buoyancy carries them up and out of the bed. They then rise more slowly through the solids settling zone and are collected by a device located above the settling zone.

Fluidized Bed Reactor

The fluidized bed reactor resembles the expanded bed system in physical appearance and differs principally in three respects: (1)

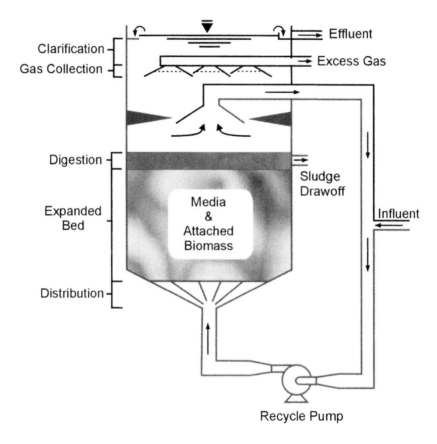

Figure 7-59   Anaerobic attached film expanded bed reactor (AAFEB).

the media used are typically much heavier (sand rather than diatomaceous earth), (2) the velocity of wastewater upflow is significantly higher (needed to fluidize the heavier media), and (3) the amount of expansion is significantly greater (20% rather than 5%).

Figure 7-60 shows a diagram of an anaerobic fluidized bed wastewater treatment system. As shown in Figure 7-60, raw or pretreated wastewater, combined with recycled effluent, is pumped through a distribution system at the bottom of the reactor and up through the media, which might be sand. The upflow velocity serves to expand the bed, which allows any given molecule of water or organic material to take a tortuous path through the bed, contacting many grains of media and, consequently, the microbial film attached thereto. The upflow velocity of the wastewater also causes the grains of the

media to constantly rub against each other, and the abrasive action keeps the microbial film from growing too thick and from bridging between grains to form blockages to flow. Biological and other solids that are knocked loose from the media are carried up through the media by the relatively high velocity of the upflowing wastewater.

As the wastewater emerges from the top of the bed of sand (or other media), the effective cross-sectional area for the flow increases greatly due to the absence of solid media. Consequently, the upflow velocity decreases so that it no longer can carry solids upward. The solids thus "settle" in the zone just above the bed.

As methane and carbon dioxide gases are produced within the microbial films on the surfaces of the media, bubbles form and grow in size until their buoyancy carries

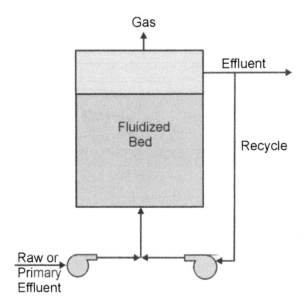

Figure 7-60 Anaerobic fluidized bed reactor.

them up through and out of the bed. These gas bubbles then rise through the solids settling zone and are collected at the top of the reactor. The type of gas collection device is one of the characteristics that distinguishes between different proprietary fluidized bed systems.

Packed Bed Reactor

An anaerobic packed bed reactor consists of a vessel (round steel tank, rectangular concrete basin, earthen lagoon, etc.) filled with a stationary solid medium to which microbes can attach and through which wastewater flows and comes into contact with the microbes. An anaerobic packed bed reactor can be operated in either the upflow or downflow mode, and effluent recycle can be used to even out variations in raw wastewater flow rate. Many different types of packing can be used, but the most common are (1) stones and (2) plastic devices of various shapes. The design objectives of the plastic media include high surface-to-volume ratio, structural strength, and being nonreactive with any chemical that might be in the wastewater. Figure 7-61 shows a diagram of a packed bed reactor.

Anaerobic Lagoons

Anaerobic lagoons are subjected to such a high organic loading that anoxic or anaerobic conditions prevail throughout the entire volume. The biological treatment processes that take place are the same as those that take place in anaerobic digesters. However, there is no mixing, no heating, and no attempt to control or manage either the size or location of the "clumps" of biological solids that develop. Consequently, the progress of treatment is relatively slow, although highly cost effective in many applications.

Basically, the typical anaerobic lagoon is a relatively deep earthen basin with an inlet, an outlet, and a low surface-to-volume ratio. If the basin is not excavated from soil of very low permeability, it must be lined to protect the groundwater below. Because of the ever-present potential for problems due to odors, it is usually necessary to cover anaerobic lagoons. Covers have been manufactured from synthetic membranes and Styrofoam. In some cases, the layer of solids that formed on the surface of the lagoon due to floating greases, oils, and the products of microbial metabolism (the scum layer) has successfully prevented intolerable odor problems.

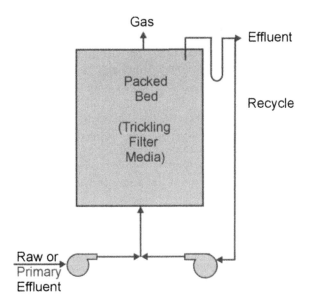

Figure 7-61   Anaerobic packed bed reactor.

Organic loading rates reported for anaerobic lagoons have varied from 54 to 3,000 pounds of BOD_5 per acre per day. BOD_5 removal performance has varied from 50% to 90%. Lagoon depths have varied from 3.5 to 7 feet, and hydraulic detention times have varied from 4 to 250 days. More typical organic loadings have been in the range of 1,000 pounds of BOD_5 per acre per day, with BOD_5 removals in the range of 70% to 80%. Depths have been typically in the range of five to seven feet. More typical hydraulic detention times have been in the range of 30 to 50 days, and have varied with climate, being longer in colder climates.

Physical Methods of Wastewater Treatment

Physical methods of wastewater treatment accomplish removal of substances by use of naturally occurring forces, such as gravity, electrical attraction, and van der Waal forces, as well as by use of physical barriers. In general, the mechanisms involved in physical treatment do not result in changes in chemical structure of the target substances. In some cases, physical state is changed, as in vaporization, and often dispersed substances are caused to agglomerate, as happens during filtration.

Physical methods of wastewater treatment include sedimentation, flotation, and adsorption, as well as barriers such as bar racks, screens, deep bed filters, and membranes.

Separation Using Physical Barriers

There are many separation processes that make use of a physical barrier through which the target pollutants cannot pass, simply because of their size. These physical barriers are classified according to the size of the passageways through which all but the target pollutants (and larger) can pass, and they range from bar racks to reverse osmosis. Bar racks, screens, and sieves are considered to be either part of the headworks or part of primary treatment, while filters, microscreens, dialysis processes, and reverse osmosis are normally considered either secondary or tertiary treatment, depending on specific use.

In the cases of the physical barriers with smaller pores (filters and microscreens, for instance), particles are often caught on the barrier bridge cross the openings and form a filter themselves. For some substances, this surface filter is very effective with respect to degree of suspended solids removal, and builds to appreciable thickness. Often, this process contributes to longer filter or screening runs. Some substances, however, are prone to clogging the surface of the filter or screen and are thus not appropriate candidates for filtration or screening.

Racks and Screens

Screening is a physical treatment method that uses a physical barrier as the removal mechanism. Screening ranges from coarse bar racks, used to remove objects of an inch or more in size, to microscreening, used to remove particles as small as macromolecules (ten or so millimicrons in size).

The success of screening as a treatment method depends on the appropriateness of the mesh size of the screening device compared with the sizes of the target substances; the clogging characteristics of the material removed on the screen; and the self-cleaning, or nonclogging, characteristics of the screening device. Bar racks can either be equipped with mechanical self-cleaners or must be periodically cleaned by hand. Smaller screens, including microscreens, are more difficult to clean, and the clogging characteristics of the wastewater must always be taken into account.

Screening is not to be confused with filtration in regard to mechanism of removal. Screening employs the mechanism of physical barrier as the primary removal method. Filtration employs adsorption, sedimentation, and coagulation in conjunction with the physical barrier.

All screening devices that have mesh sizes smaller than one-half inch must be pilot tested before design. The objectives of the pilot test program should include the clogging characteristics of both the screen and the materials in the wastewater, the build-up of head loss over time, and the potential for damage to the screen.

Bar Racks

Bar racks, often the first treatment devices to be encountered by wastewaters en route to renovation, have the primary purpose of protecting pumps and other equipment from damage. These devices remove objects such as pieces of product or raw material, broken or dropped items of maintenance equipment, gloves, plastic wrapper material, or other foreign objects that inadvertently gain access to the industry's system of drains and sewers. A self-cleaning feature is considered well worth the extra cost, since organic and other materials that are objectionable in

Figure 7-62 Bar rack with automatic moving mechanical cleaning system (photo courtesy of U.S. Filter).

many ways, including production of odors, tend to accumulate on these types of equipment. Materials of construction, as they relate to corrosiveness of both the wastewater and the atmospheric environment in which the equipment must operate, are vitally important design considerations.

Two types of bar racks in general use include racks with stationary bars equipped with a moving automatic mechanical cleaning system, and so-called "rack conveyors," which have a slotted conveyor that passes all but the target objects. These objects are conveyed away from the wastewater by "fingers" attached to the slotted conveyor. Figures 7-62 and 7-63 show examples of a bar rack with an automatic moving mechanical cleaning system and a rack conveyor, respectively. Rack conveyors have been used with exem-plary success at fish processing plants, tanneries, woolen mills, and sugar refineries, to name a few.

Vibrating Screens

An example of a vibrating screen, used widely and successfully at vegetable processing plants, metal milling shops, and poultry processing plants, is the so-called "Sweco screen," pictured in Figure 7-64. This screen operates by receiving a wastewater stream at its center. The vibrating action of the screen causes particles or objects larger than the mesh size of the screen to migrate to the periphery, while everything else goes on through the screen. Alternative types of vibrating screens include inclined vibrating screens and drum-type vibrating screens.

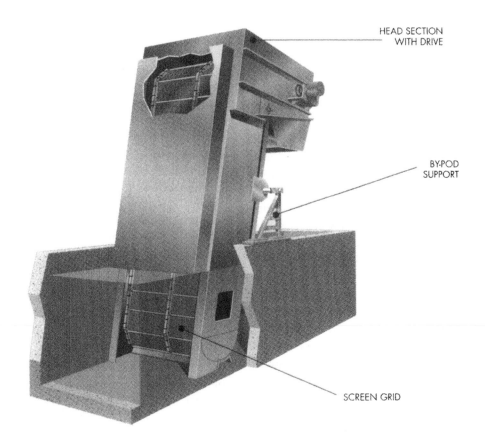

Figure 7-63   Rack conveyor (photo courtesy of U.S. Filter).

Figure 7-64   Vibrating screen (courtesy of SWECO).

Tangential Screens

Tangential screens, also known as "sidehill screens," have the advantage of having no moving parts. Various devices have been used to clean these screens while in use, including water sprays, steam blasts, and air blasts.

The tangential screen shown in Figure 7-65 operates in the following manner. Wastewater enters the reservoir at the back of the screen and, after filling the reservoir, overflows the top of the screen and flows down over the inclined face of the screen. Water and solids smaller than the mesh size of the screen proceed through the screen to the collection device, and are then conveyed to the next treatment step. All other solids slide down the incline to be collected at the bottom. The name derives from the fact that the wastewater actually approaches the screen tangentially as it overflows the reser-

voir at the top of the screen and down the incline until it proceeds through the screen.

Rotating Cylindrical Screens

Rotating cylindrical screens are designed to continuously clean themselves, using the flushing action of the screened water itself. Figure 7-66 shows a photograph of a typical rotating cylindrical screen. Figure 7-67 illustrates how the self-flushing action cleans the rotating screen on a continuous basis. As shown in Figure 7-67, wastewater enters the reservoir on one side of the rotating screen. Water and solids that are smaller than the mesh size of the screen flow on through the screen, while larger solids are carried on the surface of the rotating screen, over the top, down the side, and are removed by the doctor blade located on the side of the screen cylinder opposite the raw wastewater reservoir. Small solids that are caught in the spaces between the wedge-shaped bars that

Figure 7-65   Tangential screen (courtesy of Hycor).

make up the screen are knocked loose and reenter the wastewater (downstream of the screen itself) by the wastewater that has just gone through the screen, as shown in Figure 7-67. The wedge shape of the bars of the screen and the action of the screened wastewater flushing back through the screen in the direction opposite to that of the screening action accomplish the continuous cleaning as the screen operates.

Figure 7-66   Rotating cylindrical screen (courtesy of Hycor).

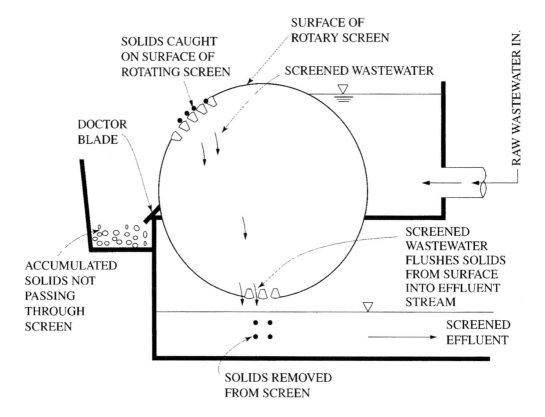

Figure 7-67 Rotating cylindrical screen (self-cleaning process).

Microscreening

Microscreening, often called microstraining, is a physical treatment process that closely resembles the rotating screen described above. The major differences are the size of the screen mesh openings and the flow path taken by the wastewater being screened. Screen mesh openings used by microscreens are in the range of 20 to 35 m. The filter medium is usually a fabric made of synthetic fiber.

As shown in Figure 7-68, the microscreen is a rotating drum with filter fabric around its periphery. The drum is mounted into one wall of a reservoir of wastewater, which enters the drum through that end. The wastewater then flows through the filter fabric from the inside to the outside of the drum. Solids that are too large to pass through the filter fabric are removed from the fabric by a mechanism at the highest part

of the rotating drum. These collected solids flow via a trough to a collection tank. The screened wastewater flows over a weir, which, as illustrated in Figure 7-68, serves to maintain a certain degree of submergence of the microscreen. Because of this submergence, a certain amount of screened wastewater flows back through the filter fabric (from the outside to the inside), dislodging particles that have been wedged into the mesh. In addition to this filter fabric cleaning action, there is usually a system of spray nozzles that cleans the fabric. In total, the fabric cleaning process consumes from 2% to 5% of the wastewater flow that has already been screened.

Microscreens have been used to remove algae and other solids from oxidation pond effluents, as well as from other types of wastewater treatment lagoons. Reported removal performance has usually been in the range of 50%. Hydraulic loading rates have

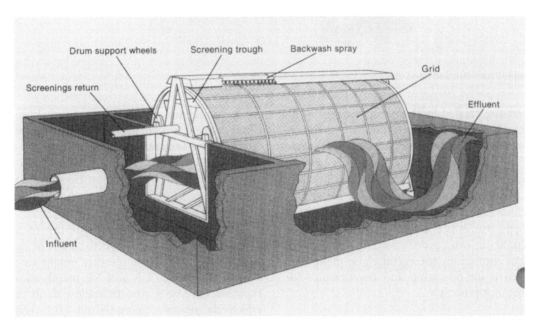

Figure 7-68   Microscreening process (courtesy of U.S. Filter/Envirex).

been in the range of 75 to 150 gallons per square foot of filter medium per minute (gal/ft^2/min.).

Plate and Frame Filters

Plate and frame filter presses, normally thought of in the context of waste treatment as sludge dewatering devices, have been used with excellent results as separation processes for precipitated metals and other substances, immediately following flocculation. A plate and frame filter can also be used, sometimes with and sometimes without precoat and/or body feed, as shown in Figure 7-69.

Figure 7-69 presents an illustration of a plate and frame filter press. The feed slurry, which may contain diatomaceous earth or another "body feed," such as "perlite," enters the center of each framed cavity. The liquid is forced by pressure created by the feed pump to proceed through the filter media, which can consist of fabric made of natural, synthetic, or metal fibers. The target particles that, by virtue of their size, cannot go through the filter media (fabric plus body feed, if any) accumulate within the framed cavities until the cavities are full, or until the resistance to flow equals or nearly equals the

Figure 7-69   Plate and frame filter press (courtesy of U.S. Filter/JWI).

pressure created by the pump. At this point, the filter run is stopped, and the plates are opened to discharge the accumulated cake.

A principal advantage of using a plate and frame filter press immediately following flocculation of a precipitated or coagulated substance is that the product is in an easily handled form. It is also a noteworthy advantage that there is no gravity-settling device used, thus eliminating the expense and manpower needed for operation and maintenance for that unit process.

Membrane Separation

Membrane separation processes include microfiltration (MF), ultrafiltration (UF), nanofiltration (NF), reverse osmosis (RO), dialysis, and electrodialysis. They make use of one or more membranes, which can be thought of as physical barriers between phases, through which only limited components of the phases can pass. These processes are used to separate molecular or ionic species from waste streams. The properties of pollutants to be removed that bear upon the appropriateness of these processes, or, indeed, that influence the selection of one of these processes over the others, are particle size, diffusivity, and/or ionic charge. Each of these membrane processes functions by way of something in the wastewater stream pass-

ing through one or more membranes and being concentrated in a stream known as the "reject." The treated effluent is less concentrated in the target pollutants than the influent by an amount that is proportional to the driving force as well as to the length of time over which the treatment takes place.

The rejection (removal) capability of a given membrane is rated on the basis of its nominal pore size, or molecular weight cutoff (MWCO). MF, UF, and NF membranes typically have pore sizes that allow them to effectively remove micron-sized (10^{-6} m) substances from water. UF membranes remove substances that are 10^{-9} m and larger.

Nanofilters are usually rated by the smallest molecular weight substance that is effectively removed. These substances include organic molecules. Molecular shape and polarity are influential, as well as molecular size, in determining removal effectiveness.

Reverse osmosis, also referred to as "hyperfiltration," effectively removes ion-sized substances such as sodium, calcium, sulfate, chloride, and nonpolar organic molecules.

Membrane materials are made from organic polymers, such as cellulose acetate, polysulfone, polycarbonate, and polyamide, and inorganics, including aluminum oxide (ceramic) and porous carbon. The preparation of a given polymer can be varied to produce membranes of different porosity, or MWCO. These membranes can be configured in stacks of plates, spiral-wound modules, or bundles of hollow fibers. Figure 7-70 shows several different configurations in which commercial membranes can be obtained.

Probably the greatest deterrent to more widespread use of membrane processes is the tendency of the membranes to become fouled, or blocked, by colloidal and other substances in wastewater. This phenomenon, sometimes referred to as concentration polarization, leads to flux inhibition, or

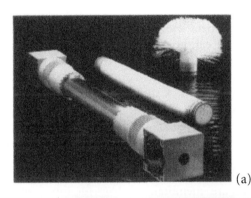

(a)

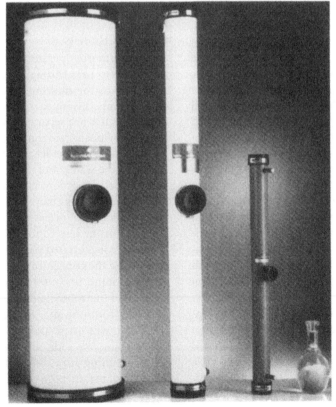

(b)

Figure 7-70 Commercial membrane configurations. (a) Courtesy of U.S. Filter/Memcore. (b) Courtesy of IDI/Infilco Degremont, Inc.

reduction in throughput. Various techniques have been, and are being, developed to combat this problem, including flow perturbations such as back-flushing and pulsing.

Another problem with use of membrane processes for wastewater treatment is that of biological fouling. Various microorganisms can become deposited on membrane surfaces, where they can flourish as a result of fresh nutrients constantly being transported to them by action of the treatment process itself. Sometimes, this problem can be controlled by disinfection of the wastewater by chlorination, ultraviolet, or other treatment, upstream of the membrane system. Care must be taken to avoid damage to the membrane by the disinfection activity.

Removal Mechanisms

As explained above, microfiltration, ultrafiltration, and nanofiltration are membrane separation processes that differ principally in the size range of the target substances. These filtration processes involve physical filtration of wastewater by using pressure to force water molecules through a membrane that is not permeable to the target substances. These target substances can vary in size from small particles to molecules. While this description can be applied equally well to reverse osmosis, the two processes differ in respect to the types of membranes used. There is only very small osmotic pressure to overcome in the case of ultrafiltration, for instance.

The membranes used for ultrafiltration are not of the "semipermeable" type associated with osmosis. Therefore, although pressure is the principal driving force for both reverse osmosis (RO) and ultrafiltration (UF), the required pressure is significantly less and the cost for power is less for UF. Typical operating pressures for ultrafiltration systems vary from 5 to 100 psig, compared with 300 to 1,500 psig for RO systems. While reverse osmosis has the ability to remove dissolved ions such as sodium and chloride, as well as organic molecules and nondissolved solids, these filtration processes do not remove low-molecular-weight ions and molecules. Ultrafiltration and nanofiltration are useful for removing higher-molecular-weight organic and inorganic molecules as well as nondissolved solids.

Ultrafiltration has been used in "direct filtration" of precipitated metal ions. This process involves addition of a precipitating agent, such as sodium hydroxide or sodium sulfide, followed directly by ultrafiltration. The coagulation and sedimentation steps are eliminated. Ultrafiltration has also been used to clarify dilute colloidal clay suspensions, as well as to remove microorganisms.

As with all other applications of treatment devices applied to industrial wastewater, an extensive pilot program must be conducted prior to final design. For the purpose of evaluating preliminary alternatives, it is probably best to use recommendations of the various manufacturers and equipment representatives for this developing treatment technology.

Reverse Osmosis

Reverse osmosis operates by allowing water molecules to pass through a membrane that will not pass the molecules or ions regarded as pollutants. In any system where two volumes of water are separated by a membrane that is pervious to water molecules but not to the particular molecules or ions that are dissolved in the water, and the concentration of those molecules or ions is greater on one side of the membrane than it is on the other, water molecules will pass through the membrane from the less concentrated volume to the more concentrated volume, in an attempt to equalize the concentration on each side of the membrane, in conformance with the second law of thermodynamics. This movement of water from one side of the membrane to the other will cause the depth of water to increase on one side and decrease on the other, resulting in a differential head, thus, differential pressure, on one side of the membrane with respect to the other. The differential pressure will counteract the tendency of water to move across the membrane until the point is reached that the differential pressure resisting movement of water through the membrane equals the "pressure" caused by the desire of the system to equalize the concentration of all dissolved substances on each side of the membrane. At that point, equilibrium is reached, and the net movement of water across the membrane will be zero. The differential pressure at equilibrium will be equal to the "osmotic pressure," and is directly proportional to the difference in concentration of dissolved substances in the two water volumes.

The process described above, of course, is "osmosis," and it takes place whenever volumes of water of different concentrations in one or more dissolved substances are separated by a membrane (known as a "semipermeable membrane") that is permeable to water but not to the dissolved substances. If the water on one side of a semipermeable membrane is very low in dissolved solids, and wastewater is placed on the opposite side, there will be a relatively strong osmotic pressure tending to drive water molecules from the clean water side to the dirty water side. Now, if pressure greater than the osmotic pressure is imposed on the dirty water side, the osmotic pressure will be overcome. Because the membrane is permeable to water molecules, the pressure will force water through the membrane, from the wastewater into the clean water compartment, against the osmotic pressure, which increases in proportion to the increasing dissolved solids concentration differential. The result is concentration of the wastewater and production of clean water.

The membrane's ability to pass water molecules but not other ions and molecules accounts for the wastewater treatment mechanism. The osmotic pressure that must be overcome, added to additional pressure required to force water molecules through the membrane, accounts for the relatively high cost of energy to operate a reverse osmosis wastewater treatment system. Typical operating pressures for RO systems range from 300 to 1,500 psig.

Reverse osmosis systems have been successfully applied to removing fats, oils, and greases (FOG), as well as salts and other dissolved substances, from wastewaters in order to comply with discharge limitations. In a number of cases, the substances removed by RO have been successfully recycled and reused, substantially reducing the real cost for this treatment step. In fact, since reverse osmosis is more widely used as a manufacturing process than for wastewater treatment, it should always be considered as part of a wastes reduction or pollution prevention program.

Electrodialysis

The mechanism of separating pollutants from wastewater using electrodialysis is that of electrical attraction of ions and consequent movement through a solution toward an electrode, of opposite charge, combined with selective transport of ionic species through membranes. The driving force is electrical attraction, and the selectivity of the membranes makes possible the separation of target pollutants from wastewater. Figure 7-71 presents a diagrammatic representation of an electrodialysis cell. Electrodes of opposite charge are on either end of the cell. Within the cell are placed, alternately, cation-permeable and anion-permeable membranes. When the cell is filled with wastewater and the electrodes are charged, cations migrate toward the cathode, and anions migrate toward the anode. In cell 2, cations migrate toward cell 1 and are admitted into cell 1 because the membrane between the two is cation-permeable. Anions migrate from cell 2 toward cell 3, drawn by the anode, and are admitted to cell 3 through the anion-permeable membrane. At the same time, anions migrate in cell 1 toward cell 2, drawn by the anode, but are not admitted to cell 2 because the membrane between cells 1 and 2 is not anion-permeable. Likewise, in cell 3, cations migrate toward cell 2, drawn by the cathode, but are not admitted to cell 2 because the membrane separating cells 2 and 3 is anion-permeable, not cation-permeable.

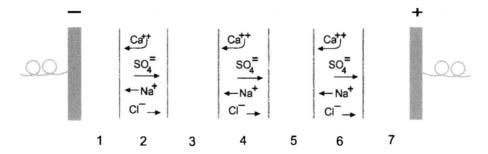

Figure 7-71 Schematic diagram of an electrodialysis cell.

As this process continues, cells 2, 4, 6, and so on lose nearly all ions, while cells 1, 3, 5, and so on gain the ions lost by the even-numbered cells. The effluent from the even-numbered cells is called the product water, and the effluent from the odd-numbered cells is called the concentrate. The concentrate may be considered as waste and may be either disposed of directly, processed further, or considered a source of valuable substances to be recycled or otherwise made use of.

The potential to reduce real costs for treatment by realizing value from pollution prevention should always be given full consideration. As with reverse osmosis and ultrafiltration, electrodialysis may have its greatest potential usefulness in purifying isolated waste streams for reuse of substances that had previously been considered pollutants and/or for simple reuse in the manufacturing process from which it came.

Electrodialysis is applicable to removing only low-molecular-weight ions from wastewater. Electrodialysis may be used in combination with ultrafiltration where dissolved ions would be removed by electrodialysis and organic molecules would be removed by ultrafiltration.

Filtration Using Granular Media

Three principal types of granular filters used for industrial wastewater treatment include deep bed granular filters, precoat filters, and slow sand filters. Filters are physical treatment devices, and the mechanisms of removal include one or more of the following: physical entrapment, adsorption, gravity settling, impaction, straining, interception, and flocculation. While slow sand filtration accomplishes removal within only the first few millimeters of depth from the surface of the sand, both deep bed granular filters and precoat filters (when body feed is used) make use of much more of the filter medium.

There are two distinct operations that characterize any granular filter: the filtering phase and the cleaning phase. With respect to these two phases, operation may be continuous or semicontinuous. In continuous operation, the filtering of wastewater and the cleaning of the filter medium or media take place at the same time. In semicontinuous operation, these steps take place in sequence.

Deep Bed Granular Filters

The intended operational process of a deep bed granular filter is that solid particles in the suspending medium are caused to follow a torturous path through the filter granules until one or more removal mechanisms results in the particle being retained somewhere within the depth of the filter. Deep bed granular filters can be of the downflow type, in which a considerable portion of removal takes place on the surface of the filter medium as well as in the initial few inches, or in the upflow mode, in which the filter bed is expanded, or fluidized, and the initial few inches of filter medium are not as significant to overall removal. In fact, the primary removal mechanism involved in an upflow filter application is likely quite different from

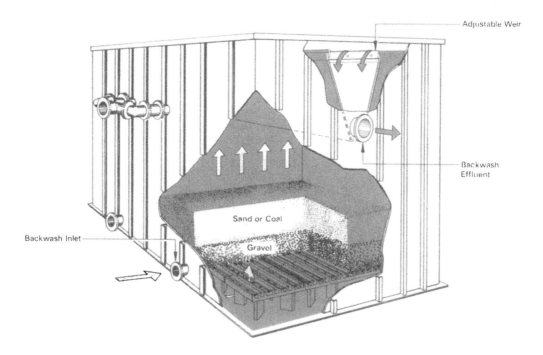

Figure 7-72   Illustration of a deep bed granular filter shown in backwash mode (courtesy of IDI/Infilco Degremont, Inc.).

that involved in a filter operated in the down-flow mode.

Deep bed granular filters use one, two, or more types of media, including sand, anthracite coal, and garnet. Figure 7-72 presents a sketch of a multimedia deep bed granular filter in which granular anthracite overlays silica sand, which overlays granular garnet. The specific weights of these three materials increase in the same order, so that during the backwashing part of the filter's operational cycle (cleaning phase) there is a natural separation of the three materials that takes place.

Upflow Granular Filters

Figure 7-73 illustrates one of the primary mechanisms by which solid particles are removed by deep bed granular filters operated in the upflow mode. As illustrated in Figure 7-73, the upflow velocity of the wastewater being treated creates a buoyant force on the granules of filter medium, causing the bed to expand in proportion to that upward velocity. Target particles within the wastewater are carried up into the expanded filter

bed, wherein there are velocity gradients. The velocities of liquid flow within the expanded bed are greatest near the middle portions of the spaces between filter medium granules, and are nearly zero in close proximity to the granules. Immediately above each filter medium granule there is a space in which the liquid velocity is either zero or is in a downward direction. The target particles are carried (up) into the "depths" of the filter by the upflow velocity gradients between particles, then become trapped in the regions adjacent to and on top of, the individual granules of filter medium. Once the particles come into contact with the granules, adsorption takes place, and the particles are removed from the wastewater. Also, as the particles are brought into contact with one another, coagulation and flocculation take place, and the resultant larger, denser flocs settle under the influence of gravity in the regions of stagnant velocity above each of the granules. The target particles are thus removed from the bulk liquid by a combination of coagulation, flocculation, gravity settling, and adsorption.

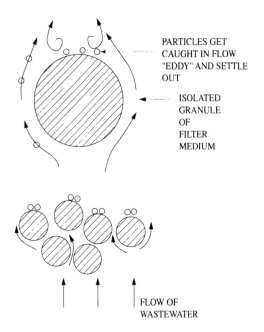

PARTICLES GET CAUGHT IN FLOW "EDDY" AND SETTLE OUT

ISOLATED GRANULE OF FILTER MEDIUM

FLOW OF WASTEWATER

Figure 7-73 Removal mechanism of an upflow granular filter.

There is no widely accepted mathematical model for use in designing an upflow deep bed granular filter as described above. Experience has shown that filter performance is a function of the filtration rate; the concentration and characteristics of the particles to be removed; and the size, surface characteristics, density, and other characteristics of the filter media. An extensive pilot program is always required to develop reliable design parameters.

Within the limits allowed by the size and numbers of filter units available and the quantity of wastewater to be treated, the operator has the most control over effluent quality by controlling the filtration rate. If the filtration rate is too high, excess solids will be swept up and out of the filter. If the filtration rate is too low, there will not be adequate bed fluidization, and the removal mechanisms described above will not be able to manifest themselves. The most desirable filtration rate is that which results in the greatest quantity of satisfactory filtrate, on a long-term basis, per unit time of filter use.

Downflow Granular Filters

Deep bed granular filters operated in the downflow mode remove solid particles by mechanisms that are quite different from those that take place in the upflow mode. In the downflow mode, particles that are smaller than the sizes of the pores and passageways between filter medium granules are carried down into the depths of the filter until they contact, and are adsorbed onto, one of the granules of filter medium. Figure 7-74 illustrates the currently accepted model of filtration mechanisms within downflow granular filters.

Figure 7-74 illustrates that particles are transported to the vicinity of filter granules (collectors) more or less along the paths of fluid streamlines, where attachment, and sometimes detachment, occur. Figure 7-74 shows that several transport mechanisms are operative, including hydrodynamics, diffusion, gravity-settling, inertia, and interception.

As a deep bed granular filter is operated in the downflow mode, many of the larger particles are intercepted and removed by sieve action at the surface of the filter bed. Also, to some extent, these particles bridge across each other and form their own filter, which further filters out new particles brought to the filter surface. This is the primary removal mechanism in the case of slow sand filters, precoat-body feed filters, and other "surface-type" filters. However, in the case of deep bed granular filters, it is necessary that the filtration rate be sufficiently high that hydrodynamic forces carry most of the particles past the media surface and into the depths of the filter, where the mechanisms of diffusion, sedimentation, interception, etc., as illustrated in Figure 7-74, lead to attachment and, thus, removal.

In the case of precoat-body feed filters, described in more detail below, the filter bed continually builds in depth as the filter run progresses, and there is a continually renewed "surface" on which to collect particles by sieving action.

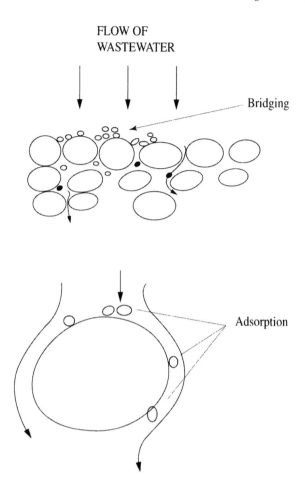

FLOW OF
WASTEWATER

Bridging

Adsorption

Figure 7-74   Removal mechanism of a downflow granular filter.

Filtration rate, then, is a critical operational parameter in downflow filters as well as upflow filters. In the case of downflow filters, too low a filtration rate will result in too much removal of solids in the upper layers of the filter. The head loss will become too high for further operation before the deeper portions of the filter are made use of. Too high a filtration rate, on the other hand, will result in flushing too many solids through the filter, causing the filtrate to be unacceptable.

Head loss is a major operational factor in the case of downflow granular filters, since it is hydraulic head, or pressure, that causes flow through the filter. As particles are removed on and in the filter, resistance to

flow through the filter increases. This resistance to flow manifests itself in hydraulic head loss. Eventually, head loss builds to the point where the rate of flow through the filter (filtration rate) is unacceptable, and the particles must be removed by backwashing. Usually, previously filtered water is used for backwashing. The backwash water is pumped through a manifold and distribution system at the bottom of the filter. The backwashing rate of flow is sufficient to expand the bed so that the scouring action of hydrodynamic forces and the filter medium granules bumping and rubbing against each other dislodge the particles removed during filtration. The upflowing backwash then carries the removed solids up and out of the filter media and into a trough or other collection device above the expanded filter bed.

Auxiliary scouring is sometimes necessary, and is normally provided by one or a combination of high-pressure sprays onto the surface layers of the filter medium, mechanical stirring or racking within the filter bed, and the introduction of air bubbles at the bottom of the filter.

Downflow deep bed granular filters are operated either as constant head with declining rate of flow or as constant rate of flow with either increasing head or constant head loss. Constant head loss is achieved by use of an artificial head loss device, such as a valve on the outlet that gradually decreases its effect to maintain a constant total head loss across both the filter and the valve during the filter run. Constant rate of flow with increasing head is achieved by either gradually increasing the depth of water over the filter throughout the filter run or by applying pressure with air or a pump to an enclosed filter influent chamber. It is common practice to attach a manifold that applies water at a constant head to each of several parallel filters that then operate in the declining rate of flow mode. As the head loss in one filter increases, more flow proceeds to the other filters until the flow rate through the entire group of filters becomes unacceptable, either because of inadequate total filtration rate or because of unacceptable filtrate quality. At this point, backwashing takes place.

Often, filter performance can be significantly improved by use of chemical coagulants. It is critically important to provide adequate and appropriate chemical feed and rapid mixing, but it is not necessary to provide slow mixing (flocculation), since flocculation takes place very effectively within the filter in either the upflow or downflow mode. This mode of operation, referred to as direct filtration, can be used when the load to the filter is not such that it would be less expensive in the long term to provide a solids removal step consisting of chemical feed, rapid mix, slow mix (flocculation), and gravity settling prior to filtration. One technique for removal of phosphorus, as explained in the section on chemical methods of wastewater treatment, is to add alum prior to the filter.

Deep bed granular filters, both upflow and downflow, can be operated in configurations where the filtration media are continuously washed. Sometimes referred to as "moving bed filters," these systems have worked well in certain applications. Usually, a reasonably constant solids loading is desirable for successful continuous washing applications.

Figure 7-75 shows four types of deep bed granular filters.

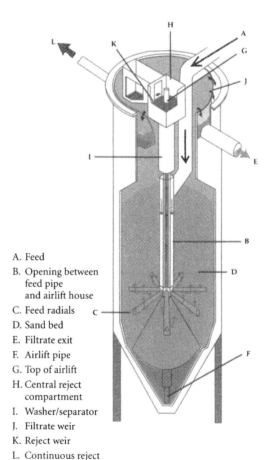

A. Feed
B. Opening between feed pipe and airlift house
C. Feed radials
D. Sand bed
E. Filtrate exit
F. Airlift pipe
G. Top of airlift
H. Central reject compartment
I. Washer/separator
J. Filtrate weir
K. Reject weir
L. Continuous reject exit.

Figure 7-75 Granular filters for industrial waste (courtesy of Parkson Corporation)

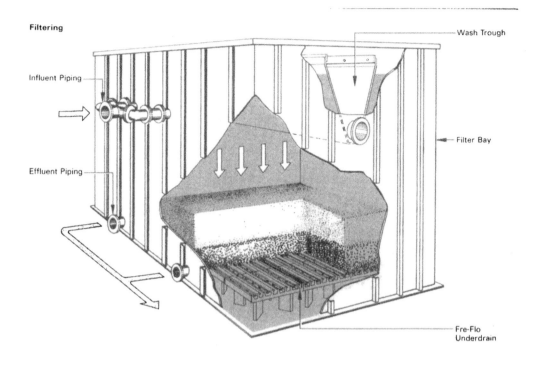

Filtering

Influent Piping

Effluent Piping

Wash Trough

Filter Bay

Fre-Flo
Underdrain

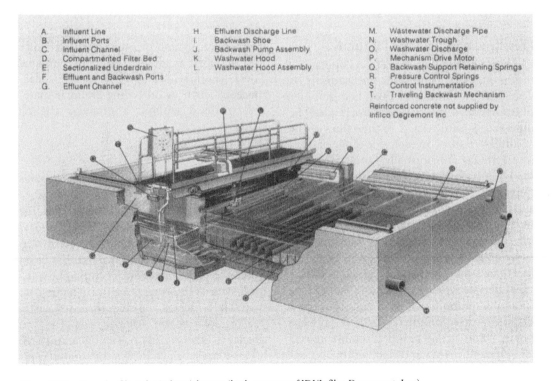

A.	Influent Line	H.	Effluent Discharge Line	M.	Wastewater Discharge Pipe
B.	Influent Ports	I.	Backwash Shoe	N.	Washwater Trough
C.	Influent Channel	J.	Backwash Pump Assembly	O.	Washwater Discharge
D.	Compartmented Filter Bed	K.	Washwater Hood	P.	Mechanism Drive Motor
E.	Sectionalized Underdrain	L.	Washwater Hood Assembly	Q.	Backwash Support Retaining Springs
F.	Effluent and Backwash Ports			R.	Pressure Control Springs
G.	Effluent Channel			S.	Control Instrumentation
				T.	Traveling Backwash Mechanism

Reinforced concrete not supplied by
Infilco Degremont Inc

Figure 7-75   Granular filters for industrial waste (both courtesy of IDI/Infilco Degremont, Inc.).

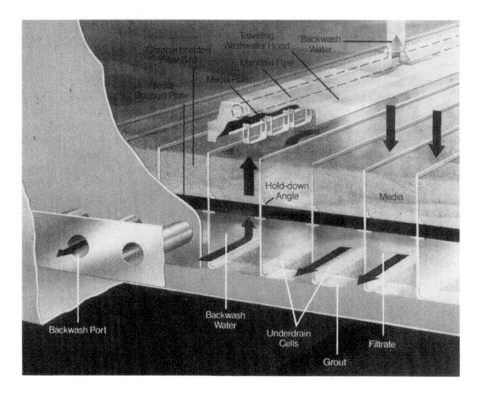

Figure 7-75   Granular filters for industrial waste (courtesy of IDI/Infilco Degremont, Inc.).

Pressure and Vacuum Filtration

Pressure filters and vacuum filters operate by means of exactly the same mechanisms as described previously for deep bed granular filters operated in the downflow mode. The difference is with respect to the values of the absolute pressures on each side of the barrier. These values are normally atmospheric on the low-pressure side of a pressure filter and greater than atmospheric on the high-pressure, or feed, side. In the case of the vacuum filter, the pressure on the feed side of the barrier is normally atmospheric, while the pressure on the low-pressure side of the barrier is less than atmospheric. The principal differences between the equipment, then, are that a liquid pump and air-tight conveying devices are used on the feed side of pressure filters, while an air pump (vacuum pump) and air-tight piping and duct work are used on the low-pressure side of the barrier in the case of a vacuum filter. The physical barrier can consist of sand or other granular material, or fabric made of natural or synthetic fibers. The fabric can be coated with diatomaceous earth or other fine granular material, or used without a granular coating material.

Precoat

If the fabric is coated with diatomaceous earth, perlite, or other granular material of low specific gravity just prior to each use, the filter is said to be "precoated." This "precoat" is applied by simply mixing a predetermined quantity of the precoat material in clean water, then passing the slurry through the filter just as in a normal filter run. The precoat material, which is the diatomaceous earth, perlite, or other granular material as described above, will collect in a uniform layer on the fabric to form a thin granular filter bed. A valve is then activated, and the precoat slurry is replaced with wastewater. It is important that the hydraulic pressure used to precoat be maintained continuously from the start of the precoat process to the end of the filter run to maintain the integrity of the precoat.

Body Feed

After the precoat has been applied, and the process of filtering the wastewater through the precoat has commenced, a prescribed concentration of the same material as used for precoat (diatomaceous earth, perlite, or other granular material) is added to the wastewater. Figure 7-76(a) shows that a slurry of filter media can be metered in to the stream of wastewater as it is being pumped to the filter. Figure 7-76(b) shows that, alternately, a predetermined amount of the filter medium can be mixed into the wastewater that is to be filtered by use of a mixing tank. The mixing tank can be of a size that holds one day's contribution of wastewater, one filter run, or another convenient cost-effective size.

Figure 7-77 illustrates the way in which the granules of body feed add to the thickness of the precoat to develop a filter of increasing depth as the filtration process proceeds in time. As shown in Figure 7-77, the granules of body feed, which are evenly dispersed throughout the wastewater, collect on the outside of the precoated filter support surface as the wastewater is forced through. These granules thus build a filter of ever-increasing thickness, preventing the original precoat from blinding over with suspended solids from the wastewater. Rather, the suspended solids, which are the target pollutants intended to be removed by the filter, are removed in the ever-increasing depth of the body feed filter as it builds.

Pressure or vacuum filters (examples of which are shown in Figure 7-78) are routinely used to remove solids from dilute industrial wastewaters, usually with chemical conditioning, as well as to dewater sludges. One common use of pressure or vacuum filters is to dewater precipitated metals.

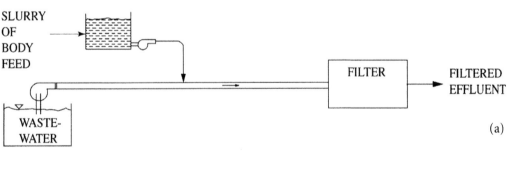

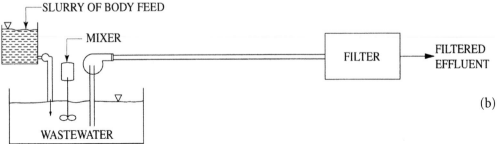

Figure 7-76   (a) Body feed added to precoat filter. (b) Body feed added to mixing tank.

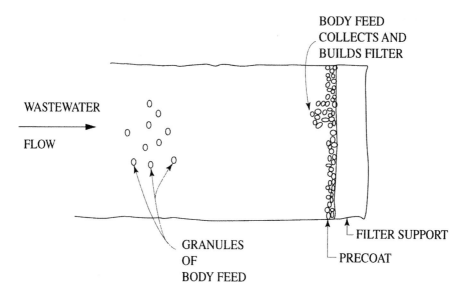

Figure 7-77   Body feed increases filter depth.

(a)

Figure 7-78   Pressure and vacuum filters. (a) Courtesy of IDI/Infilco Degremont, Inc.

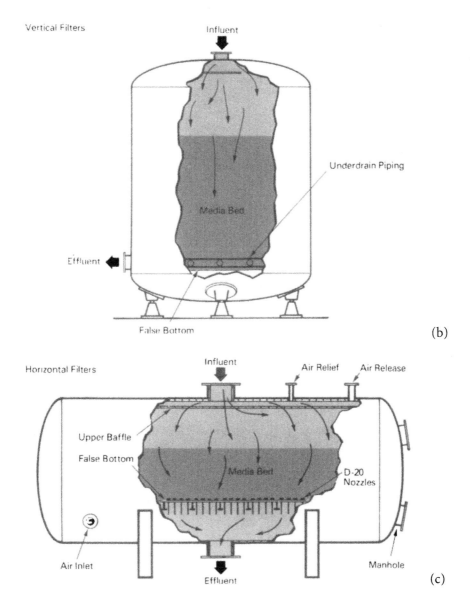

Figure 7-78   Pressure and vacuum filters. (b, c) Courtesy of IDI/Infilco Degremont, Inc.

Swimming Pool Filters

In some cases of short-term need, an off-the-shelf packaged, automated deep bed granular or septum-type filter assembly, intended for swimming-pool use, is a cost-effective alternative to achieve treatment by filtration. These items are inexpensive and are designed and constructed to provide reliable, automated service for a few years, assuming a nonaggressive water mixture.

Packaged Water Treatment Systems

There are many packaged treatment systems that were designed and manufactured for the purpose of treating potable water. Many of these systems can be used to treat many different industrial wastewaters, with little or no modification. Some wastewaters that are candidates for these systems include dilute, small-volume process wastes, stormwater

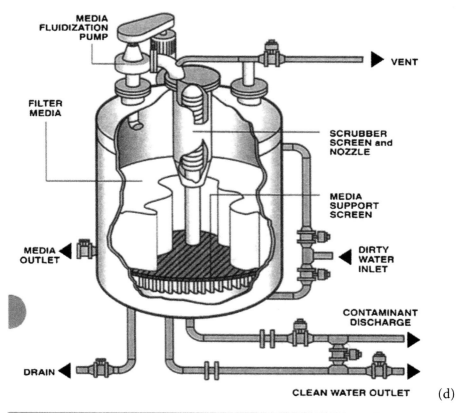

Figure 7-78   Pressure and vacuum filters. (d, e) Courtesy of Eimco Process Equipment, a Baker Hughes Company.

(f)

Figure 7-78 Pressure and vacuum filters. (f) Courtesy of Waterlink Separators, Inc.

runoff, contaminated groundwater, and cooling water.

Several issues must be thoroughly addressed during the concept or preliminary design phase of such a project, including corrosivity, effects of any hazardous materials; or other properties, such as heat, intermittent flow, color, susceptibility to weather, and variability of wastewater characteristics.

Slow Sand Filter

Slow sand filters are characterized by a loading rate that is significantly lower than the more conventional (sometimes referred to as "rapid sand filter") filter described earlier. In general, slow sand filters are useful for polishing small quantities of treated wastewater, water that has been used for scrubbing gases and is to be either discharged or renovated for another use, or simply as a final "emergency" barrier before discharge of water that is normally very clean.

Slow sand filters are loaded at rates between two and four gallons per day per square foot of filter surface area. Bed depths are typically two to four feet.

The mechanisms of removal in slow sand filtration include entrapment, adsorption, biological flocculation, and biological degradation. In fact, one of the reasons for the very low hydraulic loading rate of slow sand filter is to allow aerobic conditions within the first inch or two of the sand filter depth. The aerobic conditions allow for aerobic and, thus, relatively complete, biological removal of the organic portion of the material removed from the waste stream. Sand filters are cleaned periodically by removal of the top one-half to two inches of sand from the surface.

Plain Sedimentation

Plain sedimentation can be described as the separation of particulate materials from wastewater as a result of the influence of gravity. The more quiescent (nonmoving) the hydraulic regime, the more effective is this removal process; therefore, a principal objective of plain sedimentation equipment is to produce "quiescent conditions." The process of plain sedimentation is often referred to as "clarification," and the devices used to accomplish plain sedimentation are called "clarifiers" or "settling tanks."

There are three modes by which particles undergo the plain sedimentation process: discrete settling, flocculent settling, and zone settling. Discrete settling is that process by which individual particles proceed at a steady velocity, governed principally by the specific gravity of the particle and the viscosity of the wastewater, toward the bottom of the settling device. The settling of sand in water is an example of discrete settling. Flocculent settling is undergone by particles, often organic, that agglomerate to larger, faster-settling particles as the settling process takes place over time. The settling of activated sludge in the upper several feet of a secondary clarifier (a clarifier that follows, hydraulically, an activated sludge aeration tank or a fixed growth treatment process) is an example of flocculent settling. Zone settling is typically undergone by a suspension of relatively concentrated solid particles (sludge) as it concentrates even more. The vertical distance an individual particle moves decreases with increasing depth in the sludge mass itself.

The overall objective of any physical wastewater treatment device, including clarifiers, is to remove as much solid material as possible, as inexpensively as possible, with a resulting sludge that has as low a water content as possible. The final residuals of the treatment processes must undergo final disposal, and it is usually the case that the cost of final residuals disposal is proportional to some degree to the water content.

Discrete Settling

Under the influence of gravity, a particle immersed in a body of fluid will accelerate downward until the forces relating to viscosity that resist its downward motion equal the force due to gravity. At this point, the particle is said to have reached terminal velocity, described by equation 7-67:

$$V_s = \left[\frac{4g(\rho_s - \rho_l)\ D}{3\ C_D \rho_l} \right]^{1/2} \qquad (7\text{-}67)$$

where

V_s = Velocity of sedimentation (ft/sec)
g = Acceleration due to gravity (ft/sec^2)
$_s$ = Density of the particle (lb/ft^3)
$_l$ = Density of the fluid (lb/ft^3)
D = Diameter of particle (ft)
C_D = Coefficient of drag

The coefficient of drag, C_D, is dependent upon a number of factors, including the diameter of the particle, if hydraulic turbulence is significant within the settling device. For the small size of particles usually encountered in clarification of industrial wastewater, and the relatively low turbulence encountered in a well-operating clarifier, the relationship presented in equation 7-68 is considered to hold:

$$C_D = \frac{24}{N_R} \qquad (7\text{-}68)$$

where

N_R = Reynolds number

Because Reynolds number increases with increasing turbulence of the fluid, it is seen that the settling velocity, analogous to the rate of clarification, increases with decreasing turbulence.

There are interesting relationships, as shown by Hazen and Camp, between particle settling velocity, the surface loading of a clarifier, and the theoretical effect of depth of the clarifier on particle removal efficiency. These relationships are based on the conditions of a uniform distribution of particles throughout the entire clarifier at its influent end, and of whether a particle can be considered to be removed as soon as it reaches the bottom of the clarifier. If the rate of settling, expressed as vertical distance settled per unit of horizontal distance of forward flow (in ft/ft) of a particle in a clarifier is multiplied, top and bottom (numerator and denominator), by the rate of wastewater flow, in cubic feet per

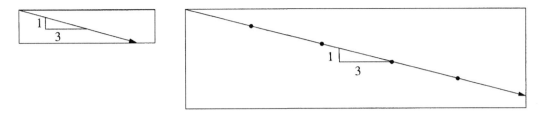

Figure 7-79   Illustration of the effect of increasing clarifier depth.

day, the units of gallons per day can be developed in the numerator, and the units of area, or square feet, can be developed in the denominator.

Therefore, equation 7-69 can be written:

$$V_s\,(ft/ft)$$
$$= surface\ loading\ rate\,(GPD/ft^2) \qquad (7\text{-}69)$$

which states that particle settling velocity, in ft/ft, is numerically equivalent to a term called "surface loading rate," having the units, gallons (of wastewater) per square foot (of clarifier surface area) per day, indicating that only clarifier surface area, to the exclusion of clarifier depth, is of significance to the performance of a clarifier.

Figure 7-79 illustrates that the major effect of increasing clarifier depth is simply to increase the time it takes to remove a given particle, rather than to have any effect on removal effectiveness or efficiency.

Figure 7-79 illustrates that all particles with settling velocities equal to or greater than V_s will be completely removed. All particles having settling velocities less than V_s will be removed as described by the ratio V/V_s.

Flocculent Settling

Many solid particles found in wastewaters tend to agglomerate as they settle under quiescent hydraulic conditions, and because they become denser as they coagulate and agglomerate, their terminal velocities increase over time. The paths followed by solid particles undergoing flocculent settling, then, are illustrated in Figure 7-80.

Often, a given industrial wastewater will contain a mixture of solids; some settle as discrete particles and others undergo flocculent settling. When sufficient particles reach the lower portion of the clarifier, the accumulated mass ("sludge blanket") undergoes zone settling. Normally, either discrete or flocculent settling predominates in the upper portion of the clarifier and, therefore, becomes the basis for design. Those gravity clarification devices that combine the processes of primary sedimentation and sludge thickening are called "clarifier-thickeners." For these devices, the thickening characteristics of the settled sludge are also a basis for design.

Sludge Thickening

Once solids have settled under the influence of gravity, they must be removed from the bottom of the clarifier and ultimately disposed of. In order to reduce the cost of handling this material, it is desirable to reduce the moisture content, and thereby the volume, to a minimum as soon as practicable. To this end, "gravity-thickeners" are often designed and built as integral components of wastewater treatment systems. Some clarifiers incorporate a thickening capability into the sludge zone and are thus called "clarifier-thickeners."

Kynch, Talmage and Fitch, Behn and Liebman, and Edde and Eckenfelder have developed various methods for predicting performance of thickeners based on results of laboratory bench-scale tests. These thickening tests have the common procedure of using standard two-liter graduated cylinders, which

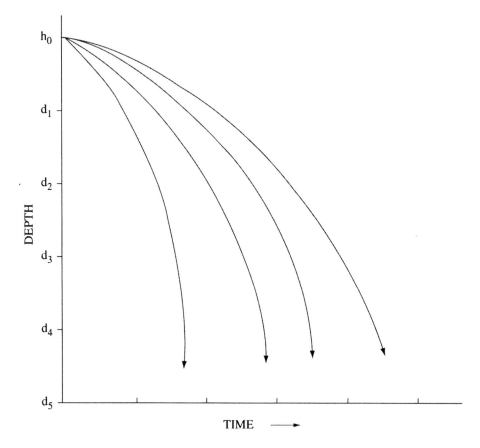

Figure 7-80   Settling path taken by particles undergoing flocculent settling.

are filled with primary clarifier sludge, and then determine the rate of downward progress (under the influence of gravity) of the sludge-water interface. The above-mentioned researchers have published different methods for analyzing the data collected using the two-liter graduated cylinders.

Kynch's method makes use of the assumption that the settling velocity of a particle is a function of only the concentration of particles around that particle. A conclusion that resulted from this analysis was that waves of constant concentration move upward from the bottom of a settling solids concentration, each wave moving at a constant velocity. Talmage and Fitch used this analysis to show that the area of a gravity-thickener is a function of the "solids handling capacity" of a given zone of constant concentration. The

unit area (U.A.) required to pass a unit weight of solids through a given concentration (C_i) can be expressed as:

$$U.A. = \frac{1}{V_i}\left(\frac{1}{C_i} - \frac{1}{C_u}\right) \qquad (7\text{-}70)$$

where

V_i = settling velocity of the particles at concentration C_i
C_u = underflow solids concentration

The "underflow solids concentration" is the suspended solids concentration of the sludge as it is withdrawn from the bottom of the clarifier.

Behn and Liebman developed a relationship for the depth of the sludge compression zone as a function of the initial sludge concentration, rate of elimination of water from the mass of compressing sludge, dilution ratio, specific gravities of the solids and water, underflow concentration, and thickener area.

Edde and Eckenfelder developed a procedure for scale-up from bench-scale results to design parameters for prototype thickeners. Thickener design is based upon the following mathematical model relating thickener mass loading (ML) to underflow concentration (C_u):

$$\left(C_u/C_O\right) - 1 = D/(ML)^n \tag{7-71}$$

The depth of sludge in a batch test (D_b) is related to depth of sludge in the prototype (D_t) based on conditions observed in the field.

The design procedure presented below makes use of a "settling column" (for discrete and/or flocculant settling) and a two-liter graduated cylinder (for sludge thickening) to obtain laboratory data for use in design of an industrial wastewater clarifier-thickener.

Design Procedure

Figure 7-81 shows a standard laboratory settling column used to generate data for use in designing a clarifier. Figure 7-82 shows a photograph of a settling column in use. Usually, two or more six-inch diameter transparent plastic pipe sections are bolted together to make a column that is 12 to 16 feet long. The column is fitted with sampling ports at each foot of depth when the pipe is standing on end. A plate seals the bottom, and a cleanout port is located as close to the bottom as possible.

The settling column is filled with mixed, raw wastewater and allowed to achieve quiescent conditions. At intervals of 5 to 15 minutes, samples are withdrawn from each of the sample ports for suspended solids analysis.

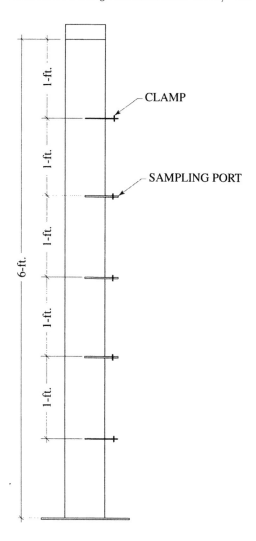

Figure 7-81   Settling column.

Figure 7-83 presents a plot of the results of suspended solids analyses obtained at progressive time intervals from each of the sampling ports in the settling column. Lines that connect the data points from each sampling port represent given fractions of TSS removal and describe the approximate path that would be taken by particles settling in an ideal settling tank (no physical or other influences on the settling particles other than gravity and viscous resistance to motion). The data plotted in Figure 7-83 are analyzed, and criteria for design of a clarifier are developed, as shown in the following example.

Figure 7-82   Settling column.

Example 7-6: Development of design criteria for a clarifier-thickener

Using a standard settling column and the procedures described by Eckenfelder and Ford develop design criteria for the settling portion of a combination clarifier-thickener to remove and concentrate suspended solids from wastewater from a paper mill. Using standard two-liter graduated cylinders, develop design criteria for the sludge-thickening portion of the combination clarifier-thickener.

Solution: Raw wastewater was collected from the paper mill effluent at an integrated kraft pulp and paper mill. The standard settling column shown in Figure 7-81 was filled with the (well-mixed) wastewater, and quiescent conditions were allowed to develop. Then, samples were withdrawn from each of the ports simultaneously at various time intervals. Each sample was analyzed for suspended solids concentration. The BOD_5 was determined on samples from the port located three feet below the top of the column. Two of these procedures were carried out, one at a temperature of 3.5°C and another at 18.5°C.

The data thus developed were plotted as shown in Figures 7-84 and 7-85. Figure 7-84 presents the results of data from the procedure performed at the temperature of 3.5°C. Figure 7-85 presents the results of data taken at 18.5°C. Comparison of these results shows

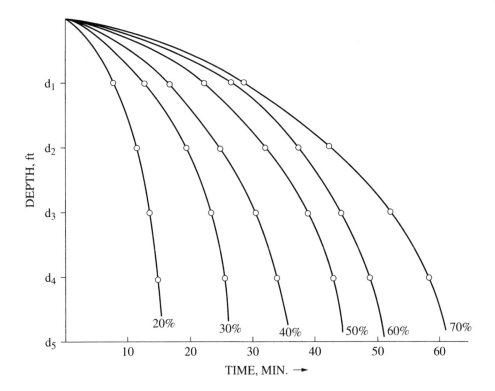

Figure 7-83   Plot of percent removal of TSS at given depths versus time (settling column analysis).

that higher settling rates were obtained at the higher temperature, a result to be expected due to the higher viscosity, thus greater drag, and thus slower settling of particles through the more viscous medium at the lower temperature.

Figure 7-86 presents suspended solids removal versus detention time for the wastewater at 3.5°C and at 18.5°C. The corresponding overflow rates are presented in Figure 7-87. BOD_5 removal as a function of suspended solids removal is presented in Figure 7-88. These results, as expected, show that BOD_5 removal increased with increasing suspended solids removal but at a decreasing rate, as the limit of BOD_5 contributed by suspended solids (as opposed to dissolved organics).

These three figures, then, are the principal product of the settling column analysis procedure. The next step is to determine the cost for the different sizes of clarifier, as deter-

mined by overflow rate and the desired percent removal of suspended solids and/or BOD_5.

Thickener Portion of Clarifier-Thickener
Samples of primary clarifier sludge were collected at a paper mill that was very similar in type of manufacturing process to the mill for which the column-settling analyses presented above were being performed. The samples of sludge were diluted (with primary effluent) to the desired initial suspended solids concentration, then placed in the two-liter graduated cylinders. The initial suspended solids concentration, temperature, and specific weight of the water were determined for each cylinder. The liquid-solids interface level was recorded at regular intervals until a constant level was reached. The results of these data are presented in Figure 7-89, which presents plots of solids-liquid interface height versus time for four initial sludge concentrations.

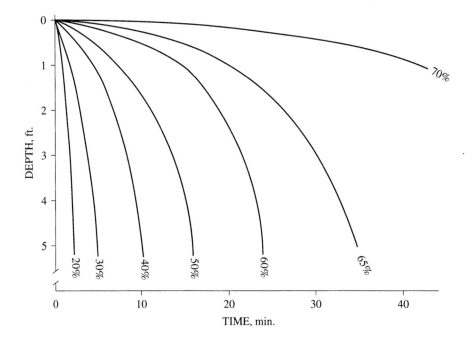

Figure 7-84   Solids removal for depth versus time (3.5°C).

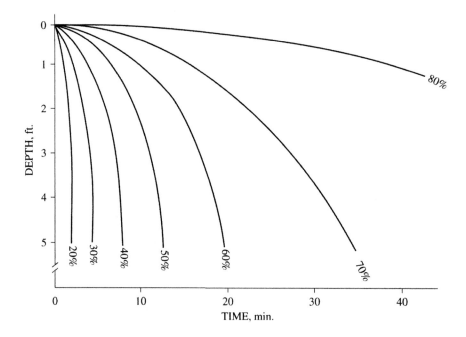

Figure 7-85   Solids removal for depth versus time (18.5°C).

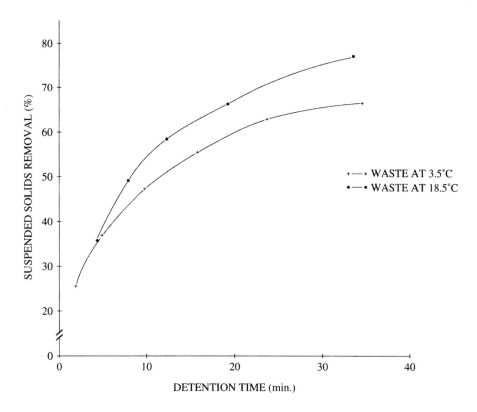

Figure 7-86   Suspended solids removal versus time.

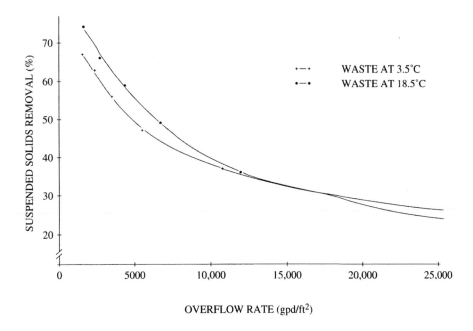

Figure 7-87   Suspended solids removal versus overflow rate.

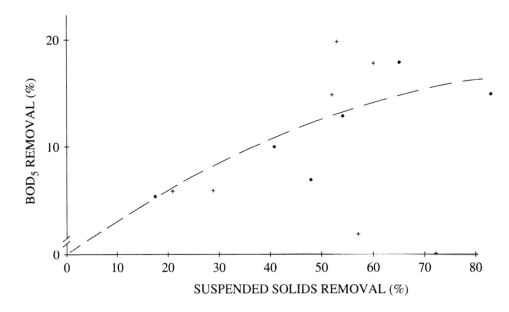

Figure 7-88   BOD$_5$ removal versus suspended solids removal.

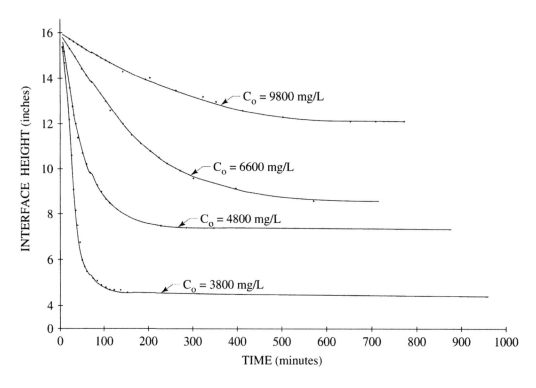

Figure 7-89   Liquid-solids interface height versus time.

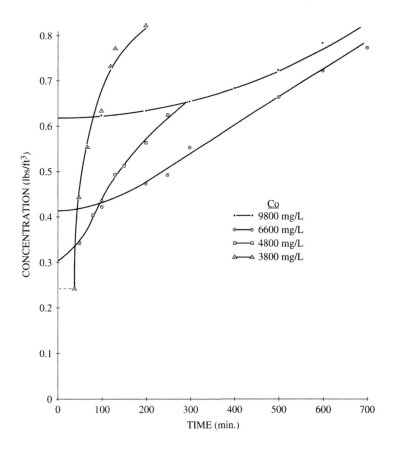

Figure 7-90   Sludge concentration versus time.

Using the mathematical models presented by Talmage and Fitch and Behn and Liebman, the solids concentration, C_i, at the interface at time, t_i, and the settling velocity, V_i, of the particles at concentration C_i can be determined.

Figure 7-90 presents solids concentration, C_i, versus time relationships for the four sludge slurries examined, and Figure 7-91 presents the concentration-velocity relationships.

The next step is to determine the unit area (ft^2/lb/day) required for given values of sludge-liquid interface settling velocities for three or more different values of initial sludge concentrations. These results are presented in Figure 7-92, which shows that for given values of initial sludge concentration there is a maximum unit area (high point on each curve). As discussed by Behn and Liebman and Talmage and Fitch, the maximum unit area for a given underflow (initial sludge) concentration is the theoretical cross-sectional area required for a sludge-thickener.

Figure 7-93 is then constructed in order to determine the depth required for thickeners for given values of underflow (initial sludge) concentration. As discussed by Behn and Liebman, the depth of a thickener is a fraction of the volume represented by a given "dilution" of the solids, or "dilution factor." Figure 7-93 presents plots of dilution factors versus time for the appropriate values of

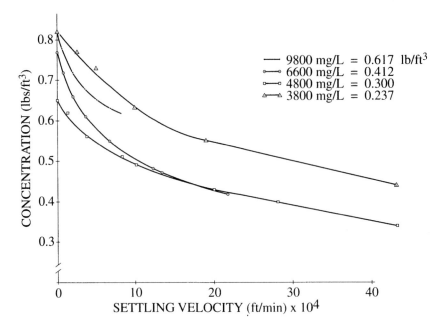

Figure 7-91   Sludge concentration versus settling velocity.

sludge underflow concentration. The slopes of the appropriate lines are equal to the values of k in equation 7-72:

$$Depth = H = \frac{Q_0 C_0 \left(t_u - t_c\right)\left(\dfrac{1}{\rho_s} + \dfrac{D_\infty}{\rho_w}\right)}{A - \dfrac{\left(D_C - D_U\right)\left(Q_0 C_0\right)}{\rho_w\left(kH_0\right)}} \quad (7\text{-}72)$$

Figure 7-94 can be constructed to illustrate the analysis of data described by Edde and Eckenfelder. In this analysis, the results of sludge-liquid interface height versus time and unit area versus settling velocity are plotted as presented in Figures 7-89 and 7-92. Then, the values of maximum unit area for different feed and underflow concentrations are constructed, as shown in Figure 7-94. The next step is to construct the semilog plot of the depth in the batch test versus the original solids concentration, as presented in Figure 7-95. Figures 7-94 and 7-95 can be interpolated to obtain values needed for design calculations that are not identical to those used in the laboratory analysis.

Design Calculations
by Kurt Marston

Having constructed the curves presented in Figures 7-84 through 7-95, the appropriate size of clarifier-thickener for wastewaters of given solids settling characteristics can be determined, as illustrated in the following demonstrative calculations.

Example 7-7

Design a clarifier-thickener for the pulp and paper wastewater using the testing data described in Figures 7-84 through 7-95 to accommodate an influent design flow of 10 million gallons per day (MGD) and to achieve 60% suspended solids removal at 3.5°C with an underflow concentration of 10,000 mg/L. Influent suspended solids (the raw wastewater) were measured at approximately 300 mg/L.

Clarifier requirements can be derived from Figures 7-86 and 7-87. Based upon the data curves for wastewater at 3.5°C, Figure 7-86 shows a minimum detention time of approximately 20 minutes, and Figure 7-87

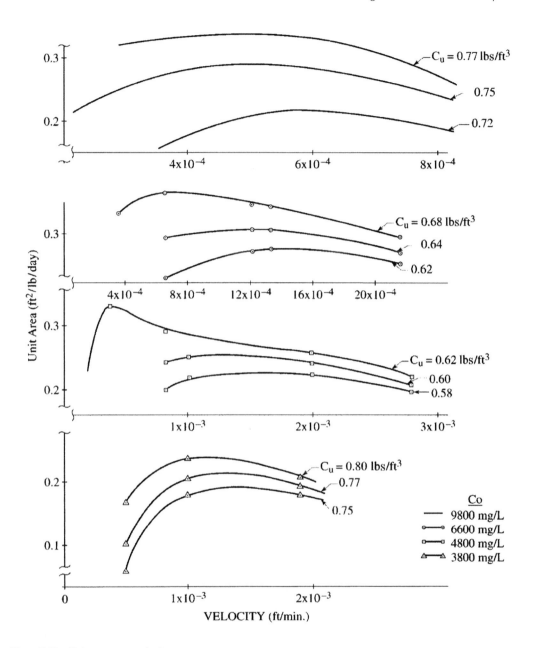

Figure 7-92   Unit area versus velocity.

shows a minimum overflow rate of approximately 2,500 gpd/ft^2 required to achieve 60% suspended solids removal.

Minimum area for clarification

= (10,000,000 gpd) / (2,500 gpd/ft^2)

= 4,000 ft^2

Minimum depth for clarification

= (Flow)(detention time)/Clarifier area

= 4.6 ft

These values represent results obtained under ideal (laboratory) settling conditions. It is necessary to apply scale-up factors

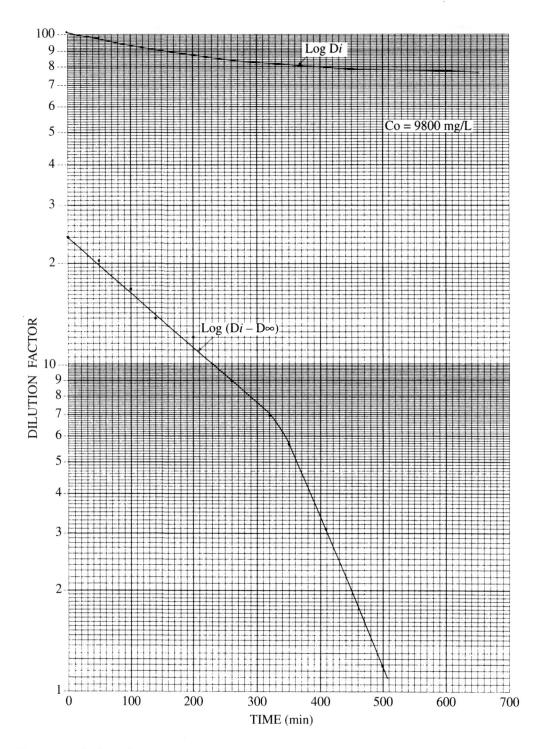

Figure 7-93   (a) Dilution factor versus time.

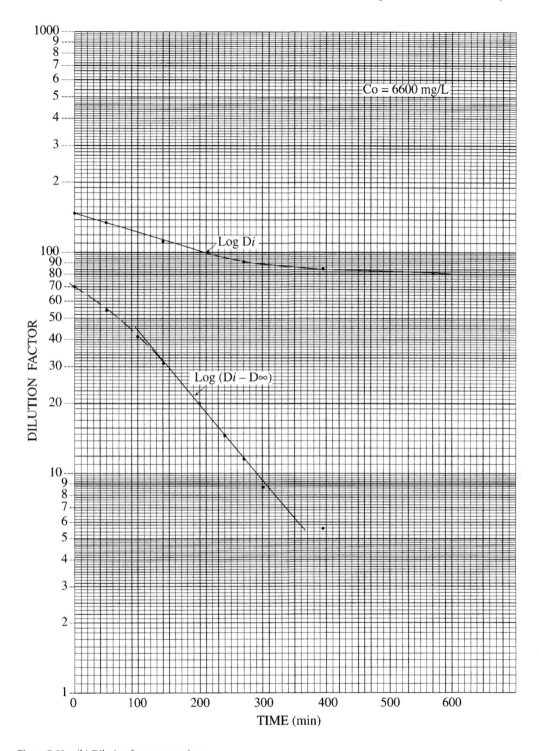

Figure 7-93   (b) Dilution factor versus time.

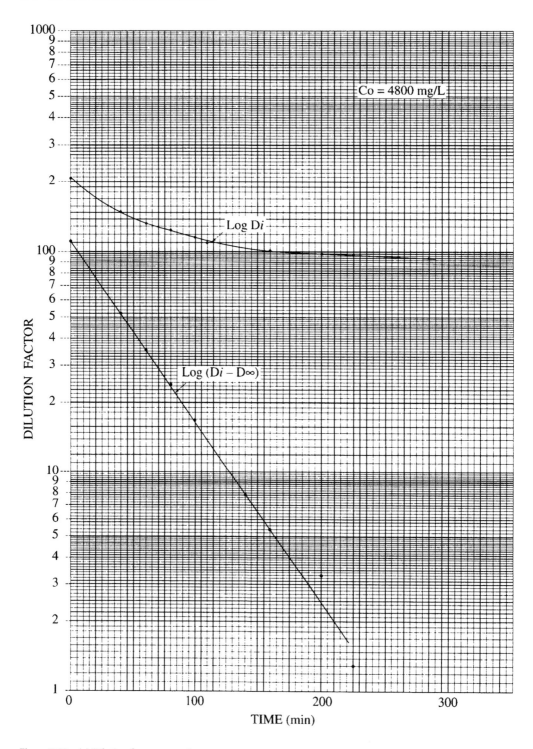

Figure 7-93   (c) Dilution factor versus time.

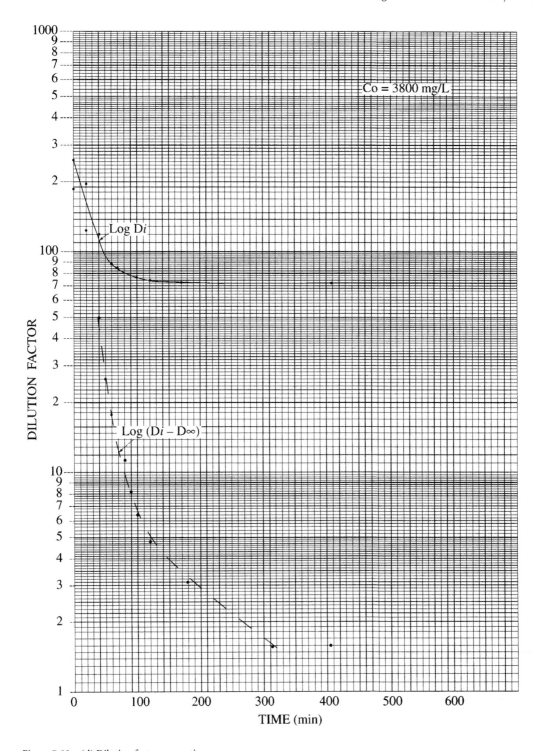

Figure 7-93 (d) Dilution factor versus time.

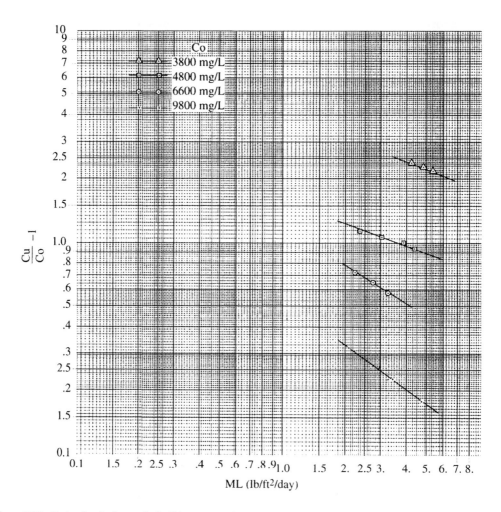

Figure 7-94   Ratio of underflow to feed solids concentration versus mass loading.

appropriate to each type of clarifier. Scale-up factors are normally selected based upon experience reported in the literature and through discussions with manufacturers. Typical scale-up factors are in the range of 1.5 to 2.0.

Thickener requirements calculated using the Behn & Liebman methods:

$$Cu = \text{Underflow Concentration} = 10,000 \\ \text{mg/L} = 0.624 \text{ lb/ft}^3$$

From Figure 7-90, the maximum time for reaching a concentration of 10,000 mg/L (t_u)

is the data curve for the 6,600 mg/L test sample with a value of approximately 440 minutes.

Figure 7-92 shows a maximum unit area of approximately 0.28 ft²/lb/day for the family of curves for the 6,600 mg/L test sample.

Solids loading to the thickener would be 60% of the 300 mg/L entering the clarifier = 180 mg/L in 10 MGD = 15,200 lb/day.

Minimum area for thickening
$$= (15,200 \text{ lb/day}) (0.28 \text{ ft}^2/\text{lb/day})$$
$$= 4,260 \text{ ft}^2$$

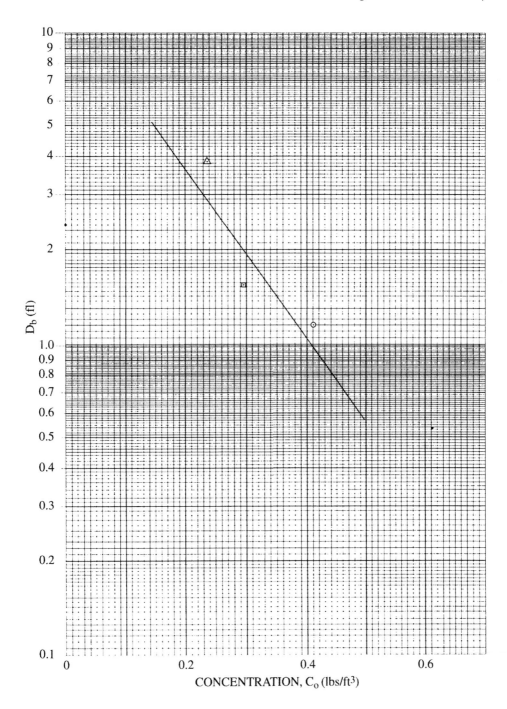

Figure 7-95   Depth versus concentration.

Sludge Depth (H) required for thickening is calculated using equation 7-72.

From Figure 7-93, the value for k is the slope of the lower curve or 7.76×10^{-3} and, using the upper curve, $D_u = 85$ for $t_u = 440$ minutes, $D_c = 118$ for $t_c = 120$.

$$H = \dfrac{\dfrac{15,200}{1440}(440-120)\left(\dfrac{1}{(1.35)62.4} + \dfrac{81.2}{61.91}\right)}{4260 - \dfrac{(118-85)\left(\dfrac{15,200}{1440}\right)}{61.91(1.33)7.76\left(10^{-3}\right)}}$$

$$= 1.2 \; feet$$

Plate, Lamella, and Tube Settlers

Examination of Figure 7-80 reveals that a particle settling at terminal velocity, V_S in an ideal settling tank of hydraulic retention time, Q/V (where Q is the flow rate and V is the tank volume), will not become "removed" until it reaches the end of the tank, if it starts at the surface. Furthermore, the deeper the tank, the longer the required HRT to remove the particle. There is a temptation, then, to imagine a very shallow tank, in which particles would reach the bottom very quickly, thus achieving removal in a very short amount of time and with relatively lit-

tle cost (owing to the small size of the clarifier). Common sense reveals, however, that turbulence resulting from such a shallow flow path would prevent effective settling (the clarifier would not be able to attain quiescent conditions). The theoretical implication still remains, however, that the shallower the settling tank the quicker the removal.

What, then, is the optimum depth for a clarifier, given, on the one hand, that surface loading is the only parameter that affects removal in an ideal clarifier and, on the other hand, that hydraulic turbulence affects removal in a negative way in a "real" clarifier?

Looking again at Figure 7-80, it can be seen that, no matter what the depth of the ideal settling tank, removal (by virtue of having reached the "bottom") of solids having settling velocities much slower than v_s could theoretically be achieved if a series of false bottoms were placed within the tank, similar to a stack of trays. Figure 7-96 illustrates such a concept.

While there is no disputing the theoretical validity of the concept illustrated in Figure 7-96, the actual occurrence of hydraulic turbulence imposes limits on application. Several devices based on the theoretical advantage illustrated in Figure 7-96 have been developed and are in widespread use. The

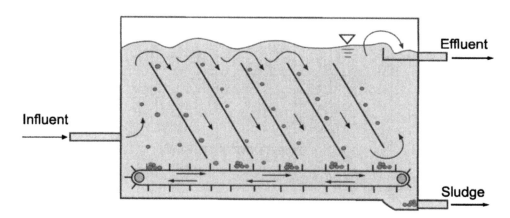

Figure 7-96 Illustration of conceptual increase in removal efficiency by adding a series of false bottoms to an open, ideal settling tank.

device illustrated in Figure 7-97, known as a "plate settler," is an example.

Figure 7-97 shows a tank that contains a stack of flat plates that nearly fills the tank. The plates are inclined at an angle of 60°. The influent end of the tank is equipped with a device to distribute incoming flow uniformly across the tank so that, to the extent possible, equal amounts of raw wastewater are caused to proceed through the spaces between each of the plates.

It is a fundamental principle of hydraulic flow, in both theory and fact, that fluids flowing in contact with a solid surface establish a velocity gradient. The velocity of flow is very small close to the solid surfaces, and increases with distance from solid surfaces. Within the spaces between the plates shown in Figure 7-97, then, the velocity of flow from the influent end of the tank to the effluent end is greatest near the midpoints between the plates and is very small close to the plates. Therefore, if the velocity of flow of wastewater flowing through the "plate settler" shown in Figure 7-97 is less than the settling velocity of some of the solid particles contained in the wastewater, these particles will "settle." As these particles settle, they quickly approach one of the plate surfaces, where the flow velocity steadily decreases. Eventually, the solid particles become very close to the

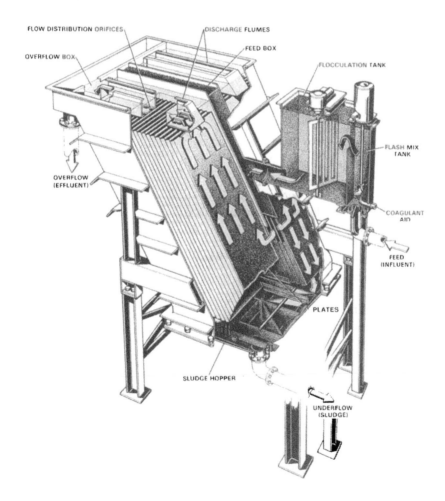

Figure 7-97   "Plate" or "lamella" separator (courtesy of Parkson Corporation).

plates, where the velocity of flow is almost zero. These settled particles will then slide down the plates and into the sludge hopper, as illustrated in Figure 7-97.

It is seen, then, that the treatment effectiveness of the simple, open clarifier can be improved dramatically by the addition of a series of "false bottoms," which have the effect of greatly decreasing the distance through which solid particles must settle before they are "removed" from the bulk liquid. Practical problems such as clogging of the spaces between the plates, the need for periodic cleaning before resuspension of organic solids by the action of denitrification or other phenomena, distribution of the flow at the influent end of the tank, collection of clarified flow at the effluent end, and the cost of the plate system become the factors that influence design. These practical considerations also influence the selection procedure between alternative types of false bottom systems or, indeed, whether it is more desirable to use a simple, open clarifier.

A great deal of experience has shown that, in general, plate settlers or similar alternatives are the clarifiers of choice for inorganic solids, such as those associated with metals precipitation. Because of bridging and clogging between the plates, however, plate settlers have not been shown to be advantageous for organic solids, such as activated sludge. For metals precipitation, for instance, it has been shown that the required HRT for satisfactory removal can be less for plate-type settlers than for open clarifiers by factors of six to eight.

Centrifugation

Centrifuges, which are devices that amplify the forces at work in gravity separation, are of several different types. Centrifuges are used for separation of grit and other relatively heavy solids at the head end of wastewater treatment facilities, as well as for sludge thickening and/or dewatering at the "other end" of such facilities.

So-called "swirl separators," an example of which is illustrated in Figure 7-98, are devices that use a relatively low centrifugal force to separate grit. The source of energy for these devices is normally a pump. The pump causes wastewater to flow through the swirl separator at a velocity sufficient to cause grit, or other high specific gravity material, to be thrown to the outside of the circular path within the separator (by centrifugal force). The solids are then collected at the outside of the separator, while the degritted liquid is collected at, or near, the center.

There are three basic types of centrifuges used for concentration of sludges: the solid bowl type, the basket type, and the disk-nozzle type. The essential differences are in the method by which separated solids are harvested. Figure 7-99 presents illustrations

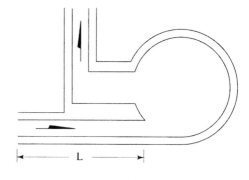

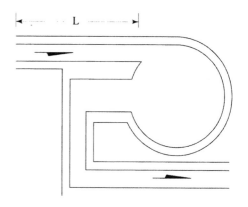

Figure 7-98 Swirl separator (courtesy of Smith & Loveless).

(b)

(a)

(c)

Figure 7-99   (a) Disk-stack centrifuge. (b) Basket-type centrifuge. (c) Solid-bowl centrifuge. (All courtesy of Alfa-Laval, Inc.)

of the three types of sludge concentrating centrifuges.

Centrifuges have also been used to a limited extent for separation of lighter-than-water substances, such as oil. In these cases, the aqueous discharge is collected at the opposite outlet from that where solids having a specific gravity greater than one are the target substances for separation.

A solid bowl-type centrifuge is diagrammed in Figure 7-100. The horizontal cylindrical-conical bowl rotates so as to cause solid particles to be thrown against its wall as they enter the bowl, with the liquid to be dewatered at the center of the cylinder. A helical scroll within the bowl rotates at a slightly slower rate than the bowl itself, operating in a manner similar to a screw conveyor to scrape the solids along the wall of the bowl to the solids discharge area. Centrate (the liquid from which solids have been removed) is discharged at the opposite end of the bowl.

The effectiveness of solids concentration is somewhat proportional to the speed of rotation of the bowl and inversely proportional, to a certain degree, to the feed rate. In some

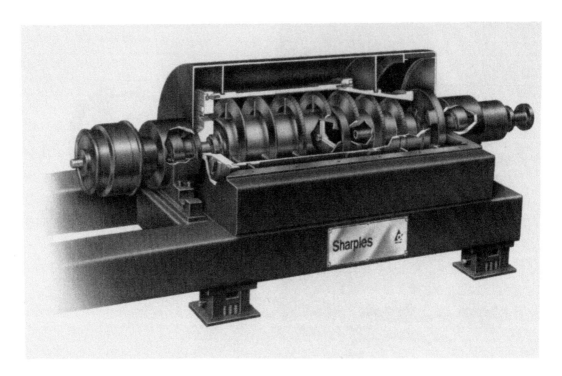

Figure 7-100   A solid-bowl centrifuge. (Courtesy of Alfa-Laval, Inc.)

instances, chemical conditioning of the feed enhances operational effectiveness. Various polymers, as well as lime, ferric chloride, and other chemical conditioning agents, have proven to be cost-effective performance enhancers in various applications. Variations in operating parameters of the solid bowl centrifuge can also affect performance, such as scroll rotation rate and pool volume.

Figure 7-101 shows a diagram of a basket-type centrifuge. Here, feed (liquid plus the solids to be removed) enters the basket at the bottom. The basket rotates about its vertical axis, and solid particles are caused by centrifugal force to be thrown against the sides of the basket. The basket gradually fills up with solids as liquid spills out over the top of the basket. When the basket becomes sufficiently full of solids, the feed is stopped, and a set of knives removes the solids from the basket. Machine variables that affect performance include basket spin rate, solids discharge interval, and skimmer nozzle dwell time and travel rate. Performance is normally inversely

proportional, to some degree, to feed rate, and, often, conditioning chemicals are not used.

The disc-stack centrifuge, diagrammed in Figure 7-102, is composed of a series of conical disks and channels between the cones, which rotate about a vertical axis. Feed enters the top, and solid particles are forced by centrifugal force against the cones. Centrate is discharged at the top of the unit just below the feed port, and solids exit the bottoms of the conical discs to a compaction zone. A small fraction of the wastewater is used to flush out the compacted solids. The moisture content of the dewatered solids is proportional to the feed rate. Normally, conditioning chemicals are used to enhance operational performance.

Flotation

Gravity Flotation

Gravity flotation is used, sometimes in combination with sedimentation and sometimes

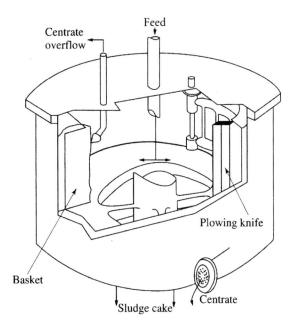

Figure 7-101   Diagram of a basket-type centrifuge (after EPA 625/1-74-006)..

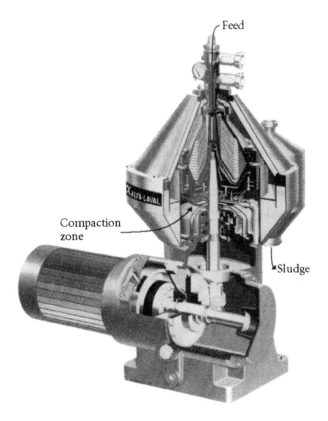

Figure 7-102   Disc-nozzle-type centrifuge (courtesy of Alfa Laval, Inc.).

alone, to remove oils, greases, and other flotables, such as solids, that have a low specific weight. Various types of "skimmers" have been developed to harvest floated materials, and the collection device to which the skimmers transport these materials must be properly designed. Figure 7-103 shows different types of gravity flotation and harvesting equipment.

Dissolved Air Flotation

Dissolved air flotation (DAF) is a solids separation process, similar to plain sedimentation. The force that drives DAF is gravity, and the force that retards the process is hydrodynamic drag. Dissolved air flotation involves the use of pressure to dissolve more air into wastewater than can be dissolved under normal atmospheric pressure and then releasing the pressure. The "dissolved" air, now in a supersaturated state, will come out of solution, or "precipitate," in the form of tiny bubbles. As these tiny bubbles form, they become attached to solid particles within the wastewater, driven by their hydrophobic nature. When sufficient air bubbles attach to a particle to make the conglomerate (particle plus air bubbles) lighter than water (specific gravity less than 1.0), the particle will be carried to the water surface.

A familiar example of this phenomenon is a straw in a freshly opened bottle of a carbon-

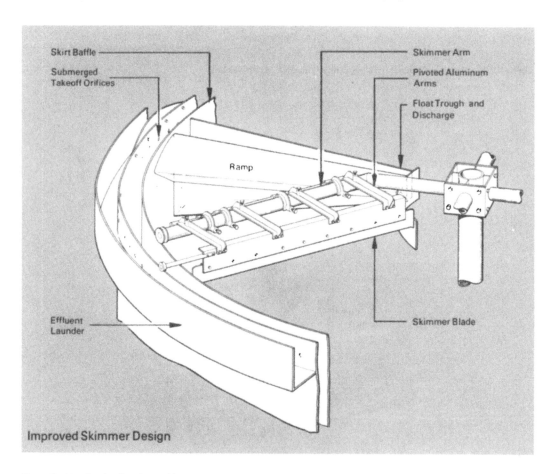

Figure 7-103   Gravity flotation and harvesting equipment. (a) Courtesy of IDI/Infilco Degremont, Inc.

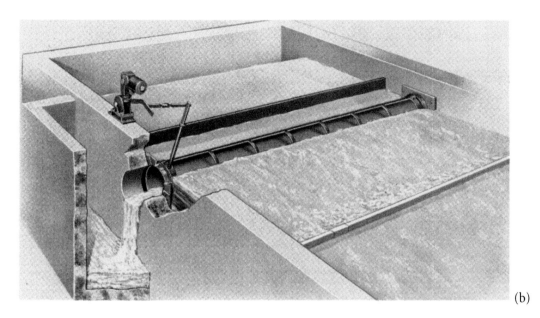

(b)

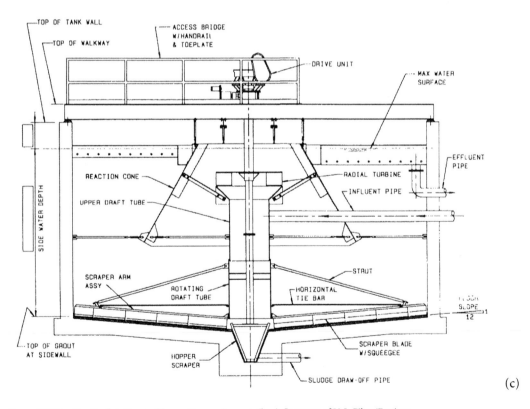

(c)

Figure 7-103   Gravity flotation and harvesting equipment. (b, c) Courtesy of U.S. Filter/Envirex.

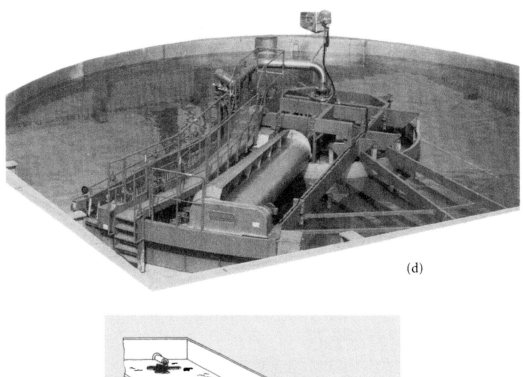

(d)

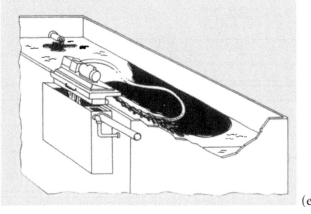

(e)

Figure 7-103   Gravity flotation and harvesting equipment. (d) Courtesy of KWI. (e) Courtesy of Oil Skimmers, Inc.

ated beverage. Before the bottle is opened, its contents are under pressure, having been pressurized with carbon dioxide gas at the time of bottling. When the cap is taken off, the pressure is released, and carbon dioxide precipitates from solution in the form of small bubbles. The bubbles attach to any solid surface, including a straw, if one has been placed in the bottle. Soon, the straw will rise up in the bottle.

In a manner similar to the straw, solids with a specific gravity greater than one can be caused to rise to the surface of a volume of wastewater. As well, solids having a specific gravity less than one can be caused to rise to the surface at a faster rate by using DAF than without it. Often, chemical coagulation of the solids can significantly enhance the process, and, in some cases, dissolved solids can be precipitated chemically and then separated from the bulk solution by DAF.

Dissolution of Air in Water

Examination of the molecular structures of both oxygen and nitrogen reveals that neither would be expected to be polar; therefore, neither would be expected to be soluble in water.

$$:O::O: \qquad\qquad :N:::N:$$

Molecular oxygen   Molecular nitrogen

The question arises, then, as to how non-polar gases like oxygen and nitrogen can be dissolved in water. The answer, of course, is that they cannot. Nonpolar gases are "driven" into a given volume of water by the mechanism of diffusion. Molecules of nitrogen and oxygen in the gaseous state create pressure as a result of constant motion of the molecules. The generalized gas law:

$$pV = nRT \qquad\qquad (7\text{-}73)$$

where

  p = Pressure, psi
  V = Volume, in^3
  n = Number of moles of gas in a given volume of gas
  R = Universal gas constant 0.082 liter atmospheres/mole/° Kelvin
  T = Temperature, ° Kelvin

rewritten as

$$p = \frac{nRT}{V} \qquad\qquad (7\text{-}74)$$

states that the pressure exerted by a given number of molecules of a gas is directly proportional to the temperature and inversely proportional to the size of the volume in which those molecules are confined.

Dalton's law of partial pressures states further that, in a mixture of gases, each gas exerts pressure independently of the others, and the pressure exerted by each individual gas, referred to as its "partial pressure," is the same as it would be if it were the only gas in the entire volume. The pressure exerted by the mixture, therefore, is the sum of all the partial pressures. Conversely, the partial pressure of any individual gas in a mixture, such as air, is equal to the pressure of the mixture multiplied by the fraction, by volume, of that gas in the mixture.

Henry's law, which mathematically describes the solubility of a gas in water in terms of weight per unit volume, is written as

$$C = Hp \qquad\qquad (7\text{-}75)$$

where

  C = concentration of gas "dissolved" in water (mg/L)
  H = Henry's constant, dimensionless, specific for a given gas

The consequence of the above is that, by way of the process of diffusion, molecules of any gas in contact with a given volume of water will diffuse into that volume to an extent that is described by Henry's law, as long as the quantity of dissolved gas is relatively small. For higher concentrations, Henry's constant changes somewhat. This principle holds for any substance in the gaseous state, including volatilized organics. The molecules that are forced into the water by this diffusion process exhibit properties that are essentially identical to those that are truly dissolved. In conformance with the second law of thermodynamics, they distribute themselves uniformly throughout the liquid volume (maximum disorganization), and they will react with substances that are dissolved. An example is the reaction of molecular oxygen with ferrous ions. Unlike dissolved substances, however, they will be replenished from the gas phase with which they are in contact, up to the extent described by Henry's law, if they are depleted by way of reaction with other substances or by biological metabolism. The difference between a substance existing in water solution as the result of diffusion and one that is truly dissolved can be illustrated by the following example.

Consider a beaker of water in a closed space—a small, airtight room, for instance. An amount of sodium chloride is dissolved in the water, and the water is saturated with oxygen; that is, it is in equilibrium with the air in the closed space. Now, a container of sodium chloride is opened, and, at the same time, a pressurized cylinder of oxygen is released. The concentration of sodium chloride will not change, but because the quantity of oxygen in the air within the closed space increases (partial pressure of oxygen increases), the concentration of dissolved oxygen in the water will increase. The oxygen molecules are not truly dissolved; that is, they are not held in solution by the forces of solvation or hydrogen bonding by the water molecules. Rather, they are forced into the volume of water by diffusion, which is to say, by the second law of thermodynamics. The molecules of gas are constantly passing through the water-air interface in both directions. Those that are in the water are constantly breaking through the surface to return to the gas phase, and they are continually replaced by diffusion from the air into the water. An equilibrium concentration becomes established, described by Henry's law. All species of gas that happen to exist in the "air" participate in this process: nitrogen, oxygen, water vapor, volatilized organics, or whichever other gases are included in the given volume of air.

The concentration, in terms of mass of any particular gas that will be forced into the water phase until equilibrium becomes established, depends upon the temperature, the concentration of dissolved substances such as salts, and the "partial pressure" of the gas in the gas phase. As the temperature of the water increases, the random vibration activity, "Brownian motion," of the water increases. This results in less room between water molecules for the molecules of gas to "fit into." The result is that the equilibrium concentration of the gas decreases. This is opposite to the effect of temperature on dissolution of truly soluble substances in water, or other liquids, where increasing temperature results in increasing solubility.

Some gases are truly soluble in water because their molecules are polar, and these gases exhibit behavior of both solubility and diffusivity. Carbon dioxide and hydrogen sulfide are examples. As the temperature of water increases, solubility increases, but diffusivity decreases. Also, since each of these two gases exists in equilibrium with hydrogen ion when in water solution, the pH of the water medium has a dominant effect on their solubility, or rather their equilibrium concentration, in water.

In the example above, where a beaker of water is in a closed space, if a flame burning in the closed space depletes the oxygen in the air, oxygen will come out of the water solution. If all of the oxygen is removed from the air, the concentration of "dissolved oxygen" in the beaker of water will eventually go to zero (or close to it), and the time of this occurrence will coincide with the flame extinguishing, because of lack of oxygen in the air.

Dissolved Air Flotation Equipment

The dissolved air flotation process takes advantage of the principles described above. Figure 7-104 presents a diagram of a DAF system, complete with chemical coagulation and sludge handling equipment. As shown in Figure 7-104, raw (or pretreated) wastewater receives a dose of a chemical coagulant (metal salt, for instance) and then proceeds to a coagulation-flocculation tank. After coagulation of the target substances, the mixture is conveyed to the flotation tank, where it is released in the presence of recycled effluent that has just been saturated with air under several atmospheres of pressure in the pressurization system shown. An anionic polymer (coagulant aid) is injected into the

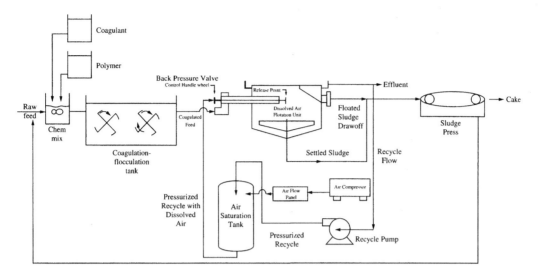

Figure 7-104   Dissolved air flotation system.

coagulated wastewater just as it enters the flotation tank.

The recycled effluent is saturated with air under pressure as follows: a suitable centrifugal pump forces a portion of the treated effluent into a pressure holding tank. A valve at the outlet from the pressure holding tank regulates the pressure in the tank, the flow rate through the tank, and the retention time in the tank, simultaneously. An air compressor maintains an appropriate flow of air into the pressure holding tank. Under the pressure in the tank, air from the compressor is diffused into the water to a concentration higher than its saturation value under normal atmospheric pressure. In other words, about 24 ppm of "air" (nitrogen plus oxygen) can be "dissolved" in water under normal atmospheric pressure (14.7 psig). At a pressure of six atmospheres, for instance (6 × 14.7 = about 90 psig), Henry's law would predict that about 6 × 23, or about 130 ppm, of air can be diffused into the water. In practice, dissolution of air into the water in the pressurized holding tank is less than 100% efficient, and a correction factor, f, which varies between 0.5 and 0.8, is used to calculate the actual concentration.

After being held in the pressure holding tank in the presence of pressurized air, the recycled effluent is released at the bottom of the flotation tank, in close proximity to where the coagulated wastewater is being released. The pressure to which the recycled effluent is subjected has now been reduced to one atmosphere, plus the pressure caused by the depth of water in the flotation tank. Here, the "solubility" of the air is less, by a factor of slightly less than the number of atmospheres of pressure in the pressurization system, but the quantity of water available for the air to diffuse into has increased by the volume of the recycle stream.

Practically, however, the wastewater will already be saturated with respect to nitrogen, but may have no oxygen, because of biological activity. Therefore, the "solubility" of air at the bottom of the flotation tank will be about 25 ppm, and the excess air from the pressurized, recycled effluent will precipitate from "solution." As this air precipitates in the form of tiny, almost microscopic, bubbles, the bubbles attach to the coagulated solids. The presence of the anionic polymer (coagulant aid), plus the continued action of the coagulant, causes the building of larger solid conglomerates, entrapping many of the adsorbed air bubbles. The net effect is that the solids are floated to the surface of the flotation tank, where they can be collected by some means and thus be removed from the wastewater.

Some DAF systems do not have a pressurized recycle system, but, rather, the entire forward flow on its way to the flotation tank is pressurized. This type of DAF is referred to as "direct pressurization" and is not widely used for treatment of industrial wastewaters because of undesirable shearing of chemical flocs by the pump and valve.

Air-to-Solids Ratio

One of the principal design parameters for a DAF wastewater treatment system is the so-called air-to-solids ratio (A/S). The mass of air that must be supplied per day by the compressor and air delivery system is calculated by multiplying the mass of solids to be removed each day by the numerical value of the A/S. If the A/S is too low, there will not be sufficient flotation action within the treatment system. If, on the other hand, the A/S value is too high, there will be many more air bubbles than can attach to the solids. Many of the unattached fine bubbles will coalesce, and large bubbles will result, causing turbulence in the flotation tank. Experience has shown that A/S values in the range 0.02 to 0.06 result in optimal flotation treatment.

Calculations of Recycle Ratio and Quantity of Air

The quantity of air needed for a given industrial waste treatment application can be calculated directly from the air-to-solids ratio, once its value has been determined by laboratory and pilot experimentation:

Weight of air, lb/day
$$= (\text{lb solids/day}) \times \text{A/S} \qquad (7\text{-}76)$$

Optimum Pressure for the Air Dissolving System

The optimum pressure for the air dissolving system is determined by balancing the cost for operating the water pressurizing pump and the compressor against the capital costs for the individual components of the system, which include the pressurization tank. Generally, the higher the pressure, the greater the cost for power but the smaller the recycle flow and, therefore, the smaller the size of the components. However, as shown by the equation for pump horsepower:

$$HP = \frac{Q \rho H}{550} \times \text{ pump efficiency} \qquad (7\text{-}77)$$

where

HP = Horsepower required to operate pump
Q = Rate of flow being pumped (ft³/sec)
ρ = Specific weight of water (lb/ft³)
H = Pressure or "head" against which the pump is pumping. In this case, H equals the pressure of the pressurization tank, in feet of water

As shown in equation 7-77, the horsepower required for the pressurization pump increases directly with increasing pressure and decreases with decreasing rate of flow. Therefore, there is a compensating tradeoff. As the pressure is increased to dissolve more air in a given flow rate of water or to dissolve a given quantity of air in a smaller flow rate of water, power requirement increases. Power requirement decreases, however, as the rate of flow decreases due to less water needed by the higher pressure. The fact that higher pressure results in the need for more heavy-duty equipment, coupled with the fact that higher pressure requires more power consumption by the air compressor (the air requirement is governed by the solids load, which is not related to the pressure), results in increased costs for increasing pressure.

Because of these counteracting effects on costs, it is not a simple matter to select an operating pressure that will minimize overall costs. In most applications, operating pressures between 60 and 120 psig have been used.

Design of the air-dissolving portion of a DAF treatment system includes the following:

1. Determination of the size (i.e., air flow rate, working pressure, and, consequently, the motor horsepower of the compressor)
2. Determination of the water flow rate, total dynamic head, and, consequently, the motor horsepower of the recycle pump
3. Determination of the size and pressure rating of the pressure holding tank
4. Selection of the appropriate valve for control of flow rate from the pressure holding tank, as well as the pressure to be maintained
5. Selection of the appropriate sizes of piping and materials of construction

Competing considerations that concern the working pressure for the pressurization-air dissolving system are that the higher the pressure, the more air can be dissolved but the more horsepower is required, and therefore the operating costs will be higher for the recycle pump motor and the compressor motor.

Adsorption

Adsorption can be defined as the accumulation of one substance on the surface of another. The substance undergoing accumulation, and thus being adsorbed, is called the adsorbate, and the substance on which the accumulation is taking place is called the adsorbent. The adsorbate can be dissolved, in which case it is called the solute, or it can be of the nature of suspended solids as in a colloidal suspension. Colloidal suspensions of liquids or gases can also be adsorbed. The discussion presented here addresses substances such as ions or organic compounds dissolved in water adsorbing onto solid adsorbents. However, the principles hold for solid, liquid, or gaseous adsorbates adsorbing onto solid, liquid, or gaseous adsorbents.

The mechanism of adsorption can be one or a combination of several phenomena, including chemical complex formation at the surface of the adsorbent, electrical attraction (a phenomenon involved in almost all chemical mechanisms, including complex formation), and exclusion of the adsorbate from the bulk solution, resulting from stronger intermolecular bonding between molecules of solvent (hydrogen bonding, in the case of water) than existed between molecules of solvent (water) and the solute. As is almost always the case, the driving force is explained by the second law of thermodynamics. The sum total energy of all bonds is greater after adsorption has taken place than before.

As a wastewater treatment process, adsorption is of greatest use when the substance to be removed is only sparingly soluble in water. Many organic substances have groups that ionize weakly or have very few ionizable groups per unit mass of substance. These groups form hydrogen bonds with water molecules that are of sufficient strength to hold the organic molecules in solution, but a good adsorbent will reverse the solvation process because, as stated above, the sum total of all chemical bonds is greater after adsorption than before.

The Adsorbent

Activated carbon is the most common adsorbent in use for industrial wastewater treatment. Other adsorbents include synthetic resins, activated alumina, silica gel, fly ash, shredded tires, molecular sieves, and sphagnum peat. Because adsorption is a surface phenomenon, a desirable characteristic of an adsorbent is a high surface-to-volume ratio. Surface-to-volume ratios are increased in two ways: by decreasing the size of particles of adsorbent and by creating a network of pores or "tunnels" within the particles of adsorbent.

Activated carbon is manufactured by pyrolyzing organic materials such as bones, coconut shells, or coal andthen oxidizing residual hydrocarbons using air or steam. The result is a granular material with an intricate, interconnected network of pores ranging in size from a few angstroms to several thousand angstroms in diameter. The pores that are 10 to 1,000 angstroms in diameter are referred to as micropores and are responsible for most of the adsorptive capability of the material. The larger pores are relatively unimportant in adsorptive treatment capability, but are important passageways through which ions and dissolved organics can diffuse to reach the innermost adsorptive surfaces.

Most adsorbents have weak negative charges on their surfaces; therefore, the pH of the wastewater being treated has an influence on the adsorptive process. This is because hydrogen ions repress the negative surface charges at lower pH values. Also, at low pH conditions, there is an increase in adsorption of dissolved or suspended substances that tend to have a negative surface charge.

Adsorption Equilibria

When wastewater is successfully treated by adsorption, the substances being removed adsorb onto and thus "coat" the adsorbent until equilibrium is established between the molecules of substance adsorbed and those still in solution. Depending on the characteristics of both the adsorbate and the adsorbent, equilibrium will be reached when either a complete monolayer of molecules coats the adsorbent or when the adsorbent is coated with layers of adsorbate several molecules thick.

A standard laboratory procedure that is used to determine the effectiveness with which a given wastewater can be treated with a given adsorbent is described as follows: Several flasks containing samples of the wastewater (containing the adsorbate) are dosed with different quantities of adsorbent, which is in either granular form or has been ground in a ball mill to powder form. The flasks are shaken using a device that shakes

all the flasks together, and the temperature is held constant. Shaking continues until the contents of all flasks have reached equilibrium, as determined by the concentration of adsorbate remaining constant with time. The adsorbate may be a specific chemical compound, such as trichloroethylene, or a group of compounds measured as COD, for instance. Then a plot is constructed showing the relationship between the concentration of adsorbate in solution after equilibrium has been reached and the quantity of adsorbate adsorbed per unit mass of adsorbent. Such a plot is called an isotherm, owing to the conditions of constant temperature held throughout the test.

Several standardized isotherm models have been developed to provide the ability to calculate estimates of adsorbent efficiency and costs after generating a relatively small amount of laboratory data. The model developed by Langmuir is one of the most widely used, and is stated as follows:

$$q = \frac{q_m K_a C}{1 + K_a C} \qquad (7\text{-}78)$$

where

q = Mass of adsorbate adsorbed/mass of adsorbent

q_m = Mass of adsorbate adsorbed/mass of adsorbent if a complete layer, one molecule thick, were adsorbed

K_a = Constant (related to enthalpy of adsorption)

C = Concentration of adsorbate present in solution at equilibrium

Langmuir's model is based on (1) the assumption that only a single layer of molecules of adsorbate will adsorb to the adsorbent, (2) the immobility of the adsorbate after being adsorbed, and (3) equal enthalpy of adsorption for all molecules of adsorbate. It has long been accepted as a good, generalized model for use in making estimates based on very little data. It is most useful when linearized as follows:

$$\frac{C}{q} = \frac{1}{K_a q_m} + \frac{C}{q_m} \qquad (7\text{-}79)$$

or

$$\frac{1}{q} = \frac{1}{K_a q_m C} + \frac{1}{q_m} \qquad (7\text{-}80)$$

Laboratory data can be plotted as shown in Figures 7-105 and 7-106.

Values for the constants K_a and q_m can then be determined from the slopes and intercepts. These constants can then be used in the original Langmuir equation (equation 7-79) to estimate the quantity of adsorbent

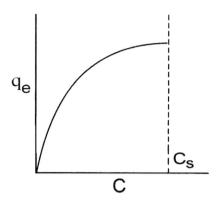

Figure 7-105   Plot of laboratory data for Langmuir isotherms.

needed given a known quantity of adsorbate to be removed (inverse of q).

If adsorption is known or suspected to take place in multiple layers, a more appro-

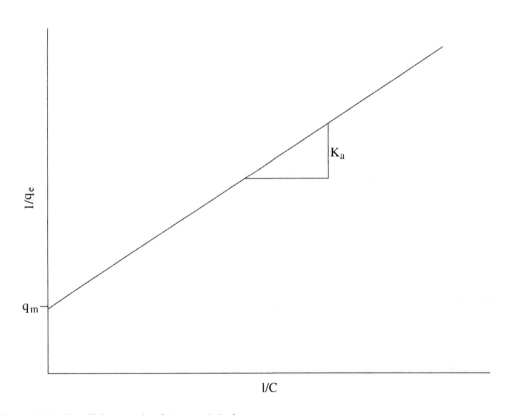

Figure 7-106   Plot of laboratory data for Langmuir isotherms.

priate model is given by the Brunauer-Emmett-Teller (BET) equation (7-81):

$$q = \frac{q_m K_b C}{(C_S - C)\left[1 + (K_b - 1)\left(\frac{C}{C_S}\right)\right]} \quad (7\text{-}81)$$

where

C_S = Concentration of adsorbate in the wastewater when all layers of adsorbate on the adsorbent are saturated

K_b = Constant related to energy of adsorption

Figure 7-107 shows the approximate relationship between the concentration of adsorbate in the wastewater after adsorption has taken place to the point of equilibrium and the amount adsorbed.

As shown in Figure 7-107, this relationship is not linear; however, the BET model can be rearranged to the following linear form:

$$\frac{C}{(C_S - C)q} = \frac{1}{K_b q_m} + \frac{K_b - 1}{K_b q_m}\left(\frac{C}{C_S}\right) \quad (7\text{-}82)$$

and the plot of laboratory data shown in Figure 7-108 can be used to determine the mathematical value of the constant K_b.

Perhaps the most widely used model for estimating adsorbent efficiency and costs is an empirical model developed by Freundlich, stated as follows:

$$q = \frac{\frac{1}{n}}{K_f C} \quad (7\text{-}83)$$

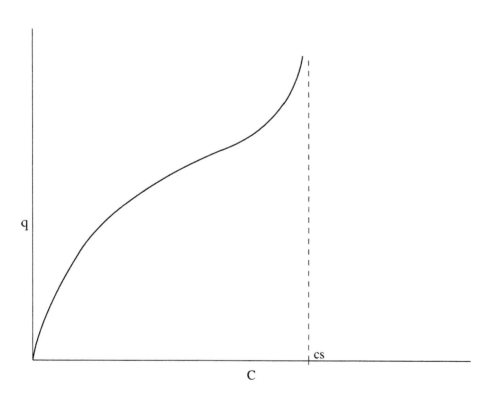

Figure 7-107   Concentration of adsorbate in bulk liquid versus amount adsorbed.

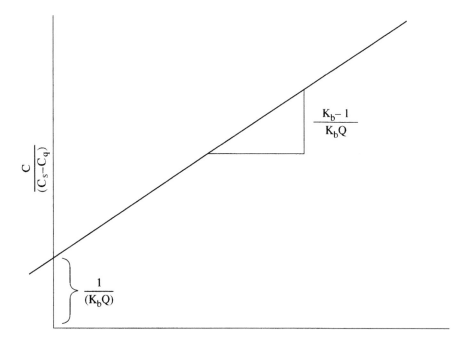

Figure 7-108   Plot of laboratory data for BET isotherms.

where

    K_f = Constant
    n = Empirical constant greater than one

Freundlich's model requires no assumptions concerning the number of layers of adsorbed molecules, heats of adsorption, or other conditions. It is strictly a curve-fitting model and works very well for many industrial wastewater applications.

Freundlich's model is made linear by taking the logarithm of both sides:

$$\log\ q = \log\ K_f + (\tfrac{1}{n}) \times \log\ C \qquad (7\text{-}84)$$

Plotting q vs. C on log-log paper yields a straight line, the slope of which is the inverse of n, and the vertical intercept is the value of K_f, as shown in Figure 7-109.

In practice, candidate adsorbent materials (different types of activated carbon, for instance) are evaluated for effectiveness in treating a given industrial wastewater by constructing the Freundlich, Langmuir, or BET isotherms after obtaining the appropriate laboratory data. The isotherms provide a clear indication as to which candidate adsorbent would be most efficient in terms of pounds of adsorbent required per pound of adsorbate removed, as well as an indication of the quality of effluent achievable. Example 7-8 illustrates this procedure.

Example 7-8

Wastewater from a poultry processing plant was treated to the extent of greater than 99% removal of COD and TSS using chemical coagulation, dissolved air flotation, and sand filtration. The question then arose as to the economic feasibility of using activated carbon to further treat the water for (re)use in the plant as cooling water or for initial washdown during daily plant cleanup. Table 7-15 presents the data obtained from the isotherm tests using the adsorbate designated Carbon A.

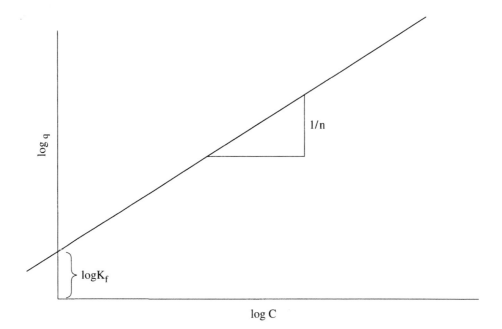

Figure 7-109   Log-log plot of q vs. C (Freundlich isotherm).

The data shown in Table 7-15 were plotted on a log-log scale, as shown in Figure 7-110.

Figure 7-110 shows that the slope of the isotherm was about 0.899, and the vertical intercept was about 0.007. Therefore, the value for the constant, n, in the Freundlich model is calculated as:

$$1/n = 0.899 \; ; \; n = 1.112 \tag{7-85}$$

and the value for the constant, K_f, is calculated as:

$$intercept = \log K_f = \log\ 0.007;$$
$$K_f = 0.007 \tag{7-86}$$

It is now possible to use the Freundlich model to calculate the quantity of Carbon A required to produce treated effluent of whichever quality is deemed appropriate, balancing cost against effluent quality:

Table 7-15   Data Obtained from Isotherm Tests—Carbon A

Flask No.	Carbon Dose (Grams)	TOC of Effluent at Equilibrium (mg/L)	TOC on Carbon (mg/mg Carbon)
1	3.2	2.2	18
2	1.8	7.6	0.028
3	1.3	20.1	0.038
4	0.92	22.3	0.052
5	0.55	28.1	0.076
6	0.41	30.2	0.096

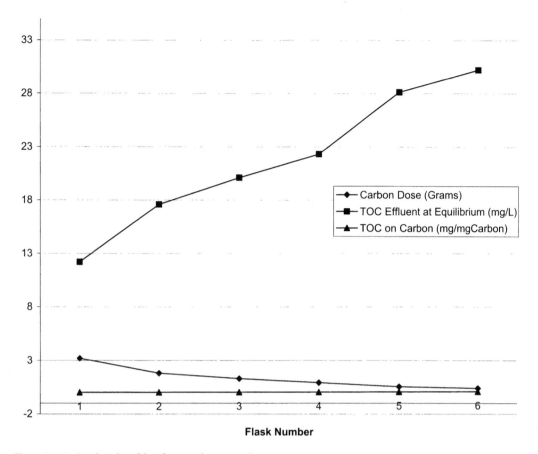

Figure 7-110   Log-log plot of data from poultry processing wastewater.

For a desired effluent quality of TOC = 10 mg/L = C:

$$\frac{1}{n} = \frac{1}{1.112}$$

$$q = K_f C \qquad (7\text{-}87)$$

$$q = 0.007(10)$$

$$= 0.06 \text{ mg TOC adsorbed/mg adsorbent}$$

or, 1/0.06 = 18 pounds carbon adsorbent required for each pound of TOC removed from the sand filter effluent.

Evaluating Relative Effectiveness between Different Carbon Products

Two characteristics of activated carbon can be evaluated from the Freundlich isotherms.

One is the relative capability for removing impurities from the wastewater to a sufficiently low level to satisfy requirements for reuse or for discharge. The second characteristic is the relative efficiency of several different carbon products, that is, the quantity of activated carbon required to remove a given quantity of pollutants, as illustrated in Figure 7-111.

Figure 7-111 shows Freundlich isotherms for three different activated carbon products plotted on the same graph. In such a composite graph, the further to the right the isotherm, the less efficient the activated carbon product in terms of pounds of the carbon product required to remove one pound of TOC. Figure 7-111 further shows that if a TOC concentration of no more that 10 mg/L (for instance) can be tolerated, Carbons A

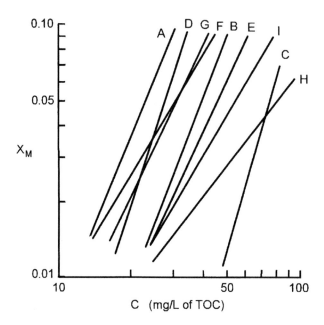

Figure 7-111   Composite plot of isotherms of three different carbon products.

and C have potential use, while Carbon B need not be considered further.

As additional notes, it has been found that in general, high-molecular-weight (greater than 45) organics that are not readily biodegradable tend to be good candidates for removal by adsorption on activated carbon or other adsorbates. Also, certain inorganics that are not highly soluble in water (some heavy metals, reduced sulfur gases, and chlorine, are examples) are good candidates for removal by adsorption.

Activated carbon can be very effective in removing metal ions from wastewater by first chelating the metal ions with an organic chelant, such as citric acid or EDTA. When the chelated mixture is treated using activated carbon, the organic chelant will adsorb to the carbon, removing the chelated metals along with it.

Ion Exchange

Ion exchange is a physical treatment process in which ions dissolved in a liquid or gas interchange with ions on a solid medium.

The ions on the solid medium are associated with functional groups that are attached to the solid medium, which is immersed in the liquid or gas.

Typically, ions in dilute concentrations replace ions of like charge and lower valence state; however, ions in high concentration replace all other ions of like charge. For instance, calcium ions or ferric ions in dilute concentrations in water or wastewater replace hydrogen or sodium ions on the ion exchange medium. The divalent or trivalent ions move from the bulk solution to the surface of the ion exchange medium, where they replace ions of lesser valence state, which, in turn, pass into the bulk solution. The ion exchange material can be solid or liquid, and the bulk solution can be a liquid or a gas.

However, if a bulk solution has a high concentration of ions of low valence state and is brought into contact with an ion exchange material that has ions of higher valence state, the higher valence ions will be replaced by the lower valence state ions. For instance, if a strong solution of sodium chloride is brought into contact with an anion exchange material that has nitrate ions associated with its functional groups, the chloride ions will replace the nitrate ions.

In these examples, the ions that exchange are referred to as mobile, and the functional groups are referred to as fixed. Ion exchange materials occur extensively in nature; ion exchange materials of high capacity can be manufactured. Clays are examples of naturally occurring ion exchange materials. Approximately one-tenth of a pound of calcium (expressed as calcium carbonate), for instance, per cubic foot of clay is a typical cation exchange capacity for a naturally occurring clay. So-called zeolites are naturally occurring materials of much higher exchange capacity. About one pound of calcium per cubic foot of zeolite is a good average value for this material. Synthetic ion exchange materials are produced that have capacities of over 10 pounds of calcium per cubic foot of exchange material.

Naturally occurring materials that have ion exchange capability include soils, lignin, humus, wool, and cellulose. The ion exchange capacity of soils is made use of by land treatment systems. Synthetic ion exchange resins of high capacity are manufactured, typically, as illustrated in Figure 7-112.

As shown in Figure 7-112, a "good" ion exchange material consists of a foundation of an insoluble, organic, or inorganic three-dimensional matrix having many attached functional groups. In the example shown in Figure 7-112, the three-dimensional matrix is formed by polymerization, in three dimensions, of styrene and divinylbenzene molecules. The functional groups are soluble ions that are able to attract ions of opposite charge; they are attached by reacting various chemicals with the basic matrix material. In the example shown in Figure 7-112, the soluble ions are sulfonate ions, attached to the three-dimensional matrix by reacting the matrix with sulfuric acid. The term *resin* is used for the foundation matrix material.

Materials that exchange cations have acidic functional groups such as the sulfonic group, $R\text{-}SO3^-$; carboxylic, $R\text{-}COO^-$; phenolic, $R\text{-}O^-$; or the phosphonic group, $R\text{-}PO_3H^-$. In each case the R represents the foundation matrix. Materials that exchange anions have the primary amine group, $R\text{-}NH^+$; the secondary amine group, $R\text{-}R'N^+$; or the quaternary amine group, $R\text{-}R'3N+$. The tertiary amine group can also be used. Cation exchange resins with a high degree of

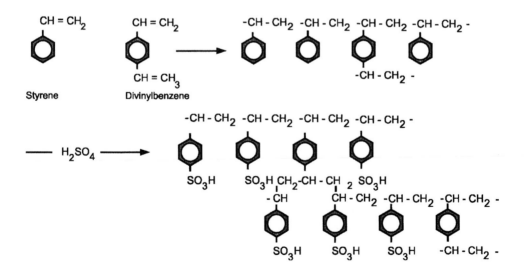

Figure 7-112   Typical process for manufacturing ion exchange resins.

ionization are referred to as "strongly acidic," and those with a low degree of ionization are called "weakly acidic exchangers." Strongly basic and weakly basic anion exchangers are named in the same way.

Strongly acidic or strongly basic ion exchangers normally have greater attraction for the target ions and are therefore more efficient in terms of effluent quality for a given wastewater loading rate than are weakly acidic or weakly basic ion exchangers. The strongly acidic or basic exchangers, however, require more regenerant than the weakly ionized exchangers. In all cases, exchange capacity, in terms of mass of ions exchanged per mass of exchange material, is determined by the number of functional groups per unit mass of the material.

Higher numbers of functional groups per unit mass of material are made possible by higher surface-to-volume ratios for the matrix material. As with all sorption materials, higher surface-area-to-volume ratios result from the existence of pores throughout the basic material. However, when compared with simple exchange at the outer surface of a bead of ion exchange material, the ion exchange process is somewhat slower when ions must diffuse into and out of pores.

The density of cross-linking within the latticework, or matrix, of the exchange material is what governs the number and size of pores and, consequently, has a strong influence on exchange capacity. Greater cross-linking produces a stronger, more resilient resin; however, if the degree of cross-linking is too high, pores will be too small. Larger ions will be physically blocked from entering or moving through them.

Mechanisms of Ion Exchange

The ion exchange process takes place as illustrated in Figure 7-113.

Figure 7-114(a) shows a cross-section of a bead of ion exchange material immersed in wastewater containing zinc ions. The functional groups (sulfonic groups) are associated with hydrogen ions at the start of the exchange process. There is a strong gradient of concentration of zinc ions between the bulk solution and the interior of the pores, which are filled with water. There is also a strong gradient of concentration of hydrogen ions between the surface of the exchange resin throughout its porous structure and the

(A) Initial State Prior to Exchange
Reaction with Cation B^+

(B) Equilibrium State After Exchange
Reaction With Cation B^+

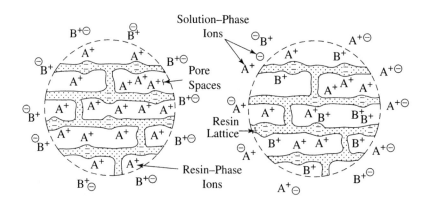

Figure 7-113   Schematic diagrams of a cation exchange resin framework with fixed exchange sites prior to and following an exchange reaction. (a) Initial state prior to exchange reaction with cation B^+. (b) Equilibrium state after exchange reaction with cation B^+.

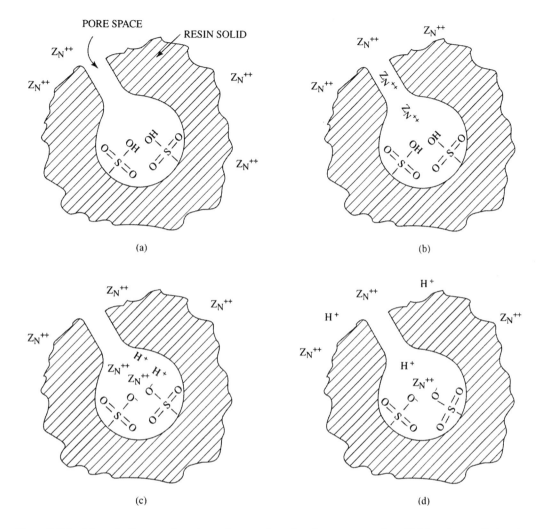

(a)

(b)

(c)

(d)

Figure 7-114   Schematic diagram of a cation exchange resin exchanging two monovalent (H^-) ions for one divalent ion (Z_N^{++}).

water within the pores. Because of this, hydrogen ions have a strong tendency to leave the functional groups and diffuse throughout the pores, but the stronger tendency to maintain electrical neutrality prevents this from taking place. However, as the zinc ions are driven into the interior of the pores by molecular diffusion, due to their concentration gradient (Figure 7-114[b]), they are available to interchange with the hydrogen ions (Figure 7-114[c]). Thus, the force tending to distribute hydrogen and zinc ions uniformly throughout the aqueous system (entropy) is satisfied, and the tendency

to maintain electrical neutrality is also satisfied (Fig 7-114[d]). The tendency to maintain electrical neutrality results in stoichiometric exchange (i.e., one divalent ion exchanges with two monovalent ions, etc.).

Kinetics Of Ion Exchange

Although the description given above of the example illustrated in Figure 7-114 is useful, there is more to the overall ion exchange process. In general, the rate-controlling process is either the rate of diffusion of ions through the film (the region of water molecules surrounding the ion exchange resin material) or the rate of diffusion of the interchanging ions within the pores. The first of these processes is called film diffusion and the second is termed pore diffusion. If the exchange treatment process is of the batch type, higher rates of stirring will minimize the retarding effects of film diffusion. In a continuous flow column system, higher flow rates minimize these effects. Larger pores can minimize the retarding effects of pore diffusion.

Ion Selectivity

Ion exchange materials exist that are selective for specific ions. For instance, Gottlieb has reported on resins that have a high selectivity for nitrate ions. However, for most ion exchange materials, there is a common selectivity sequence that is based on fundamental chemical properties and/or characteristics.

In general, the smaller the mobile ion for a given charge (valence state), the more strongly attracted and, thus, selective it is toward a given ion exchange material. Also, in general, the lower the atomic weight of an ion, the smaller its size. However, ions are dissolved in water as a result of water molecules surrounding the ion, attracted by electrolytic forces and held by hydrogen bonding, as described in Chapter 2. The "solvated" ion (ion plus attached layers of water molecules) constitutes the mobile unit. Since ions of higher atomic weight hold solvated water layers more tightly than ions of lower atomic weight, the heavier ions make up mobile units that are of smaller radius than ions of lower atomic weight. The end result is that, in general, the higher the molecular weight for a given ionic charge, the more highly selective most ion exchange materials are for that ion. Monovalent cations, consequently, exhibit selectivity as follows:

$$Ag^+ > Cs^+ > Rb^+ > K^+ > Na^+ > Li^+$$

For divalent cations, the order of selectivity is:

$$Ba^{++} > Sr^{++} > Ca^{++} > Mg^{++} > Be^{++}$$

and for monovalent anions, the order of selectivity is:

$$CNS^- > C104^- > I^- > NO_3^- > Br^- > CN^-$$
$$> HSO_4^- > NO_2^- > Cl^- > HCO_3^-$$
$$> CH_3COO^- > OH^- > F$$

Application of Ion Exchange to Industrial Wastewater Treatment

Ion exchange can be used to remove undesirable ions from industrial wastewaters as a final or tertiary treatment step, as treatment for isolated process streams as part of a waste minimization program, or as a polishing step prior to recycle and reuse of process water or wastewater. Ion exchange can also be used to recover valuable metals or other exchangeable substances.

As a process, ion exchange can be operated in either the batch mode or the continuous flow mode. In the batch mode, a container of fluid to be treated is dosed, then mixed with an appropriate quantity of ion exchange "beads." Beads of 2 to 4 mm diameter are the physical form in which ion exchange resins are normally used. After the desired amount of exchange of ions has taken place, the beads of resin are separated from the treated fluid by sedimentation, filtration, or other solids separation process. The exchange resin is then rinsed, then "recharged" with an appropriate solution of acid, base, or salt.

Far more widely used than the batch process is the continuous flow ion exchange process, illustrated in Figure 7-115.

As shown in Figure 7-115, continuous flow ion exchange is carried out by passing the fluid to be treated through one or more cylindrical containers (referred to as "columns") packed with exchange resin beads, 2 to 8 feet or more in depth. The resin is first "charged" with an appropriate solution (as an example, 5% to 20% sulfuric acid could be used to charge a strong acid cation exchange resin). When essentially all of the functional groups have charging ions associ-

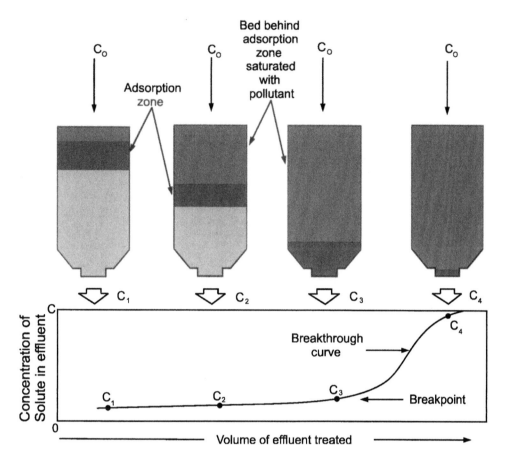

Figure 7-115   Schematic of the continuous flow ion exchange process.

ated with them, the charging cycle is stopped, and the material is rinsed with water of very low ion content. Then the removal cycle is initiated by passing 2 to 5 gpm/ft^3 (gpm of wastewater per ft^3 of ion exchange resin) through the bed. During this cycle, the target ions diffuse into the pores of the beads of resin and replace the charging ions on the functional groups. The replaced charging ions diffuse out through the pores and into the bulk solution, then exit the column as a waste that, in its own right, may need to be managed.

As wastewater passes through the column, most of the exchange activity takes place in a zone referred to as an "active front" or "zone of active exchange." As resin within the front becomes saturated with target ions, the front

progresses from the top to the bottom of the bed. It is this active front that characterizes the continuous flow exchange process. Because the resin ahead of the active front is fully charged, there is always a concentration gradient in terms of the target ions, enabling maximum removal. Batch treatment, on the other hand, is limited by the equilibrium phenomena between the exchange resin and the bulk solution.

It is important to note that maximum removal is not necessarily synonymous with complete removal. In practice, it is usually the case that complete removal of target ions occurs during the initial portion of the treatment phase. How long this complete removal lasts depends upon the hydraulic loading rate of the exchange columns, as well as on the

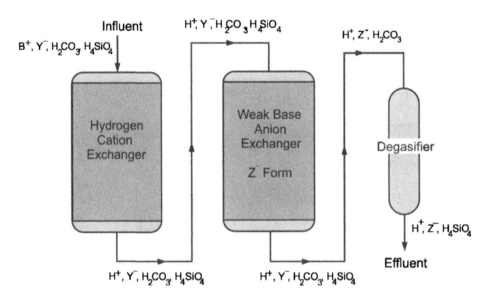

Figure 7-116   Treatment train for removal of both cations and anions.

strength of attraction between the exchange sites and the target ions. Sooner or later, target ions begin to bleed through and appear in the effluent (referred to as "breakthrough") until, at some point, the effluent becomes unacceptable. At this point, the treatment run is stopped, and the exchange resin is backwashed with water to remove debris and to regrade the resin.

If both cations and anions are to be removed, the treatment train is set up as shown in Figure 7-116, with the cation exchangers preceding the anion exchangers. Otherwise, metal cations that might appear during noncomplete removal would precipitate with hydroxide ions within the anion exchanger and cause fouling.

A familiar example of cation-anion removal is the "demineralization" of water. In this process, the cation exchanger is charged with hydrogen ions and is said to be on the acid cycle or on the hydrogen cycle. The anion exchanger is charged by passing sodium or potassium hydroxide solution through it and is said to be on the hydroxide cycle. Cations (metals, calcium, etc.) are exchanged for hydrogen ions, which then react with hydroxide ions exchanged for

anions (sulfate, chloride, etc.) to form water, the only intended substance in the effluent from the anion exchanger.

In many cases, a mixed bed process is advantageous. The cation and anion exchange resins are mixed and packed into a single column. The two resins must be separated, however, prior to recharging and then mixed again for the next ion exchange cycle.

Often, ions that have just been exchanged are soon replaced by ions for which the exchange material is more highly selective. This process results in the less strongly selected ions being pushed along at the front of the zone of active exchange. When breakthrough occurs, the effluent will be enriched in the less strongly selected ions.

Practical limitations to the ion exchange process include the following:

1. The fluid to be treated must be reasonably free of nondissolved solids. The economics of ion exchange are such that it is cost effective to install and operate solids removal facilities prior to the process, rather than lose capacity due to fouling.

2. Corrosion-resistant materials of construction are required for the column

containers as well as the pumps and pip-ing.

3. Disposal of spent regenerant (which contains the target ions that have been removed), as well as rinse waters, may pose an expensive problem if they can-not be discharged to the municipal sewer or to the industry's main wastewater treatment system.

Significant advantages of ion exchange as a process, compared with chemical precipita-tion, include:

1. No significant sludge disposal problem.
2. No chemical feeders, mixers, etc., other than what is required to make up and feed regenerant.
3. The systems are simple to operate and do not require much attention.

Design Criteria

Design criteria, including brand and type of exchange resins, volumes of resins, treat-ment train configuration, charging sub-stances, and hydraulic loading rates, are nor-mally generated in the laboratory. Also, it is sometimes cost effective to perform a series of batch studies wherein varying quantities of a fully charged exchange resin are placed into Erlenmeyer flasks containing the liquid to be treated and then mixed in a standard-ized manner. Plots are then made in the same manner as they are made for carbon adsorption isotherms, discussed previously. While these plots may not be able to be used exactly as adsorption isotherms are used, they show important trends needed in the design process.

Removal of Specific Organic Substances

In general, ion exchange is useful for remov-ing inorganic substances, but not organic substances, and the reverse is normally the case for activated carbon. However, activated carbon can be very effective in removing metal ions from wastewater by simply chelat-ing the metals with an organic chelant, such as EDTA, and certain organics can be removed by ion exchange–type resins manu-factured for that specific purpose. These macroreticular resins are available for use in removing specific nonpolar organic materi-als. Passing a liquid that is a solvent for the target substance regenerates these resins.

Stripping

Stripping is a physical treatment technology, in the sense that there are no chemical reac-tions involved. Stripping is a method of mov-ing one or more chemical substances from one medium, either liquid or gas, to another, also either liquid or gas, but usually the opposite of the first medium. That is, if the first medium is liquid, the second is gas, and vice versa. An example of stripping as a treat-ment technology is the stripping of acetone from water with air, described in the follow-ing text and illustrated in Figure 7-117.

Acetone is highly soluble in water, but can be removed from water solution by bubbling air through an acetone-water mixture or by causing droplets of the liquid mixture to pass through the air as takes place during mechanical aeration.

The mechanism of removal is accounted for by a combination of Henry's law of solu-bility and Dalton's law of partial pressures. Acetone is volatile compared with water; therefore, molecules of acetone will pass from a container of the mixture, through the acetone-water surface, into the air until the partial pressure of acetone in the air is equal to the vapor pressure of the acetone. If the container is covered, and there is enough ace-tone in the mixture, equilibrium will be reached and the concentration of acetone in the mixture will then remain constant. If the vessel is not covered, equilibrium will never be reached. This fact alone will result in all the acetone eventually being removed from the water mixture, given enough time. How-ever, because of the length of time required, this arrangement hardly qualifies as a waste-water treatment method. The process can be

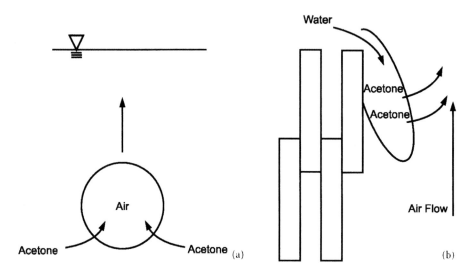

Figure 7-117   Stripping as a waste treatment technology. (a) Air bubble stripping molecules of Acetone as it rises through the Acetone-bearing water. (b) Acetone being stripped out of Acetone-bearing water cascading over packing.

greatly accelerated by increasing the surface area of the acetone-water mixture. This can easily be accomplished by introducing a host of air bubbles to the bulk volume of the acetone-water mixture or by converting the bulk volume of the mixture to a host of tiny droplets.

Either way, the surface area of the acetone-water mixture will be greatly increased; molecules of acetone will pass from the mixture, through the surface films of either the air bubbles or the liquid droplets, into the air phase, in an attempt to raise the partial pressure of acetone in the air phase to a value equal to the vapor pressure of acetone (at the prevailing temperature). The result will be essentially complete removal of the acetone from the water.

In the case of volatile substances, which are toxic or otherwise objectionable, the substance must be contained after stripping and subjected to appropriate reuse or final disposal. In many cases, activated carbon has been used to capture organic substances that have been removed from aqueous solution (polluted groundwater, for instance). In other cases, stripped organics have been disposed of directly, by incineration using a

flame in a conduit leading from the enclosed stripping reactor to the atmosphere.

Scrubbing

Scrubbing, like stripping, is a physical treatment technology, because no chemical reactions are involved. Also, like stripping, its mechanism is accounted for by a combination of Henry's law of gas solubility and Dalton's law of partial pressures. Usually, scrubbing is used to remove one or more target gases from a stream of mixed gases, whereas stripping is used to remove one or more substances that have a higher vapor pressure than water from a wastewater stream.

An example of scrubbing as a treatment method is that of the removal of hydrogen sulfide from air using chlorine in a water solution of high pH. Figure 7-118 shows a stream of waste air being passed through a "scrubber." The scrubber resembles a discharge stack and operates as follows: a fan draws the air stream through the duct work to the scrubber, then forces it up through the scrubbing apparatus, then into the atmosphere. The scrubbing apparatus consists of two parts. The first is a chamber filled with "telerets," known as the "packing," which are

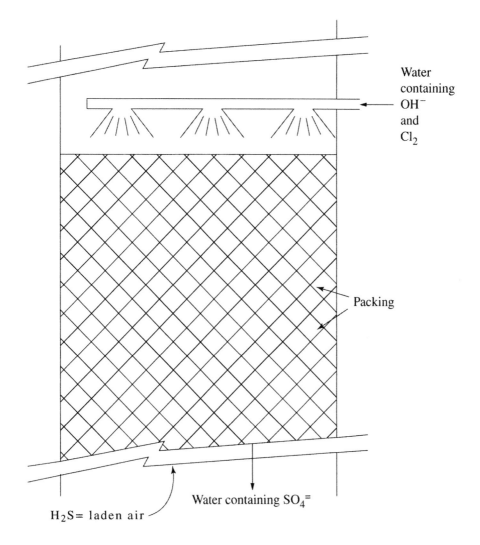

Figure 7-118   Illustration of the air "scrubbing" process.

solid objects with a high surface-to-volume ratio. The second is a system for circulating a solution of dissolved chlorine plus dissolved sodium or potassium hydroxide (in water) through the scrubbing chamber by pumping from the reservoir through spray nozzles down onto the telerets. The caustic chlorine solution flows over the surface of the packing, where it contacts the air to be treated. During this contact period the hydrogen sulfide gas dissolves in the water, where it exists in equilibrium according to the following:

$$H_2S \leftrightarrow HS^- + H^+ \qquad (7\text{-}88)$$

Because the solution is of a high pH, OH^- ions "remove" almost all of the H+ ions:

$$H^+ + OH^- \rightarrow H_2O \qquad (7\text{-}89)$$

causing the equilibrium (equation 7-88) to displace to the right. In so doing, essentially all of the hydrogen sulfide that was dissolved out of the stream of contaminated air exists as dissolved sulfide ion, which reacts with dissolved chlorine as follows:

$$HS_2 + 4Cl_2 + 4H_2O$$
$$\rightarrow SO_4^- + 9H^+ + 8Cl^- \quad (7\text{-}90)$$

Thus, hydrogen sulfide is "scrubbed" from the air and oxidized to soluble, nonodorous sulfate ion. Of course, it can be reduced back to the sulfide state and come out of solution as hydrogen sulfide under favorable conditions of neutral to low pH. Therefore, it is necessary to dispose of the waste scrubber solution in an appropriate manner. In many applications, the "spent" scrubber solution is simply discharged to the industrial wastewater sewer, after which it is treated in the industry's wastewater treatment system.

Magnetically Enhanced Solids Separation (CoMag Process)

This patented process, referred to as the "CoMag Process," makes use of finely granulated magnetite as a nucleation site for precipitation and then coagulation of suspended solids or dissolved substances, such as phosphorus. The granular magnetite is added to the (usually treated) wastewater, mixed vigorously for a short period of time (five seconds), then mixed slowly for a somewhat longer time (five minutes). The rapid mix accomplishes complete dispersion of the magnetite, and the slow mix accomplishes coagulation of the chemical precipitates, as well as agglomeration into large flocs. Coagulants and coagulant aids (polymers) are typically used, as well.

The flocculated material then flows to a settling chamber where a magnet, acting upon the entrained magnetite, greatly aids the process of separating the flocs from the bulk liquid. Use of the magnetite and magnet allows loading of the clarifier at rates that are more than ten times greater than is typical for a conventional physical-chemical process. Therefore, the cost of construction of the clarifier is very low, and the footprint is far smaller than would be required for a conventional clarifier.

Figure 7-119 is a process flow diagram of the CoMag system.

Other Wastewater Treatment Methods

Land Application

There are several versions of land application as a wastewater treatment method, including spray irrigation, wetlands treatment, overland flow, hyponics, and a relatively new proprietary process called "snowfluent," in which pretreated wastewater is made into snow by use of the same equipment used at ski areas. In all land application systems, the treatment mechanisms include evaporation, evapotranspiration, microbiological metabolism, adsorption, and direct plant uptake. As well, land application systems rely upon the groundwater for final disposal of the water after treatment has taken place. For this reason, land application is appropriate for use with only wastewaters that have all biodegradable organics. Bacteria and other microorganisms living in the soil use the organics as food for energy and reproduction. To this extent, a land treatment site is regenerative and can be used for many years, with appropriate rest periods. To the extent that some substances are removed by adsorption to soil particles, as is the case for phosphorus, a land application site has a limited life.

Land application systems are subject to limitations, some of which are as follows:

1. Land application systems must not be used if it is unacceptable, or even undesirable, for the wastewater, raw or treated, to enter the groundwater. The groundwater is the ultimate disposal destination of all the wastewater that does not evaporate, transpire, or find its way via overland flow to a surface water body (a form of system failure).

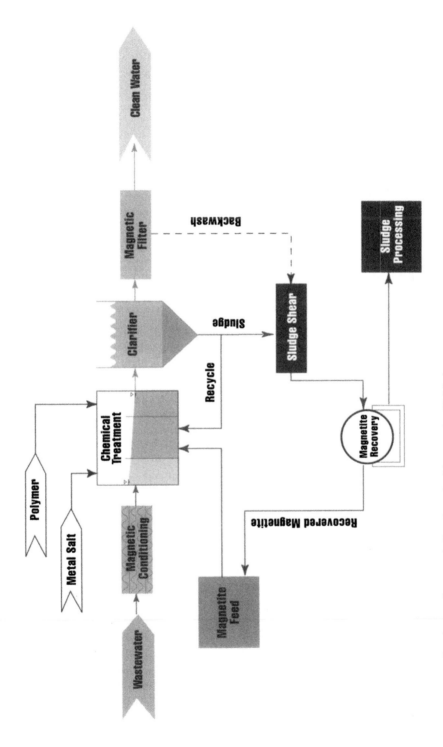

Figure 7-119   Process flow diagram for magnetically enhanced solids separation (CoMag).

2. The hydraulic application rate must be more than the evapotranspiration rate. Otherwise, salts will be left in the soil, rather than washed down into the groundwater. The consequence will be plugged soil, i.e., a failed wastewater treatment system.

3. There are many laws, rules, and regulations governing rates of application. Factors that must be considered include:

   a. The hydraulic conductivity of the soil
   b. The rates of evaporation and transpiration in the geographical area
   c. The allowable nitrogen loading in the soil
   d. The type of cover crop used and whether or not it is to be harvested on a regular basis (subnote: There MUST be a cover crop)
   e. The substances in the wastewater

With respect to allowable application rates, each state has its own regulations. Typically, hydraulic application rates are limited to 5,000 to 10,000 GPD/acre (1/4 to 1/2 inch per day). Nitrogen loading rates are typically limited to 400 lb/acre/yr if the cover crop is harvested, or to 200 lb/acre/yr if the cover crop is not harvested.

Spray Irrigation

Spray irrigation sites are used in rural areas for food processing wastewaters generated on a seasonal basis, as well as for many other industrial wastewaters that contain no non-biodegradable substances. In arid regions, the wastewater treatment process serves also to supply the water needs of appropriate crops. In cold regions, sufficient storage must be provided during the nongrowing season.

Design criteria for spray irrigation systems include wastewater application rate, distribution system pipe sizing, and storage facility sizing.

Wastewater Application Rate

The water-balance equation (7-91) is used to determine the acceptable rate of application of wastewater, based on soil hydraulic conductivity considerations. In many cases, one of the substances contained in the wastewater (nitrogen, for instance) might govern the rate of wastewater application. The water balance equation is stated as:

$$L_w(p) = ET - p + W_p \qquad (7\text{-}91)$$

where

$L_w(p)$ = wastewater application rate, in./month

ET = Rate of evapotranspiration, in./month

p = Rate of precipitation, in/month

W_p = Rate of percolation into soil, in./month

Notice that if the rate of precipitation exceeds the combined rates of evapotranspiration and percolation, spray cannot take place.

Wetlands Treatment

Wetlands treatment can be described as biological and microbiological treatment of wastewater resulting from the use of pollutants as food for living organisms in a natural or artificial wetlands. Sedimentation (in regions of slow flow and eddy currents), adsorption onto soils and the root systems and other parts of plants, photooxidation, direct plant uptake, cation exchange, and photosynthesis are also important mechanisms of pollutant removal. In wetlands treatment, wastewater, usually pretreated to a rather high degree in the case of natural wetlands, is allowed to flow, very slowly, through the wetlands system. Bacteria, fungi, and many other types of organisms inhabit the aqueous medium and use pollutants contained in the wastewater for food. Wetlands systems are usually no more than two feet deep, in order to enable photosynthesis and diffusion from the air to maintain aerobic conditions. Area loading rates must be managed so as to avoid the development of anaerobic zones. Flow through them is very slow, so that hydraulic retention times are very long, due to the lack of input of energy of the types used with mechanical treatment systems. There is no return of sludge or other attempt to increase the numbers of organisms.

Artificial or constructed wetlands are sometimes used for raw industrial wastewaters, or wastewaters that have had only minimal pretreatment. The organic strength of these wastewaters is typically low. Constructed wetlands are of two types: those with a free water surface (FWS), as with natural wetlands, and those with subsurface flow systems (SFS), also known as "rock-reed filters" and "root zone filters." These systems have highly porous media, such as sand or rocks, that support the growth of plants and through which the wastewater flows at a slow rate.

Wetlands treatment makes use of all of the biological treatment mechanisms that are involved in any of the conventional biological treatment processes, and, as such, these systems require nutrients and trace elements. Wintertime operation requires special consideration.

The fact that most natural wetlands occur in areas of groundwater discharge or are underlain by impermeable material makes them unlikely sources of future groundwater contamination. However, protection of the groundwater must be positively ensured during the design and construction process. In the case of artificial or constructed wetlands, protection of groundwater by installation of an appropriate liner is a major design consideration.

Reeds, cattails, sedges, and bullrushes are types of emergent plants that have performed well in wetlands treatment systems. Floating plants such as water hyacinths and duckweed are also effective. In particular, the root systems of water hyacinths project a mesh into the flow that effectively adsorbs both dissolved and suspended matter. Microbial degradation, however, is responsible for the bulk of the treatment of dissolved and suspended

organic matter. The plant life participates in reaeration. If the plant life is used for nutrient removal, nitrogen and/or phosphorus, for instance, there must be a periodic harvesting of these plants to prevent reintroduction of these nutrients to the water flow when the plants die. The bioaccumulation of trace elements such as heavy metals must be considered when selecting final disposal methods for the plants.

Design parameters for wetlands systems include system geometry; hydraulic retention time; type, size, and porosity of media; hydraulic loading rate; organic or nutrient loading rate; temperature; and slope.

The area and length-to-width ratio of the basin are determined by either hydraulic retention time or organic or nutrient loading rate, and whether the system is FWS or SFS. Temperature is a very important consideration, closely paralleling lagoons and stabilization ponds in that respect.

Bibliography

American Public Health Association, American Society of Civil Engineers, American Water Works Association, and Water Pollution Control Federation. *Glossary, Water and Wastewater Control Engineering.* 1969.

American Society of Civil Engineers. *Standard Guidelines for In-Process Oxygen Transfer Testing.* Publication No. ASCE-18-96, 1997.

American Society of Civil Engineers. "Wastewater Treatment Plant Design." *Water Pollution Control Federation Manual of Practice,* 8; *ASCE Manuals and Reports on Engineering Practice,* No. 36, 1977.

Amirtharajah. *Journal of the American Water Works Association.* Dec. 1988.

Behn, V. C., and J. C. Liebman. "Analysis of Thickener Operation." *Journal of the Sanitary Engineering Division, American Society of Civil Engineers* 89 (1963).

Burd, R. S. "A Study of Sludge Handling and Disposal." Federal Water Pollution Control Administration. WP-20-4. Washington, DC: 1968.

Butler, J. N. *Ionic Equilibrium—A Mathematical Approach.* Reading, MA: Addison-Wesley, 1964.

Cummins, M. D., and J. J. Westrick. "Packed Column Air Stripping for Removal of Volatile Compounds." *Proceedings of National Conference on Environmental Engineering, American Society of Civil Engineers.* Minneapolis, MN: 1982.

Eckenfelder, W. W. "Trickling Filter Design and Performance." *Transcript of the ASCE* 128, pt. III (1963): 371.

Eckenfelder, W. W., and W. Barnhart. "Performance of a High-Rate Trickling Filter Using Selected Media." *Journal of the Water Pollution Control Federation* 35, No. 12 (1963).

Eckenfelder, W. W., and D. L. Ford. *Water Pollution Control: Experimental Procedures for Process Design.* New York: Pemberton Press, 1970.

Edde, H. J., and W. W. Eckenfelder. "Theoretical Concept of Gravity Sludge Thickening; Scaling-up Laboratory Units to Prototype Design." *Journal of the Water Pollution Control Federation* 40 (1968).

Fair, G. M., J. C. Geyer, and D. A. Okun. *Elements of Water Supply and Wastewater Disposal.* New York: John Wiley & Sons, 1971.

Gailer, W. S., and H. B. Gotaas. "Analysis of Biological Filter Variables." *Journal of the Sanitary Engineering Division, ASCE* 90, No. 6 (1964).

Germain, J. E. "Economical Treatment of Domestic Waste by Plastic-Medium Trickling Filters." *Journal of the Water Pollution Control Federation* 38, No. 2 (1966).

Great Lakes-Upper Mississippi River Board of State Sanitary Engineers. *Recommended Standards for Sewage Works.* Albany, NY: Health Education Service, Inc., 1978.

Hawkes, H. A. *Microbial Aspects of Pollution.* London: Academic, 1971.

Kavanaugh, M. C., and P. R. Trussell. "Design of Aeration Towers to Strip Volatile Contaminants from Drinking Water." *Journal*

American Water Works Association 72, No. 12. (1980).

Kynch, G. J. "A Theory of Sedimentation." *Transactions of the Faraday Society* 48 (1952).

Lanouette, K. H. "Heavey Removal," *Chemical Engineering Journal* 84, No. 10 (1977): 73–80.

McKinney, R. E. *Microbiology for Sanitary Engineers.* New York: McGraw-Hill, 1962.

Metcalf & Eddy, Inc. *Wastewater Engineering—Treatment, Disposal, and Reuse.* Edited by G. Tchobanoglous and F. L. Burton. New York: McGraw-Hill, 1991.

National Fire Protection Association. "Recommended Practice for Fire Protection in Wastewater Treatment and Collection Facilities." *NFPA Standard* 820 (1992).

Novotny, V., and A. J. Englande, Jr. "Equalization Design Techniques for Conservative Substances in Wastewater Treatment Systems." *Water Research* 8, No. 325 (1974).

Novotny, V., and R. M. Stein. "Equalization of Time Variable Waste Loads." *Journal of Environmental Engineering, ASCE* 102, No. 613 (1916).

Schulze, K. L. "Load and Efficiency of Trickling Filters." *Journal of the Water Pollution Control Federation* 32, No. 3 (1960): 245.

"Sewage Treatment at Military Installations." National Research Council. *Sewage Works Journal* 18, No. 5 (1946).

Shinskey, F. G. *pH and pION Control.* New York: John Wiley & Sons, 1973.

Smith, J. M. *Chemical Engineering Kinetics.* 2nd ed. New York: McGraw-Hill, 1972.

Smith, R., R. G. Eilers, and E. D. Hall. "Design and Simulation of Equalization Basins." For the U.S. Environmental Protection Agency. Advanced Waste Treatment Research Laboratory. Cincinnati, OH: 1973.

Stenstrom, M. K., and R. G. Gilbert. "Effects of Alpha, Beta, and Theta Factor Upon the Design, Specification, and Operation of Aeration Systems." *Water Research* 15 (1981).

Talmage, W. P., and E. B. Fitch. "Determining Thickener Unit Areas." *Industrial and Engineering Chemistry* 47 (1955).

The Soap and Detergent Association. *Phosphorus and Nitrogen Removal from Municipal Wastewater: Principles and Practice.* 2nd ed. Edited by R. Sedlak. New York: Lewis Publishers, 1991.

U.S. Congress. Office of Technology Assessment. *Technologies and Management Strategies for Hazardous Waste Control.* Washington, DC: U.S. Government Printing Office, 1983.

U.S. Environmental Protection Agency. *Chemical Aids Manual for Wastewater Treatment Facilities.* EPA 430/9-79-018, MO-25. Washington, DC: U.S. Government Printing Office, 1979.

U.S. Environmental Protection Agency. *Design Manual—Phosphorus Removal.* EPA/621/1-87/00 1. Washington, DC: U.S. Government Printing Office, 1987.

U.S. Environmental Protection Agency. *Disinfection of Wastewater Task Force Report.* EPA-430/9-75-012. Washington, DC: U.S. Government Printing Office, 1976.

U.S. Environmental Protection Agency. *Electroplating Industry.* EPA 625/5-79-016. Washington, DC: U.S. Government Printing Office, 1979.

U.S. Environmental Protection Agency. *Evaluation of Sludge Management Systems.* EPA 430/9-80-00 1. Washington, DC: U.S. Government Printing Office, 1980.

U.S. Environmental Protection Agency. *Flow Equalization.* Technology Transfer. EPA 625/4-74-006. Washington, DC: U.S. Government Printing Office, 1974.

U.S. Environmental Protection Agency. *Lining of Waste Impoundment and Disposal Facilities.* SW-870. Washington, DC: U.S. Government Printing Office, 1983.

U.S. Environmental Protection Agency. *Manual for Nitrogen Control.* EPA/625/R-93/010. Washington, DC: U.S. Government Printing Office.

U.S. Environmental Protection Agency. *Meeting Hazardous Waste Requirements for Metal Finishers.* EPA 625/4-87/018. Cincinnati, OH: 1987.

U.S. Environmental Protection Agency. *Nitrification and Denitrification Facilities, Wastewater Treatment.* Technology Transfer. Washington, DC: U.S. Government Printing Office, 1973.

U.S. Environmental Protection Agency. *Operations Manual for Anaerobic Sludge Digestion.* EPA 430/9-76-001. Washington, DC: U.S. Government Printing Office, 1976.

U.S. Environmental Protection Agency. *Oxygen-Activated Sludge Wastewater Treatment Systems.* Technology Transfer Seminar Publication. Washington, DC: U.S. Government Printing Office, 1973.

U.S. Environmental Protection Agency. *Process Control Manual for Aerobic Biological Wastewater Treatment Facilities.* EPA III—A-524-77. Washington, DC: U.S. Government Printing Office, 1977.

U.S. Environmental Protection Agency. *Process Design Manual for Carbon Adsorption.* Technology Transfer. EPA 625/1-71/0022. Washington, DC: U.S. Government Printing Office, 1973.

U.S. Environmental Protection Agency. *Process Design Manual for Phosphorus Removal.* Technology Transfer. EPA 625/1-76-001a. Washington, DC: U.S. Government Printing Office, 1976.

U.S. Environmental Protection Agency. *Process Design Manual for Sludge Treatment and Disposal.* Technology Transfer. EPA 625/1-74-006. Washington, DC: U.S. Government Printing Office, 1974.

U.S. Environmental Protection Agency. *Process Design Manual for Suspended Solids Removal.* Technology Transfer. EPA 62511-75-003a. Washington, DC: U.S. Government Printing Office, 1975.

U.S. Environmental Protection Agency. *Process Design Manual for Upgrading Existing Wastewater Treatment Plants.* Technology Transfer. EPA 625/1-71-004a. Washington, DC: U.S. Government Printing Office, 1974.

U.S. Environmental Protection Agency. *Process Manual for Carbon Adsorption.* Technology Transfer. Washington, DC: U.S. Government Printing Office, 1971.

U.S. Environmental Protection Agency. *Physical-Chemical Nitrogen Removal, Wastewater Treatment.* Technology Transfer. Washington, DC: U.S. Government Printing Office, 1974.

U.S. Environmental Protection Agency. *Upgrading Trickling Filters.* EPA 430/ 9-78-004. Washington, DC: U.S. Government Printing Office, 1978.

U.S. Environmental Protection Agency. *Wastewater Filtration, Design Considerations.* Technology Transfer. EPA 62514-74-007a. Washington, DC: U.S. Government Printing Office, 1974.

U.S. Environmental Protection Agency. Industrial Environmental Research Laboratory. *Control and Treatment Technology for the Metal Finishing Industry.* EPA 625/ 8-82-008. Cincinnati, OH: 1982.

U.S. Environmental Protection Agency. Industrial Environmental Research Laboratory. *Control and Treatment Technology for the Metal Finishing Industry—Sulfide Precipitation.* EPA 625/8-80/003. Cincinnati, OH: 1980.

U.S. Environmental Protection Agency. Municipal Environmental Research Laboratory. *Innovative and Alternative Technology Assessment Manual.* EPA 430/9-78-009, MCD 53. Cincinnati, OH: 1978.

U.S. Environmental Protection Agency. Office of Research and Development. *Control of Organic Substances in Water and Waste Water.* EPA 600/8-83/01 1. Cincinnati, OH: 1983.

U.S. Environmental Protection Agency. Office of Research and Development. *Estimating Costs of Granular Activated Carbon Treatment for Water and Wastewater.* Washington, DC: U.S. Government Printing Office, 1983.

U.S. Environmental Protection Agency. Office of Research and Development. *Research Summary—Industrial Wastewater.* EPA-60018-80-026. Washington, DC: U.S. Government Printing Office, June 1980.

U.S. Environmental Protection Agency. Office of Wastewater Enforcement and Compliance. *Ultraviolet Disinfection Technology Assessment.* EPA 832-R-92-004. Washington, DC: U.S. Government Printing Office, 1992.

Velz, C. J. "A Basic Law for the Performance of Biological Filters." *Sewage Works Journal* 20 (1948): 607.

Wachinski, A. M., and J. E. Etzel. *Environmental Ion Exchange—Principles and Design.* New York: Lewis Publishers, 1997.

Wallace, A. T. "Analysis of Equalization Basins." *Journal of the Sanitary Engineering Division, ASCE* 94, No. 1161 (1968).

Water Environment Federation. *Design of Municipal Wastewater Treatment Plants; WEF Manual of Practice.* No. 8. Alexandria, VA: 1992.

Water Environment Federation. *Odor Control in Wastewater Treatment Plants; WEF Manual of Practice.* No. 22. Alexandria, VA: 1995.

Wei, J., and C. D. Prater. "A New Approach to First Chemical Reaction Systems." *AIChE Journal* 9, No. 77 (1963).

Wei, J., and C. D. Prater. *Advances in Catalysis,* Vol. 13. New York: Academic Press, 1962.

8 Treatment of Air Discharges from Industry

Air Discharges

The discharge, or release, of substances to the air, no matter how slight, is regarded as air pollution. Discharges can be direct, by means of a stack, vent, hood, or the like, or indirect, by way of leaks from a building's windows, doors, or other openings. These indirect emissions are referred to as fugitive emissions. Fugitive emissions must be considered to evaluate a facility's total emissions; they include any emission that is generated from an outlet not specifically designed and built to discharge substances to the air.

For example, when evaluating air discharges at a paper mill, discharges from boiler stacks and laboratory hoods would be considered direct emissions. If the same facility used calcium carbonate in the process, and dust generated in its handling entered the atmosphere, this would be considered a fugitive emission.

Other examples of fugitive emissions include volatilization of organic compounds such as solvents and gasoline from storage containers, transfer equipment, and even points of use.

Discharges can be in only one of two categories: within compliance or out of compliance. Major sources of criteria pollutants and hazardous air pollutants (HAPs) are regulated by the Clean Air Act, as amended in 1990, and administered by state or local air-quality management agencies. Some industrial establishments may be regulated by one or more state or agency requirements that are more restrictive than the CAA. The federal government, through the EPA, serves two functions:

1. Issue regulations that must be followed by the administrating authorities
2. Oversight of the administration

The following sections summarize the regulations governing discharges of air. Chapter 3 provides a more comprehensive discussion of air pollution control laws.

Air Pollution Control Laws

Federal involvement in air pollution control began in 1955 with the passage of the Air Pollution Control Act of 1955, Public Law 84-159. While very narrow in scope, the Act served as a wake-up call to states that air pollution control was to be taken seriously.

Subsequent amendments eventually led to the Clean Air Act of 1963, Public Law 88-206, which, since its inception, has been amended several times, resulting in today's prevailing law, the 1990 amendments.

The *1963 Clean Air Act* designated six pollutants as "criteria pollutants," thought to be the most important substances affecting the public's health and welfare. These criteria pollutants, still regarded as such in the year 2005, are as follows:

- Sulfur dioxide (SO)
- Nitrogen oxides (NO_x)
- Carbon monoxide (CO)
- Lead (Pb)
- Ozone (O_3)
- Particulate matter (Pm)

The 1970 amendments established the basic framework under which the Clean Air Act exists today. In addition to authorizing the EPA to establish National Ambient Air

Quality Standards (NAAQS), the 1970 amendments authorized the EPA to establish emission standards for categories of sources. Such categories include major sources and nonmajor sources. There are significantly different requirements for sources that qualify as a "major source" compared with those that do not. A "major source" is any stationary source that emits in excess of ten tons per year of any listed hazardous substance, or 25 tons per year of any combination of those substances. The emission standard that an industrial facility is held to is dictated, in part, by whether it is categorized as a major source or a nonmajor source. This facet of the Clean Air Act is discussed in greater detail in Chapter 3.

The 1990 amendments to the Clean Air Act made some of the most notable changes with regard to industrial sources. The amendments established a new operating permitting system that had the effect of permanently changing the way environmental managers in industry must do their job. Monitoring and reporting requirements were greatly expanded compared with previous requirements. Furthermore, the requirement was established to identify all "regulated pollutants" emitted by a facility, to monitor emissions (continually or periodically), to operate the equipment in compliance with standards written into the permit, and to certify compliance with all standards in the permit on an annual basis .

In addition to the new permitting, monitoring, and reporting requirements, the number of designated hazardous air pollutants (air toxics) was increased to 189 and then subsequently (in 1999) reduced by one to 188.

The list of 188 "air toxics" includes pesticides, metals, organic chemicals, coke oven emissions, fine mineral fibers, and radionuclides.

With a clear understanding of the laws governing a facility, and the air emissions standards it is required to meet, a facility can begin the process of controlling air emissions.

Air Pollution Control

After all potential sources of air pollutants have been identified, the process of developing an air pollution control plan can begin. Control of the discharges of air pollutants from industries can be organized into three categories:

1. Waste minimization and reduction at the source
2. Containment
3. Removal by use of one or more treatment technologies

In the sections that follow, each of these categories will be discussed in greater detail.

Waste Minimization and Reduction at the Source

As explained in Chapter 4, waste minimization is an important aspect to any pollution prevention program, and as a means of controlling pollution, is it the preferred method. Reduction at the source can be accomplished using the following strategies.

- Identification of each and every source within the entire industrial facility. This must include the full range of air pollutant generation activities, from the boilers used for hot water and space heating, to each separate manufacturing process step, to dust blowing around the parking lot.
- Analysis of each source to determine whether it can be eliminated.
- If elimination is not an option, substitute nonobjectionable process materials for those that are hazardous or otherwise objectionable.
- Determination of whether or not a change in present operations can reduce pollutant generation. For example, maintaining vigorous maintenance and preventive maintenance programs to ensure that all process equipment is generating a minimum of pollutants, all

containment facilities are performing as designed so as to minimize fugitive emissions, and all treatment equipment is operating at top efficiency in terms of pollutant removal.

- Maintenance of accident and spill prevention procedures and facilities to ensure that they are up-to-date, well-known to those who should know them, and reviewed and revised at regular intervals.
- Maintenance of emergency response procedures and facilities to ensure that they are up-to-date and well known.
- Maintenance of a rigorous program of analyzing past spills and emergencies with the objective of determining (1) how to prevent each and every spill event from ever happening again, and (2) how to improve responses to emergency events over past responses.

A formal Air Pollution Management Plan (APMP) is a necessity. The APMP must be a living, active document, used often, and reviewed and revised frequently. The APMP is truly the key to minimizing the cost for air pollution control.

Containment

After prevention at the source, containment is the most cost-effective method of air pollution control. Containment refers to the absence of leaks, or of any type of breach in structural integrity in buildings, ductwork, storage tanks, or any location from which air pollutants could enter the environment without such entrance resulting from the express intention of the environmental managers. Such unwanted discharges are called "fugitive emissions" and are the primary target of a pollutant containment program.

One of the most effective methods of containment to prevent release of fugitive emissions is to maintain a "negative pressure" inside buildings—that is, to maintain a lower pressure inside the building than the atmospheric pressure outside the building.

This can be accomplished by use of one or more wet scrubbers installed in such a way that air from any location within the plant must go through one of the scrubbers before reaching the outside. Induced air fans continually evacuate air from the building and force this air to pass through the scrubbers. If the fans and scrubber systems are sized properly, occasional opening of doors can take place without significant release of air pollutants.

Hoods and Isolation Chambers

Appropriate and effective use of hoods is important to the objective of containing air pollution. Containing pollutants with hoods and isolation chambers can prevent air pollutants from contaminating large volumes of air. Furthermore, it is always less expensive to treat a smaller volume of more highly concentrated pollutant than to remove the same mass of pollutant from a larger volume of air or other gas.

Fans and Ductwork

Properly designed, operated, and maintained fans and ductwork are key to successful containment of air pollutants. Fans with inadequate capacity allow air pollutants to drift out of the containment system at the source. Fans with unnecessarily abundant capacity will dilute the target pollutants excessively, leading to increased cost of treatment.

Stabilization

After a prudent waste minimization program has been carried out, a stabilization period should occur before sizing the equipment. If air pollutant flow rates and concentrations are calculated based on improved maintenance and operational procedures, and these procedures are not maintained, the handling and treatment equipment designed on the basis of the improved procedures will be overloaded and will fail.

Once the sources of air discharges have become stabilized, each of the sources should be subjected to a characterization program to determine flow rates and target pollutant concentrations (flows and loads) for the purpose of developing design criteria for handling and treatment facilities. Examples of handling facilities are hoods, fans, and ductwork. Examples of treatment equipment are electrostatic precipitators and fabric filters (bag houses, for instance). The characterization study consists of developing estimates of emission rates based on either the historical records of the facility under consideration or those of a similar facility. For instance, materials balances showing amounts of raw materials purchased and products sold can be used to estimate loss rates.

In Chapter 5, the waste characterization procedure is discussed in detail. A summary of the process is provided in the sections that follow.

Characterization of Discharges to the Air

There are three categories of air pollutant characterization work: (1) stack discharge characterization, (2) fugitive emissions characterization, and (3) ambient air–quality characterization. An effective air discharge characterization includes each of these categories. The three involve quite different sampling procedures but similar (in most cases, identical) analysis methods. Stack discharge and fugitive emission characterization work are done primarily to determine the state of compliance with one or more discharge permits. Ambient air–quality characterization is done primarily to determine the quality of air in a given area and establish a baseline for comparison of air discharge quality.

Stack Sampling

The most common reason for conducting a stack sampling program is to determine the state of compliance with regulatory requirements. As such, the substances sampled are usually dictated by the list of substances included in the air discharge permits issued to the facility.

The purpose of stack sampling is to determine, with as much accuracy as is practicable, the quantity (magnitude) of the total gas source and the quality (types and amounts of air contaminants) of the total source gas discharge. The equipment included in a typical stack sampling station includes devices to measure characteristics from which gas flow rate can be calculated, devices to measure certain characteristics directly, and equipment to collect and store samples for subsequent analyses in the laboratory.

In general, the equipment used to obtain data from which to calculate gas flow rate includes pitot tubes to measure gas velocity; a device to measure the static pressure of the stack gas; and devices to measure barometric pressure, moisture content, and temperature. Equipment used to characterize stack discharges in terms of specific substances is classified in two broad categories: particulate and gaseous. The objective of equipment in both categories is to quantitatively remove air contaminants in the same condition as they occur when they are discharged to the air. Many individual devices as well as integrated systems are commercially available to accomplish this objective.

Sample Collection

Ambient air or gas streams (including stack emissions and fugitive emissions) are sampled to determine the presence of, and concentrations of, gaseous pollutants by use of the following equipment and mechanisms:

- Vacuum pumps, hand operated or automatic
- Vacuum release of an evacuated collection container
- Tedlar bags
- Adsorption on a solid
- Condensation (freeze-out) in a trap

Equipment used to collect particulate matter and to determine the concentration, in ambient air as well as stack emission and other gas streams, uses one or more of the following mechanisms:

- Filtration
- Electrostatic impingement
- Centrifugal force
- Dry impingement
- Wet impingement
- Impaction

Each of the mechanisms for particulate and gaseous pollutant sampling is described further in Chapter 5.

Sample Analysis

Once a sample is collected, it is subject to change as a result of chemical reaction, chemical degradation, absorption or adsorption onto the walls of the container or to other substances in the container, or other phenomena. There is uncertainty about how much change will take place in a sample once it is collected. For this reason, it is extremely important to perform the chemical or other analyses as soon as possible after the samples are collected.

Ambient Air Sampling

The objective of ambient air sampling is determining the quality of ambient air as it relates to the presence and concentration of substances regarded as pollutants. The specialized devices and techniques for carrying out this task have been developed over half a century. Obtaining representative samples is a major objective of the work plan. Decisions to be made must strike a balance between the cost of the characterization program and the value of the data and include the duration of the sampling period, number of discrete samples taken, the size of each sample, and the number of substances sampled.

Particulate matter in ambient air is measured by use of a "high-volume sampler," which is an integrated filter holder–vacuum pump. A glass fiber filter is held in the filter holder, and a high flow rate of ambient air is drawn through it over a measured period of time. Calculations of particulate matter concentration in the ambient air are carried out using the weight of particulate matter collected on the filter and the flow rate (or total volume) of air drawn through the filter.

Treatment

Treatment of air discharges and, therefore, removal of pollutants from gas streams prior to discharge to the air are the major subjects of this chapter. Often the most expensive, but also the most important in terms of compliance with the law, treatment is considered a last resort after minimization and containment.

Treatment Systems for Control of Particulates

In general, there are five methods used for controlling particulate emissions:

1. Gravity separators
2. Inertial separators
3. Electrostatic precipitators
4. Fabric filters
5. Wet scrubbers

Gravity and inertial separators, including so-called "cyclones," are dry, "no-moving-parts" devices. They take advantage of the relatively high specific gravity of certain types of particulate matter, including fly ash, dust, cement particles, and organic solids. Electrostatic precipitators take advantage of the electrostatic charge on the surface of particles, either present from natural phenomena or induced. Fabric filters make use of physical blocking and adsorption, and wet scrubbers make use of a liquid to entrap particulates, thus removing them from a gas stream.

Gravity Separators

Gravity separators are devices that provide quiescent conditions, thus counteracting the tendency of a particulate-laden gas stream to sweep the particulates along with it as a result of aerodynamic drag. Most gravity separators are simple, open chambers, sometimes with a mechanism at the bottom to remove collected material. As is the case with other,

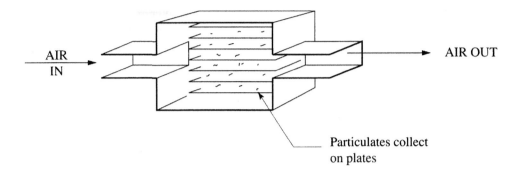

Figure 8-1 An example of a gravity settling device for removal of particulates as an element in an air pollution control system.

rather passive particulate removal devices, gravity separators are normally used as pretreatment devices, upstream of more sophisticated equipment. The greatest value of gravity separators is to prolong the life of the more sophisticated equipment. Figure 8-1 shows an example of a gravity settling device for removal of particulates as an element in an air pollution control system.

Inertial Separators

Inertial separators make use of the differential specific gravity between particulates and the gas that contains them. Inertial separators cause the stream of flowing gas to change directions. The inertia of the particulates, being directly proportional to the weight of each particle, causes them to resist the change in direction; thus, they are propelled out of the stream. Cyclones of various designs are the most common examples of inertial separators. Some dry venturi devices make use of the inertia of particulates moving along in a gas stream to effect removal. Figure 8-2 illustrates typical inertial separators.

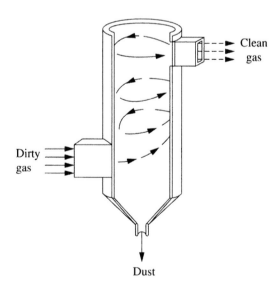

Figure 8-2 Cyclone separator: example of a typical inertial separator (from Buonicore and Davis, © 1992; reprinted by permission of John Wiley & Sons, Inc.).

Electrostatic Precipitators

Electrostatic precipitators (ESPs) are more active (as opposed to passive) than gravity separators but still have no moving parts. There is, however, a constant input of energy. Electrostatic precipitators consist of a series of elements having an electrostatic polarity that attracts particulates because of an electrostatic charge of opposite polarity on the surfaces of the particulates. Figure 8-3(a) through (c) shows, schematically, the principle of operation of electrostatic precipitators.

Even though there is a naturally occurring surface charge on essentially all particles in nature, the corona that is generated in the chambers of ESPs induces an even greater surface charge.

There are five types of electrostatic precipitators in general use:

* The plate-wire precipitator
* The flat-plate precipitator
* The two-stage precipitator
* The tubular precipitator
* The wet precipitator

The wet precipitator is a variation of any of the other four types.

The Plate-Wire Precipitator

The plate-wire precipitator is shown schematically in Figure 8-3(b). This device consists of parallel metal plates and wire electrodes of high voltage. Voltages range between 20,000 and 100,000 volts, as required. As the particulate-laden gas flows between the parallel plates, the particulates are attracted to, and adhere to, the plates. The plates must be cleaned periodically, usually by "rapping" and in some cases by water spray, and the wires must be cleaned of collected dust, as well. The plate-wire precipitator is the most commonly used ESP by industry.

The Flat-Plate Precipitator

The flat-plate precipitator is shown schematically in Figure 8-3(c). The flat-plate ESP consists of a series of parallel flat plates in which some of the plates serve as the high-voltage electrodes. Corona, needed to

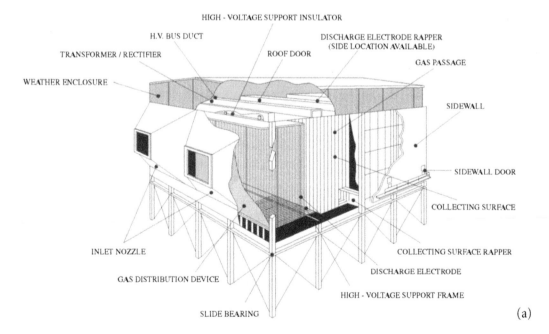

Figure 8-3 (a) Electrostatic precipitator components (from Buonicore and Davis, © 1992 by Van Nostrand Reinhold. Courtesy of the Institute of Clean Air Companies and reprinted by permission of John Wiley & Sons, Inc.).

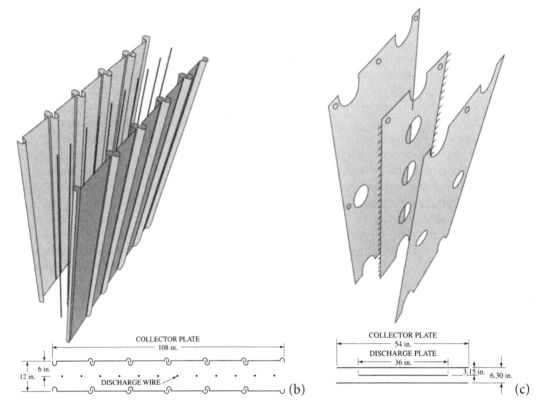

Figure 8-3 (b) Plate-wire precipitator components; (c) flat-plate precipitator components (from Buonicore and Davis, © 1992 by Van Nostrand Reinhold; courtesy of the Institute of Clean Air Companies and reprinted by permission of John Wiley & Sons, Inc.).

increase the surface charge on the particles to be removed, is generated in chambers preceding the ESP itself. As with wire-plate ESPs, the plates have to be cleaned periodically.

The Two-Stage Precipitator

In the two-stage precipitator, the high-voltage electrodes precede the collector electrodes, as opposed to their being arranged in parallel, as is the case with the wire-plate ESP and the flat-plate ESP. The plates are normally cleaned by spraying with water. In some instances, detergents are added to the water.

The Tubular Precipitator

The tubular precipitator is named for its shape and has its high-voltage wire coincid-ing with the axis of the tube. Tubular ESPs can fit into the stack because their shape is similar to that of the stack. Tubular ESPs are cleaned with a water spray

The Wet Precipitator

If an ESP, wire-plate, flat-plate, two-stage, or tubular precipitator, is operated with wet walls, it is properly referred to as a wet precipitator. Wet walls are used where the increase in particulate removal efficiency warrants the added cost for capital, operation, and maintenance.

Fabric Filters

Fabric filters make use of the "physical barrier" mechanism and are aided by a degree of adsorption. Fabric woven from a wide variety

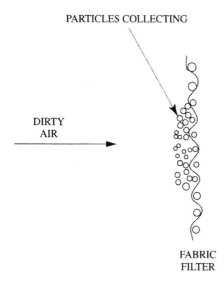

PARTICLES COLLECTING

DIRTY
AIR

FABRIC
FILTER

Figure 8-4   The process of "developing the filter."

of materials collects particulates from a gas that is forced through it. As a coating of filtered-out particles builds up on the surface of the filter fabric, this layer becomes an additional filter, usually more effective than the fabric alone. This process is referred to as developing the filter (as shown in Figure 8-4). Eventually, the layer becomes so thick that the filter as a whole is too restrictive to pass the desired flow rate of gas undergoing treatment. At this point the filter must be taken off line and either cleaned or disposed of and replaced.

Fabric filters can be obtained in many different configurations. Figure 8-5 shows a "bag house," which is a type of fabric filter that is used for a large number of industrial applications.

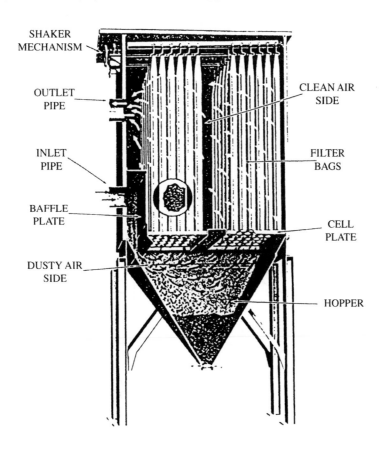

SHAKER
MECHANISM

OUTLET
PIPE

INLET
PIPE

BAFFLE
PLATE

DUSTY AIR
SIDE

CLEAN AIR
SIDE

FILTER
BAGS

CELL
PLATE

HOPPER

Figure 8-5   Bag house air filter installation (from Cooper and Alley, *Air Pollution Control: A Design Approach*, 2nd ed., 1986; reprinted by permission of Waveland Press, Inc., Prospect Heights, IL).

Wet Scrubbers

Wet scrubbers, used for removal of gases and other chemicals, as well as particulates, are the most common type of air pollution control in use by industries. They are also the most extensive, in terms of complexity of equipment, moving parts, requirement for controls, and operation and maintenance requirements. Wet scrubbers can be designed for a single target pollutant—for instance, particulates—but while in operation they will remove, to some degree, any other pollutant that will react with, or dissolve in, the scrubber fluid. As well, wet scrubbers can be designed for multipurpose removal. For instance, a scrubber can be designed to remove both particulate matter and sulfuric acid fumes by using a caustic solution as the scrubbing fluid in a system also configured to remove particulates.

Wet scrubbers that are intended for different target pollutants have several design and construction features in common. Figure 8-6 presents a schematic drawing of a basic wet "scrubbing tower," which has physical features that are common to wet scrubbers used for many different target pollutants.

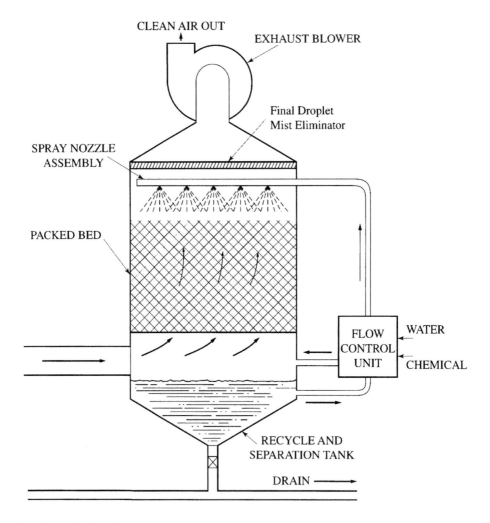

Figure 8-6   Schematic of a basic wet scrubber.

As diagrammed in Figure 8-6, the components of a basic scrubber include a vessel, some type of packing (of which there are many different types), a fan or blower, a reservoir for the scrubber fluid, and a pump for the fluid. There are many options for additional features, and many optional configurations for the system as a whole.

Venturi Scrubbers

Venturi scrubbers are intended only for removal of particulates. Figure 8-7 shows a diagram of a venturi scrubber, which consists of a restriction in the air transport duct work and spray nozzles located either in the restricted zone, referred to as the "throat" of the venturi, or just upstream of the throat. The throat is preceded by a "converging section" and is followed by a "diverging section." These two sections together make up the venturi.

The mechanism for removing particulates from the air stream with a venturi scrubber is as follows: as the stream of gas approaches the throat of the venturi, its velocity increases dramatically. The kinetic energy imparted by the high velocity has the effect of shearing the fluid sprayed into the gas stream into tiny droplets. Particulates and other substances in the gas stream become adsorbed onto the extremely large surface area of the fluid droplets. Then, as the gas stream slows down to its original velocity in the diverging section and beyond, the droplets coalesce, become too heavy to be propelled along with the gas stream, and consequently settle out of the stream under the influence of gravity.

Venturi scrubbers have the advantage of being relatively inexpensive in terms of both initial cost and costs for operation and have few, if any, moving parts other than the blower. On the other hand, venturi scrubbers are relatively low-efficiency treatment devices. They are most often used as pretreatment devices, upstream from devices with higher removal efficiency, such as packed wet scrubbers.

Tray Scrubbers

Tray scrubbers, shown in Figure 8-8, are somewhat more extensive than venturi scrubbers. Tray scrubbers may be used for pretreatment or as the only treatment for certain air streams. Figure 8-8 shows a schematic drawing of a tray scrubber.

As shown in Figure 8-8, the basic components of a tray scrubber include a set of orifices, or nozzles, and a set of trays, which may be perforated, in addition to the basic components. The mechanism by which tray scrubbers remove particulates from gas streams is as follows: The orifices or nozzles direct the gas stream onto the trays, which are covered with the scrubber fluid. As the stream of gas proceeds through the orifices, the velocity of the gas increases rather suddenly. The particulates are carried by inertia into the scrubber fluid and are thus removed from the air stream. The high velocity of the gas as it strikes the fluid-covered plates serves a second purpose, which is to cause the fluid to become frothy, greatly increasing the fluid surface area and thus particle-capture efficiency.

Treatment Objectives

Treatment objectives are needed to complete the development of design criteria for handling and treatment equipment. The air discharge permit, either in hand or anticipated, is one of the principal factors used in this development. Another principal factor is the strategy to be used regarding allowances, i.e., whether or not to buy allowances from another source or to reduce emissions below permit limits and attempt to recover costs by selling allowances. This strategy and its legal basis are discussed in Chapter 3. Only after all treatment objectives have been developed can candidate treatment technologies be determined. However, it may be beneficial to employ an iterative process whereby more than one set of treatment objectives and their appropriate candidate technologies are com-

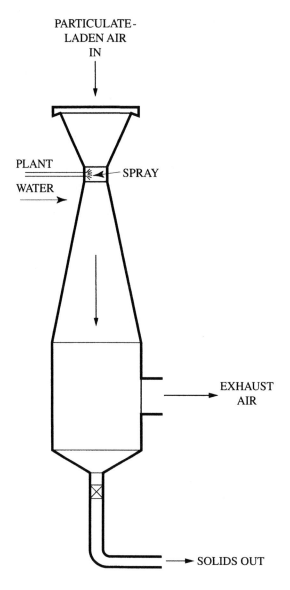

PARTICULATE-
LADEN AIR
IN

PLANT
WATER

SPRAY

EXHAUST
AIR

SOLIDS OUT

Figure 8-7   Diagram of a venturi scrubber.

pared as competing alternatives in a financial analysis to determine the most cost-effective system.

Treatment Systems for Control of Gaseous Pollutants

There are five methods in general use for removing gaseous pollutants from gas streams.

1. Adsorption
2. Absorption
3. Condensation
4. Incineration
5. Biofiltration

Activated carbon is the technology most used for treating gas streams by the mechanism of adsorption. Those that employ the mechanism of absorption, condensation, or

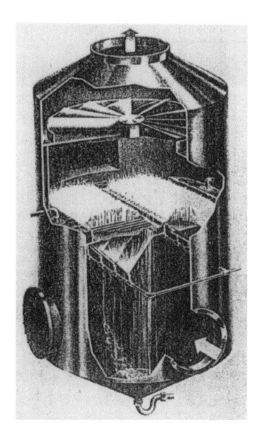

Figure 8-8   Schematic drawing of tray scrubber (courtesy of Sly Manufacturing Co.).

both often use wet scrubbers as the basic equipment. Incineration is normally accomplished in a chamber of more sophisticated design than either carbon columns or scrubbers and normally requires a high degree of safety assurance technology.

Adsorption

Removal of air pollutants by adsorption onto granules of activated carbon is an extremely effective technology for volatile organic compounds (VOCs) and other organic pollutants. It is not effective for removing most inorganic substances. Activated carbon is a nonreactive material having an extremely high surface-to-volume ratio. Activated carbon is normally manufactured in a two-step process: The first is to char the raw material (bituminous coal and coconut shells are

examples) to eliminate hydrocarbons; the second is to heat the charred material to 750°C to 950°C in the presence of steam and the absence of oxygen. The result is a very highly porous residual. Many activated carbon products have surface areas of 1,000 to 1,500 m^2/gram. The highly developed system of pores accounts for the extremely large surface area. The large surface area accounts for the highly effective adsorptive characteristic of activated carbon.

The mechanism by which adsorption works as a treatment process is explained in detail in Chapter 2. In being removed from a gas stream by adsorption, the pollutant moves from a gas phase to a solid phase and must now be managed as a solid waste. An advantage of activated carbon is that the spent carbon can be reheated so that the adsorbed pollutants are incinerated (converted to carbon dioxide, water [vapor] and ash), and the activated carbon is regenerated for reuse. With each regeneration cycle, however, a certain amount of adsorptive capacity is lost. There is always the requirement for makeup with some portion of new activated carbon.

Activated carbon treatment systems for treating gas streams are usually configured as illustrated in Figure 8-9. Cylindrical containers referred to as "carbon columns" are filled ("packed") with beds of activated carbon granules, through which the gas stream to be treated is forced to flow by use of a blower. Very often, several containers are connected in series. The multiple container arrangement allows for a factor of safety, as well as provides a means to remove one or more columns for maintenance or bed replacement without stopping the flow of gas for longer than the time required to shunt out the column to be removed.

As the contaminated gas stream travels through the bed, adsorption of the pollutants takes place. The purified effluent gas exits the last column in the series. The portion of the bed that is closest to the inlet receives a continuous dose of concentrated pollutants and

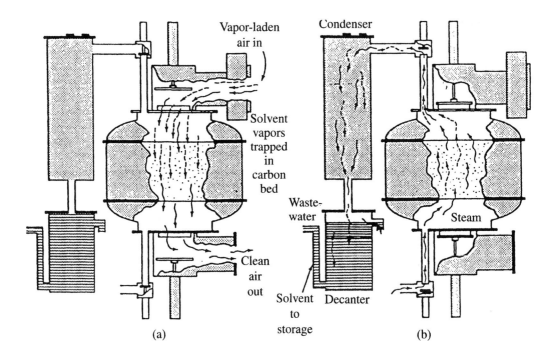

Figure 8-9   Activated carbon adsorption system for removal of gaseous pollutants and other chemicals from gas streams (from Bounicor and Davis © 1992; reprinted by permission of John Wiley & Sons, Inc.).

is thus driven to the point of saturation by the highest possible driving force (the concentration gradient). Portions of the bed that are downstream of the inlet receive progressively less concentrated amounts of the pollutants. As the portions of the bed closest to the inlet become progressively saturated, a "front of saturation" moves steadily toward the outlet end. In this manner, a "concentration profile" develops and progressively moves toward the outlet end of the system, as illustrated in Figure 8-10.

Eventually, dilute concentrations of pollutants appear in the effluent. When these concentrations increase to the point of non-acceptability, "break-through" is said to have occurred, and it is necessary to remove the columns from service.

One strategy for operation of a gas stream treatment system as described above is to remove the most upstream column from the series system at the occurrence of break-through, and, at the same time, to install a fresh column as the most downstream col-

umn. The carbon from the spent column can then be regenerated or disposed of and replaced with virgin activated carbon.

Adsorbents Other Than Activated Carbon

Certainly, activated carbon is the most commonly used adsorbent for treatment of gas streams for removal of gaseous pollutants. Other adsorbents that have been successfully used include a variety of resins, activated alumina, silica gel, and so-called molecular sieves. One of the primary characteristics of a good adsorbent, of course, is a high surface area per unit weight. While no commercially available products compare to activated carbon in this respect, other characteristics in combination with reasonably high surface area per unit weight make some adsorbents useful for certain applications.

Resins

Resins are produced by inducing controlled cross-linking between certain organic sub-

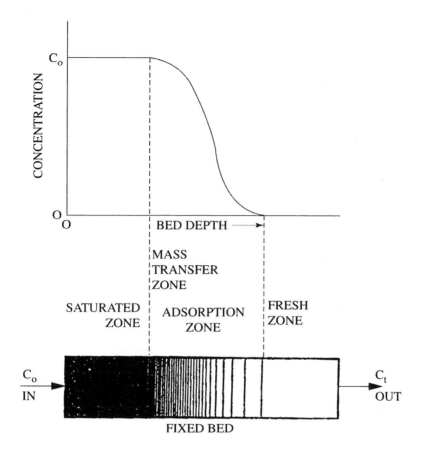

Figure 8-10 Concentration profile along adsorption column.

stances. Resins with a surface area of 100 to about 700 m² per gram can be produced so as to exhibit a high selectivity for certain substances. For instance, phenolic resins have been successfully used to remove odorous substances from air streams.

Resins can be produced in granular form such that they resemble activated carbon in physical size and shape. Resins can therefore be used in a packed bed configuration, using the same vessels and equipment as are used for activated carbon.

Activated Alumina

Activated alumina is produced by a specialized heat treatment of aluminum trihydrate, a primary ingredient of bauxite as it is mined. It can be obtained in granular form similar in size and shape to activated carbon

granules. Therefore, it can be directly substituted for activated carbon so as to use the same physical setup and equipment. Although activated alumina is most often used to remove moisture from air, it has been used, and has potential use, for removal of certain air pollutants from gas streams that are either being discharged to the air or are being recycled. A potential use is in series with another adsorbent. Surface areas of activated alumina products are in the range of 300 m² per gram.

Silica Gel

Silica gel has been used to remove sulfur compounds from a gas stream and to remove water from gas. It is produced by neutralizing, washing, drying, and roasting sodium silicate. It can be obtained in granular form, and, as is the case with resins and activated alumina, it can be used in the same physical setup as activated carbon. Silica gel products have surface areas of about 700 to 800 m^2 per gram.

Molecular Sieves

Molecular sieves have been effectively used for removing odorous chemical substances, such as hydrogen sulfide, and methyl and ethyl mercaptans from gas streams. They are crystalline substances manufactured from aluminosilicate and can be obtained in granular form and used as a substitute for activated carbon, using the same equipment in the same configuration. Molecular sieves have surface areas comparable to activated carbon. Typically, the surface area of molecular sieve products averages about 1,200 m^2 per gram.

Absorption

The chemical mechanism of absorption is that of dissolution. In a gas stream treatment system that employs absorption as the treatment technology, the stream of gas to be discharged to the air or recycled for reuse is brought into intimate contact with a liquid. Substances dissolve into the liquid and are thus removed from the gas stream. In some cases the removed substance changes in character; in other cases it does not. Either way, the removed substances have been converted from an air pollutant to a potential water pollutant and must be dealt with further. Absorption systems, then, are not complete as treatment systems in themselves but are components of treatment systems.

The primary purpose of absorption equipment is first to contain the pollutants and then to maximize the opportunity for pollutants to move from the gas phase to the liquid phase. This purpose is accomplished by maximizing the surface area of the liquid absorbent and causing the gas stream to move past as much of the liquid surface as possible. Time of contact, of course, is a major parameter.

Where the target pollutants are highly soluble in water, the liquid absorbent can be water. However, it is the usual case that a chemical substance present in the liquid absorbent readily reacts with the target pollutant to form a product that is either highly soluble in the liquid absorbent or forms a precipitate. For instance, sulfur dioxide, a gas at ambient temperatures, can be removed from a stream of air by contacting it with a solution of sodium hydroxide. Soluble sodium sulfate will quickly form and remain in the liquid. As another example, a stream of air containing silver sulfate in aerosol form can be contacted with an aqueous solution of sodium chloride. Insoluble silver chloride will form and remain suspended in the liquid until it is removed by an additional treatment step.

The "packed tower," an air pollution treatment system that resembles the wet scrubber system used for removal of particulates and discussed earlier, is the most common technology used for removal of gaseous (and aerosol) pollutants. The basic components of packed tower technology are illustrated in Figure 8-11.

The packed tower system, also called a "packed column," consists of the following elements:

- A vessel (tower), usually cylindrical, usually constructed of steel and coated as needed to prevent corrosion or other form of destruction
- Packing to promote intimate contact between molecules of target pollutants and the liquid absorbent
- A spray distribution system to apply the liquid absorbent evenly over the entire top surface of the packing
- A reservoir, usually at the bottom to the tower, to serve as a wet well for the pump
- A pump to transfer liquid absorbent from the reservoir to the spray system
- A blower to force the gas stream from its source to the packed tower and up through the packing
- A support floor, highly perforated, to perform several functions: holding the packing above the reservoir so as to pro-

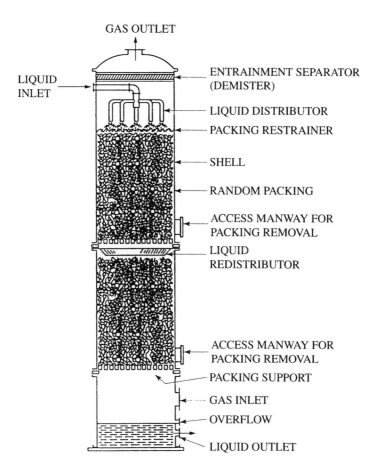

GAS OUTLET

LIQUID
INLET

ENTRAINMENT SEPARATOR
(DEMISTER)

LIQUID DISTRIBUTOR

PACKING RESTRAINER

SHELL

RANDOM PACKING

ACCESS MANWAY FOR
PACKING REMOVAL

LIQUID
REDISTRIBUTOR

ACCESS MANWAY FOR
PACKING REMOVAL

PACKING SUPPORT

GAS INLET

OVERFLOW

LIQUID OUTLET

Figure 8-11   Schematic drawing of a packed tower absorber (from Alley © 1998; reprinted by permission of McGraw-Hill, Inc.).

vide a space for incoming gas (influent) to distribute itself evenly across the cross-section of the tower; serving as an inlet device to promote even application of the influent gas to the bottom of the column of packing; and allowing the liquid absorbent to readily drain away from the packing

Additional elements that are often included as components of a packed tower absorption system are the following:

- A packing restrainer to prevent the individual units of packing material from being carried up by the gas as it passes through the packing

- A demister to prevent droplets of liquid from exiting the tower with the exiting treated gas stream
- An overflow device to maintain the proper depth of liquid absorbent in the reservoir
- A liquid redistributor located within the depth of the packing to collect liquid absorbent after it has flowed through a portion of the packing and redistribute it over the top surface of the next portion of packing

A description of the operation of the packed tower is as follows. The liquid absorbent is pumped continuously from the reservoir to the spray distribution system. After being applied evenly over the top surface of the packing material, the liquid absorbent flows slowly down over the surfaces of the packing. As the gas stream, which has entered the tower in the space between the reservoir and the bottom of the packing, flows up through the packing, substances that can dissolve in the liquid do so. These substances have thus been removed from the gas stream, which continues its upward flow and exits the tower at the top. Excess moisture in the form of aerosol-size droplets or larger are trapped by the demister as the gas stream passes through.

Design parameters for a packed tower system include the quantity of packing material and the flow rate capacity of the blower. These parameters, in addition to the flow rate capacity of the liquid absorbent pump, determine the time of contact between the gas stream and the absorbent. The physical characteristics of the packing material have a great effect on the mass transfer efficiency, since the thinner the film of liquid absorbent as it flows down over the packing, and the more turbulent the flow of this thin film, the greater the opportunity for each molecule of target pollutant to contact nonsaturated absorbent in which it can dissolve or otherwise interact. In the same manner, the physical characteristics of the packing material influence the characteristics of flow of the gas stream up through it. The more torturous the flow paths and turbulent the flow of the gas, the greater will be the opportunity for individual molecules of target pollutant to actually physically contact the liquid absorbent. The greater contact will lead to higher removal efficiency, all other influences being equal.

Condensation

Gases can be changed to liquids by decreasing temperature or increasing pressure or both. Thus, the mechanism by which condensation technology accomplishes air pollutant removal is based on the generalized gas law, stated as:

$$PV = nRT \qquad (8\text{-}1)$$

or

$$V = \frac{nRT}{P} \qquad (8\text{-}2)$$

where

V = Volume of a given weight of gas or volume/unit wt (m^3/g)

n = Number of moles of gas in the volume V

R = Universal gas constant

P = Pressure exerted on the volume of gas

T = Absolute temperature of the gas in volume V

which states that the volume that a given weight of gas will occupy decreases as temperature decreases and pressure increases. At some point, the gas will change from the gaseous state to the liquid state, after which it no longer obeys the gas laws.

The most commonly used equipment that employs condensation technology uses temperature decrease as the mechanism. Often, water-cooled condensers are used as pretreatment devices to remove easily condensed substances (such as vapors of sulfuric acid) to protect or prolong the operating cycle times of downstream equipment. Figure 8-12 presents schematic drawings of three types of condensers used to remove gaseous pollutants from gas streams.

Incineration

The fundamental mechanism on which incineration technology for air pollution control is based is combustion. Combustion of organic pollutants entails conversion to carbon dioxide, water, and ash. Some inorganic materials, such as sulfur and nitrogen, are often oxidized to problematic substances during the combustion process. Other inorganic materials—for instance, heavy metals,

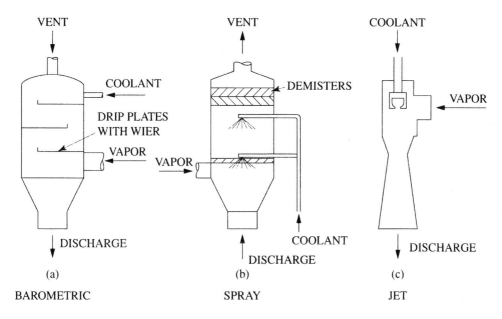

Figure 8-12   Three types of condensers for air pollution control.

become incorporated in the ash and can in this way add to ultimate disposal problems.

As an air pollution abatement technology, incineration is used for many purposes, including odor control, reduction of releases of hydrocarbons to the air (flares at petroleum refineries, for example), and destruction of volatile organic compounds (VOCs).

In the context of air pollution control, incinerators are of two types: thermal oxidizers and catalytic oxidizers. The difference between the two is that thermal oxidizers accomplish combustion by use of heat alone. Catalytic oxidizers use a catalyst to decrease the activation energy of the combustion process, or to otherwise effect acceleration of the combustion process, and are thus able to accomplish reasonably complete combustion at significantly lower temperatures.

Thermal Oxidizers

The basic components of a thermal oxidizer for air pollution control are illustrated in Figure 8-13.

A typical thermal oxidizer consists of a combustion chamber in which the combus-

tion process takes place; a burner for the purpose of combusting a support fuel needed to elevate the temperature in the combustion chamber as necessary; an injection device, used to inject what is to be treated into the combustion chamber; and a flue, used to transport the treated gas stream to the discharge location.

Additional components that can be added to a thermal oxidizer include heat recovery equipment and a system to "preheat" the gas stream containing pollutants.

Catalytic Oxidizers

As explained above, the process of catalytic oxidization is essentially the same as thermal oxidization, except that a catalyst enables the combustion process to take place at lower temperatures. The advantages include less expensive construction costs for the equipment and reduced use of auxiliary fuel.

A schematic drawing of a typical catalytic oxidizer is shown in Figure 8-14. A photograph of a catalytic oxidizer in operation is presented in Figure 8-15.

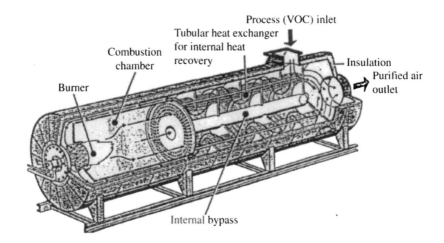

Figure 8-13 Basic components of a thermal oxidizer for air pollution control (from Freeman © 1989; reprinted by permission of McGraw-Hill, Inc.).

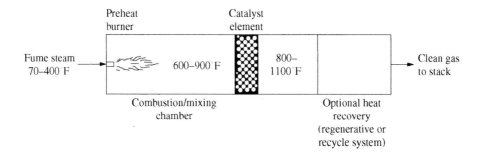

Figure 8-14 A typical catalytic oxidizer (from Bounicor and Davis © 1992; reprinted by permission of John Wiley & Sons, Inc.).

Figure 8-15 An operating catalytic oxidizer (courtesy of Stealth Industries, Inc.).

Biofiltration

General

The removal of air contaminants from air streams by means of dissolution of those contaminants into water, followed by biodegradation of the contaminants, has been a useful concept for control of air pollutants for many years. When a solid phase filter is used as the medium to contain both the liquid into which the contaminants dissolve and the microorganisms to effect the biodegradation, the process is known as "biofiltration." Biofiltration has been used since the mid-1970s for treatment of malodorous compounds as well as volatile organic compounds (VOCs). This technology is applicable to the treatment of contaminated air that contains the normal concentration of oxygen and dilute concentrations of biodegradable organic gases.

The solid phase filter is normally compost, sphagnum peat, or soil. The solid media are surrounded by a film of aqueous liquor, which is teeming with microorganisms. As the stream of contaminated air flows through the filter, the contaminants dissolve into the aqueous liquor (driven by entropy). These dissolved contaminants are then consumed as food by the microorganisms. Carbon dioxide, water, oxidized organic compounds, and more microorganisms are the end products. When reduced sulfur compounds, such as hydrogen sulfide, dimethyldisulfide, and mercaptans, are present as contaminants to be removed from the stream of contaminated air, mineral acids are among the end products.

Oxygen contained in the stream of contaminated air is necessary to maintain aerobic conditions within the biofilter. Also, nitrogen, phosphorus, and other nutrients are required for maintenance of a robust microbial population. Typically, a solution of water containing sufficient nutrients is sprayed onto the biofilter to supply sufficient moisture as well as nutrients. In addition, the stream of contaminated air is humidified before it is applied to the biofilter. The drying action (by means of evaporation) of the stream of air flowing through the biofilter must be counterbalanced by the humidifier and the applied liquid spray.

Figure 8-16 presents a schematic of a typical biofilter system, which consists of a bed of filter material, an air distribution system, a system to supply nutrients, a humidifier, a blower, and ductwork. The filter material can be composed of compost material, sphagnum peat, or highly porous soil. The filter bed is typically about one meter (three feet or so) in depth.

Design

A primary design parameter of a biofilter system is residence time, i.e., the stream of contaminated air must remain within the filter bed for sufficient time for the contaminants to dissolve into the liquid that surrounds the filter media. A second design parameter relates to the quantity of liquid to be maintained within the filter bed. If too much liquid is supplied to the bed, the pores, or passageways, for air will be blocked, with the consequence that anoxic or anaerobic conditions will develop. If insufficient liquid is present within the filter bed, the target contaminants will not dissolve out of the airstream, with the consequence that treatment will not take place. Other design parameters

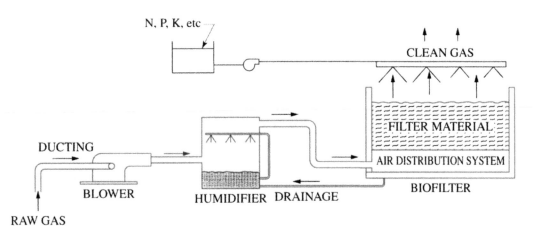

Figure 8-16   Schematic of a typical biofilter system.

include quantity and type of nutrients to be supplied, temperature control, pH control, and removal of excess liquid.

The residence time for the biofilter is a function of rate of flow of contaminated air as well as volume of filter media. Since it has been determined through operating experience that a filter bed depth of about one meter is optimum, the parameter of filter volume reduces to one of filter area, that is, the horizontal dimensions of the filter bed. Successful biofilter systems have filter areas that range from 100 to 22,000 ft^2 (10 to 2,000 m^2). Flow rates of contaminated gases have ranged from 600 to 90,000 cfm (1,000 to 150,000 m^3/hr). These parameters are compatible with air pollutant treatment rates of 10 to 100 g/m^3/hr and surface loads of up to 16 scfm/ft^2 (300 m^3/hr/m^2). The optimal values for these parameters for any given application depend upon the filter media, the concentration of pollutants, and the nature of the pollutants, as well as the temperature, pH, and alkalinity.

The nature of the filter media has a major influence on the economics of construction and operation of a biofilter system, in that the cost of operation is highly dependent on the amount of electrical energy used to blow contaminated air through the filter. The amount of electrical energy, in turn, is a function of the backpressure created by the filter. The backpressure increases with rate of flow, decrease in filter media porosity, and increase in filter moisture content, and is affected by temperature. Also, increased biological growth on the filter media has the effect of increasing the backpressure.

Sphagnum peat has proved to be an excellent medium for biofilters. This material, when handled and installed properly, provides for extremely large surface area, enables a relatively free flow of air, holds moisture well, and provides an excellent medium for the growth and maintenance of a robust microbial population. In particular, sphagnum peat has been found to harbor microbial populations that have a high percentage of fungi species, which are very effective in breaking down certain air pollutants. Sphagnum peat can be obtained in large quantities through agricultural supply outlets.

Compost material has also proved to be an effective medium for biofilters. Use of compost derived from municipal solid waste, bark, tree trimmings, and leaves has the additional benefit of avoiding the creation of more solid waste.

Eitner and Gethke have developed values for parameters for filter material as follows:

- Pore volume $\geq 80\%$
- pH = between 7 and 8
- $d_{60} \geq 4$ mm
- Total organic matter content $\geq 55\%$

Operation and Maintenance

Over time, biofilter beds have undergone settling and consolidation to such an extent that "fluffing" is required to avoid excess backpressure and channeling of air flow. Typically, beds have required reworking at two-year intervals, with complete replacement every five to six years.

Routine and periodic maintenance of biofilters includes:

- Daily check of major operating parameters:

  — Off-gas temperature
  — Off-gas humidity
  — Filter temperature
  — Backpressure

- Periodic check of filter moisture content and pH

When designed, operated, and maintained properly, biofilter technology offers an effective, low-cost solution to control air pollution caused by dilute concentrations of malodorous reduced sulfur compounds, certain VOCs, and many other biodegradable organics. Industries that have made successful use of biofilters worldwide include:

- Fish rendering
- Print shops

- Pet food manufacturing
- Flavors and fragrances
- Residential wastewater treatment plants
- Landfill gas extraction
- Tobacco processing
- Chemical manufacturing
- Chemical storage
- Industrial wastewater treatment plants

Bibliography

Alley, R. E., & Associates, Inc. *Air Quality Control Handbook.* New York: McGraw-Hill, 1998.

Avallone, E. A., and T. Baumeister III (eds.). *Marks' Standard Handbook for Mechanical Engineers.* 9th ed. New York: McGraw-Hill, 1987.

Billings, C. E. (ed.) *The Fabric Filter Manual.* Northbrook, IL: McIlvaine Co..

Billings C. E., and J. Wilder. *Handbook of Fabric Filter Technology, Vol. 1: Fabric Filter Systems Study.* Bedford, MA: GCA Corp., PB-200-648; Springfield, VA: NTIS, December 1970.

Bilotti, J. P. and T. K. Sutherland. In *Air Pollution Control Equipment: Selection, Design, Operation and Maintenance.* L. Theodore and A. J. Buonicore (eds.). Englewood Cliffs, NJ: Prentice Hall, 1982.

Borgwardt, R. H. "Limestone Scrubbing of S02 at EPA Pilot Plant." *Progress Report 12.* St. Louis, MO: Monsanto Research Company, 1973.

Buonicore, A. J., and W. T. Davis, *Air Pollution Engineering Manual.* New York: Van Nostrand Reinhold, 1992.

Cheremisinoff, P. N., and R. A. Young (eds.) *Pollution Engineering Practice Handbook.* Ann Arbor, MI: Ann Arbor Science, 1975.

Cooper, C. D., and F. C. Alley. *Air Pollution Control: A Design Approach.* 2nd ed. Prospect Heights, IL: Waveland Press, Inc., 1986.

Crawford, M. *Air Pollution Control Theory.* New York: McGraw-Hill, 1976.

Davis, W. T., and R. F. Kurzynske. "The Effect of Cyclonic Precleaners on the Pressure Drop of Fabric Filters." *Filtration Separation* 16, No. 5.

Davis, W. T., and W. F. Frazier. "A Laboratory Comparison of the Filtration Performance of Eleven Different Fabric Filter Materials Filtering Resuspended Flyash." *Proceedings of the 75th Annual Meeting of the Air Pollution Control Association,* 1982.

Dennis, R., and H. A. Klemm. *Fabric Filter Model Format Change, Vol. 1: Detailed Technical Report.* EPA-600/7-79-0432. U.S. Environmental Protection Agency, Research.

Dennis, R., R. W. Cass, D. W. Cooper, et al. *Filtration Model for Coal Flyash with Glass Fabrics.* EPA-600-7-77- 084. August 1977.

Donovan, R. P., B. E. Daniel, and J. H. Turner. *EPA Fabric Filtration Studies, 3. Performance of Filter-Bags Made from Expended PTFE Laminate.* U.S. Environmental Protection Agency. EPA-600/2-76-168c. NTIS PB-263-132. Research Triangle Park, NC: December 1976.

Durham, J. F. and R. E. Harrington. "Influence of Relative Humidity on Filtration Resistance and Efficiency of Fabric Dust Filters." *Filtration Separation* 8 (July/August 1971): 389–393.

Eitner, D., and Gethke, H. G. "Design Construction and Operation of Bio-filters for Odor Control in Sewage Treatment Plants." Presented at the 80th Annual Meeting of the Air Pollution Control Association, New York, 1987.

Ellenbecker, M. J., and D. Leith. "Dust Removal from Non-Woven Fabrics." Presented at the Annual Meeting of the Air Pollution Control Association, Montreal, June 22–27, 1980.

Fernandes, I. H. "Incinerator Air Pollution Control." *Proceedings of National Incinerator Conference.* New York: ASME, 1970.

Frazier, W. F., and W. T. Davis. *Effects of Flyash Size Distribution on the Performance of a Fiberglass Filter.* "Symposium on the Transfer and Utilization of Particulate Control Technology." Vol. II: *Particulate*

Control Devices. EPA-600/9-82-0050c. July 1982: 171–180.

Frederick, E. R. *Electrostatic Effects in Fabric Filtration: Vol. 11, Turboelectric Measurements and Bag Performance* (annotated data). EPA-600/7-78-142b NTIS PB-287-207. July 1978.

Freeman, H. M. (ed.) *Standard Handbook of Hazardous Waste Treatment and Disposal.* New York: McGraw-Hill, 1989.

Greiner, G. P., J. C. Mycock, and D. S. Beachler. "The IBFM, a Unique Tool for Troubleshooting and Monitoring Baghouses." Paper 85-54.2, Annual Meeting of the Air Pollution Control Association. Detroit, MI: June 1985.

Hotchkiss C. B., and L. F. Cox. "Fabric and Finish Selection, Manufacturing Techniques and Other Factors Affecting Bag Life in the Coal-Fired Boiler Applications." *Proceedings: Second EPRI Conference on Fabric Filter Technology for Coal-Fired Power Plants.* EPRI CS-3257. Palo Alto, CA: Electric Power Research Institute, November 1983.

Katz, J. *The Art of Electrostatic Precipitation.* Munhall, PA: Precipitator Technology, Inc., 1979.

Ketchuk, M. M., A. Walsh, O. F. Fortune, et al. "Fundamental Strategies for Cleaning Reverse-Air Baghouses." *Proceedings, Fourth Symposium on the Transfer and Utilization of Particulate Control Technology, Vol. 1: Fabric Filtration.* U.S. Environmental Protection Agency. EPA-600/9-84-025a. Research Triangle Park, NC: November 1984.

Larson, R. I. "The Adhesion and Removal of Particles Attached to Air Filter Surface." *AIHA Journal* 19 (1958).

Licht, W. *Air Pollution Control Engineering.* New York: Marcel Dekker, 1980.

Liptak, B. G. (ed.) *Environmental Engineers' Handbook, Vol. II: Air Pollution.* Radnor, PA: Chilton Book Company, 1974.

Lund, H. F. (ed.) *Industrial Pollution Control Handbook.* New York: McGraw-Hill, 1971.

Magill, P. L., F. R. Holden, and C. Ackley (eds.) *Air Pollution Handbook.* New York: McGraw-Hill, 1956.

Ottmers, D. M. et al. "Evaluations of Regenerable Flue Gas Desulfurization Processes." Vol. 11. EPRI RP-535-1. Palo Alto, CA: Electric Power Research Institute, 1976.

Perry, R. et al. (eds.) *Perry's Chemical Engineers' Handbook.* New York: McGraw-Hill.

Rafson, H. J. (ed.) *Handbook of Odor and VOC Control.* New York: McGraw-Hill, 1998.

Reisinger A. A., and W. T. Grubb. "Fabric Evaluation Program at Coyote Unit #1, Operating Results Update." *Proceedings, Second EPRI Conference on Fabric Filter Technology for Coal-Fired Power Plants.* EPRI CS03257. Palo Alto, CA: Electric Power Research Institute, November 1983.

Robinson, J. W. R. E. Harrington, and P. W. Spaite. "A New Method for Analysis of Multicompartmented Fabric Filtration." *Atmospheric Environment* 1, No. 4 (1967) 499.

Siebert, P. C. *Handbook on Fabric Filtration.* Chicago: ITT Research Institute, 1977.

Stern, A. C. (ed.) *Air Pollution.* 3rd ed. New York: Academic Press, 1977.

Strauss, W. *Industrial Gas Cleaning.* 2nd ed. New York and Oxford: Pergamon Press, 1975.

Theodore, L., and A. Buonicore. *Industrial Air Pollution Control Equipment.* Cleveland, OH: CRC Press, 1976.

U.S. Department of Health, Education and Welfare. *Recommended Industrial Ventilation Guidelines.* Publication No. (NIOSH) 76-162. Washington, DC: U.S. Government Printing Office, 1976.

U.S. Department of the Interior. Bureau of Mines. *Methods and Costs for Stabilizing Fine-Sized Mineral Wastes.* RI 7896. Washington, DC: 1974.

U.S. Environmental Protection Agency: National Research Council. *Odors from Stationary and Mobile Sources.* Contract 68-01-4655. Washington, DC: U.S. Government Printing Office, 1979.

U.S. Environmental Protection Agency. *Evaluation of the Efficiency of Industrial Flares: Test Results.* EPA 600/2784/095. Washington, DC: U.S. Government Printing Office, 1984.

U.S. Environmental Protection Agency. *Flue Gas Desulfurization System Capabilities for Coal-Fired Steam Generators, Vol. 11.* EPA 600/7-78/032b. Washington, DC: U.S. Government Printing Office, 1978.

U.S. Environmental Protection Agency. *Flue Gas Desulfurization: Lime/Limestone Processes.* Technology Transfer. EPA 625/8-81/006. Research Triangle Park, NC: 1981.

U.S. Environmental Protection Agency. *Fugitive Emissions from Integrated Iron and Steel Plants.* EPA 600/2-78/050. Washington, DC: U.S. Government Printing Office, 1978.

U.S. Environmental Protection Agency. *Manual of Electrostatic Precipitator Technology.* APTD 061 0. Washington, DC: U.S. Government Printing Office, 1970.

U.S. Environmental Protection Agency. "Menard Cleaning of a Fabric Filter Dust Collector." *Proceedings, Fourth Symposium on the Transfer and Utilization of Particulate Control Technology, Vol. 1: Fabric Filtration.* EPA-600/9-84-025a. Research Triangle Park, NC: November 1984.

U.S. Environmental Protection Agency. "Standards of Performance for New Stationary Sources." *Code of Federal Regulations.* Title 40, Part 60. Washington, DC: U.S. Government Printing Office, 1971–1997.

U.S. Environmental Protection Agency. *Adipic Acid-Enhanced Lime/Limestone Test Results at the EPA Alkali Scrubbing Test Facility.* Technology Transfer. EPA 625/2-82/029. Research Triangle Park, NC: 1982.

U.S. Environmental Protection Agency. *Air Pollution Engineering Manual.* 2nd ed. AP-40. Research Triangle Park, NC: 1973.

U.S. Environmental Protection Agency. *An Update of the Wellman-Lord Flue Gas Desulfurization Process.* EPA 600/2-76/136a. Washington, DC: U.S. Government Printing Office, 1976.

U.S. Environmental Protection Agency. *Assessment of the Use of Fugitive Emission Control Devices.* EPA 600/7-79/045. Washington, DC: U.S. Government Printing Office, 1979.

U.S. Environmental Protection Agency. *Capital and Operating Costs of Selected Air Pollution Control Systems.* EPA 450/5-80/002. Washington, DC: U.S. Government Printing Office, 1978.

U.S. Environmental Protection Agency. *Compilation of Air Pollutant Emission Factors.* EPA AP-42. Research Triangle Park, NC: 1973.

U.S. Environmental Protection Agency. *Control of Volatile Organic Compound Emissions from Manufacture of High-Density Polyethylene, Polypropylene, and Polystyrene Resins.* EPA 450/3-83/008. Research Triangle Park, NC.

U.S. Environmental Protection Agency. *Control of Volatile Organic Compound Leaks from Synthetic Organic Chemical and Polymer Manufacturing Equipment.* EPA 450/3-83/006. Research Triangle Park, NC: 1983.

U.S. Environmental Protection Agency. *Control of Volatile Organic Emissions from Existing Stationary Sources, Vol. 1: Control Methods for Surface Coating Operations.* EPA 450/2-76/028. Research Triangle Park, NC: 1976.

U.S. Environmental Protection Agency. *Control Techniques for Particulate Emissions from Stationary Sources, Vol. 1.* EPA 450/3-8 IA/05a. Research Triangle Park, NC: 1982.

U.S. Environmental Protection Agency. *Evaluation of Control Technologies for Hazardous Air Pollutants—Appendices.* EPA 600/7-86/0096. Washington, DC: U.S. Government Printing Office, 1985.

U.S. Environmental Protection Agency. *Evaluation of the Efficiency of Industrial Flares: Test Results.* EPA 600/2-84/095. Washington, DC: U.S. Government Printing Office, 1984.

U.S. Environmental Protection Agency. *Flue Gas Desulfurization and Sulfuric Acid Production Via Magnesia Scrubbing.* Technology Transfer. EPA 625/2-75/077. Washington, DC: U.S. Government Printing Office, 1975.

U.S. Environmental Protection Agency. *Flue Gas Desulfurization: Dual Alkali Process.* Technology Transfer. EPA 625/8-80/004. Research Triangle Park, NC: 1980.

U.S. Environmental Protection Agency. *Fugitive and Fine Particle Control Using Electrostatically Charged Fog.* EPA 600/7-79/078. Washington, DC: U.S. Government Printing Office, 1979.

U.S. Environmental Protection Agency. *Fugitive Emissions from Iron Foundries.* EPA 00/7-79/195. Washington, DC: U.S. Government Printing Office, 1979.

U.S. Environmental Protection Agency. *Fugitive Emissions Sources of Organic Compounds.* EPA 450/3-82/010. Research Triangle Park, NC: 1982.

U.S. Environmental Protection Agency. *Guidelines for Development of Control Strategies in Areas with Fugitive Dust Problems.* EPA 45012-77/029. Washington, DC: U.S. Government Printing Office, 1977.

U.S. Environmental Protection Agency. *Handbook—Control Technologies for Hazardous Air Pollutants.* EPA 625/6-86/014. Cincinnati, OH: 1986.

U.S. Environmental Protection Agency. *Handbook—Permit Writer's Guide to Test Burn Data—Hazardous Waste Incineration.* EPA 625/6-86/012. Cincinnati, OH: 1986.

U.S. Environmental Protection Agency. *Industrial Guide for Air Pollution Control.* Technology Transfer. EPA 625/6-78/004. Washington, DC: U.S. Government Printing Office, 1978.

U.S. Environmental Protection Agency. *Inspection Manual for Evaluation of Electrostatic Precipitator Performance.* EPA 340/1-79/007. Washington, DC: U.S. Government Printing Office, 1981.

U.S. Environmental Protection Agency. *Investigation of Fugitive Dust—Vol. I: Sources, Emissions and Control.* EPA 450/3-74/036a. Washington, DC: U.S. Government Printing Office, 1974.

U.S. Environmental Protection Agency. *Organic Chemical Manufacturing, Vol. 4: Combustion Control Devices.* EPA 450/3-80/026. Washington, DC: U.S. Government Printing Office, 1980.

U.S. Environmental Protection Agency. *Physical Coal Cleaning for Utility Boiler SO, Emission Control.* EPA 600/7-78/034. Washington, DC: U.S. Government Printing Office, 1978.

U.S. Environmental Protection Agency. *Procedures Manual for Fabric Filter Evaluation.* EPA 600/7-78/113. Washington, DC: U.S. Government Printing Office, 1978.

U.S. Environmental Protection Agency. *Regulatory Options for the Control of Odors.* EPA 450/5-85/003. Research Triangle Park, NC: 1980.

U.S. Environmental Protection Agency. *Standards Support and Environmental Impact Statement, Vol. I: Proposed Standards of Performance for Grain Elevator Industry.* EPA 450/2-77/001 a. Washington, DC: U.S. Government Printing Office, 1977.

U.S. Environmental Protection Agency. *Technical Guidance for Development of Control Strategies in Areas with Fugitive Dust Problems.* EPA 450/3-77/0 1 0. Washington, DC: U.S. Government Printing Office, 1977.

U.S. Environmental Protection Agency. *Technical Guide for Review and Evaluation of Compliance Schedules.* EPA 3401 I-73/00 la. Washington, DC: U.S. Government Printing Office, 1973.

U.S. Environmental Protection Agency. *Wet Scrubber Handbook.* EPA R/72/1 18a. Research Triangle Park, NC: 1972.

U.S. Environmental Protection Agency. *Wet Scrubber Performance Model.* EPA 600/2-77/127. Washington, DC: U.S. Government Printing Office, 1977.

U.S. Environmental Protection Agency. *Wet Scrubber System Study.* EPA R2/ 72/118a. Research Triangle Park, NC: 1972.

U.S. Environmental Protection Agency. Emission Standards and Engineering Division. *Control Techniques for Sulfur Oxide Emissions from Stationary Sources.* 2nd ed. EPA 45013-811064. Research Triangle Park, NC: 1981.

U.S. Environmental Protection Agency. Office of Research and Development. *Control Technologies for Hazardous Air Pollutants.* EPA/625/6-91/014. Washington, DC: U.S. Government Printing Office, 1991.

U.S. Environmental Protection Agency. Office of Air Quality and Standards. Emission Measurement Branch. *Enhanced Monitoring Reference Document,* Washington, DC: U.S. Government Printing Office, 1993.

U.S. Environmental Protection Agency. *Handbook of Fabric Filter Technology.* APTD-0690. Research Triangle Park, NC: 1970.

U.S. Environmental Protection Agency. Air and Energy Engineering Research Laboratory. *Operation and Maintenance Manual for Electrostatic Precipitators.* EPA 625/1-85/017. Research Triangle Park, NC: 1985.

U.S. Environmental Protection Agency. Air and Energy Engineering Research Laboratory. *Operation and Maintenance Manual for Fabric Filters.* EPA 625/1-86/020. Research Triangle Park, NC: 1986.

U.S. Environmental Protection Agency. Center for Environmental Research Information. *Organic Air Emissions from Waste Management Facilities.* EPA/625/R-92/003. Cincinnati, OH: 1992.

U.S. Environmental Protection Agency. Center for Environmental Research Information. EPA 625/6-86/ 014. Cincinnati, OH: 1986.

U.S. Environmental Protection Agency. Energy Engineering Research Laboratory. EPA 625/1-85/019. Research Triangle Park, NC: 1985.

U.S. Environmental Protection Agency. Emission Standards and Engineering Division. *Control Techniques for Nitrogen Oxides Emissions from Stationary Sources.* Rev. 2nd ed. EPA 450/3-83/002. Research Triangle Park, NC: 1983.

U.S. Environmental Protection Agency. Industrial Environmental Research Laboratory. *Particulate Control by Fabric Filtration on Coal-Fired Industrial Boilers.* EPA 625/2-79/02 1. Research Triangle Park, NC: 1979.

U.S. Environmental Protection Agency. Office of Air Quality Planning and Standards. *Fugitive VOC Emissions in the Synthetic Organic Chemicals Manufacturing Industry.* EPA 625/10-84/004. Research Triangle Park, NC: 1984.

U.S. Environmental Protection Agency. Office of Research and Development. *Handbook-Control Technologies for Hazardous Air Pollutants.* EPA 625/6-86/014. Cincinnati, OH: 1986.

U.S. Environmental Protection Agency. Office of Research and Development. *Continuous Emission Monitoring Systems for Non-criteria Pollutants Handbook.* EPA 1625/R97/001. Cincinnati, OH: 1997.

Vesilind, P. A., and J. J. Pierce. *Environmental Engineering.* Ann Arbor, MI: Ann Arbor Science: 1982.

Zimon, A. D. *Adhesion of Dust and Powder.* New York: Plenum Press: 1969, 112.

9 Solid Waste Treatment and Disposal

In an industrial facility, solid waste is generated in a number of ways. For example, if a manufacturing process generates "scrap" that cannot be reused, it may be treated as solid waste. Think of a shoemaking facility; it is easy to imagine scraps of leather, rejected from the cutting equipment, that cannot be reused. As another example, a facility might purchase components of the manufacturing process from a third party; those components may be received in boxes. The boxes may in turn be treated as solid waste. The ways a facility can generate solid waste are virtually limitless.

Another contribution to the solid waste that a facility generates is derived from the wastewater and air treatment processes. For many of these processes, sludges are generated that can be a large percentage of the waste a facility generates. It is important to recognize that a sludge generated by a wastewater treatment process, for example, may not represent the end of the treatment train. Thickening and dewatering techniques are available to further reduce the volume of sludge that will eventually be disposed. Those treatment technologies will not be discussed here, as there are texts devoted solely to that subject. The Bibliography provides recommended reading resources.

Likewise, in Chapter 4, the principles of waste minimization are discussed. The easiest way to treat and ultimately dispose of solid waste is by not generating it in the first place. However, when there are no further means of waste treatment and no options for waste reduction, solid waste must be handled. The discussion that follows is intended to address that scenario.

When a final residue is produced that cannot be further treated or disposed of economically on-site, it must be shipped off-site for disposal. The residues are often sludges or solid wastes, which may be regulated by state, local, and federal law. This waste material can take a variety of forms, from highly toxic hazardous waste to bulk material for the dumpster, but it all constitutes an expense and liability for the environmental manager. Because the cost of storing and disposing of a final residue varies widely in accordance with the characteristics of the waste, a knowledge of the regulatory framework governing final disposal is important for both the manager and the engineer, since low treatment costs can be outweighed by high disposal costs and liabilities.

From a disposal standpoint (which doesn't necessarily correspond to the physical form of the waste) final residues generally fall into one of three categories: normal solid waste (trash), industrial or "special" waste, or federally regulated hazardous waste. Knowledge of the differences and requirements of each is important when designing any waste treatment system. This chapter provides a background for effective disposal decisions by describing the various categories of wastes, strategic and technical considerations for disposing of nonhazardous industrial wastes, and the major types of disposal facilities currently in use. Although a comprehensive treatment of each of these topics is outside the scope and focus of this book, a working knowledge of the issues associated with each is necessary to make sound and cost-effective choices.

Background

Before the Industrial Revolution in the mid-1800s, almost all wastes—industrial, commercial, and domestic—were derived from natural substances and were, therefore, biodegradable. Pollution of any portion of the environment was caused by pathogens from human wastes or was the result of simply overwhelming the ability of the environment to undergo self-purification. Either way, the pollution was temporary, and the self-purification capability of "Mother Earth" would eventually prevail.

The Industrial Revolution was attended by the production of industrial residuals that could never be removed by the natural processes of microbiological degradation, chemical oxidation and reduction, adsorption, absorption, or other natural phenomena. In consequence, some former dumps are now sites of very large amounts of heavy metals pollution. Other sites, where coal gasification took place, are still polluted with tarry residuals that are classified as hazardous because of toxic organics.

The problem of soil and groundwater pollution greatly accelerated with the development of synthetic organics, beginning a few years before World War I and greatly increasing during and after World War II. New synthetics were developed and produced in great quantity during that period of time, including chlorinated solvents; synthetic rubber; a number of pesticides and herbicides; polychlorinated biphenyls (PCBs); and plastics such as polyethylene, polyurethane, and a host of others. Microorganisms did not have the ability to produce enzymes necessary to degrade these synthetic organics, since the organics were not present on the Earth during the millions of years of development of the microorganisms. Consequently, the Earth was unable to undergo self-purification, and these substances were carried via percolating rainwater and other precipitation to the groundwater.

To prevent the recurrence of such problems, extensive and sophisticated safeguards have become standard and required components of landfills, incinerators, composting facilities, or any alternative solid waste disposal technology. As described in later sections, modern landfills have double liners with leachate collection and treatment and leak detection. Some landfills have triple liners. All landfills must be closed, according to strict regulations, upon reaching the end of their useful life. Typically, an impermeable cap equipped with gas collection and erosion control means are required.

Incinerators have stack emission controls and are subject to strict regulations regarding handling and disposal of ash and residuals from stack emission controls. Composting facilities have extensive safeguards to protect against groundwater pollution. All three disposal technologies are subject to strict and extensive regulations regarding what is allowed to be disposed of using their technology. As well, all three technologies are subject to the requirement of groundwater monitoring at their sites.

Of paramount importance to the area of solid wastes handling and disposal is the mandate for pollution prevention, discussed in Chapter 4. The pollution prevention mandate requires that industries develop, plan for, and provide for handling and disposal of all wastes associated with any product throughout its life cycle, including minimizing the sum total of those wastes. Minimization involves source reduction as well as recycling and reuse.

Among the most desirable methods of waste reduction is the production of a salable substance or material, either by modification of the waste itself or by combining it with another waste material. Chipboard is an example: used as a substitute for plywood by the construction industry, it was originally developed using wastes from woodworking. Composted food processing wastes, including wastes from the processing of fish, poultry, meat, and vegetables, have been extensively used for soil conditioning and fertilization. Rendering plants have developed capabilities to process many organic wastes into animal feed supplements that were formerly disposed of. Potato processing wastes are further processed to produce starch. Many industries have constructed and used facilities to store solid wastes that contain metals or other substances of value until an improved recovery process is developed or until market conditions become more favorable. These are more examples of a continuing effort on the part of industry to reduce the quantity of solid wastes that requires disposal.

Categories of Wastes

Disposal of final treatment residues and plant wastes in general depends on the source and chemical characteristics of the waste material itself. Thus, the first step is to determine into which category the waste belongs. From a functional standpoint, wastes can generally be categorized into the following types:

1. Hazardous waste, which meets the explicit criteria for a hazardous waste as defined by federal and state regulations
2. Nonhazardous solid waste

Nonhazardous waste is further categorized as industrial or "special" waste or solid waste. Industrial or "special" wastes do not meet the definition of hazardous waste, but they are excluded from most municipal landfills because of physical or chemical characteristics. Examples include ash and some tannery sludge. Solid waste is general trash and refuse. This material goes into the dumpster and can be disposed of at a municipal landfill or incinerator.

Industrial residues and process wastes can fall into any of these categories, and identifying which category the waste belongs in is solely the responsibility of the facility producing the waste. Depending on the advice of a waste contractor is risky, since the generator of a waste is always liable for the consequences of its disposal regardless of the role of a waste contractor. Criminal charges can be levied for the improper disposal of hazardous waste, yet significant unnecessary costs will be incurred if a nonhazardous waste is mistakenly disposed of as hazardous. Thus, a working knowledge of these waste types is essential for the environmental manager, and the services of an environmental attorney or consultant may be necessary in tricky cases.

General information about each waste type is provided below; however, managers should consult their own state regulations for specific and current requirements for their waste type.

Hazardous Wastes

Hazardous waste is typically the most toxic, expensive, and regulated type of industrial waste. Hazardous wastes are governed at the federal level, primarily by the Resource Conservation and Recovery Act (RCRA), which both defines which wastes are hazardous and includes extensive requirements for their management and disposal. Most states are authorized to implement RCRA on behalf of the EPA and so will have their own requirements, which may be more stringent than the federal ones. Determining whether a waste is hazardous is the responsibility of the waste generator and can be based on a review of the regulations, actual waste testing, or simple generator knowledge. Consulting these regulations to determine whether a waste is hazardous and thus subject to RCRA regulation is the first step in the waste disposal process. Regulations for the identification of hazardous waste are included in the federal regulations at 40 C.F.R., Part 261, or at the appropriate corresponding section in state regulations.

Under RCRA, solid wastes (which can include liquids) are categorized as hazardous if they meet one of two conditions:

1. The waste is included on one of three "lists" of hazardous wastes included in the regulations.

2. The waste has one or more specified "characteristics" of a hazardous waste.

"Listed hazardous wastes" include waste commercial products, wastes from specific industrial processes, and wastes (e.g., spent solvents) from nonspecific sources. For the generator, the first step is thus to review these regulations to see if the waste, or the process producing the waste, is listed in the regulations. If a waste is listed on one of these tables, it must be managed as a hazardous waste. Also, if a nonhazardous waste is mixed with a listed hazardous waste, the whole quantity must be regarded as hazardous.

The three types of listed hazardous wastes are as follows:

1. Hazardous wastes from specific sources: Wastes produced by specific, listed industrial sources are automatically regulated as hazardous. They are designated as K wastes (40 C.F.R. 261.32) and are highly industry-specific. Examples are as follows:

   - K004: wastewater treatment sludge from the production of zinc yellow pigments

   - K083: distillation bottoms from aniline production

2. Hazardous wastes from nonspecific sources (40 C.F.R. 261.31): These wastes are produced by many industries and are associated with common industrial processes. They are designated as F wastes. Examples are as follows:

   - F001: the following spent halogenated solvents used in degreasing: tetrachloroethylene, trichloroethylene, methylene chloride, 1,1,1-trichloroethane, carbon tetrachloride, and chlorinated fluorocarbons; all spent solvent mixtures/blends used in degreasing that contain, before use, a total of 10% or more (by volume) of one or more of the above halogenated solvents or those solvents listed in F002, F004,

and F005; and still bottoms from the recovery of these spent solvents and spent solvent mixtures.

   - F008: Plating bath residues from the bottom of plating baths from electroplating operations where cyanides are used in the process.

3. Discarded chemical commercial products, off-specification species, container residues, and spill residues thereof. These are specific chemicals that are hazardous when disposed of, regardless of the industrial source or activity. They are considered acute wastes (designated as P wastes) or toxic wastes (designated as U wastes). Many common feedstock and product chemicals are included on these lists, as follows:

   - P051: Endrin

   - P076: Nitric oxide

   - U002: Acetone

   - U019: Benzene

   - U051: Creosote

If a waste is not included on one of these lists, it may still be regulated as hazardous because it has a hazardous waste characteristic. "Characteristic" hazardous wastes are those that meet specific criteria for ignitability (low flashpoint), reactivity (as determined by a cyanide reaction), corrosivity (very acid or alkaline), or toxicity. The toxicity characteristic is a measure of how much specific contaminants leach from the waste. For this test, the waste (a solid) is subject to a Toxicity Characteristic Leaching Procedure (TCLP) analysis, and the concentration of specified contaminants present in the leachate (in mg/L) is compared with RCRA standards (40 C.F.R. 261.24). Both organic and metals are included as compounds that, when leached in sufficient quantities, may cause a waste to be designated as a characteristic hazardous wastes.

Many routine, and especially nonroutine, industrial wastes fall into this category, and must be tested to determine whether they meet hazardous waste characteristics. Most commercial laboratories offer an analytical package that includes all these "hazardous waste characteristics" parameters for a single price. Samples must be representative of the waste, and a generator may use his or her own knowledge to specify test analytes, based on knowledge of the waste. For instance, a waste need not be tested for TCLP organics if the generator knows that no organics are in the waste.

If a waste is determined to be hazardous, it must be labeled, stored, inspected, shipped, and disposed of in accordance with strict RCRA regulations. Only licensed hazardous waste transporters may be used to transport hazardous wastes, which must only be disposed of at a licensed treatment, storage, and disposal facility (TSDF). An appropriate waste code (as listed in the RCRA regulations) is applied, and the waste is shipped under a special hazardous waste manifest. If the waste stream will remain consistent and the initial sampling was representative, it need not be sampled again for every shipment; the same waste code applies.

RCRA regulations are among the most complex of all environmental regulations, and a thorough understanding of the many requirements is essential for those involved with managing these wastes. While this section presents a brief overview of the process for identifying hazardous wastes, it is no substitute for the regulations themselves, which provide significant additional detail, qualifications, and exemptions.

Nonhazardous Industrial or Special Wastes

In many cases, an industrial waste may not be hazardous but may still be barred, at least in significant quantities, from municipal landfills and incinerators because of the composition of the waste. Examples of these "special" wastes vary by state and facility but may include tannery leather scraps, feathers and other wastes from poultry processing,

ash, nonhazardous sludge, and some construction debris. These materials are normally disposed of in an industrial landfill, which is generally more strictly regulated and managed than municipal landfills. As with hazardous waste, prior waste testing and approval are necessary before an industry can ship waste to the site. The disposal facility will provide a list of required testing parameters.

Industrial wastes are normally regulated on the state and local levels, and most facilities are licensed to accept only certain kinds of waste. Special state approval is often necessary for unusual waste streams. While a reputable waste disposal contractor usually knows the requirements and limitations of a number of special waste facilities, direct communication with the facility regarding the acceptability of your waste is prudent to ensure that the waste will indeed be managed appropriately.

While nonhazardous industrial wastes are less tightly regulated than hazardous waste, they can still be the source of significant disposal expense. Recycling, waste exchange with other industries, and waste minimization are useful and wise approaches to reducing these expenses, especially for wastes that have been produced and disposed of in the same manner for a long time: new technologies and recycling opportunities may now be available for these materials. These and other options for disposing of industrial wastes are described in the sections that follow.

Nonhazardous Solid Waste

Solid waste (i.e., trash) includes such routine wastes as office trash, unreusable packaging, lunchroom wastes, and manufacturing or processing wastes that are not otherwise classified as "hazardous" under RCRA. These wastes are normally deposited in trashcans and dumpsters and collected by a local trash hauler for disposal in a municipal landfill or treatment at a municipal incinerator. For the environmental manager or design engineer, this is the ideal category for a final treatment residue to fall into, since these wastes are typically of relatively low toxicity and thus their disposal is relatively inexpensive. Although RCRA contains design and other standards for municipal waste management facilities, these facilities are normally governed primarily by state and local regulation.

For the waste generator, the challenge is usually keeping industrial or hazardous waste out of the solid waste dumpster. When establishing a contract with a waste hauler, be sure to get a clear understanding (preferably a written list) of what is and is not allowed in the trash receptacles. Many RCRA violations arise from improper disposal in the dumpster of common shop items, such as solvent-soaked rags or absorbent material. Clear guidelines for workers, ready access to proper disposal areas, and constant vigilance of workers and subcontractors are some of the tools that work.

The major methods of solid waste disposal are described in further detail in the following sections. For some larger facilities, on-site disposal of waste at on-site landfills or incinerators is a viable option. As with industrial waste, however, a focus on recycling and reuse can help reduce the costs of solid waste disposal and should be an ongoing component of a facility's waste management practice.

The distinctions between each of the waste categories (hazardous, industrial, and solid waste) described above are not always clear, and the onus is on the individual industry, or appropriate facility, to make the correct determination. Some states, for instance, consider waste oils and PCBs to be hazardous waste, even though federal law does not. Cans of dried paint are generally regarded as a normal solid waste that can go in a dumpster; cans of wet paint, especially those that contain lead or chromate, are usually designated hazardous. Tannery wastes with trivalent chrome usually can go to an industrial landfill, but some states consider these materials hazardous.

Waste management and disposal often represent significant and constantly increasing costs for industry. To minimize these costs and reduce the likelihood of enforcement actions by regulators, environmental managers must ensure that a sound program is in place and that all personnel, from laborers to top managers, are vigilant in carrying it out. The following guidelines are often helpful:

- *Know the facility waste streams.* These are seldom the same for different plants. As a first step, facilities must know how much of each type of solid waste they are producing.
- *Keep wastes segregated.* Heavy fines, as well as criminal sentences, are the penalties for improper waste disposal. Facilities must ensure that hazardous wastes are not put in the trash dumpster, that listed hazardous wastes are not mixed with other nonhazardous materials, and generally that wastes are handled as they are supposed to be.
- *Choose waste disposal firms carefully.* Since facilities can be held responsible for cleanup costs of the waste facilities they use, waste transporters and facilities should be chosen carefully. Check with your state's environmental enforcement division to uncover any violations or tour and interview the facility or transporter directly.
- *Institute a pollution prevention program that includes a vigorous wastes minimization effort.* Although, as described in Chapter 1, this is one of the first steps in a waste treatment system design, implementing a waste minimization study as a stand-alone task is a common and effective means of reducing disposal costs. Reducing the quantity or toxicity of materials used in production reduces both disposal and health and safety–related liabilities.
- *Keep areas clean.* Frequent spills or routine drippage not only present safety hazards but also increase the amount of facility decontamination necessary at closure.

- *Keep good records.* Industry-wide, a great deal of money is wasted on testing and disposing of unknown materials or in investigating areas with insufficient historical data. Good recordkeeping is essential to keep both current and future waste management costs to a minimum.

Characterization of Solid Wastes

Each significant solid waste stream should undergo characterization in order to determine the following:

- Opportunities for waste reduction
- Volume rate of waste generation
- Whether or not the waste is hazardous
- Suitability of the waste for landfilling
- Physical properties as they relate to suitability for landfilling
- Chemical properties as they relate to suitability for landfilling
- Estimation of leachate characteristics
- Suitability of the waste for incineration
- Estimated requirement for auxiliary fuel
- Estimated characteristics of stack emissions
- Estimated characteristics of ash
- Suitability of the waste for composting

Opportunities for Waste Reduction

The first and most important order of business in a solid waste management program is to identify any and all opportunities for reducing the volume, strength, and hazardous nature of wastes, always with the goal of elimination. For instance, the solid wastes manager should constantly be seeking opportunities for given wastes to be used as raw material in another manufacturing process, even if there is a cost. The cost should be compared with the cost of processing and disposing of the waste, and an appropriate amount should be added as insurance against having to deal with the disposed waste in the future for one reason or another. If the waste can be safely incorporated into

another product, a permanent solution will have been implemented.

As a very simple example, a certain cardboard manufacturing plant had a daily production capacity of 1,000 tons per day of heavy brown paper. The plant had a wastewater treatment plant that produced about three tons (dry basis) of waste biosolids per day. After years of landfilling this sludge, the suggestion was made to simply incorporate the waste biosolids into the brown paper. The relatively tiny amount of biosolids, which had been a sizable solid waste disposal problem, "got lost" in the 1,000 tons of brown paper, forever solving a previously expensive problem.

As another example, during the 1970s, the number of poultry farms and poultry processing facilities increased dramatically in response to a growing desire on the part of the American public to reduce intake of cholesterol. Many of the poultry processing plants made use of dissolved air flotation (DAF) for wastewater treatment. The residual from the DAF process is a sludge that is very difficult and expensive to dispose of. A solution that was developed and used by some of the processing plants was to incorporate the DAF residual into the raw materials that were being used to produce feed for the poultry. Since the DAF residual was essentially all poultry parts and pieces, it was perfect as a feed additive, to be fed right back to the chickens. Again, an expensive and problematic solid waste disposal problem was resolved. In this case, the result was a significant savings of money for both waste disposal and feed production.

Volume Rate of Waste Generation

Often rigorous attention paid to managing production processes so as to "do it right the first time," and having little or no off-spec product to dispose of, can significantly reduce the volume of solid wastes. Also, improved preventive maintenance and improved operation and maintenance (O&M) practices, so as to reduce leaks, spills, and accidents, can have the effect of significantly reducing the volumes of solid wastes, as well as wastewater and air pollutants.

Whether or not the waste is hazardous has a major effect on both the cost of solid waste handling and disposal and the risk of future liability. It is always a good idea to substitute nonhazardous materials for hazardous materials used in production, if the resulting wastes become nonhazardous. An example is to use hot water and detergent or a hot caustic solution as a degreaser fluid, rather than one of the chlorinated organic solvents. In many cases, the hot detergent or caustic solution outperforms the chlorinated solvent. The principal value of this substitution, however, is that the sludge from the hot detergent or caustic process will likely not require disposal as a hazardous substance.

As another example, certain sludges from industrial processes or industrial waste treatment facilities are required to be handled and disposed of as hazardous wastes because, on being subjected to the Toxic Characteristic Leaching Procedure (TCLP) test, the quantity of metals leached is more than is allowed. It has been possible, in some instances, to mix "superphosphate" fertilizer into the sludge, with the result that the sludge is then able to "pass" the TCLP test. The treatment mechanism at work here is that the metal ions become tied up as extremely insoluble metal phosphates. The acid solution used for the TCLP test is unable to leach the metal ions from the sludge, and rainwater percolating through the landfilled sludge will certainly not be able to leach the metals and become a groundwater pollution threat.

Suitability of the Waste for Landfilling

Whether or not a particular solid waste is suitable for landfilling depends upon its physical characteristics, chemical characteristics, and its probable leachate characteristics, i.e., whether or not there will be hazardous substances in the leachate.

Physical Properties

Physical properties that influence suitability for landfilling include those that influence structural stability. Often, water content is used to determine suitability for landfilling in the case of waste sludges. Another property is physical size. Still another relates to the ability of landfill machinery to handle the waste at the landfill site.

Chemical Properties

Chemical properties that influence suitability for landfilling, in addition to those properties that determine whether or not a waste is hazardous (corrosivity, toxicity, ignitability, and reactivity) include foaming agents (methylene blue active substances), iron and manganese, odor, and odor generation potential, for instance, sulfate.

Formal testing procedures for determining whether or not a waste is hazardous are presented in Chapter 5.

Suitability of the Waste for Incineration

Whether or not a waste can be destroyed by incineration (often referred to as "combustion"), and the quantity of auxiliary fuel that would be required to maintain a sufficiently high temperature in the combustion chamber to accomplish complete combustion, depends on several characteristics of the waste. Water content and chemical composition are major factors. Water content is important because of the demand that water places on auxiliary fuel usage and thus the cost of incineration. Chemical composition influences the following:

- The fraction of the total waste stream that can be converted to carbon dioxide and water
- The characteristics of the stack emissions
- The characteristics of the ash
- The quantity of auxiliary fuel required

Within a total waste stream, the only elements that will be converted to carbon dioxide and water are carbon and hydrogen. All other elements will be converted to other substances or will remain unchanged. Some elements will be converted to anions of high oxidation state; for instance, sulfur will be converted to sulfur dioxide or sulfur trioxide. Some elements will be liberated as the free cation; for instance, mercury that has been incorporated into organic material (methyl mercury) will most likely exit the stack as free Hg+. Chunks of ferrous metals will tend to remain unchanged. Sand and other inerts will end up in the ash. Some heavy metals will end up in the ash, while a portion will exit the stack.

It is critically important, then, to be aware of the chemical composition of any waste stream for which treatment by incineration is being contemplated. The word "treatment" is used here rather than "disposal" because incineration yields residuals from stack emission treatment (for instance, electrostatic precipitation) as well as ash. These residuals then become the subjects of evaluation of alternatives for disposal method.

Suitability of the Waste for Composting

Certain industrial solid wastes—vegetable processing, for instance—may be excellent candidates for composting. Composting is a method of converting waste material to usable material and is thus highly desirable. A very large composting facility has been operated for many years in the U.S. midwest for the purpose of converting meat processing wastes to a soil enhancement product.

Solidification and Stabilization of Industrial Solid Wastes

Solid wastes from industries, including both fly ash and bottom ash from combustion processes, can sometimes be conveniently, safely, and cost-effectively stabilized by one or more of the processes that are characterized as "solidification and stabilization" (S/S). In this context, "safely" refers to compliance with all laws and regulations, as well as with respect to potential harm to people or the environ-

ment. "Stabilized" refers to transformation of a substance from a form in which it is leachable, hazardous, or otherwise objectionable to a form in which it cannot be leached, vaporized, or enter into a reaction with another substance.

Solidification and stabilization technologies are used with the objective of converting the solid waste, with or without any free-flowing liquid that may be associated with it, to a state such that it can be landfilled or otherwise applied to the land without danger of forming unacceptable leachate or gas. There are a number of technologies that can be used, depending on the characteristics of the waste material.

Laws and regulations regarding S/S methods of solid waste treatment before disposal are ever changing, and this poses a problem. It can be said that if the S/S product can "pass" all applicable tests, such as the Paint Filter Liquids Test (PFLT) and the Toxic Characteristic Leaching Procedure (TCLP), it is "safe" to be disposed of in an approved landfill. However, the criteria for passing those tests, as well as the procedures of the tests themselves, have been changing over time—thus, the moving target problem. Nevertheless, there are certain industrial solid wastes, with certain hazardous or otherwise objectionable characteristics, that can be treated using certain S/S techniques that would seem essentially certain to comply with any present or future acceptance testing procedure. For instance, the objectionable characteristic of certain industrial solid wastes that contain formaldehyde is that the formaldehyde, which is toxic, can be leached out by water and become a groundwater pollution threat. The solid waste mass may be treated by mixing it with phenol (another toxic substance) to produce (stoichiometrically) a "phenol-formaldehyde resin," which is extremely insoluble—resin that does dissolve is relatively inert and nontoxic. Whether or not such a treatment process can be reliably and cost effectively (in the context of alternatives) carried out is largely dependent on the characteristics of the solid waste.

A second problem regarding S/S methods for treatment of solid wastes prior to landfilling is that additional volume is added. As "air space" at approved landfills increases in its already high value, this consideration will increase in importance.

Notwithstanding the moving target problem, a number of S/S technologies exist that warrant serious consideration, which means a very close examination of the characteristics of the solid waste and the characteristics of the S/S product. It is to be emphasized that the feasibility of using any given S/S technology for treatment prior to disposal of a given solid waste stream must be closely evaluated using a four-step process:

- Evaluate the likelihood of successful application of each alternative S/S technology on a theoretical basis (basic chemistry and physics). This evaluation should include a thorough search to determine whether or not any of the available alternative S/S technologies have been used successfully on a similar waste stream at another location.
- Perform bench-scale tests using those technologies that appear promising. Testing must include all product evaluation tests such as TCLP and PFLT procedures, where applicable.
- Perform pilot plant evaluations using S/S technologies with the most favorable bench-scale tests. Include product evaluation tests.
- Perform a detailed preliminary design of the prototype treatment system(s). A thorough and detailed cost opinion analysis must be included.

There are at least three systems by which S/S technologies are categorized:

- By process system
- By binder used
- By S/S mechanism

S/S Process Systems

There are at least four systems in use to accomplish S/S treatment of industrial waste streams: in the 55-gallon drum in which the waste materials to be treated have been collected; in a specially designed and constructed facility to carry out the S/S process on site; in a mobile facility; or in a sludge pit, lagoon, or other collection facility of the solid waste stream.

In the case of the "in-drum" system, treatment chemicals or binders are added directly into the 55-gallon steel drums used to collect the solid wastes. Stirring and reaction are carried out, and the stabilized product is often placed in a landfill or other facility still in the drum.

Specially built S/S plants, similar to incinerators, are operated either on site or at central locations. The process can be batch-type or continuous. These facilities are appropriate for large quantities of solid wastes.

In some cases, the reaction chemicals or binders are added directly into a collection tank, storage tank, or lagoon in which the target solid wastes have collected. In the case of lagoons, it might be feasible, safe, and cost effective to simply close the lagoon, under RCRA or other approved procedures, with the stabilized solid wastes remaining. This type of procedure is referred to as "*in situ* treatment and disposal."

Binders Used for S/S Technologies

Solidification agents that have been successfully used to stabilize industrial solid wastes include both organic and inorganic substances, as listed below:

Organic Binding Agents

- Epoxies
- Asphalt
- Urea-formaldehyde
- Polyesters
- Polyolefins

In all of these technologies, appropriate pretreatment of the solid wastes is very important.

Epoxies

Several different epoxy mixes have been used as binders for the S/S treatment of industrial wastes, including proprietary substances and procedures. Epoxies have been successfully used for S/S treatment in 55-gallon drums, as discussed above. Epoxies make use of at least two reactants; therefore, mixing must be rapid and thorough. If executed properly, the hardened epoxy will incorporate the solid waste substances within its polymerized matrix, effectively reducing, by orders of magnitude, the surface area of the target material that can be contacted by water or other potential leaching liquid. Epoxies are characteristically nonreactive and are expected to prevent leaching or other reaction of the target substances for extremely long periods of time.

There are different types of epoxy, using different reactants and producing polymerized matrices having different chemical properties. Some are more resistant to certain organic solvents or strong acids or caustic substances. For these reasons, the most appropriate epoxy for a given application should be the subject of considerable research before a given S/S technology is selected or rejected. As well, the results of bench-scale trials should be evaluated in light of these characteristics.

Asphalt

Many industrial solid wastes have been successfully solidified and stabilized by simply incorporating them into an asphalt mix at a conventional asphalt batch plant. In other cases, special asphalt mixes using specially designed and constructed mixing equipment have been used. Factors that must be considered include reactivity potential between the asphalt binder and any of the components of the solid waste stream; the future effects of heat on the physical integrity of the asphalt; the potential for substances that are solvents for the asphalt coming in contact with the disposed products; and the potential for wearing or abrading of the asphalt product.

Urea-formaldehyde

An epoxy-like matrix can be formed by the polymerization that occurs when urea and formaldehyde are mixed in proper proportions. As with the epoxy S/S treatments discussed above, target substances in solid wastes can be incorporated into this matrix, effectively solidifying and thus stabilizing the target substances. The choice between urea-formaldehyde and epoxy as the S/S treatment depends largely on potential reactions between substances in the solid waste stream and the urea-formaldehyde mixture.

Polyesters

Several polyesters can be used for S/S treatment of certain solid waste target substances. As with epoxy and urea-formaldehyde technologies, the binding effect that polyesters can have on solid waste target substances comes about as a result of incorporating the target substances (and other substances present with the target substances) in the inner mass of the matrix that forms when the polymerization process takes place, in which small organic ester molecules combine to produce a relatively huge monolithic mass. It is not (necessarily) that the target substances become chemically incorporated into the matrix but rather that the target substances become trapped within the matrix. Because the target substances are trapped, it is impossible for water or other dissolving or reacting substance to come in contact with them. Thus, the goal of the S/S technology is accomplished.

Polyolefins

Polyethylene and polyethylene-butadiene are two polyolefins that can be used in the same way as is described for polyesters, above, to solidify and thus stabilize certain solid waste streams. The choice between polyesters and polyolefins is, again, based on the chemical properties of the solid waste substances and the chemical and physical properties and characteristics of the environment of the disposal site. The mechanisms of stabilization are essentially the same. For example, if a solid waste is first, conditioned by evaporation to dryness and grinding to a fine granular consistency, it can be mixed together with appropriate proportions of ethylene and butadiene, plus a catalyst. In the presence of the catalyst, ethylene and butadiene polymerize to form a monolithic mass. As the polymerization process takes place, the granules of solid waste material become entrapped within the polymeric mass.

Inorganic Binding Agents
- Portland cement
- Pozzolan substances
- Lime
- Gypsum
- Silicates

Portland Cement

One of the first S/S techniques to be used was to simply mix conditioned solid wastes with Portland cement, sand, and water and then allow the mixture to harden into concrete. In many cases, fly ash has been mixed in. The fly ash contains silica and thus enters into a pozzolan reaction that adds to the structure of the concrete mass. Also, the fly ash may have been a disposal problem in itself. In this case, including fly ash in the mixture with cement and another waste material has solved two hazardous waste disposal problems concurrently.

Solidification with Portland cement achieves stabilization of solid wastes by preventing water from any source(e.g., percolating rainwater or groundwater) from being able to leach substances from the solid waste, because the water can no longer come into contact with the waste material. In this case, the waste material has been trapped within a monolithic block of relatively high strength. Even if cracks eventually develop in the concrete mass, the surface area of the waste material that can be contacted by water will be orders of magnitude less than before the solidification process.

There are at least two stabilization mechanisms involved when Portland cement is used as the solidification agent:

1. Entrapment within a monolithic mass
2. Incorporation into the chemical structure of the monolithic substance

Any water that is included in the solid waste stream will be taken into the structure of the concrete as the curing process takes place.

Use of Portland cement as the solidification agent or "binder" is relatively safe, but relatively expensive. The decision to use Portland cement rather than a less expensive solution is usually made to reduce perceived future risk to a minimum.

Pozzolan Substances

Fly ash, from the burning of fossil fuels and other substances, contains fine grains of noncrystalline silica. When fly ash is mixed with lime and water, the calcium in the lime reacts with the silica from the fly ash to produce a low-strength cement-like solid substance. This reaction is referred to as a "pozzolan" reaction. Conditioned solid wastes can be mixed into the lime–fly ash–water mixture, as described for the Portland cement S/S technology above. The result is much the same, in that the target substances in the solid waste stream have become immobilized and are protected form the leaching process. The reactants are less expensive, and one of them, the fly ash, is a solid waste in its own right. This technology, then, accomplishes treatment for safe disposal of two, or at least portions of two, solid waste streams.

As was the case with Portland cement as the binder, there are at least two stabilization mechanisms involved when the pozzolan-type solution is used:

1. Entrapment within a monolithic mass
2. Incorporation into the chemical structure of the monolithic substance

Any water that is included in the solid waste stream will be taken into the structure of the monolithic mass as it forms via the "curing process."

Lime

Some solid waste streams, having certain characteristics of chemical composition and moisture content, can simply be mixed with lime and allowed to dry. On drying, a low structural strength solid mass with low solubility in water is formed. Many metal ion species that were present in the waste stream will have been precipitated as the highly insoluble hydroxide, except for those metal ion species, such as lead and zinc, that have the lowest solubilities in the lower pH range. pH, of course, has no meaning in a solid substance; however, a metal that is precipitated as the hydroxide and is then placed in a dry environment will not dissolve into water that eventually comes into contact with it to a degree higher than its theoretical solubility limit.

If there are significant quantities of lead and/or zinc, or other metal whose hydroxide is more soluble than desired in the solid waste stream (along with those metals that do precipitate best at high pH levels), there are at least two alternatives. One is to add a source of water-soluble sulfide ions to the lime (e.g., sodium sulfide); the second is to add a source of soluble phosphate ions (e.g., triple-super-phosphate fertilizer). The sulfide or phosphate ions will precipitate those metal ions that remain in water solution after the pH has been raised to the 12 to 13 range by the solubilized lime.

The security of a lime-treated solid waste mass can be enhanced significantly by encapsulating the final product in an impermeable membrane, either by spraying it onto the surfaces and allowing it to cure or by wrapping the final product in an already formed membrane.

Gypsum

Gypsum is another relatively inexpensive substance, like lime, that will form a solid mass on drying. Unlike lime, however, gypsum is not an alternative for treating wastes in which metal ions are among the target substances. Whereas the anion associated with lime is hydroxide, the anion associated with gypsum is sulfate. Metal sulfates are soluble in water. Gypsum is therefore an alternative substance for treating certain solid waste streams where incorporation into a low structural strength solid mass is the principal objective. Importantly, gypsum itself (calcium sulfate) is very sparingly soluble in water; therefore, the substances that were components of the original solid waste stream and are incorporated into the dried monolithic mass product are "sheltered" from contact with water. If disposed of properly, by placing in a suitable burial facility, the problem of future leachate will have been solved.

Silicates

Silicates represent another category of substances that can be mixed as a liquid with properly conditioned solid wastes and then, on drying, form a monolithic matrix that is:

- Insoluble in water and water repellent
- Relatively inert regarding reaction with most chemical substances
- Stable over time
- Capable of incorporating the target substances into its monolithic mass

An advantage of the silicates compared with pozzolans, lime, and gypsum is that the silicates form a gel that is pliable and, therefore, resistant to cracking if physically deformed. The basic objective of treatment is the same as for pozzolans and gypsum; that is, to make it extremely difficult, if not impossible, for water to produce leachate from the treated solid waste in the future. Treatment using silicates is different, however, from treatment with lime or Portland cement, in that metal ions will not be precipitated as a water insoluble substance by using silicates. For this reason, silicate technology would likely not be a good choice for treatment prior to disposal of a solid waste stream in which metal ions are target compounds.

Mechanisms Involved in S/S Treatment Procedures

Some of the mechanisms by which individual S/S technologies accomplish stabilization of target substances in solid waste streams have been discussed in the preceding paragraphs. A more thorough discussion of these and other mechanisms is presented below.

To restate the overall objective of S/S treatment: it is to produce a product that can be reliably and safely disposed of, usually in an appropriate landfill, without danger of generating an objectionable leachate in the future. The mechanisms employed by the individual solidification technologies discussed above, are as follows:

- Incorporation into a monolithic solid mass
- Precipitation of target substances to form nonleaching products
- Adsorption of target substances to form nonleaching products
- Absorption of target substances to form nonleaching products
- Adsorption of water from the solid waste mass
- Absorption of water from the solid waste mass
- Encapsulation

Incorporation into a Monolithic Solid Mass

In order for leachate to form, water must come into contact with a soluble material, dissolve an amount of that material, and then continue to flow (under the influence of gravity) away from the solid waste material. One effective way to prevent leachate formation, then, is to make it impossible for water from outside a mass of solid waste material

to contact target substances. One way this objective can be accomplished is to incorporate the target substances within a monolithic mass of an insoluble solid. Any water approaching this solid mass will be forced to flow around it and thus will never achieve contact with the target substances.

The mechanism is simple, but the objective is not. The most difficult problem has to do with all of the counterobjective activities that take place over a very long period of time. Destruction of the structural integrity of the monolithic solid mass, either slowly over time, or as a result of physical damage, must be prevented. The long-term effects of "natural" phenomena, such as acid rain, must be carefully considered and designed into the overall technological solution.

There are basically two ways by which target substances can be incorporated into a formed monolithic mass: (1) by chemical reaction and actual incorporation into the lattice structure of the monolithic solid as it is forming, for instance, by polymerization, and (2) by entrapment within the physical structure of the monolithic solid as it is being formed.

Precipitation of Target Substances to Form Nonleaching Products

There are several alternative technologies that make use of precipitation of target substances. As discussed above, using lime as a binder is one of them. Other alternatives include mixing with soluble carbonate, sulfide, or phosphate compounds, usually combined with a substance to form a monolithic solid mass.

Adsorption of Target Substances to Form Nonleaching Products

Certain adsorbents, such as activated carbon, can be mixed into a slurry of solid waste materials with the result that water is not able to desorb the target substances over time. Before this technology can be deemed appropriate for a given waste stream, however, extensive testing must be carried out to determine the edegree of adsorption, as well as resistance to desorption. Moreover, it is seldom that the optimum adsorbent is selected early in the evaluation procedure. There are many different activated carbons as well as alternatives to activated carbon. Also, there are many different chemical characteristics regarding the potential leaching solution that must be evaluated—for instance, different values of pH, acidity, alkalinity, ORP, and TDS content, to name a few.

Absorption of Target Substances to Form Nonleaching Products

Regarding S/S technologies, there are certain materials that can absorb and hold target substances in a solid waste stream in the manner that a sponge absorbs and holds water. This technology type has potential application when the target substances within a solid waste stream are part of the liquid portion of the stream.

Adsorption of Water from the Solid Waste Mass

In some cases, activated carbon, or another adsorbent, has been used to adsorb water from a solid waste mass to change its physical characteristics to a more manageable state. Substances that are dissolved in the water are often incorporated into the bulk activated carbon (or other adsorbent) product and must be dealt with appropriately. One alternative for this requirement is incineration to destroy the adsorbed target substances and to reactivate and recover the activated carbon for extended use.

Absorption of Water from the Solid Waste Mass

Materials having the capability of absorbing water from a solid waste stream have been used for two purposes: (1) to prepare the solid waste for further treatment in a less wet state, and (2) to isolate target substances that are dissolved in the water phase of the solid waste stream. Once absorbed into the absorbing material, that material can be fur-

ther processed by treating the material, the water, and the dissolved target substances as a whole, or by physically removing the absorbed water phase and then treating that liquid for recovery (for reuse), fixation, or destruction of the target pollutants.

Encapsulation

Encapsulation has the objective of preventing contact between water and target substances by forming a physical, impermeable barrier around a volume of solid waste, or a volume of solid waste that has been treated using one of the technologies discussed above. Alternative methods for encapsulation include wrapping a volume of the treated or "raw" solid waste in a membrane of high-density polyethylene (HDPE) or other material; spraying it with one of the commercially available products that then polymerizes to form a seamless, impermeable coat; or dipping a volume of treated or "raw" solid waste in a solution that will then polymerize to form a seamless, impermeable coating.

The Solid Waste Landfill

The Conventional Landfill

Modern landfills are highly secure facilities with sophisticated features to prevent pollution of groundwater or soil. Modern landfills also have safeguards against blowing trash problems, odor problems, rodent and fly problems, fires, and contamination of surface waters.

There are alternatives, but most modern landfills are constructed as described below.

First, the site is cleared and grubbed. The topsoil is removed and stored. Access roads are built, and physical facilities such as an office building, maintenance and storage building, scales, hot load pit with firefighting equipment, and laboratory facility are constructed and/or installed.

After excavation to provide a basin in which the solid wastes will be placed has been completed, the liner system is constructed.

Landfill Liner System

Figure 9-1 presents a cross-section of a landfill liner system constructed in 1999 and is herewith discussed as a typical modern landfill liner system. The landfill liner system shown in Figure 9-1 can be described as a "double composite with primary and secondary leachate collection." This particular design can also be described as having leak detection, since any leachate collected by the secondary leachate collection is direct and positive evidence of a leak in the primary liner.

It is convenient to describe the landfill liner system illustrated in Figure 9-1 by describing how it is constructed. After excavating the landfill "basin" to the subgrade, (shown in Figure 9-1), an 18-inch thick layer of "secondary barrier soil" is placed. This material constitutes a liner in its own right, since the hydraulic conductivity (the rate at which water can move through the soil under the influence of gravity) can be no greater than 10^{-7} centimeters per second, equivalent to about 1 foot in 30 years. Federal regulations (Subtitle D) require that a soil liner of not less than two feet of soil having a coefficient of permeability ("k") equal to or less than 1×10^{-7} be an integral component of the liner system for all landfills. In the case of the landfill used for this example, the state regulatory agency as well as the EPA agreed that 18 inches of soil with a "k" value of 1×10^{-7} plus 12 inches of soil with a "k" value of 1×10^{-5}, plus the geosynthetic clay liner as discussed below, more than satisfied this requirement.

Some states will not accept a substitute for a soil liner at least two feet thick with a "k" value of 1×10^{-7} or less. One solution for this situation, if it is not possible to obtain enough soil of that low hydraulic conductivity within a reasonable distance of the landfill site, is to augment native soil (of higher hydraulic conductivity [larger value of "k"]) with powdered sodium bentonite. Bentonite is a natural clay that exhibits a high degree of swelling on contact with water. In fact, it will swell to 25 times its volume when dry. When relatively small amounts of this material are mixed into soil of higher hydraulic conductivity, the particles of bentonite fit into the void spaces. Then, when a water solution (for

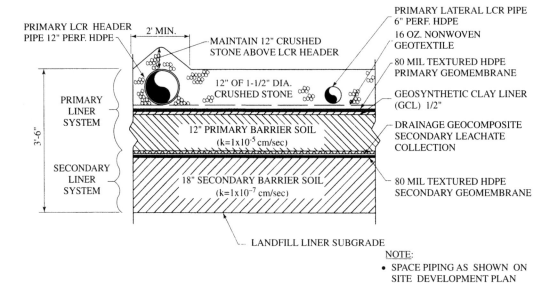

Figure 9-1   Double composite landfill liner system.

instance, leachate) comes into contact with the mixture, the particles of bentonite swell. This swelling action closes off the void spaces in the more highly permeable soil, with the result that the mixture as a whole has a "k" value that satisfies the Subtitle D mandate of 1×10^{-7} or less.

Next, an impermeable membrane is placed over the secondary soil barrier. In the system illustrated in Figure 9-1, that membrane consisted of 80-mil (thick) textured high-density polyethylene (HDPE). This barrier is referred to as a "synthetic membrane" and constitutes the secondary liner for the landfill.

Over the top of the synthetic membrane secondary liner is placed a fabric that is referred to as a "geocomposite." This fabric, which is loosely woven using threads of synthetic material, performs two functions. One is to protect the HDPE membrane from being punctured by anything in the layer of soil above it. The other is to act as a leachate collection and conduit device. Its openly woven texture allows it to perform this function.

Over the top of the geocomposite fabric is placed a one-foot layer of compacted soil referred to as the "primary soil barrier." This layer of compacted soil, like the secondary barrier soil, is strictly regulated as to characteristics of aggregate size and uniformity coefficient, and its hydraulic conductivity must be no higher than 10^{-5} centimeters per second.

Over the top surface of the primary barrier soil is placed a sheet of a manufactured material called a "geosynthetic clay liner" (GCL). This material, which is delivered to the site in large rolls and is about one-half inch thick, consists of a mat-like fabric that is filled with clay. The clay is used for its impermeability and resistance to chemical attack, which is a possibility in the case of HDPE, though a very remote one. The synthetic fabric has the function of providing the GCL as a whole with structural strength and the capability of constructing a layer of clay that is quite thin.

The next layer, placed over the top of the GCL, is the primary liner for the landfill, a sheet of 80-mil-thick textured HDPE. This membrane is constructed by rolling out very large rolls of the HDPE material as it is delivered to the site and welding sheets edge to edge (actually, the edges overlap) to form a contiguous membrane underlying the entire landfill. The same procedure is used to construct the secondary liner. If everything were as intended throughout time, there would be no passage of leachate through the primary membrane, and everything below it would be superfluous.

The final layer of fabric in the landfill liner system shown in Figure 9-1 consists of a "nonwoven geotextile" placed over the top of the primary synthetic liner. The primary function of this layer of synthetic fabric is to protect the primary liner from being punctured or otherwise breached by the crushed stone, or anything inadvertently put in the crushed stone, above it.

A bed of crushed stone consisting of aggregates of one to one-and-a-half inches in diameter is placed on top of the nonwoven geotextile, along with the primary leachate collection system. The crushed stone bed has the multiple functions of (1) holding the leachate collection and transport pipe system in place, (2) providing a high degree of hydraulic conductivity to enable leachate to flow to the collection and transport system, and (3) serving as a bed for the solid wastes placed in the landfill.

The primary leachate collection and transport system is a grid of plastic pipe (HDPE in the example illustrated in Figure 9-1), perforated at all locations within the landfill and solid outside the landfill. The layout of the pipe grid, the size and spacing of the laterals and headers, and the material of construction of the pipes themselves are subject to guidelines and regulations issued by each state or solid waste authority. The treatment and disposal of leachate, once collected, is addressed in Chapter 7.

There are landfill liner designs that have a layer of groundwater protection in addition

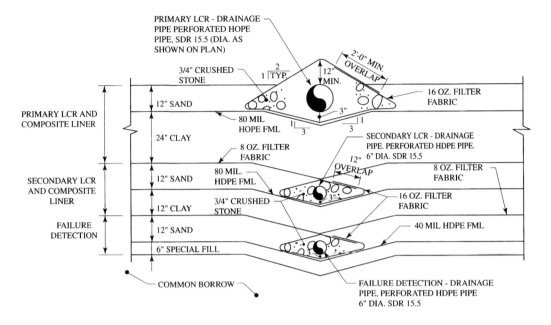

Figure 9-2   Landfill liner with three synthetic membranes and two leachate collection systems.

to the two that are included in the design illustrated in Figure 9-2. The landfill liner system shown in Figure 9-2, which was installed in 1999 in Augusta, Maine, has primary and secondary synthetic membrane liners with primary and secondary leachate collection, as shown in Figure 9-2, plus a third synthetic membrane liner below the primary and secondary systems. There is a system of liquid collection and transport piping just above the third liner, which is correctly referred to as "leak detection" since its function is to show evidence of leaks in the "conventional" landfill liner system above it.

Landfill Cover and Cap Systems

Daily and Intermediate Cover

An important feature of any landfill is a practice that began with the earliest landfills in the 1940s—that of daily, intermediate, and final cover. Daily cover consists of soil or other approved material that is placed over the deposited wastes as soon as is practical throughout, and without fail at the end of, each day of landfill operation. The purposes

of daily cover include control of flies and rodents, fire, odors, VOC emissions, and scavenging, as well as the prevention of items being blown out of place by wind. A six-inch depth of soil is mandated by the EPA (Title 40, Subpart C, 258.21) for this purpose. However, because of the high value placed on landfill "air space," the six inches taken up by each lift when soil is used for the daily cover has been looked upon as a candidate for replacement by much thinner alternative materials. Although each time a substitute material is used it must be approved on a case-by-case basis, a significant number of landfills have been successful in doing so. Alternative materials have included specially constructed membranes (referred to as "tarps"), sludge from primary clarifiers, proprietary fiber slurries, foam-in-place products, and spray-on asphalt membranes.

Intermediate cover is soil that was earlier removed from the landfill excavation and stockpiled. The same is true for daily cover. In some instances, another (approved) material has been used for daily cover. An example is sludge from primary clarifiers at paper mill wastewater treatment plants. The frequency of placement of intermediate cover depends on the characteristics of the material being landfilled but is typically required to be placed on any portion of the landfill if it is to be inactive for 60 or more days. The thickness of intermediate cover is typically required to be no less than 12 inches.

Final Cover and/or Cap

When a landfill, or a discrete segment of a landfill, has reached capacity, it must be "closed" in accordance with applicable regulations. Typically, a constructed "cap" is required, an example of which is illustrated in Figure 9-3. The primary purpose of the cap is to stop the formation of leachate by preventing rainwater or other precipitation from being able to percolate into the landfilled waste. Therefore, the design objective is to provide an impermeable barrier that is able to dissipate methane and other gases into the ambient air, thus preventing an explosion or fire hazard from developing.

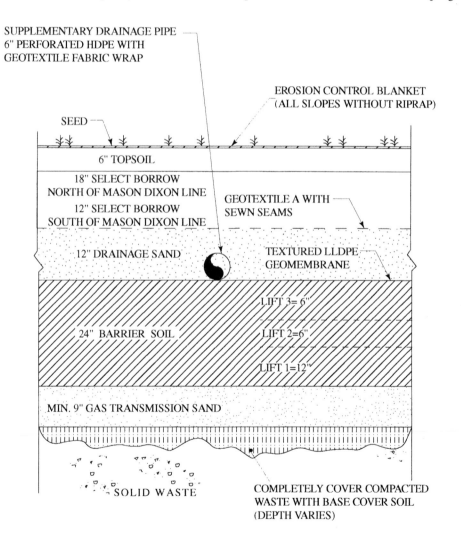

Figure 9-3   Typical cover section for a landfill.

As shown in Figure 9-3, it is first necessary to completely cover the (compacted) waste with soil of approved characteristics. This first layer is the base cover soil. Normally, a minimum thickness of 12 inches is required.

The base cover soil is then overlain with a layer of sand, 9 to 12 inches thick. The primary purpose of this layer is to collect methane and other gases that are generated within the wastes as a result of microbiological or other activity, and to provide for free flow of the gas to the pipe grid that constitutes the gas collection and relief system. The objective of the combined sand layer and pipe grid is to allow generated gas to pass continuously into the ambient air and not build up beneath the impermeable cap. Therefore, the gas transmission sand must be clean and properly graded (high uniformity coefficient) so as to have high gas conductivity.

Above the layer of gas transmission sand is placed a two-foot layer of "barrier soil." The barrier soil is placed in three lifts, the first of 12-inch thickness, the second and third of 6 inches each. This layer of soil is then overlain with a synthetic, impermeable membrane. The primary purpose of the combined barrier soil–synthetic membrane system is to prevent water from any source from percolating down through the landfilled waste. If the cap is successful, leachate formation will cease.

What is placed on top of the impermeable membrane system should support the growth of grass and conduct away rainwater and water from any other source. In order to support the growth of grass, the system constructed over the top of the impermeable membrane must be able to allow rainwater to percolate through it and then be collected and carried harmlessly off the site. To this end, the synthetic membrane is overlain with a 12-inch layer of "drainage sand." The purpose of the drainage sand is to provide an easy flow path for the percolated water to reach the drainage piping network. The drainage pipes then conduct this water to the stormwater collection and retention system so as to prevent erosion or other harmful activity.

Above the drainage sand is a geotextile sheet, and above that is a 12- to 18-inch layer of common borrow. The geotextile prevents percolating water from carrying fines from the common borrow into the drainage sand, thus causing plugging. The purpose of the common borrow is to support the topsoil above it in a way that helps the topsoil retain moisture. If the topsoil were placed directly on the sand, it would dry quickly and not support the growth of grass between periods of rain events.

Topsoil is placed in a six-inch-thick layer over the borrow. It has been found to be good practice to place an erosion control blanket over the topsoil before, or just after, seeding. The erosion control blanket is a fabric of synthetic or natural material that resembles fine netting. It allows ready penetration of water and allows grass to grow up through it.

Gas Venting

The organic material that has been placed in a landfill will undergo anaerobic degradation, with the consequent generation of methane, hydrogen, hydrogen sulfide, and other gases, possibly for many years. Explosions have occurred when one or more of these gases, principally methane, have migrated through the soil and entered a building. To prevent such an occurrence, it is necessary to provide for easy and continuous escape of these gases into the ambient air. It is for this purpose that a well-designed and installed gas venting system is a necessary component of a landfill or portion of a landfill that has been closed and provided with final cover or an impermeable cap.

Figure 9-4 presents a drawing of a cross-section of a landfill cap, including a portion of the gas venting system. The standard installation procedure for a gas venting system is as follows. A grid system of trenches is excavated a minimum of five feet into the final layer of deposited and compacted solid

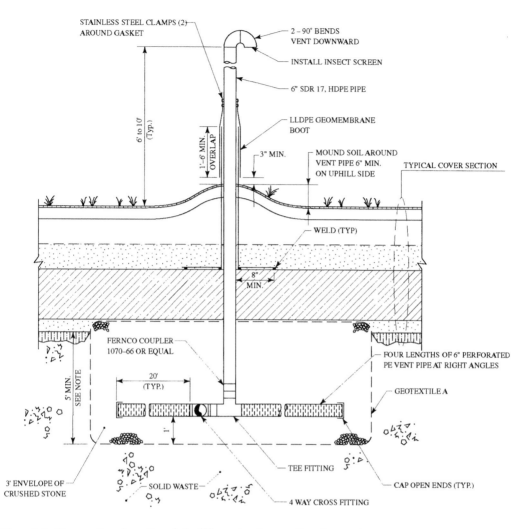

STAINLESS STEEL CLAMPS (2)
AROUND GASKET

2 – 90° BENDS
VENT DOWNWARD

INSTALL INSECT SCREEN

6" SDR 17, HDPE PIPE

LLDPE GEOMEMBRANE
BOOT

6' to 10' (Typ.)

1'-6 MIN.
OVERLAP

3" MIN.

MOUND SOIL AROUND
VENT PIPE 6" MIN.
ON UPHILL SIDE

TYPICAL COVER SECTION

WELD (TYP)

8"
MIN.

FERNCO COUPLER
1070–66 OR EQUAL

FOUR LENGTHS OF 6" PERFORATED
PE VENT PIPE AT RIGHT ANGLES

GEOTEXTILE A

20'
(TYP.)

5' MIN.
SEE NOTE

1'

TEE FITTING

CAP OPEN ENDS (TYP.)

3' ENVELOPE OF
CRUSHED STONE

SOLID WASTE

4 WAY CROSS FITTING

Figure 9-4   Cross-section of a portion of a landfill cap, showing a portion of gas venting system.

waste material. Perforated plastic pipe, six inches in diameter, is installed in each trench, surrounded by crushed stone. The crushed stone completely fills each trench, except for the plastic pipe. The purpose of the crushed stone is to provide for ready migration of the gas generated in the landfill toward and into the plastic (polyethylene [PE], in the case shown if Figure 9-4) pipe. The purpose of the plastic pipe, of course, is to collect the gas and allow it to disperse into the ambient air via the vertical vent pipes shown.

The mechanism of gas collection and dispersion is accounted for by the low specific weight, compared with air, of the mix of gases, which is predominantly methane. When the mixed gas reaches the vertical vent pipe, it rises, under the influence of gravity, thus creating a decrease in pressure (partial vacuum) behind it. This lowered pressure draws gases within the solid waste mass toward and into the vent pipe system. These gases rise up the vent pipe and out the 90° bend, as shown, and the process perpetuates itself.

In general, there are guidelines and regulations issued by each state or solid waste authority regarding the layout of the gas collection grid and the number and spacing of vertical vents. Also, the characteristics of the industrial solid waste that is landfilled must be taken into account regarding how readily gas can move through it to be collected by the venting system, when designing the system.

Stormwater Management

The objectives of stormwater management at a landfill site are to prevent runoff from contacting the landfilled waste material and to prevent erosion. Stormwater management involves the use of berms, grading, catch basins, a storm sewer system, and retention basins. Rip-rap-lined ditches and channels are also used for stormwater management and erosion control, during active use of the landfill as well as after closure and capping of the landfill. There is always a slope to the cap to prevent standing water at a minimum but usually as the result of placing the maximum possible amount of waste in the landfill.

Rip-rap-lined ditches and channels are usually designed as an integral component of the final cover or cap for control of erosion. Detention basins are used to hold collected stormwater for a period of time to allow solids to settle out before discharge to the receiving water body.

Essentially all solid waste management facilities, including transfer stations, incinerators, compost facilities, and certainly landfills, should include a well-designed system of groundwater sampling wells, sometimes referred to as "sentinel wells," to detect the occurrence of groundwater contamination. There should be at least one upgradient well, to show background or uncontaminated groundwater quality, and a sufficient number of downgradient wells to detect contamination from any reasonably possible location. The sentinel wells should be monitored, and records maintained, throughout the life of the facility, as well as for a reasonable period of time after closure.

Discharges from Landfills

There are two types of discharges from sanitary landfills: liquid, in the form of leachate,

and gaseous, i.e., that which is discharged from the vent system. Both are regulated by federal and state statutes. In the case of leachate, there are strict requirements to contain, collect, and treat it before ultimate discharge to the environment, which would necessarily be either a surface water body or the ground water.

Leachate is contained and collected by the mandated landfill liner and leachate collection system described above. Regarding treatment, there are three alternatives: (1) traditional trucking or piping to the industrial facility's wastewater treatment plant or to a publicly owned treatment works (POTW), (2) treatment using a facility designed specifically for the leachate, and (3) evaporating the leachate in a burner fueled by gas collected from the landfill. Of course, there is also the alternative of combining two or more of these three alternatives. For instance, pretreatment facilities have been used to render leachate from industrial solid waste landfills compatible with either the industrial facility's wastewater treatment plant or a POTW. Alternative technologies for treating or pretreating landfill leachate are discussed in Chapter 7.

Regarding evaporation by use of a burner fueled by gas from the landfill itself, Figure 9-5 shows an example of a system that was developed for this purpose during the mid-1990s. Landfill gas (LFG), being primarily methane, is fed in to the bottom of the combustion chamber shown on the right. Air is fed in very close to the gas inlet nozzle, so as to provide the correct quantity of oxygen for complete combustion of both the landfill gas (including the methane and the other gases such as hydrogen sulfide) and the vaporized leachate, which enters this combustion chamber above the flame. A temperature of 1,600°F, achieved in the combustion chamber, ensures complete destruction of VOCs or other hazardous substances from either the landfill gas or the leachate.

The leachate is vaporized in the evaporation vessel shown on the left in Figure 9-5. Landfill gas enters the burner at the top of

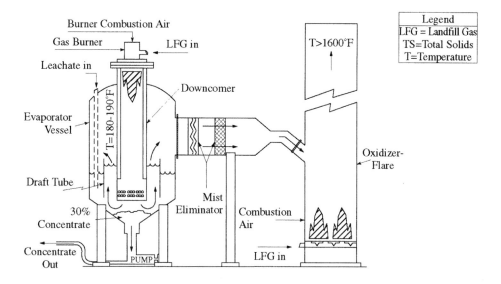

Figure 9-5 Evaporation vessel (from White et al., 1996).

this vessel, and the hot gases from the burning of the LFG are directed down and into the leachate, which enters the evaporation vessel just outside the draft tube within the vessel. Evaporated leachate, plus completely and incompletely oxidized LFG, exits the liquid surface within the draft tube and then proceeds to the combustion chamber of the right. Those substances in the leachate that will not evaporate sink to the bottom of the evaporator vessel to form a concentrate, reportedly amounting to less than 5% of the original leachate volume. This concentrate must be dealt with, and options include returning it to the landfill. In order to do this, however, the restrictions and provisions of 40 C.F.R. 264.314 and 265.314 (solid waste regulations) must be complied with.

Landfill gas discharges are regulated by Subtitle D (see Chapter 3) and the Clean Air Act, primarily to control greenhouse gas emissions. Methane is 21 times more potent in creating climate change than carbon dioxide, on a carbon-to-carbon basis, according to the EPA. Moreover, it is the EPA's estimate that 40% of the methane emitted to the atmosphere on a yearly basis is from active and closed landfills. Consequently, the EPA

has been charged by Congress to enforce controls on methane emissions from landfills. The result is an ever-increasing requirement for landfill gas collection, with subsequent management to minimize the effect on global climate change.

Gas collected from industrial waste landfills usually requires a certain amount of cleaning before use as fuel in a boiler or other burner. Reasons for cleaning include the following:

- Limiting pollutants released to the atmosphere
- Removing corrosive substances
- Increase the fuel value in terms of BTUs per cubic foot

Additionally, to limit emissions of NO_x and carbon monoxide (CO), it may be necessary to use catalytic converters to remove these substances from postcombustion gas, as well as other technologies to address VOCs, sulfur compounds, and silicon-based compounds. The latter have been the source of problems with the catalytic converters used to control emissions of NO_x and CO.

Alternative Landfills

The conventional modern landfill, with leachate containment, collection and treatment, and impermeable cap, is referred to after closing as a "dry tomb." The combination of impermeable cap, which prevents moisture from reaching the landfilled material via percolating precipitation, and the landfill liner system, which carries away leachate resulting from the period of time when the landfill was active, ensures that the landfill interior will be dry for a long period of time. The dry conditions effectively prevent microbial activity. The advantage of this condition, of course, is that neither gas nor leachate will be produced within the landfill in significant quantity. The disadvantage is that the landfilled material will remain "entombed" for those many years. While this may be a good thing regarding future recycle and reuse of some of the landfilled substances, or for future beings from outer space to examine to reconstruct our civilization, there is a growing sentiment that, on balance, the disadvantages outweigh the advantages.

Probably the greatest disadvantage of the dry tomb approach is that the closed landfill remains a potential threat to the groundwater for as long as it exists. Whenever the integrity of the cap and liner are breached, for whatever reason—earthquakes, tornadoes, accidents, or simply the passage of time—water will reach the landfilled wastes and create leachate.

An alternative to the dry tomb approach is to manage the closed landfill to encourage microbial degradation of the landfilled material (i.e., manage the landfill as a biochemical/chemical reactor). One way to do this is to continually recycle leachate back to the top of the landfilled wastes and to add water to the leachate in order to keep the moisture within the landfill at a level that will result in maximum gas generation. Then, the option of collecting the gas for use as fuel is available. The decision to manage an industrial waste landfill as a biochemical/chemical reactor, possibly for the purpose of generating gas for fuel as well as to address the "time

bomb" problem where the landfill sits as a threat to groundwater pollution "in perpetuity," must be based in large part on the characteristics of the leachate. If hazardous substances are in hazardous concentrations in the leachate, the cost for proper (including legal) management becomes of overbearing importance.

Solid Waste Incineration

There are many different types of incinerators, and several different conventions are used to classify them. One convention is to classify incinerators as either "mass burn" or "refuse derived fuel" (RDF). Mass burn technology involves combusting, as completely as possible, all (combustible) substances in a solid waste stream. RDF systems attempt to separate, by volatilization, as much of the organic portion of a solid waste stream as possible and then to clean and otherwise process the resulting mixture of gases to produce a low-, medium-, or even high-grade fuel. The remaining portion of the solid waste stream (char and ash) is then disposed of or further processed and then disposed of.

Another convention used to classify incinerators (or "combustors"—the terms *incinerator* and *combustor* are used synonymously) is based upon the equipment technology itself. The following is a list of incinerator technologies in use:

- Fluidized bed
- Rotary kiln
- Hearth-type
- Liquid injection

Another convention is to classify incinerators as either "excess air" or "starved air," and still another is to classify incineration systems as either "hazardous waste incinerators" or "conventional waste incinerators."

Under any classification system, all incinerators have certain characteristics in common:

- All solid waste substances must be converted to the vapor state before they can be ignited and burned.
- Two types of ash leave the system: fly ash and bottom ash. Both must be managed to prevent them from becoming environmental pollution problems. Incineration is therefore regarded as a treatment and volume reduction process, rather than a disposal process.
- Solid wastes almost always have to be subjected to a conditioning process before entering the combustion chamber. Conditioning may include one or more of grinding, mixing, blending, dewatering, or other treatment.

Fluidized Bed Technology

Fluidized bed technology is considered to have high potential for treating industrial wastes because of the capability of complete combustion and history of relatively low air emissions. Fluidized bed combustion can be combined with recovery of heat; therefore, it has potential as a viable waste-to-energy (WTE) alternative, depending on the overall BTU value and moisture content of the waste.

Fluidized bed incineration systems make use of a bed of sand or sand-like material that is fluidized (suspended against the force of gravity) by the drag force of air and other gases rising up through the bed, as illustrated in Figure 9-6. The bed is heated to incineration temperature, which can range from 650°C to more than 1,200°C (1,200°F to 2,200°F or higher), depending on the characteristics of the waste. The heat content of the fluidized bed material, typically sand, provides a substantial heat reservoir to maintain temperature and combustion as new waste material is injected into the fluidized bed volume. For instance, the heat content of the bed on a unit volume basis is about three orders of magnitude greater than that of flue gas at the same temperature. Fluidized bed technology is used for hazardous as well as conventional solid wastes.

A generic fluidized bed combustor system is illustrated in Figure 9-6(a). The characteristic component of a fluidized bed combustion system is the reaction chamber, illustrated in greater detail in Figure 9-6(b). Figure 9-6(b) shows that a grate supports the sand or sand-like bed material at the bottom of the primary combustion chamber. A cone-shaped collector-distributor beneath the grate has the dual purpose of distributing air that is blown into the section below the combustion chamber, as well as collecting inerts that are not further combustible and too heavy to be buoyed up by the rising air column. The inerts are collected and removed from the lower chamber by a conveyor device that locks out air.

The air that is blown into the lower chamber is distributed evenly across the cross-section of the primary combustion chamber by the cone-shaped collector-distributor. This air then rises up through the bed, supplying oxygen for combustion and drag force to fluidize the sand bed. The fluidized bed of sand (or sand-like material) fills most of the primary reaction chamber. Waste materials are fed directly into the primary reaction chamber, along with auxiliary fuel, needed to start up the combustion process and to maintain desired temperature, if necessary. Typically it is not necessary to supply auxiliary fuel to maintain desired temperature while combustion is taking place, because the heat value of the organic solid wastes being incinerated is more than enough for this purpose. In fact, it is usually necessary to supply excess air to prevent the temperature from rising too high, causing damage to the system.

As the waste is injected into the lower part of the fluidized bed within the combustion chamber, the heat from the sand grains first gasifies the waste material and then ignites it. Complete combustion of this material follows. The grains of sand are greatly agitated by the column of air passing up through the bed, and this agitation has beneficial effects. The agitation tends to break up the heated particles of solid waste, enhancing the gasification process and resulting in quicker

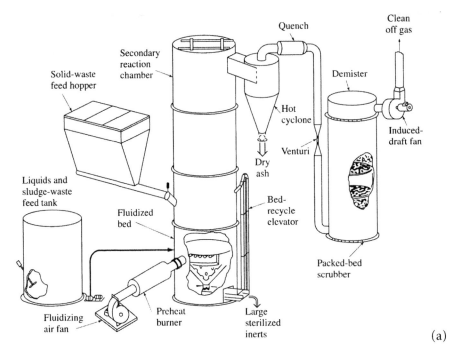

(a)

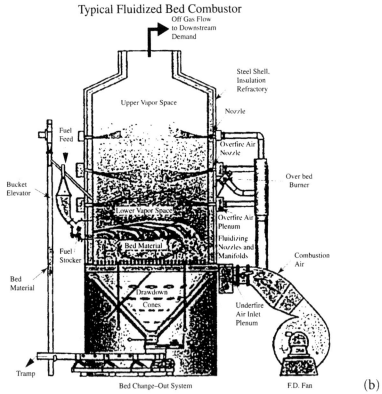

(b)

Figure 9-6   (a) Fluidized bed combustor system (from Freeman © 1989; reprinted by permission of McGraw-Hill, Inc.). (b) Typical fluidized bed combustor (from Pope, 1999).

ignition. The agitation also tends to abrade the bed material itself, and this tends to keep it clean. Without this self-cleaning action, the particles of bed material would tend to become coated with soot and other products of the combustion of the solid waste material. The agitation also tends to mix and homogenize the injected waste material, promoting faster and more thorough combustion. As the solid waste material undergoes combustion, liberated heat is transferred back into the bed material (typically sand) to perpetuate the gasification-ignition-combustion process.

The products of combustion rise out of the fluidized bed, toward the top of the reactor. Some systems have a secondary combustion chamber at this point to ensure complete destruction of the waste material. This, of course, is especially important when hazardous materials are being treated.

From the secondary combustion chamber, which may or may not have auxiliary fuel capability, the products of combustion, along with the oxygen-depleted air stream, flow into and through a system that cleans and cools them before discharge to the ambient air. The gas treatment devices included in this system can include cyclones for collecting particulates; heat exchangers or quenchers to lower the temperature; one or more packed column scrubbers; activated carbon adsorbers; venturi scrubbers; or other devices. Chapter 8 presents discussions regarding the mechanisms and uses of these air pollution control devices.

Fluidized bed reactors are used in some waste-to-energy (WTE) systems. In these systems, typically, only gasification takes place in the fluidized bed reactor. The volatilized organic substances from this process are either burned immediately in a boiler or other energy conversion device or are captured and stored for later use. In some cases, the volatilized product (gases) from the fluidized bed reactor are processed to remove substances that degrade the fuel value; in other cases, other substances of highest fuel value are removed as a valuable by-product.

The remaining substances are then burned as a lower-grade fuel.

Advantages

Many years of development and operating experience with fluidized bed combustors have resulted in the following advantages, compared with other combustion systems for treating solid wastes from industries:

- Air emissions are relatively low.
- Generation of nitrogen oxides (NO_x) tends to be significantly less, due to lower excess air and lower temperatures.
- Low carbon monoxide (CO) emissions due to relatively quick and relatively complete combustion.
- Solids, liquids, and gases can all be burned simultaneously.
- Few moving parts are used.
- Maintenance costs are relatively low.
- The large surface area of bed grains and waste particles enhances the gasification-ignition-combustion process.
- The large heat capacity of the bed tolerates fluctuations in solid waste feed rate.
- The size of the facility is relatively small.
- The high degree of agitation in the combustion chamber practically eliminates hot spots and cold spots.

Disadvantages

- There is a low tolerance for items such as wire that get stuck in the grate.
- Substances that tend to agglomerate can plug up the sand bed.
- Inerts are difficult to remove from the bed if they are not heavy enough to fall through the grate against the air upflow.

Rotary Kiln Technology

Rotary kilns use an inclined cylinder as the combustion chamber, which rotates slowly to accomplish mixing of the materials and to attempt to expose all surfaces and substances to oxygen within the chamber. Auxiliary fuel, mixed with air, can be injected into either end of the rotating combustion chamber or at one or more injection nozzles along the side. Rotary kilns have been successfully used to treat hazardous as well as conventional wastes; solid, semisolid sludges, and liquids; alone or simultaneously. The kilns have been operated at temperatures as high as 1,500°C (3,000°F). Many rotary kiln combustors have been used to incinerate hazardous materials at temperatures in the 1,100°C (2,000°F) degree range.

As with all incinerator technologies, residence time, temperature, and quantity of oxygen made available are important parameters. Among other factors, residence time is influenced by the angle of incline, or slope, of the rotating combustion chamber and the rate of rotation. The slope typically ranges from 0.02 ft/ft to 0.04 ft/ft. Speed of rotation ranges from 0.5 revolutions per minute (rpm) to 3 rpm. Figure 9-7 illustrates characteristics of the rotation combustion chamber. Figure 9-8 illustrates a typical rotary kiln solid waste incinerator system.

Figure 9-7 shows that the conditioned solid waste is fed into the higher end of the inclined, rotating combustion chamber. Conditioning can include any or all of grinding, mixing, or dewatering. As the kiln rotates, the solid waste tumbles from the up-rotating side back toward the middle of the bottom of the cylinder. As the wastes tumble, they become heated, vaporize (gasify), mix with oxygen, and ignite. They also progress toward the lower end of the rotating combustion chamber. Finally, ash discharges from the lower end.

Figure 9-7(c), (e), and (f) also illustrates that the auxiliary fuel-air mixture can be injected into the combustion chamber at alternative locations, and that exhaust gases exit at either (or both) the higher end or the lower end of the chamber.

Figure 9-8 is an example of a complete rotary kiln incinerator system. As shown, after ignition, the gasified solid waste substances burn in the upper portion of the rotating cylinder (i.e., the combustion chamber), and the exhaust gases proceed to a post-combustion chamber, sometimes referred to as the afterburner. The postcombustion chamber is the primary air pollution control device and is especially important if hazardous materials are contained in either the solid waste stream or the products of combustion, or both. The postcombustion chamber is followed by a heat exchanger, which, in turn, is followed by a particle collector. The cleaned exhaust gas is then discharged.

The heat exchanger is required to cool the exhaust gases in order to protect the particle collector. It may also be the functional component of a waste-to-energy (WTE) system. In its simplest form, the heat exchanger can be a water spray or radiator, to cause heat to be wasted to the ambient air. In its most useful form, it can be a waste heat boiler or other energy conversion device.

The ash, of course, must be disposed of in an appropriate manner, and there are alternatives to be evaluated. Landfilling in a secure facility designed and approved for either "special" or "hazardous" solid wastes has been most often used. Other alternatives include solidification with cement before burial, incorporation into asphalt, or solidification with epoxy.

Advantages

- Proven technology, having been used for incineration of many different industrial wastes over many years
- Flexible operation; the rate of rotation can be varied to suit the needs of a large variety of solid waste characteristics
- Has been used to incinerate solids, liquids, and gases, in any combination
- Requires less preparation (conditioning) such as sorting, grinding, and mixing than certain other technologies
- Not subject to problems such as plugging of grates or from substances melting before gasification
- Adaptable to many alternatives for air emission control, including quenchers, venturi scrubbers, wet gas scrubbers,

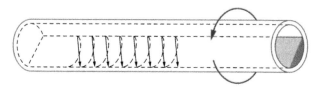

Axial movement through the reactor.

The slope of the kiln produces a slight forward movement toward the discharge end.   (a)

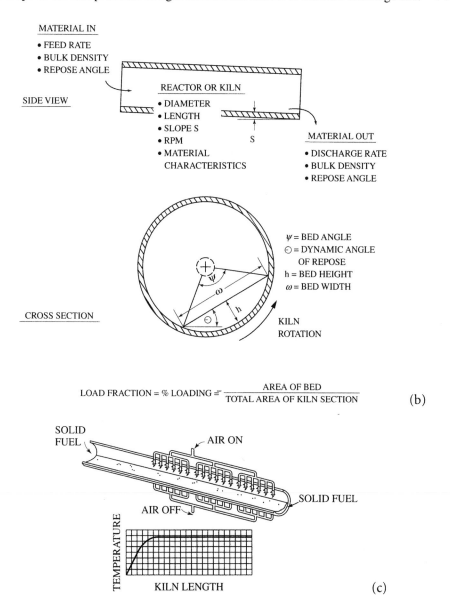

LOAD FRACTION = % LOADING = $\dfrac{\text{AREA OF BED}}{\text{TOTAL AREA OF KILN SECTION}}$ (b)

Figure 9-7   Elements of a rotary kiln incinerator. (a) Axial movement through the reactor. (b) The kiln solids flow system; definition of terms. (c) Ported-kiln concept. (All from Freeman, © 1989; reprinted by permission of McGraw-Hill, Inc.)

electrostatic precipitators, and bag houses
- Provides mixing within the combustion chamber for exposure to heat and oxygen
- Can receive bulk containers in the feed stream

Disadvantages

- Subject to formation of hot spots and cold spots
- High load of particulates on air pollution control devices
- Relatively high capital cost

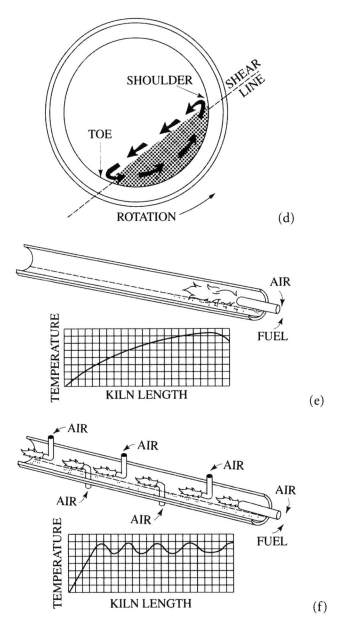

Figure 9-7   Elements of a rotary kiln incinerator. (d) Solids flow in cross-section. (e) Primary burner concept. (f) Axial burner concept. (All from Freeman, © 1989; reprinted by permission of McGraw-Hill, Inc.)

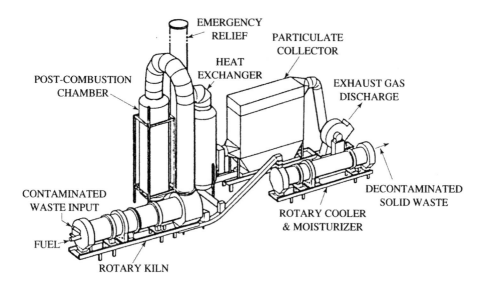

Figure 9-8   Hazardous waste incinerator system (from Freeman, © 1989; reprinted by permission of McGraw-Hill, Inc.).

- High amounts of excess air lead to high demand for auxiliary fuel
- Subject to damage to refractory lining of kiln due to tumbling solids

Hearth Incinerator Technologies

Hearth incinerators are mass-burn systems in which conditioned solid waste materials are spread out on a (nearly) horizontal surface and burned. Auxiliary fuel is used for startup and as necessary to maintain desired temperatures. In many cases, of course, excess air is used to maintain desired temperatures. In these situations, the excess air prevents the burning solid wastes from developing excessively high temperatures that would damage the combustion chamber and downstream equipment.

The most common type of hearth incinerator is the multiple hearth incinerator, illustrated in Figure 9-9. The multiple hearth incinerator system shown in Figure 9-9 has five hearths, stacked vertically. New, conditioned, solid waste is fed onto the uppermost hearth, where hot gases from materials burning on lower hearths heat the new material, volatilizing water and the most volatile of the organic materials in the solid waste stream. A

system of rakes slowly forces the wastes toward the middle of the uppermost hearth, and down through a passageway to the next lower hearth. The temperature on this hearth is much higher, and active combustion takes place. The rake system forces burning wastes on this hearth toward the outside, then down onto the next lower hearth, and so on, until the residual ash is raked from the lowest hearth to a collection and transport device that removes the ash from the combustion chamber.

As the solid wastes burn on the middle hearths, the gaseous products of combustion flow up to heat materials on the hearths above; then they exit the combustion chamber and pass through one or more devices to either finish off combustion, especially of toxic or otherwise hazardous substances, or to clean the exhaust gas stream of other emissions such as particulates, SO_x, malodorous gases, or other substances. In the case of completing the combustion process, a secondary combustion chamber (or "postcombustion chamber" or "afterburner") with its own fans and auxiliary fuel system can be used. In the case of cleaning the exhaust gases before release to the ambient air, alternative processes include a quenching system to cool

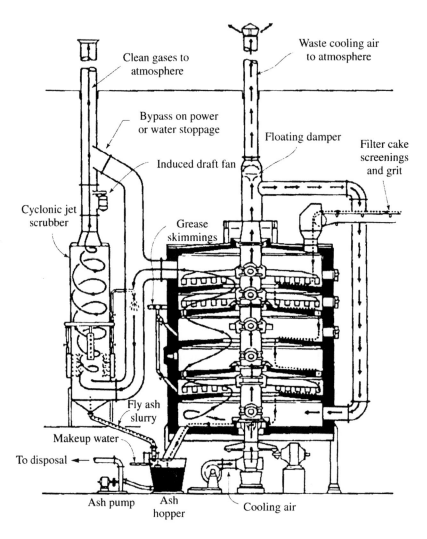

Figure 9-9   Typical multiple hearth incinerator (from Metcalf and Eddy, © 1991; reprinted by permission of McGraw-Hill, Inc.).

the exhaust gases to prevent damage to particulate collection devices; the particulate collection devices themselves; electrostatic precipitators; and various types of wet scrubbers.

Advantages

- Useful when the water content of the solid waste stream is relatively high
- Extensive operating history with conventional (nonhazardous) wastes

Disadvantages

- Lack of sufficient control for use with hazardous wastes
- Subject to formation of cold spots and hot spots
- Occupies relatively large area

Modular Systems

A variety of incinerator systems available for industrial use are referred to as "modular," because they are assembled in sections at the factory for final assembly on site. Some are completely mobile after assembly. The principal advantage of modular combustor systems is lower cost due to preassembly under favorable conditions at the factory.

Starved Air Technologies

Starved air technologies, which include pyrolysis technologies, have been used as the first step in a two-step system for processing both hazardous and conventional industrial solid wastes. In this first step, the solid wastes are heated in a low-oxygen atmosphere to accomplish volatilization of all components except char and ash. Then the volatilized material is burned, as in a conventional incinerator. At some facilities, the volatilized substances are processed to recover one or more components before the rest is incinerated. The volatilization process is carried out at significantly lower temperatures than are used for more conventional incineration technologies, with the advantage that equipment life is prolonged and maintenance costs are lower. More important in the case of hazardous materials, however, is the fact that the incineration step can be carried out under very precisely controlled conditions, since only gases are being handled in the combustor phase of the system.

The generally accepted definition of pyrolysis is chemical decomposition by heating in the absence of oxygen. Some of the starved air industrial waste treatment systems do not fit this definition, since they are carried out in atmospheres of up to 2% oxygen by volume (compared with 21% in ambient air). However, there are systems in operation that are called pyrolysis systems, even though "gasification" or simply "starved air" would be more appropriate.

Figure 9-10 is an example of a solid waste starved air system. In Figure 9-10, the conditioned, wet solid wastes enter the dryer, which uses recovered heat from the combustion of waste materials downstream in the same system. The dryer solid wastes are stored for a period of time (approximately one day) and then fed into the gasification chamber, which is maintained within a temperature range of 425°C to 750°C (800°F to 1,400°F). Residence times within the gasification chamber depend on the characteristics of the material in the waste stream. For instance, chemical makeup and particle size, as well as bed thickness, all have a significant effect on residence time. Typical residence times are as short as two minutes and as long as five hours. The concentration of oxygen in the gasification chamber is maintained near zero. Gasification chambers are maintained under slightly negative pressure to prevent release of fugitive gases.

After completing the required residence time in the gasification chamber, the nonvolatilized solids (char and ash) are removed and are either further processed or landfilled. Further processing can be by high temperature incineration (to further burn the char) or by solidification and stabilization using lime, cement, epoxy, asphalt, or other substance; chemical oxidation; or other process. The volatilized material can proceed to incineration in a one- or two-stage combustor or, preferably, to a cleaning and conditioning process for the preparation of a fuel of reasonably high value. This fuel can be burned on site or off site in a gas engine, it can be used in a gas turbine to generate electricity, or it can be burned in a boiler to generate heat or steam to generate electricity. In the case of the system illustrated in Figure 9-10, the heat given off by the combusting volatilized gas stream is recovered and used to dry incoming solid wastes as the first step in the system.

It is also possible to separate by condensation and thus recover certain components of the volatilized gas stream. Some of these substances have considerable value, representing a source of cost recovery for the system as a whole.

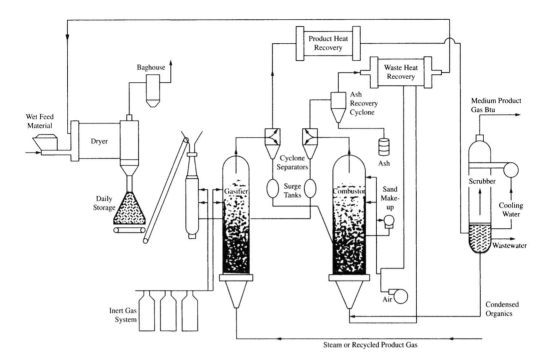

Figure 9-10   Battell gasification process (from Niessen, 1996).

Another significant advantage of the starved air process is that it results in a char or ash that is concentrated in potentially valuable nonvolatile substances, such as metals. For instance, solid waste streams that contain a significant quantity of lead and/or zinc can cause problems in conventional combustors, because at normal incinerator temperatures lead and zinc and certain other substances can volatilize and cause fouling of heat exchangers and spalling of the refractory linings of the combustion chamber.

Advantages
- The volatilized gas stream can be used as fuel for engines, turbines, or burning.
- Valuable substances such as lead and zinc can be recovered from the char and ash residue.
- The vapor stream can be processed to recover substances of value.
- Lower temperatures lead to longer equipment life and lower maintenance costs.

- Process control is relatively easy, because the gasification process is endothermic. There is no danger of self-overheating.

Disadvantages
- One disadvantage of starved air technologies is that the process takes place in a reducing atmosphere, with the result that the products are highly corrosive. Therefore, the materials of construction are relatively expensive.
- In the case of certain highly toxic solid waste streams, carcinogenic substances are sometimes generated. This problem may require construction and operation of very highly reliable and effective volatilized gas stream incineration.

Design Considerations
There are choices to make regarding the technology to employ for the gasification step, as well as the subsequent processing and ultimate disposal of the volatilized gas stream, and the solid residuals. The gasification step

can be accomplished in any of several furnace types, including:

- Rotary kiln
- Rotary hearth
- Roller hearth
- Car bottom

Rotary Kiln

The rotary kiln reactor for carrying out the gasification step is essentially the same as is discussed above for the rotary kiln combustor. Differences include provision to exclude oxygen from the gasification furnace and different materials of construction due to the oxidative environment in the combustor furnace versus the reducing environment in the gasification furnace. Also, auxiliary fuel must always be used in the gasification furnace, due to the endothermic nature of the process, whereas provision must be made to supply and control excess air for cooling in the case of the combustor furnace.

Rotary Hearth

A typical rotary hearth reactor, illustrated in Figure 9-11, consists of a doughnut-shaped hearth that rotates through a stationary heated gasification chamber. Industrial solid wastes, conditioned as needed, are fed onto the continuously rotating hearth ahead of the heated chamber. As the waste enters the heated chamber, the pyrolysis process begins and continues after the wastes leave the heated chamber. Vaporized organics and other gaseous substances are continuously extracted from the top of the enclosure that covers the entire hearth and proceed to the next step, which, in the example shown in Figure 9-11, is a reactor that conditions and ultimately uses the fumes for fuel to fire the waste heat boiler.

Roller Hearth

The roller hearth gasification system is used by individual industries and by centralized hazardous waste processors to gasify hazardous wastes that are brought to the facility in containers such as 55-gallon steel drums. The containers are opened at the top, usually by removing the top, and then fed into the heated chamber upright, so as not to spill the contents. Rollers cover the bottom of the heated chamber and allow the open drums to be pushed along through the heated chamber from the entrance to the exit. In some cases, the rollers are motorized so they can propel the containers through the heated chamber.

As the containers pass through the heated chamber, the target substances in those containers volatilize or "gasify;" the char and ash remain in the containers. The vapors are collected, processed, and stored and then used as either fuel or as a source of mixed substances, mostly organic, from which to recover certain substances of value. The remainder is either incinerated or, if possible, used as a (lower grade) fuel.

Car Bottom

Car bottom gasification furnaces are small two-part systems. One part is a mobile "car" on wheels, on which containerized wastes, often of a hazardous nature, are placed. The car is then rolled into the second part, which is a furnace without a bottom. The car becomes the bottom of the furnace. After the car and the furnace have been joined to form the complete furnace, the furnace is heated, and the gasification process takes place.

As fumes exit the containers (typically 55-gallon drums), they are collected and then usually combusted as a means of disposal in a furnace of higher temperature. The char and ash that remain in the bottom of the containers can then be processed to recover certain substances. The remaining char and ash must then be disposed of using appropriate means, which might include incinerating (the char) in a combustor of high temperature, landfilling in an appropriate landfill, or solidification.

Car bottom gasification technology is normally operated on a batch basis. The advantage, as with many other batch technologies, is that the residence time, or time allowed for the gasification process to take

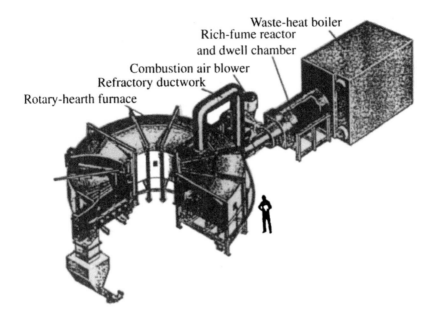

Figure 9-11   Rotary hearth pyrolyzer (from Freeman © 1989; reprinted by permission of McGraw-Hill, Inc.).

place, can be extended for as long as needed for the process to go to completion.

In the United States, there are two incinerator technologies in wide use for treatment (volume reduction and characteristic transformation) of two classifications of solid wastes. The two incinerator technologies are hearth incinerators and rotary kiln incinerators. The two classifications of solid wastes are hazardous and conventional. Of all incinerators in use for all purposes, those based on liquid-injection technology account for about half, but they are not used for incinerating solid wastes.

Generally, incinerators for hazardous materials are required to maintain significantly higher temperatures in the combustion chamber than are incinerators for conventional wastes. For this reason, hazardous waste incinerators are required to be built using more expensive materials and techniques.

The Process of Composting Industrial Wastes

Some industrial solid wastes are amenable to composting as a treatment process to prepare the waste for future use as a soil conditioner. Typically, these solid wastes are close to 100% organic in composition, are readily biodegradable, contain no hazardous materials, contain a moderate amount of moisture, and can be handled with conventional equipment such as front-end loaders and belt conveyors. The basic composting process is illustrated schematically in Figure 9-12.

As illustrated in Figure 9-12, microorganisms use the organic material within a pile of mixed solid wastes for food. Initially, the temperature will be close to the temperature of the ambient air, and oxygen present in the air will occupy the void spaces between the solid waste materials. Also, there must be enough moisture present for the microorganisms to live in. Bacteria can live only in an aqueous environment and can metabolize only substances that are dissolved in water. As the bacteria and other microorganisms metabolize the food dissolved in the moisture content of the compost pile, heat is generated. The source of the heat is as follows: as the microorganisms disassemble the proteins, carbohydrates, lipids, and other materials that make up the solid wastes in the

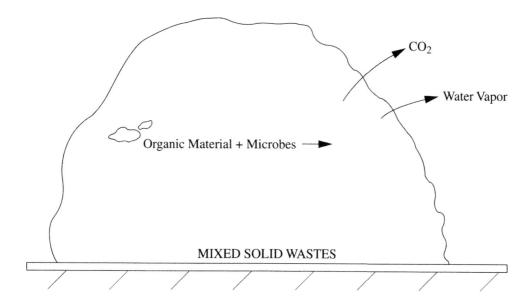

Figure 9-12   Basic composting process.

compost pile, energy from the breaking of chemical bonds is liberated. The microorganisms use some of this liberated energy to reassemble atomic and molecular parts and pieces into new cell protoplasm and other materials to make new cells, i.e., growth; however, the process is less than 100% efficient, and the leftover energy is lost as heat. As heat is generated, the temperature rises. During the increasing temperature phase, there is a continual natural selection process that favors microorganisms that thrive at higher temperatures, and then, as the temperature rises more, still other microorganisms are favored that thrive at still higher temperatures, and so on until the range of thermophilic microorganisms is reached, at about 70°C. The composting mass stays at about this temperature, because heat is now lost about as fast as it is generated.

Regarding the fact that microorganisms cannot ingest "food" unless it is dissolved in water, the reason composting of solid wastes proceeds successfully is due to the ability of some microorganisms to produce "exoenzymes" (enzymes that are manufactured inside the microorganism, but are then sent outside, into the aqueous environment) that

can attack undissolved organic solids and cause them to dissolve into the water, in parts and pieces. This relatively slow process continues until all of the solid substances have been "disassembled and dissolved," with the exception of that fraction of the original solid waste mass that is resistant to biodegradation. This fraction is called "humus."

All the foregoing can, and will, take place naturally; no management is required. However, if too much moisture evaporates and is lost from the system, the microorganisms will be unable to live and the process will be arrested. If there is too much moisture, water will fill the void spaces between solid waste materials, the dissolved oxygen in the water will soon be depleted by the respiring microorganisms, and the system will become anoxic or anaerobic. If the system becomes anoxic or anaerobic, the temperature will not rise to the thermophilic range, bad odors will be created, and the microbial degradation process will be very slow. For these reasons and others, a successful process for treating solid wastes from industries by use of the composting process to achieve volume reduction, chemical and biochemical stability, and an end product that can be safely and beneficially disposed of requires a high degree of management.

Regarding the temperature within the composting mass, there are two major benefits to achieving and maintaining a thermophilic condition (about 70°C). The first is that microbial degradation takes place very fast. In general, the rate of microbial metabolism doubles for each 10°C rise in temperature. The benefit to solid waste treatment, of course, is a shorter time required to reach a stable condition. Organic material that has been biologically stabilized will not undergo significant further biodegradation, and therefore will not become an odor or other nuisance or health problem. The second major benefit of achieving and maintaining the thermophilic range for a significant period of time is that pathogenic organisms, including bacteria and viruses, are killed. This is especially important if human wastes, usually in the form of sludge from wastewater treatment facilities, are included in the mix of solid wastes. In the context of treating solid wastes from industries, this ability to kill pathogens may or may not have value. What is always of value, however, is the ability of the composting process to produce a stable, inoffensive, useful product from a putrescible material that is inherently a disposal problem. Since sludge from industrial wastewater treatment systems is one of the industrial waste candidates for treatment by composting, and since toilet wastes may be treated along with processing wastewaters at a given industrial plant, the ability to kill pathogens obviously has value in some industrial situations.

It is often beneficial to mix fresh solid wastes with a bulking agent, such as wood chips, at the start of the composting process. The bulking agent does not necessarily become involved in the composting process itself; that is, the bulking agent does not necessarily undergo biological degradation. Rather, the bulking agent helps to provide spaces for air to reside, thus providing oxygen for microbial respiration as well as channels for air to move from outside the composting mass to the inner spaces where oxygen is being depleted. In some cases, fully composted product from the process is used. When an agent other than the compost itself is used for bulking, it may or may not be separated from the finished product for reuse. Sometimes it is left in the composted product to enhance the desirability and value of the product for certain uses. In other cases it is separated from the final product by screening, stored, and then used again with new batches of solid wastes.

Because of the continual depletion of oxygen, it is necessary to continually renew the oxygen supply by providing fresh air to the entire composting mass. This is accomplished by either turning and fluffing up the composting mass periodically or by blowing or drawing air through it. Also, if the moisture content becomes too low because of evaporation, it is necessary to add more moisture by spraying or other means.

It is important to maintain the oxygen content in the void spaces within the composting mass between 5% and 15% by volume (compared with about 21% in ambient air). Less than 5% will possibly allow local pockets of anaerobic or anoxic conditions to develop, leading to an odor problem. More than 15% is indicative of overaeration, with consequent inability of developing temperatures in the thermophilic range.

Moisture content must also be maintained within a favorable range, but that range is waste-specific. It is best determined through experience. The proper procedure is to conduct, first, bench-scale laboratory tests, followed by a pilot-scale program. Once an optimal range of moisture is determined by use of these studies, the initial compost process is set up and then observed closely and compared with the pilot-scale results. The objective is to reach thermophilic temperatures within a few days and to achieve complete composting within 14 to 28 days.

It is, in some cases, good practice, and in other cases, absolutely necessary, to grind the solid waste material before composting. Grinding has the beneficial effects of greatly increasing the surface area of the organic substances, thus enhancing the composting process and rendering the material more easily mixed with the bulking agent. Grinding also produces more uniform moisture content throughout the solid waste material. Mechanical grinders for this purpose are available from several vendors.

There are three general technologies used in the United States for composting: windrow, static pile, and mechanical. Windrow composting typically makes use of a very large, specialized mobile machine that straddles a windrow of composting solid wastes and works it over by fluffing and turning the windrow. Static pile technology also typically makes use of a windrow or other type of pile and uses a blower to either maintain a partial vacuum within the pile, to cause air to flow from the ambient air into the pile, or to blow air out through the pile.

Windrow Composting Technology

Solid wastes to be composted by the windrow method are first mixed to produce a reasonably uniform composition, so that the time required for complete composting will be close to the same throughout the windrow.

In some cases a grinding procedure using specially made solid waste grinders either precedes or follows the mixing procedure. Next, the mixture is processed to adjust the moisture content to the desired range; then, typically, the composting material is mixed thoroughly with a bulking agent such as wood chips or previously composted material. The next step is to arrange the mixture in a windrow using a front-end loader or other machinery. The windrows are 5 to 7 feet high and are 10 to 20 feet wide at the base. Figure 9-13 is an illustration of typical windrow dimensions. About once per day or in some cases more often, the windrow machine travels the length of the windrows, mechanically works over the composting material, and, in some cases, blows air into it. The objective of working over or turning the composting material is to expose new portions of the mass to the open air, thus renewing the oxygen supply. As explained previously, it is necessary to maintain an average oxygen concentration in the void spaces of 5% to 15% by volume (compared with 21% in ambient air) to prevent odor problems as well as to enable the composting mass to maintain a temperature in the thermophilic

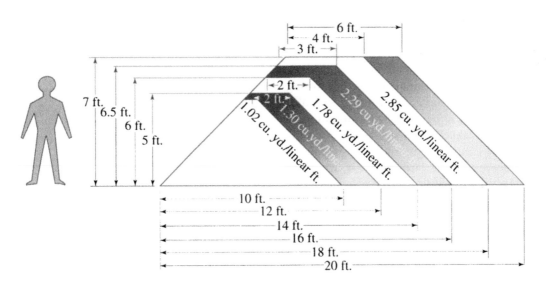

Figure 9-13   Dimensions of typical windrows (from *MSW Management*, 1996).

Figure 9-14 Windrow composting machine (courtesy of Scarab Manufacturing).

range. Figure 9-14 illustrates a windrow machine.

A large compost facility in Puyallup, Washington, has modified the windrow turning machine to include a 10,000-gallon water tank for the purpose of adjusting the moisture content of the composting mass. This facility uses grass clippings, picked up curbside throughout the city, as a bulking agent. Because the grass clippings have a tendency to settle and mat, it is necessary to blow (rather than draw) additional air through the composting mass to reestablish porosity and thus proper oxygenation. The extra air, moved by use of three 50-horsepower centrifugal blowers, removes too much moisture from the composting mass, thus the need for the water tank.

Other problems unique to the composting of a given solid waste or use of a given bulking agent or other additive can often be solved by taking advantage of simple concepts. As another example, in Rockland, Maine, a compost facility that was used to convert fish processing wastes to a soil conditioning agent made use of leather buffing dust (similar to sawdust, but of almost zero moisture content and also a troublesome industrial solid waste) to adjust from too high a moisture content to the optimum range.

The composting process is normally complete in 14 to 21 days (in some cases, as long as 28 days), after which the finished compost is stirred and allowed to "cure" for several weeks or months. During curing the microbial population reduces and stabilizes by adjusting itself to a food-poor environment. Normally, minor odor problems that may have existed at the end of the composting process disappear during the curing process, as the remaining species of microbes consume whatever is the source of the minor odors.

Static Pile Technology

Static pile technology makes use of a system that sucks air through the composting mass by drawing a vacuum at the bottom of the pile. This method, originally developed in the mid-1970s to treat sludge from sewage treatment plants, is illustrated in Figure 9-15.

As shown in Figure 9-15, a grid of perforated pipes supported on the top surface of a concrete pad is connected to a blower. The solid waste to be composted is piled over the pipe grid after first undergoing conditioning by grinding and mixing with the bulking agent. Then the pile is covered to insulate it and thus enable the temperature to build up to the thermophilic range. The composting process begins immediately, due to the presence of both bacteria and food; therefore, air flow should begin immediately to prevent the development of an odor problem. The blower draws air through the composting material from the outside in to maintain the proper concentration of oxygen within the voids in the composting material, as discussed above. Standard procedure is to discharge the exhaust air from the blower into a pile of finished compost, to filter out fine particulates and to control minor odors.

Mechanical Composting Technology

Mechanical composting technology involves the use of a container and mechanical stirring equipment. The solid waste is first conditioned by grinding and mixing with a bulking agent, as described for the windrow and

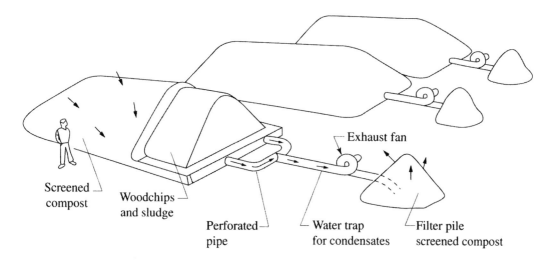

Figure 9-15   Composting with forced aeration (from Corbitt © 1999; reprinted by permission of McGraw-Hill, Inc.).

static pile technologies. Then, the mixture is placed in the container-stirring system, which maintains even distribution of oxygen, temperature, and moisture while the composting process is taking place.

Two variations of mechanical composting technology are the rotating drum system and the vertical system. As it is now used, the rotating drum system is a "precomposting" process or, more accurately a "compost starting" system. The conditioned (ground and mixed with bulking agent) solid waste is placed in one end of an inclined, rotating drum, which may or may not have additional mechanical devices inside to enhance mixing. The rotating of the drum mixes the composting material and brings about evenly distributed temperature and moisture. The rotation action also accomplishes aeration of the composting mass, by rolling the composting mass in an atmosphere of ambient air.

After two to five days, the partially composted material exits the lower end of the rotating drum. The composting process can be finished by use of either windrow technology or static pile technology. The value of the rotating drum process is that it significantly reduces the total time required for completing the composting process.

A second mechanical composting technology is referred to as a "vertical process." This technology makes use of large silos or bins that are maintained full of composting material, which is continually moving from the top to the bottom. Composted product is harvested from the bottom and is replaced with conditioned (ground and mixed with a bulking agent and, in some instances, carbonaceous material) solid waste at the top. Air is blown through, usually from the bottom to the top, to maintain the correct (5% to 15% by volume) oxygen content within the gas phase in the vertical composter. The oxygen content, as well as temperature and moisture content, are monitored by use of sampling ports in the side of the vertical composter. The principal advantage of this technology, then, is the capability to closely monitor and control the progress of the composting process.

Additional Considerations

Nutrients

Depending on the nitrogen and phosphorus content and other nutrient levels of the industrial solid waste, the nutrient content of the compost may be a significant benefit as fertilizer in the conventional sense. Some

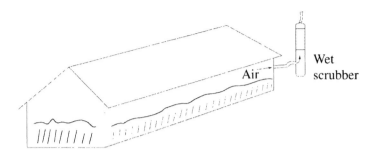

Figure 9-16 Containment for odor control.

nitrogen is lost during the composting process, however. Conversely, some industrial solid wastes might be suitable candidates for composting as a treatment process, but they lack adequate nutrients. In these cases, it might be cost effective to add nutrients such as nitrogen and phosphorus, probably in the form of agricultural fertilizer. However, the cost of the fertilizer, added to the cost of composting, minus the income from sale of the final composted product as a soil conditioner or other use, may be less than the cost of disposing of the solid waste by use of an alternate technology.

Odors

One of the primary concerns expressed when composting is under consideration, is the potential odor problem. The simple fact is that almost all composting facilities experience odor problems to some degree. The reason for the odors is simple: it is almost

impossible to maintain aerobic conditions throughout the entire composting mass at all times. When pockets of anoxic or anaerobic conditions develop, odors are produced. Whether or not the odors result in a problem depends on how odor generation is managed and the odors that are actually generated. The most important management activity, of course, is to do everything reasonably practicable to prevent odors from being generated. This comes down to managing the piles or windrows to maintain optimum oxygen, temperature, and moisture conditions. For those odors that are generated despite all management efforts, there are generally two control strategies: (1) containment and treatment, and (2) "reodorization" (masking) before release.

Containment and treatment have been successfully accomplished by conducting the composting process in an enclosure maintained under a negative pressure. The

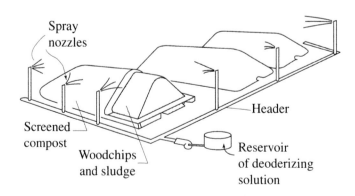

Figure 9-17 Dispensing system for odor-masking agent.

exhaust from the air withdrawal system is treated by a scrubber or flare. Figure 9-16 illustrates the containment and treatment approach to odor management.

Reodorization, or "masking," can be accomplished by using the same containment facility illustrated in Figure 9-16 for containment and treatment. The difference is that a masking agent is added to the exhaust stream when needed. Another approach is to add the masking agent to the air stream blown into the composting mass, if that is the method used to manage the oxygen concentration within the pile. There are a number of masking agents used for this purpose, most of them proprietary. Still another approach that has been successful is to use a misting system to dispense a masking agent onto the compost pile. Figure 9-17 illustrates a misting system that has been used successfully to manage odor problems at compost facilities.

Other candidates for application of composting technology for treating solid wastes from industry include meat processing wastes, feed lot wastes, vegetable processing wastes, and fish processing wastes.

Bibliography

Archinger, W. C., and L. E. Daniels. "An Evaluation of Seven Incinerators." New York: *Proceedings of National Incinerator Conference.* ASME, 1970.

American Society of Heating, Refrigerating, and Air Conditioning Engineers. *ASHRAE Handbook and Product Directory.* New York: 1977.

Baron, J. L. et al. *Landfill Methane Utilization Technology Workbook.* U.S. Department of Energy. CPE-8 101. Washington, DC: 1981.

Baumeister, T. (ed.) *Marks' Standard Handbook for Mechanical Engineers.* 8th ed. New York: McGraw-Hill, 1978.

Brunner, D. R., and D. J. Keller. "Sanitary Landfill Design and Operation." U.S. Environmental Protection Agency Report No. SW65ts. Washington, DC: 1972.

Campbell, J. A. "Waste Fuel Densification: Review of the Technology and Application." *Proceedings of the International Conference on Prepared Fuels and Resource Recovery Technology.* ANL/CNSV-TM-60. Argonne, IL: Argonne National Laboratory, 1981.

Canter, L. W., and R. C. Knox. *Groundwater Pollution Control.* Chelsea, MI: Lewis Publishers, 1985.

General Electric. *Solid Waste Management: Technology Assessment.* Schenectady, NY: 1975.

Hasselriis, F. et al. "Eco-Fuel II: The Third Generation." *Proceedings of the International Conference on Prepared Fuels and Resource Recovery Technology.* ANL/CNSV-TM-60. Argonne, IL: Argonne National Laboratory, 1981.

Hasselriis, F. *Refuse Derived Fuel Processing.* Woburn, MA: Butterworth Publishers, 1984.

Hollander, H. I., and W. A. Sanders II. "Biomass: An Unlimited Resource." *Consulting Engineer* 55 (1980).

Keitz, E. "Profile of Existing Hazardous Waste Incineration Facilities and Manufacturers in the United States." U.S. Environmental Protection Agency. EPA 600/2-84/ 052. Cincinnati, OH: 1984.

Lawrence Berkeley Laboratory. *Characterization of Solid Waste Conversion and Cogeneration Systems.* U.S. Department of Energy. DOE/EV-0105. Washington, DC: 1980.

Mitchel, G. et al. *Small-Scale and Low-Technology Resource Recovery Study.* U.S. Environmental Protection Agency. Washington, DC: 1979.

National Academy of Sciences of the National Research Council. *The Recovery of Energy and Materials from Solid Waste.,* Washington, DC: National Academy Press, 1981.

National Center for Resource Recovery. *Waste-to-Energy Compendium.* DOE/CE/

20167-05. U.S. Department of Energy. Washington, DC: 1981.

Niessen, W. R., and S. H. Chansky. "The Nature of Refuse." *Proceedings of National Incinerator Conference.* New York: ASME 1970.

Reilly, T. C., and D. L. Powers. "Resource Recovery Systems, Part 2: Environmental, Energy and Economic Factors." In *Solid Wastes Management.* New York: June 1980.

Risk Reduction Engineering Laboratory. *Landfill Leachate Clogging of Geotextile (and Soil) Filters.* EPA/600/2-91/025. Cincinnati, OH: 1991.

Schroeder, P. R., A. C., Gibson, J. M. Morgan, and T. M. Walski. *The Hydrologic Evaluation of Landfill Performance (HELP) Model, Vol. 1—Users Guide for Version I* (EPA/530-SW-84-009) and *Vol. II—Documentation for Version I* (EPA/530-SW-84-010). U.S. Army Engineer Waterways Experiment Station. Vicksburg, MS: 1984.

Schroeder, P. R. et al. "The Hydrologic Evaluation of Landfill Performance (HELP) Model." U.S. Environmental Protection Agency. U.S. Army Engineer Waterways Experiment Station. Vicksburg, MS: 1988.

Schroeder, R. L. et al. "Raytheon Service Company Experience and Programs in Resource Recovery." *Proceedings of the International Conference on Prepared Fuels and Resource Recovery Technology.* ANL/CNSW-TM-60. Argonne, IL: Argonne National Laboratory, 1981.

Shepherd, K. R. "RDF Fuels in a Waste-to-Energy System—The Teledyne National Experience." *Proceedings of the International Conference on Prepared Fuels and Resource Recovery Technology.* ANL/CNSV-TM-60. Argonne, IL: Argonne National Laboratory, 1981.

Sommerlad, R. E. "Considerations Affecting Dedicated Boiler Design for Refuse Derived Fuels (RDF)." *Proceedings of the International Conference on Prepared Fuels and Resource Recovery Technology.* ANL/CNSV-TM-60. Argonne, IL: Argonne National Laboratory, 1981.

Tchobanoglous, G., H. Theisen, and R. Eliassen. *Solid Wastes: Engineering Principles and Management Issues.* New York: McGraw-Hill, 1977.

Theodore, L., and J. Reynolds. *Introduction to Hazardous Waste Incineration.* New York: Wiley-Interscience, 1987.

Tittlebaum, M. E. "Organic Carbon Content Stabilization through Landfill Leachate Recirculation." *Journal Water Pollution Control Federation* 54, No. 5 (1982).

U. S. Environmental Protection Agency. *Development Document for Proposed Effluent Limitations, Guidelines, and Standards for the Landfills Category.* EPA 821-R-97-022, Washington, DC: U.S. Government Printing Office, 1997.

U.S. Environmental Protection Agency. "Interim Standards for Owners and Operators of New Hazardous Waste Land Disposal Facilities." *Code of Federal Regulations.* Title 40, Part 267. Washington, DC: U.S. Governmental Printing Office, 1981.

U.S. Environmental Protection Agency. *Development Document for Proposed Effluent Limitations, Guidelines, and Standards for the Landfills Category.* EPA 82 1-R-97-022. Washington, DC: U.S. Government Printing Office, 1997.

U.S. Environmental Protection Agency. *Economic and Cost-Effectiveness Analysis for Proposed Effluent Limitations, Guidelines, and Standards for the Landfills Category.* EPA 821-B-97-005. Washington, DC: U.S. Government Printing Office, 1997.

U.S. Environmental Protection Agency. *Environmental Assessment for Proposed Effluent Limitations, Guidelines, and Standards for the Landfills Category.* EPA 821-B-97-007. Washington, DC: U.S. Government Printing Office, 1997.

U.S. Environmental Protection Agency. *Environmental Assessment for Proposed Effluent Limitations, Guidelines, and Standards for the Landfills Category.* EPA 821-B-97-007. Washington, DC: U.S. Government Printing Office, 1997.

U.S. Environmental Protection Agency. *Guide to the Disposal of Chemically Stabi-*

lized and Solidified Waste. SW-872. Cincinnati, OH: 1980.

U.S. Environmental Protection Agency. *Handbook—Remedial Action at Waste Disposal Sites* (revised). EPA 625/6-85/006. Washington, DC: U.S. Government Printing Office, January 1987.

U.S. Environmental Protection Agency. *Procedures Manual for Groundwater Monitoring at Solid Waste Disposal Facilities.* EPA 530/SW-61 1. Washington, DC: U.S. Government Printing Office, 1977.

U.S. Environmental Protection Agency. *Process Design Manual for Sludge Treatment and Disposal.* EPA 625/1-74/006. Washington, DC: 1974.

U.S. Environmental Protection Agency. *Protecting Health and Safety at Hazardous Waste Sites: An Overview.* Technology Transfer. EPA 625/9-85/006. Cincinnati, OH: 1985.

U.S. Environmental Protection Agency. *Refuse-Fired Energy Systems in Europe: An Evaluation of Design Practices, An Executive Summary.* Publication SW 771. Washington, DC: U.S. Government Printing Office, 1979.

U.S. Environmental Protection Agency. *Statistical Support Document for Proposed Effluent Limitations, Guidelines, and Standards for the Landfills Category.* EPA 821-B-97-006. Washington, DC: U.S. Government Printing Office, 1997.

U.S. Environmental Protection Agency. Center for Environmental Research Information. *Protection of Public Water Supplies from Ground-Water Contamination.* EPA 625/4-85/016. Cincinnati, OH: 1985.

U.S. Environmental Protection Agency. Center for Environmental Research Information. Office of Research and Development. *Requirements for Hazardous Waste Landfill Design, Construction, and Closure.* EPA/625/4-89/022. Cincinnati, OH: 1989.

U.S. Environmental Protection Agency. Office of Research and Development. *Design and Construction of RCRA/CERCLA Final Covers.* EPA/625/4-91/025.

Washington, DC: U.S. Government Printing Office, 1991.

U.S. Environmental Protection Agency. Municipal Environmental Research Laboratory. *Management of Hazardous Waste Leachate.* SW-87 1. Cincinnati, OH: 1980.

U.S. Environmental Protection Agency. Office of Solid Waste and Emergency Response. *Design, Construction, and Evaluation of Clay Liners for Waste Management Facilities.* EPA/530/SW-88/007F. Washington, DC: U.S. Government Printing Office, 1997.

U.S. Environmental Protection Agency. Office of Solid Waste and Emergency Response. *Technical Guidance Document: Final Covers on Hazardous Waste Landfills and Surface Impoundments.* EPA/530-SW-89/047F. Washington, DC: U.S. Government Printing Office, 1999.

U.S. Environmental Protection Agency. Risk Reduction Engineering Laboratory. Center for Environmental Research Information. Office of Research and Development. *Guide to Technical Resources for the Design of Land Disposal Facilities.* EPA/625/6-88/018. Cincinnati, OH: 1988.

Velzy, C. O. "Energy from Waste Plants—An Overview of Environmental Issues." Presented at the Third U.S. Conference of Mayors/National Resource Recovery Association Conference, Washington, DC, 1984.

Velzy, C. O. "State of the Art of Emissions Control." Presented at New York Academy of Sciences, New York, 1984.

Velzy, C. O. "The Enigma of Incinerator Design." *Incinerator and Solid Waste Technology.* New York: ASME, 1975.

Velzy, C. R., and C. O. Velzy, "Incineration," Sec. 7. In *Marks' Standard Handbook for Mechanical Engineers.* 8th ed. New York: McGraw-Hill, 1978.

Vence, T. D., and D. L. Powers, "Resource Recovery Systems, Part 1: Technological Comparison." In *Solid Wastes Management.* New York: 1980.

Vesiland, P. A., and A. E. Rimer. *Unit Operations in Resource Recovery Engineering.* Englewood Cliffs, NJ: Prentice Hall, 1981.

Wilson, E. M. et al. *Engineering and Economic Analysis of Waste to Energy Systems.* U.S. Environmental Protection Agency. EPA-600/7-78-086. Washington, DC: 1978.

10 Wastes from Industries (Case Studies)

General

Wastes from industries include solid wastes, air pollutants, and wastewaters. These separate categories of wastes are regulated by separate and distinct bodies of laws and regulations. Solid wastes are regulated by RCRA, CERCLA, SARA, HSWA, and other federal laws and regulations, as well as certain state laws and regulations. Air pollutants are regulated by the Clean Air Act (as well as other federal and certain state laws and regulations). Wastewater discharges are regulated by the Clean Water Act, as amended (as well as other federal and certain state laws and regulations). However, the three categories of wastes are closely interrelated, both as they impact the environment and as they are generated and managed by individual industrial facilities. As examples, certain solid wastes handling, treatment, and disposal facilities are themselves generators of both air discharges and wastewaters. Bag houses used for air pollution control generate solid wastes; air scrubbers and other air pollution control devices generate both liquid and solid wastes streams; and wastewater treatment systems generate sludges as solid wastes and release volatile organics and aerosols as air pollutants.

The total spectrum of industrial wastes, then, must be managed as a system of interrelated activities and substances. Materials balances must be tracked, and overall cost effectiveness must be kept in focus. Moreover, as discussed more fully in Chapter 4, the principles of pollution prevention must be implemented to the most complete extent practicable. All wastes must be viewed as potential resources. In some cases, wastes can be used as raw materials for additional prod-

ucts, either on site or at other industrial facilities. In other cases, wastes can be used as treatment media for other wastes. In all cases, the generation of wastes must be minimized by employment of scrupulous housekeeping; aggressive preventive maintenance; substitution of nonhazardous substances for hazardous substances; and prudent replacement of old, inefficient process technology with technology that results in generation of less pollutants.

The objective of this chapter is to present 13 industries as representative of many more industries regarding types of manufacturing processes; generation of solid wastes, air discharges, and wastewaters; strategies for pollution prevention; and wastes handling, treatment, and disposal technologies. A general description of manufacturing processes is given so as to show the "roots" of each significant solid, airborne, and waterborne pollutant. Then techniques for wastes minimization as part of an overall pollution prevention program are discussed. Finally, methods for "end of pipe" treatment are presented and discussed.

Discussion of the 13 representative industries is preceded by a discussion of three processes that are common to many different industries: vapor degreasing, chemical descaling (pickling), and rinsing. Pickling is a process used in the metal working industry, in which acid is used to remove foreign substances from the metal surface. Vapor degreasing is used by nearly all industries that place a coating on metal as part of the manufacturing process. It is very common for these industries to require extensive cleaning of the metal surfaces before the coating is applied, and vapor degreasing is very often included in the cleaning process. As well, the process of rinsing is used by a very large number of industries to remove residual substances from one manufacturing process in preparation for another.

Chemical Descaling

Many manufacturing processes that involve metal parts include a step for removing the products of corrosion from those metal parts. A common method for doing this is to immerse the parts in a bath of aqueous solution of acid or molten alkali. If acid is used, it is often sulfuric because of its relatively low cost. If a caustic bath is used, it is usually sodium hydroxide. This process is known as "pickling," and is usually followed by a rinse to remove residual acid or caustic acid. Other chemical descaling agents include aqueous solutions of nitric and hydrofluoric acids, molten salt baths, and various proprietary formulations.

When sulfuric acid is used to descale ferrous metals, some of the iron dissolves in the acid solution and exists there as $FeSO_4$. As the quantity of dissolved $FeSO_4$ builds up over time, the solution loses its effectiveness and must be renewed, by either batch replacement or continuous makeup and overflow. The spent solution must be treated before disposal. Similar acid salts result from the use of acids other than sulfuric to descale metals other than ferrous.

As with all other immersion processes, the pickling solution that fails to drip back into the pickling bath after the piece is removed (dragout) must be dealt with. The quantity of dragout can be minimized by use of air squeegees or longer drip times, or both.

Treatment of spent pickling solutions involves neutralization and precipitation of dissolved metals. The metals precipitate as a consequence of the neutralization; however, the precipitation process must be well managed to prevent loss to the effluent due to noncompletion of the precipitation and/or solids removal process. If the precipitation/ solids removal process does not take place

before discharge, it will take place after discharge, potentially causing problems of toxicity or, at a minimum, discharge permit excursions.

Degreasing

Industries that are engaged in the working, forming, plating, or welding of metals almost always apply and then remove one or more oily or greasy substances to the metal surfaces during the manufacturing process. As a result, virtually all industries that apply a coating to metal in the course of their manufacturing activities operate one or more processes that remove greasy substances that were applied to prevent corrosion. A very commonly used device for this purpose is the vapor degreaser. Simple immersion tanks, otherwise known as "dip tanks," are also common. Vapor degreasers consist of the following elements:

- A heated tank to contain and volatilize the liquid degreaser substance
- An open chamber to contain the vapors above the heated tank
- A system to condense the vapors
- A system of hangers or baskets, typically mounted on a moving conveyor overhead, to hold the objects to be degreased ("the work")

American Society for Testing Materials (ASTM) publication No. D 3698 – 99, entitled *Standard Practice for Solvent Vapor Degreasing Operations*, defines "solvent vapor degreasing operations" as "the process by which materials are immersed in vapors of boiling liquids for the purpose of cleaning or altering their surfaces, and are subsequently removed from the vapors, drained, and dried in a solvent vapor degreaser. This publication defines a "solvent vapor degreaser" as "a solvent and corrosion-resistant tank with a heated solvent reservoir or sump at the bottom, a condensing means near the top, and freeboard above the condensing means, in which sufficient heat is introduced to boil the

Figure 10-1   A typical vapor degreaser (courtesy of Greco Brothers Incorporated).

solvent and generate hot solvent vapor. Because the hot vapor is heavier than air, it displaces the air and fills the tank up to the condensing zone. The hot vapor condenses on the cooled condensing means, thus maintaining a fixed vapor level and creating a thermal balance."

Figure 10-1 shows a photograph of a typical vapor degreaser. The system or "means" to cool the space above the heated tank typically consists of coils containing cold water or a refrigerant. As the objects to be degreased move through the open chamber, vapors of the degreasing substance condense on the (relatively cold) objects. The condensed degreasing liquid dissolves grease on the surfaces of the objects. As the degreasing liquid drips off the objects, it drops into a system of troughs that carries it to a reservoir, from which it then flows back into the tank. Anything that it dissolved or suspended from the surfaces of the objects is carried into the reservoir (and then, in the case of a portion of this material, into the tank with the condensed solvent).

Because only the vapors from the heated tank contact the objects to be degreased, foreign material from previously cleaned objects is not brought into contact with the objects being cleaned. As dirt and dissolved grease accumulate in the heated tank, they must be maintained below a desired concentration by adding fresh degreaser. Some systems use a batch process where the tank is dumped periodically and refilled with fresh degreaser; others use a continuous overflow from the tank and continuous makeup with fresh degreasing substance.

Water also condenses from the ambient air in the (relatively cool) space above the heated degreasing tank. Some of this condensation takes place as the humid shop air contacts the (relatively cold) metal objects entering the vapor degreaser. Some simply condenses in the open space of the upper portion of the degreasing chamber. This condensed water gradually accumulates in the tank, along with accumulating oil, grease, and dirt. As this water accumulates, it begins to interfere with the effectiveness of the solvent. Water must, therefore, be maintained below a certain level to achieve satisfactory operation.

Other sources of water contamination include:

- Water brought in on the surfaces of the work

- Water remaining in the degreaser after cleaning
- Leaks in the water coils or the steam coils

There are two methods by which this water is removed from the vapor degreasers. The first is by use of a separator that simply provides a quiescent volume within which water that is not dissolved in the solvent separates from it under the influence of gravity. The second is by use of a still that makes use of the large difference between the boiling temperatures of the water and the solvent. When the separated water and solvent exit the gravity separator, there is water dissolved in the solvent phase (as a contaminant), and there is solvent dissolved in the water phase (also as a contaminant). Stills can then be used to remove the water contamination from the solvent and to remove the solvent contamination from the water.

Many vapor degreaser systems include what is known as an "auxiliary still" for the purpose of removing water contamination from the solvent before returning the (reclaimed) solvent to the heated tank.

Desirable characteristics of the degreasing substance include high solubilities of the greasy materials used to preserve metals and a high degree of volatility and chemical stability. Substances used as degreasers have included methylene chloride, methyl chloroform, perchloroethylene, 1,1,1 trichloroethane, trichloroethylene, perchloroethane, n-propyl bromide, and trichlorotrifluoroethane.

Periodically, the vapor degreaser must be emptied, the sludge cleaned out, and the spent degreaser disposed of. Disposal of spent degreasing liquids is best done by fractional distillation to recover relatively pure degreasing substance for reuse. The still bottoms should be incinerated or placed in a secure landfill. Spent degreaser liquid should never be landfilled directly because of its toxic properties and the virtual impossibility of containing it "forever." As well, the sludge that is cleaned from the degreaser must be disposed of by means similar to those used for the spent solvent.

Residual degreasing substance that fails to drip back into the degreaser before the degreased object leaves the process (dragout) is a toxic contaminant that must be dealt with. It can be minimized by use of air squeegees and/or longer drip times.

Certain vapor degreasing facilities are subject to air discharge restrictions of the National Emission Standards for Halogenated Solvent Cleaning (Halogenated Solvent Cleaner NESHAP). These restrictions are contained in 40 C.F.R. Part 63, Subpart T. It is the responsibility of the user to determine the applicability of these regulations.

Rinsing

Rinsing processes are very common in many industries. Typically, parts and pieces undergoing manufacturing processes are rinsed in a water bath after each process that involves immersion in an aqueous solution. For instance, a typical electroplating process involves immersion of the object being plated (the work), typically a metal, in an aqueous solution (plating bath) of a salt of the plating material, usually a different metal. In order to dissolve the metal salt being plated, the plating bath must be highly acidic. It will, therefore, be highly corrosive. For this reason, the plating bath is normally followed immediately by two or three rinse tanks in succession.

After completing the process in the plating bath, the work is immersed in the first water bath, where most of the residuals from the plating bath are removed. Since these residuals simply dissolve in the rinse water, they will be present on the surface of the rinsed work in proportion to their concentration in the rinse water. Therefore, it is necessary to rinse the work in a cleaner bath after the first rinse, and so on, until the work is sufficiently clean. Each time an object is removed from an emersion tank it brings a certain amount of the bath solution (dragout) with it. Dragout contaminates the next emersion solution

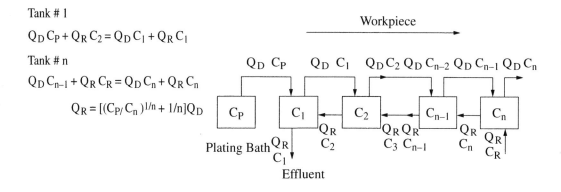

Tank # 1

$$Q_D C_P + Q_R C_2 = Q_D C_1 + Q_R C_1$$

Tank # n

$$Q_D C_{n-1} + Q_R C_R = Q_D C_n + Q_R C_n$$

$$Q_R = [(C_P/C_n)^{1/n} + 1/n]Q_D$$

Mass Balance for Counter-Current Rinse System

Figure 10-2 Rinsing sequence for a typical electroplating process (from Bishop © 2000; reprinted by permission of McGraw-Hill, Inc.).

or rinsewater bath. Use of air squeegees and/or longer drip times can minimize dragout. Figure 10-2 shows a typical rinsing sequence for an electroplating process.

It is common to have three rinse tanks in series and advisable to have them set up so that fresh makeup water is added to the final rinse tank, ensuring a concentration of "impurities" sufficiently low that the rinsed work is as clean as required. The final rinse tank should overflow into the preceding rinse tank, and so on, and the overflow from the first rinse tank should be used as makeup water for fresh plating bath solution. This setup is referred to as "countercurrent rinsing" and has largely replaced the older practice of forward flow rinsing, where fresh rinse water was added to the first rinse tank, which overflowed into the second rinse tank, and so on. Since it is the concentration of contaminants in the final rinse tank (even when there is only one tank) that governs the required flow rate of the clean makeup water, it follows, then, that the concentration of contaminants will always be lower, for any given flow rate of makeup water, in the second (or final) tank if the clean makeup water is added to the final tank rather than the first tank.

In the forward flow setup, lightly contaminated rinsewater from only the final rinse tank had to be treated before discharge within permit compliance, and spent plating bath solution, heavily contaminated, had to be treated by a separate system. With counter-current rinsing, where the overflow from the first rinse tank is used as makeup water for fresh plating solution, only the spent plating solution has to be treated.

Treatment of spent plating solution or spent rinsewater involves removal of all those substances used in the plating bath plus dissolved ions from the object being plated (the work). The same is true for treatment of spent plating solution, which is typically much lower in volume and, consequently, more concentrated. Technologies for treatment of these wastewaters before reuse or discharge are presented in Chapter 8.

Electroplating of Tin

Tin has been used for centuries as both a primary metal with which to make tools, utensils, and other useful objects and as a cover or "plate" over other metals, such as iron, to protect them from the elements. The tinning of iron, by hammering tin onto the surface of the iron, appeared around the 14th century in Bohemia. The largest simple use for tin in modern times has been as a coating for steel to make tin plate, used in the manufacture of the "tin can." Over half the tin used in the United States is for this purpose. Other uses for tin include dairy and other food handling equipment; washing machine parts; radio and electronic components; as a coating on refrigerator evaporators; as a coating on copper wire; as a component of piston rings; and for bearing surfaces. Tin is of great importance to the electronics industry, which uses coatings of tin and tin-rich (greater than 50% tin) tin-lead alloys ("solder plate"). Tin's resistance to corrosion and chemical etchants, as well as ease of soldering, are highly desirable.

The favorable characteristics of tin as a plating material or coating are the following:

- Nontoxic properties
- Resistance to corrosion
- Ease of soldering
- Ductility; ease to work with

Tin and its inorganic compounds and alloys are essentially nontoxic. This characteristic is extremely important to the food industry, the prepared foods industry, the dairy industry, and the environmental engineer or scientist. Considering that the elements that are closest to tin on the periodic chart exhibit considerable toxicity, the nontoxic nature of tin is somewhat surprising. Elements near tin include arsenic, lead, cadmium, antimony, and thallium. The only hazardous characteristics exhibited by inorganic tin substances are a result of properties other than toxicity; for instance, stannous chloride is acidic, and the various stannates are strongly alkaline. Although there are many organic compounds of tin, none is present in metal finishing operations.

Production Processes

Electroplating is by far the most widely encountered manufacturing process for tin. The basic tin electroplating process is illustrated diagrammatically in Figure 10-3. The object to be plated, called "the work," is immersed in an electrolyte (solution of metal salts). The plating metal (tin) is also immersed in the electrolyte or "bath." Using DC current, the work is made the cathode and the plating metal (tin, in this case) is made the anode. The electric current causes ions of tin to be dissolved into the bath and then to be deposited ("plated out") on the surface of the work. The quality, thickness, and other characteristics of the resulting tin plate are dependent upon the current amperage, which is a function of the electrical conductivity of the electrolyte and the applied voltage. Also affecting the quality and thickness of the tin plate product are plating time and the quality of preparation of the work.

Preplating

The surface of the work must be very clean, as well as "activated," prior to plating. Consequently, as illustrated in Figure 10-3, the first five steps in a typical tin-plating operation are for the purpose of thoroughly cleaning and activating the surface of the work. The actual mechanics of the operation of each of these five steps differ somewhat, depending on the characteristics of the work. As well, two or more of the steps may be combined for certain types of work. Of interest to the environmental engineer or scientist, each of the preplating steps (as described below) is followed by a rinse with water, using one or more rinse tanks or sprays.

Preclean

The first step in preparing the work for tin plating is to remove all gross amounts of oil, grease, and dirt. Depending on the work and its condition on arrival at the plating shop, precleaning can include wiping, air spraying, brushing, vapor degreasing, emulsifiable solvent, solvent spray, hot alkaline spray, invert emulsion cleaners, or other processes, in order to decrease, to the extent practicable, the load on the following cleaning processes. Various solvents are used for precleaning, including mineral spirits, kerosene, and chlorinated hydrocarbons such as perchloroeth-

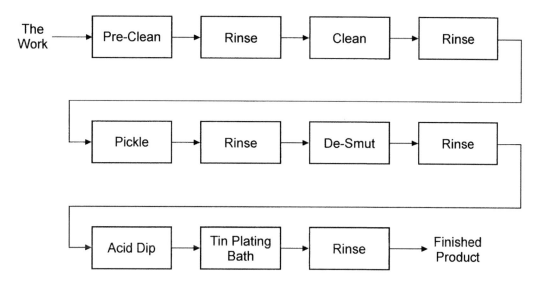

Figure 10-3 Schematic diagram of the tin-plating process.

ylene. Extreme measures are required in handling these solvents and disposing of the residuals because of their toxic nature. When vapor degreasing is used, the vapors as well as the liquids must be contained. An advantage to the volatile solvents is that they can be reclaimed on site or by use of a commercial reclaiming service, and the concentrated residuals from the reclaiming process can be destroyed or encapsulated for safe landfill disposal.

Clean

The second or "clean" step removes all residuals from the preclean step and removes any substances that were resistant to removal by vapor degreasing, solvent emulsion, or other process or processes used in the preclean step. Typically, the "clean" step makes use of alkaline solutions to remove soils that have been softened but not completely removed by the precleaning step. Cleaning solutions can be made up using various alkali soaps, detergents, chelating agents, complexing agents, and surfactants. Cleaning processes include manual brushing, mechanical brushing, soaking, spraying, or high-pressure spraying. Heat is often used. For instance,

one process involves immersion of the work in a detergent solution and heating the solution to a rolling boil, followed by soaking for a period of time. Mechanical agitation, ultrasonic energy, and high-pressure under-surface spraying are options for use in lifting the soil off the surface of the work. Another option, alkaline electrolytic cleaning, results in the production of bubbles of hydrogen and oxygen gas, which provide a scrubbing action. After rinsing, the precleaned and cleaned work should be completely free of any type of soil other than chemical derivatives of the work itself, such as oxides (rust) and scale.

Pickling

Pickling is an acid or alkaline (usually acid) dip process common to many metal cleaning operations. Pickling has the purpose of dissolving, by use of strong acids, corrosion products (rust, in the case of ferrous metals) or scale. Pickling processes vary from an acid dip to anodic treatment in a dry, alkaline salt. Sulfuric acid is widely used, due to its relatively low cost; other acids include hydrochloric, phosphoric, and nitric.

Desmutting

When steel is the material from which the work is made, a carbide film must be removed after pickling. This carbide film, referred to as "smut," must be removed prior to plating. Several processes are available for this purpose, including anodic treatment in a dry, alkaline salt. If this type of treatment is used as the pickling step, desmutting will occur simultaneously.

Acid Dip

The final step in cleaning the work is an acid dip in relatively dilute sulfuric or hydrochloric (or other) acid to remove residual alkali, remove any remaining oxides, and to "micro-etch" the surface. The microetching is referred to as "activation" because of its effect of enhancing adhesion of the plating metal.

Plating Baths

Plating baths for tin are either acid or alkaline. There are advantages and disadvantages to each, and the choice depends on the specific requirements of a given application. Acid baths are made up from stannous sulfate or stannous fluoborate. When these substances are used, the process can take place at room temperature, the current density can be relatively low, and deposition of tin onto the work takes place from the stannous, or bivalent (+2), state. Acid bath plating has the disadvantages of requiring so-called "addition agents," and there is less "throwing power." Throwing power refers to the uniformity of deposition of the plating metal over the entire surface of the work. Another characteristic of acid bath plating is the brightness of the tin plate. This represents no advantage when the tin plate is on the inside of a food can, but in other situations it might represent enough of an advantage to tip the scale in favor of the acid bath process over the alkaline bath process.

Alkaline bath plating uses sodium or potassium stannate as the base material. A wide range of concentrations can be used as a means of regulating the rate of plating, and the formulation is relatively simple. No addition agents are needed. Also, alkaline tin plating baths are characterized by having excellent throwing power. Another advantage is that insoluble anodes can be used, which means that all of the plated tin has its origin in the stannate ions. As well, hot stannate baths are known for their ability to tolerate imperfect cleaning of the work. Disadvantages of alkaline bath plating compared with acid bath plating include the fact that the tin is plated from the stannic, or trivalent (+3), state, which requires higher current density. Another disadvantage is the requirement for the bath to be heated. The hot bath is not suitable for plating delicate work, such as printed circuit boards.

DuPont has developed an alternative bath specifically for the purpose of "electro-tinning" steel strips on a continuous throughput basis. Stannous chloride is used to provide the tin, and fluoride salts provide conductivity. This bath is known as the "halogen electrolyte bath. Table 10-1 lists components of typical plating baths, including acid, alkaline, and halogen baths.

Rinse

The sequential steps of the tin-plating process must each be followed by a thorough and complete rinse prior to the next sequential step. For instance, if the work is not rinsed completely after the acid dip and prior to an alkaline plating bath, the alkalinity of the plating bath will be too quickly neutralized. As another example, if the cleaning solution is not completely rinsed from the work, the detergents and contaminants from the wash water will contaminate the following bath or baths. Some rinse steps involve only one tank. The work is dipped in the bath and then extracted from it. As the work is dipped in, any dragout from the previous process contaminates the rinse bath. When the work is extracted, it drags out a certain quantity of the rinse water with it, along with whatever contaminants are in the rinse water. Therefore, to keep the concentration of these contaminants low, fresh water must be con-

Table 10-1 Components of Typical Plating Baths[*]

Plating Bath Composition	
Alkaline Stannate Baths	
Bath	Makeup (g/L)
Potassium stannate	100
Free potassium hydroxide	15
Potassium stannate	210
Free potassium hydroxide	22
Alkaline Stannate Baths	
Bath	Makeup (g/L)
Potassium stannate	420
Free potassium hydroxide	22
Sodium stannate	100
Free sodium hydroxide	10
Acid Baths	
Stannous Sulfate Bath	
Bath	Makeup (g/L)
Stannous sulfate	72
Free sulfuric acid	50
Pheno-/cresolsulfonic acid	40
Gelatin	2
Acid Baths	
Stannous Sulfate Bath	
Bath	Makeup (g/L)
Beta-napthol	1
Stannous Fluoborate Bath	
Bath	Makeup (g/L)
Stannous fluoborate	200
Fluoboric acid	150
Gelatin	6
Beta-napthol	1
Halogen Bath	
Bath	Makeup (g/L)
Stannous chloride	63
Sodium fluoride	25
Potassium bifluoride	50
Sodium chloride	45
Addition agents	2

[*]Lowenheim, 1978.

tinually added to the rinse tank. Consequently, an amount of flow leaves the rinse tank that is equal to the inflow minus the dragout. This "outflow" water constitutes a waste stream unless a use can be found for it.

One way to reduce the quantity of outflow is to use two rinse tanks in series, rather than a single tank. With two rinse tanks in series, the first tank can be allowed to build up in concentration of contaminants to considerable strength, and the work can still be rinsed completely with much less makeup (and effluent) flow than with a single rinse tank. In fact, only if the first rinse tank builds up to a concentration equal to that in the preceding process tank will the rate of inflow of clean makeup water need to equal that of the single rinse tank.

There is a choice to be made regarding the arrangement of inflow and outflow of clean makeup and contaminated rinse waters when two or more rinse tanks are arranged in series. The clean makeup water can flow into the first tank following the process tank, as shown in Figure 10-4. Also illustrated in Figure 10-4, rinse water outflow, or wastewater, flows out from the second rinse tank. This arrangement is called "forward flow" makeup water. The second, and by far better, choice is to employ "counter-current" makeup water flow. In this arrangement, clean makeup water is added to the second of the two tanks in series (or final tank, if more than two tanks in series are used). Since it is the concentration of contaminants in the final rinse tank (even when there is only one tank) that governs the required flow rate of the clean makeup water, it follows, then, axiomatically, that the concentration of contaminants will always be lower, for any given flow rate of makeup water, in the second (or final) tank if the clean makeup water is added to the final tank rather than the first tank.

Where rinsing is the final step, it is necessary to dry the plated product completely before packing for shipping, to prevent staining, corrosion, or other discoloration.

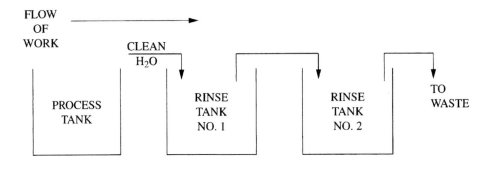

FORWARD FLOW MAKEUP

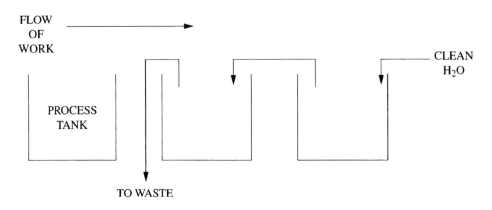

COUNTER-CURRENT FLOW MAKEUP

Figure 10-4   Alternative arrangements for rinsing.

Sources and Characteristics of Wastes

Solid Wastes

Solid wastes from the tin-plating industry include sludges from baths and from air and wastewater treatment facilities. Also included are unrecyclable trash from packaging and shipping and construction debris from general repair, rebuilding, expansion, and remodeling that take place over time. Sludges from baths include the preclean, clean, pickle, and desmut baths, and other devices, as well as the plating baths. The cleaning baths contain products of corrosion and scaling, as well as oils, greases, and general dirt.

The still bottoms (from the stills that are used to regenerate volatile organic solvents used for cleaning oils and greases from the work that is being prepared for plating) are solid wastes that must be managed as hazardous wastes.

Airborne Wastes

Air pollutants that must be managed include fumes from degreasing, etching, and plating operations, plus the air discharges from boilers. Wet scrubbers are the standard devices used to control fumes. The wet scrubbers convert air pollutants to water pollutants, with the attendant production of a certain quantity of solid wastes in the form of sludges.

Waterborne Wastes

There are three primary sources of wastewater from tin-plating facilities: (1) rinse waters, (2) cleaning and conditioning solutions, and (3) plating solutions. Rinse waters are normally produced on a continuous basis, due to continuous fresh water makeup and consequent effluent overflow. Periodically, rinse baths are dumped in order to clean the tanks.

Cleaning and conditioning solutions and plating solutions are normally discharged on a batch basis, as a result of the periodic need to dump "spent" cleaning and plating baths.

Spent Rinse Waters

As discussed previously and illustrated in Figures 10-2 and 10-3, a rinse step normally follows each cleaning, preparation, and plating step. These rinsing processes can be dip tanks, sprays, or both. Also, dip tanks can be single tanks or multiple tanks in series. Multiple (two or more) tanks can be operated as forward flow or counter-current flow, regarding the makeup of clean water and discharge of contaminated rinse water. As explained previously, for purposes of waste minimization, counter-current flow is highly preferred, to the extent that there is essentially no justification for forward flow.

Contaminated discharges from rinsing processes are basically dilute solutions of the cleaning, conditioning, or plating processes that precede the rinsing process, along with small amounts of other contaminants. For this reason, the most advantageous method for "disposing" of "spent" rinse waters is to use them, to the extent possible, as the source of water for making up new batches of the preceding cleaning, conditioning, or plating solution. The excess rinse waters must then be treated before discharge, but their quantity will be minimized.

Spent Cleaning and Conditioning Solutions

Cleaning and conditioning (conditioning is normally considered part of cleaning) are the first processes used by the tin-plating industry, as illustrated in Figure 10-3. Since the purpose of cleaning is to remove oils, grease, dirt, products of corrosion, and scaling, and to condition or "activate" the surface of the work to bond well with the tin, the spent cleaning and conditioning solutions contain all of the oils, greases, dirt, and products of corrosion and scaling as well as the spent and unspent detergents, solvents, acids, alkalis, and other substances used for cleaning and conditioning. Since many cleaning solutions are proprietary, it often takes some intensive research to determine the individual components in the waste cleaning solutions.

Although not usually present in significant amounts, some spent cleaning and conditioning solutions contain chromium and cyanide. Chromium results from dissolution from carbon steel during strong acid treatment. The chromium content of carbon steel ranges from 0.5% to 1%. Cyanide, in the form of ferocyanide, is sometimes used in tin plating to scavenge other heavy metals from the surface of the work. Also, sodium cyanide is sometimes used as a cyanide dip or as an integral part of one of the cleaning steps to help preserve work that will not be plated immediately.

Spent Plating Solutions

There is great variability in tin-plating operations; therefore, it is not feasible to set forth a "typical" or "average" wastewater. As an example of one industry's wastewater, Tables 10-2 and 10-3 are presented to illustrate the wastewater characteristics from two continuous strip electro-tinning facilities.

Spent plating baths, although considered hazardous in the as-is state because of corrosivity, or possibly other characteristics, are not normally considered toxic. This makes tin-plating waste somewhat unique as a

Table 10-2   Waste Characteristics, Continuous Strip Tin Plating[*]

Parameter	Value (g/L)
pH	3.3
Chloride	2.6
Total iron	0.04
Tin (II)	1.3
Tin (IV)	0.5
Fluoride	2.1

[*]Ellis and Whitton, 1978.

metal-plating waste, simply because, as discussed earlier, tin is not a toxic substance.

Waste Minimization

As is the case with other metal electroplating industries, the key to minimizing the cost for eventual waste treatment and disposal is wastes minimization within an overall pollution prevention program, as illustrated by the following:

- Whenever possible, nontoxic substances should be used for degreasing and cleaning.
- If toxic substances—for instance, chlorinated volatile organics—must be used

for degreasing or other cleaning process, then containment, recycle, and reuse must be practiced to the maximum extent possible.

- Biodegradable detergents should be used.
- Drips must be contained and returned to the source.
- Aggressive maintenance must be practiced to eliminate the occurrence of leaks or other "accidents" that could lead to noncontainment of chemicals and other substances.
- Reconstitution of cleaning baths, acid baths, alkali baths, and plating baths should be done on an as-needed basis according to the work performed, rather than on a regular timing or other schedule.
- Dry methods of cleanup, including brooms, shovels, and dry vacuuming, should be used to the maximum extent.
- Rinsing should be counter-current, with respect to fresh water makeup and spent rinse water overflow.
- Maximum (feasible) time should be provided for dipped work to drip back into the tank from which that work has been extracted. To this end, speed of with-

Table 10-3   Additional Waste Characteristics, Continuous Strip Tin Plating[*]

Parameter[†]	Tin Lines			Average
	4	5	6	—
pH	6.4	4.6	3.9	4.6
Chloride	39.1	207.0	45.5	97.2
Sulfate	104.6	152.00	250.7	169.1
Suspended solids	80.2	129.1	146.0	118.4
Total iron	4.4	21.6	29.0	18.3
Total chromium	20.2	15.5	2.1	12.6
Hexavalent chromium	10.4	0.34	0.0	3.6
Cyanide	0.74	1.06	1.04	0.95
Tin	48.9	122.9	12.4	61.4
Fluoride	27.0	30.5	8.6	22.0

[*]Azad, 1976.
[†]All values except pH in mg/L.

drawal should be minimized. Also, shaking techniques are options to consider.

- Air squeegees should be used to the maximum extent to increase dripping into source tanks, thus preventing dragout and consequent contamination of the next sequential bath or other process.
- Temperature and viscosity of the bath should be included in the variables that can be adjusted to minimize dragout.
- Purchasing should be guided by aggressive selection of raw materials to obtain the cleanest possible materials.
- Purchasing should be guided to demand that the packaging of materials delivered to the plant be recyclable or otherwise of low solid waste nature.
- There should be a constant and consistent program to substitute less polluting and nonpolluting substances for those that require expensive treatment and expensive disposal.
- There should be a constant and consistent program for replacing cleaning, conditioning, plating, and rinsing processes with technologies that inherently generate less wastes having even less objectionable characteristics.
- In concert with the above, there should be a constant and consistent program for replacing process controls, including sensors, microprocessors, and hardware, with the objective of decreasing waste and maximizing retention, containment, recycle, and reuse of all substances.
- Technologies for recovering and regenerating chemicals, as well as separating and removing contaminants, should be aggressively employed. Using reverse osmosis (RO) or ultrafiltration (UF) to remove oils from alkaline cleaning solutions are examples. Centrifugation has also been used for this purpose. Activated carbon can be used to remove organic impurities.

Both filtration and centrifugation produce concentrated impurities that offer the possibility of recovery. If recovery of substances is not feasible, the concentrated impurities are in a form more easily disposed of. Evaporation has also been used with success. If the total aqueous wastes or, alternatively, a side stream, can be reduced in volume by counter-current rinsing—the use of spent rinse waters as makeup for wash and/or plating baths—the cost for energy to evaporate may be less than for other treatment and disposal. The relatively pure water condensed from the vapor can be used as rinse water makeup.

Processes to recover metals from spent plating baths and concentrated rinsing baths by electrolytic techniques have been very well developed. Metals that have been deposited on the cathode are relatively easy to recover. These metals can then be reused in the plating bath. Even though complete recovery is not feasible, the process is very effective in reducing overall costs by reducing costs for the plating chemicals and reducing costs for waste treatment and disposal. Evaporation, ion exchange, and reverse osmosis are additional methods that can be used to enhance recovery by concentrating the metals prior to electrolytic recovery. As a final example, ferrocyanide has been used as a chelating agent in tin plating to effect selective scavenging of other metals in solution.

Wastewater Treatment

When wastes minimization has been implemented to the maximum extent, the contaminants that remain must be treated and disposed of. In fact, a number of the waste minimization methods discussed above can be construed to constitute waste treatment. However, there are other techniques that can be employed that are truly end-of-pipe treatments.

Although some waste streams can be combined and managed as one, regarding waste treatment processes, it is usually advisable to treat certain waste streams separately.

Treatment of tin-plating wastes usually involves removal of oils and greases from the preplating operations, and recovery of tin, and possibly other metals, from the plating bath and rinse wastes. The removal of oils and greases can be done by reverse osmosis, ultrafiltration, or chemical coagulation, followed by dissolved or dispersed air flotation in combination with simple skimming.

As discussed in Chapter 7, recovery of tin and other metals is most often accomplished by alkaline, sulfide, phosphate, or carbonate precipitation. If the wastewaters contain fluoride, use of lime as the precipitating agent will effect removal of fluorides concurrently. If the effluent contains hexavalent chromium, addition of (slightly soluble) ferrous sulfide will effect sulfide precipitation of tin and other metals and at the same time reduce the hexavalent chrome to trivalent chrome (far less toxic). If the pH is maintained between 8.0 and 9.0 during this process, the trivalent chrome will be precipitated as the hydroxide. Thus, tin, other metals, and hexavalent chrome can be removed simultaneously by addition of ferrous sulfide, pH adjustment, slow mixing, sedimentation, and filtration. These combined methods will produce an effluent having metals (tin plus other metals, including trivalent chromium) between 2 and 5 mg/L. Ion exchange can then be employed to reduce the concentrations of these substances to essentially nondetectable levels, as discussed in Chapter 7. The product water can then be returned to the process for use as either plating bath makeup water or rinse makeup water.

Insoluble starch xanthate has been used successfully as a precipitant for tin and other metals over pH levels from 3.0 to 11.0, with optimal effectiveness above 7.0. This process is effective over a wide range of metals concentration levels.

When metals, including tin, are precipitated, that is, removed by a reaction to produce an insoluble compound (for instance, stannous sulfide), the precipitation stage is normally followed by gravity sedimentation, often by use of tube or plate settlers. Because simple precipitation often results in small particles of precipitate that do not settle well, a coagulation step must be added. Coagulation (see Chapter 7) involves the addition of a metal salt or an organic polymer, followed by a very short (15 to 30 seconds) rapid mix and then by slow mixing for a period of 15 to 30 minutes prior to the gravity settling process. These processes combine to produce a large, relatively heavy floc that settles much faster and more completely than the original small precipitated particles. Three distinct processes are involved: (1) precipitation, brought about by addition of the chemical (sodium hydroxide, for instance) that reacts with the target metal ions to produce an insoluble compound (metal hydroxide); (2) coagulation, brought about by addition of the coagulant (metal salt or organic polymer); and (3) flocculation, brought about by the slow mixing process. The result is an effluent that has 5 to 15 mg/L of metal ions. Filtration can reduce the concentration to 2 to 5 mg/L. If it is desired to produce an effluent reliability lower than 5 mg/L of metals, ion exchange must be employed.

The Copper Forming Industry

Copper forming includes rolling, drawing, extruding, and/or forging copper and copper alloys. The products of copper forming vary from wires to brewery kettles.

The raw materials for the copper forming industry are copper bars, square cross-section wire bars, rectangular cakes or slabs, sheets, stripe, and cylindrical billets, all of which are cast in copper refineries. Other metals are often mixed with copper at the refinery to improve corrosion resistance, electrical conductivity, and other properties of the end product of copper forming.

The products made by the copper forming industry can be divided into six categories: plates, sheets, strips, wires, rods, and tubes. Bars and wires make up about 65% of the total, while sheets, strips, and plates account for about 20%. The remaining 15% is made up of tubes and pipes. Plates are usually greater than one-quarter inch thick and are used for the manufacture of processing vessels, heat exchangers, and printing equipment. Sheets are thin plates, and strips are basically sheets with one long dimension. Sheets and strips are both used for roof flashing, gutters, radio parts, and washers. Rods and wires have circular cross-sections and are used for springs, electrical conductors, fasteners, and cables. Tubes and pipes are used for hydraulic lines, or by the plumbing and heating industry.

The Copper Forming Process

Five different processes, plus variations within those five processes, are used to form copper.

Hot Rolling

Hot rolling is carried out at temperatures above the recrystallization temperature of the metal. The recrystallization temperature is that temperature at which the crystal lattice structure of the metal becomes reoriented. Consequently, the metal becomes more workable and ductile.

Cold Rolling

Like hot rolling, cold rolling involves passing the metal between a series of rollers, some of which are opposite others, in such a way as to make this cross-section of the metal piece become ever smaller. Since cold rolling is done at temperatures below the recrystallization temperature, the product is less ductile.

Extension

Extension involves forcing molten copper or copper alloy through an orifice or "die" at temperatures of 1,200°F to 2,000°F.

Forging

Forging involves intermittent application of pressure, as with hammering, to force the metal into a desired shape.

Annealing

Annealing involves heating the copper or copper alloy, often by the combustion of natural gas, for the purpose of reducing stresses introduced into the metal by forging or cold rolling. "Electroneal" units work by passing electrical current through the formed wire. In some cases, a quenching step follows the heating step, with the consequent production of wastewater. Figure 10-5 presents a flow diagram of a typical copper forming process.

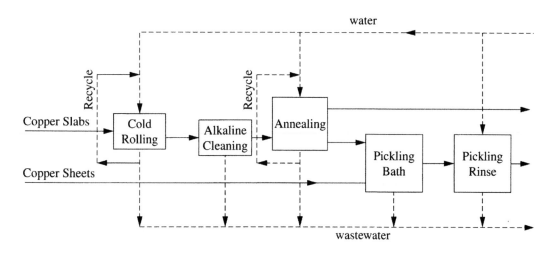

Figure 10-5   Process flow diagram for the copper forming industry.

Additional processes that may take place at a copper forming plant include the following.

Pickling

Pickling, or treatment by immersion in an acid bath, is sometimes done to remove corrosion products before or after the copper forming process.

Alkaline Cleaning

Alkaline cleaning is similar to pickling, except that the bath contains some or all of the following: caustic, sodium, polyphosphate, silicates, sodium carbonates, resin-type soap, organic emulsifiers, wetting agents, and chelating agents.

Solution Heat Treatment

Also called quenching, solution heat treatment involves cooling the copper products after forming by immersion in a continuous flow bath, or by spraying. Water is normally used. An exception is the quenching of products formed by extension, in which case an oil and water solution is normally used.

Rolling and Drawing Process

Lubricants and coolants are used in the rolling process as well as the drawing process to prevent excessive wear on the rollers and to facilitate the drawing process. Normally, dilute solutions of soluble or emulsified oil, or sometimes water alone, are used. These solutions perform the functions of cooling and lubrication. They are normally sprayed on the rolls in the cases of the hot and cold rolling process, and they are sprayed on the rod or wire as it enters the die. Alternately, the die may be immersed in the oil/water mixture, which is cooled externally, in the drawing process. These solutions are often treated (by filtration, for instance) and reused, but there is always a blow-down (constant removal of a small portion of the flow) that requires management as a waste stream, as well as periodic dumps that are replaced by fresh mixtures.

Solution Heat Treating and Annealing Quenches

There are both continuous and periodic discharges of wastewater streams, from two different types of heat treatment quenches used for cooling the copper product after it exits the forming machine and from the quenches following the annealing process. These wastewater streams are generally similar to one another in characteristics, but normally have varying substance concentrations. These waste streams are usually continuous flow as well as occasional batch dump, as described above for the rolling and drawing process lubricants and coolants.

Alkaline Cleaning Rinse and Baths

Alkaline cleaning, which normally precedes the annealing process, has the purpose of removing lubricants, tarnish, and dirt. Therefore, all of these substances, including the substances added into the alkaline cleaning baths, are present in the waste streams from those processes. The alkaline cleaning bath may contain some or all of the following: caustic, sodium polyphosphate, silicates, sodium carbonates, soaps, aquatic emulsifiers, wetting agents, and chelating agents.

Pickling Baths, Rinses, and Fume Scrubbers

Pickling is basically a cleaning process. Acid solutions are used to remove oxides from the surface of the metal. In addition, other substances on the surface of the metal will be removed, in whole or in part, and will therefore be present in the waste stream. In particular, metal sulfates accumulate in pickling baths when sulfuric acid is used as the pickling agent, as is generally the case. For instance, copper sulfate forms when copper is pickled, as follows:

$$CuO + H_2SO_4 \rightarrow CuSO_4 + H_2O \qquad (10\text{-}1)$$

Equation 10-1 shows that as the pickling process proceeds, the acid is gradually depleted and copper sulfate builds up. For this reason, makeup acid must be added, either continuously or periodically.

When the metal parts are taken out of the pickling bath and placed in a rinse tank, dragout transfers a certain amount of the pickling solution to the rinse tank. Dragout is the reason that the treated rinse waters have the same substances as the pickling bath, except that they are far more dilute, and it is the basic reason that these treated rinse waters should be used as makeup water for fresh pickling solution in order to reduce overall wastewater discharge.

Generation of Wastes

Lubrication, cooling, and blow-down from wet scrubbers are the most important sources of wastes that require management in the copper forming industry.

Solid Waste

Solid waste streams include ordinary trash from shipping and packaging, sludges from copper forming processes, and sludges from scrubbers and wastewater treatment.

Airborne Wastes

Air pollutants include fumes from furnaces, from pickling processes, and from other cleaning operations.

Waterborne Wastes, Wastes Minimization, and Wastewater Treatment

The following is a discussion of the primary types of substances that characterize wastewaters generated by the copper forming industry. Also discussed are waste minimization techniques and wastewater treatment technologies. The primary pollutants from the copper forming process are listed in Table 10-4.

Oil and Grease

Glycerides of fatty acids consisting of 16 to 18 carbon atoms are commonly used as lubricants throughout the copper forming process. The oil molecules are charged such that the water molecules reject them. Because of the specific gravity of the oils, they rise to the surface and can be removed mechanically. However, since water-soluble oils or emulsified oils are used, precipitation techniques must be used to separate them from the water.

Table 10-4   Sources of Copper Forming Wastes

Process	Waste Source	Pollutants/Parameters
Hot Rolling	Lubricant	O&G, TM, TO, SS
Cold Rolling	Lubricant	O&G, TM, TO, SS
Drawing	Lubricant	O&G, TM, TO, SS
Solution Heat Treatment (SHT)	Quench	O&G, TM, TO, SS
Extrusion Press SHT	Quench	O&G, TM, TO, SS
Alkaline Cleaning	Bath Dump	O&G, SS, high pH
Alkaline Cleaning	Rinse Water	O&G, SS, high pH
Annealing	Water	O&G, SS
Annealing	Oil	O&G, SS
Pickling	Bath	TM, low pH
Pickling	Rinse	TM, low pH
Pickling	Fume Scrubber	TM, low pH

O&G = Oil and Grease; TM = Toxic Metals; TO = Toxic Organics;
SHT = Solution Heat Treatment; SS = Suspended Solids.

Toxic Organics and Metals

Toxic organics, such as chlorinated solvents used to clean oils and greases from the surfaces of raw materials, intermediates, and products, should be replaced by nontoxic solvents, such as detergents, and thus be eliminated from the waste stream. Metals, however, cannot be eliminated. Metals can be removed from the waste stream by alkaline, sulfide, phosphate, or carbonate precipitation, followed by filtration and ion exchange, as discussed in Chapter 7. In many instances, metals can be recovered from sludges and ion exchange resins, and every opportunity to do so should be thoroughly investigated.

Suspended Solids

Particles of metals, metal oxides, and dirt are always present in metal processing wastes, due to abrasion, both intended and otherwise. Airborne particles are often scrubbed using wet scrubbers, converting these particles from air pollutants to water pollutants. It is usually a viable option to mix these wastes with wastewaters from the process itself and then to remove the combined suspended solids by chemical coagulation and filtration. Maximum recycle and reuse of treated wastewater should be a continuous objective.

Prepared Frozen Foods

Prepared frozen foods are foods that have been mixed with other ingredients, sometimes cooked or partially cooked, and packaged for easy consumer use. "TV Dinners" are examples. Prepared frozen foods vary from frozen single items to complete meals arranged in segmented aluminum or plastic plates for convenient heat and consume or microwave and consume use.

There is evidence that prehistoric people froze foods for later consumption. The rapid increase in commercialization of frozen foods (as opposed to "prepared" frozen foods) began in the 1920s, when Clarence Birdseye developed the process of quickly freezing foods in a way that preserved taste and texture. Apparently, the quick freezing results in smaller ice crystals, and the smaller the ice crystals, the more closely the product resembles fresh product at the time of consumption. The rapid growth of prepared frozen foods began with the rapid increase in time spent viewing television, starting in the 1950s.

The American Frozen Food Institute (AFFI) has divided the specialty foods industry into ten categories, the following five of which are appropriately included under prepared frozen foods:

- Prepared dinners
- Frozen bakery products
- Italian specialties
- Chinese and Mexican foods
- Breaded frozen products

Prepared frozen foods can be characterized as having value (and pollution potential) added when compared with frozen fruits, vegetables, and unprocessed meat, fish, and poultry. In general, the substances added during the (factory) preparation of prepared frozen foods are a larger source of pollutants that must be managed than are the basic foods themselves.

Preparing frozen foods, like essentially all of the food processing industry, is not associated with toxic wastes. Some of the wastes may exhibit one or more hazardous characteristics, for instance, corrosivity due to low pH or high pH resulting from one or more cleaning activities. These characteristics can readily be corrected by simple treatment techniques such as neutralization, using one or more acid or caustic substances.

Food processing of all types, including the prepared frozen foods industries, is characterized by a startup, operate, and plant cleanup sequence. It is essentially always true that plant cleanup activities produce the greatest quantities of wastes. Process startup is a distant second and, for well-operated and -maintained plants, the operation phase produces a small quantity of pollutants. This is in contrast to operations such as pulp mills, where the bulk of the pollutants are produced on a continuous basis.

The food processing industry is here to stay. Furthermore, there is every indication that the prepared frozen foods industry will enjoy steady growth in the foreseeable future.

Because plant cleanup activities produce most of the wastes from prepared frozen foods production facilities, the activities that lead to starting and stopping processing lines warrant close scrutiny regarding wastes minimization. There is an inherent economy of scale regarding wastes generation. The ideal processing plant, from a wastes minimization standpoint, is a relatively large plant that processes at near maximum capacity for two or more shifts per day and undergoes plant cleanup only once per day. The most difficult processing plant in terms of wastes minimization is a relatively small plant that changes products one or more times each shift and often operates under capacity. Each time there is a change of process on a given processing line, the line, and much of the rest of the plant, must be thoroughly washed down.

Wastes Generation

Solid Wastes

There are considerable solid wastes generated at prepared frozen food plants, including debris from fresh raw materials that make up the basic substance of the prepared food. Examples are as follows:

- Fish carcasses in the case of fish processing plants
- Feather and inedible body parts in the case of poultry processing plants
- Dirt, stems, leaves, etc., in the case of vegetable or fruit processing plants.

Some prepared frozen food plants initiate production with all "raw" materials processed to some degree at other locations. For instance, certain producers of frozen chicken potpies purchase chicken that has been processed to the degree that it is ready for cook-

ing. In these cases, the solid wastes from processing the live chickens do not have to be dealt with; however, packaging and shipping materials—boxes, wrappers, steel bands, etc.—make up an important portion of the solid waste stream. Additional solid wastes include sludges from processing tanks, sludges from wastewater treatment processes, and the material that is collected during plant cleanup operations using dry methods such as brooms, air squeegees, and shovels.

Many of the solid wastes from prepared frozen foods as well as other food processing facilities are putrescible; therefore, they are potential nuisance odor problems (and represent a source of air pollution). The best management practice to prevent such an odor problem from developing is to maintain a very clean plant throughout the premises and to dispose of the solid wastes before a problem develops.

Airborne Wastes

The only airborne wastes from prepared frozen food processing plants should be those normally associated with the boilers. Nuisance odor problems are often a potential problem but should never be allowed to develop.

Waterborne Wastes

Plant cleanup operations produce the largest percentage of waterborne wastes. The substances included in these wastes are characteristic of the products produced, as discussed in the following text.

Prepared Dinners

There are more plants producing prepared dinners than any other category of prepared frozen foods. Frozen potpies represent a large portion of the total. Some of these plants process live chickens, turkeys, swine, or cattle; most purchase processed or partially processed poultry meat or beef. Characteristically, most of the wastewater is generated by plant cleanup operations, usually during a

Table 10-5 Prepared Dinners' Average Pollutants Contained in Wastewater per Unit Production

	Constituent (kg/kkg finished product)[*]							
Plant Code	COD	BOD	SS	VSS	Total P	TKN	Grease and Oil	Volume (L/kkg)
A	69	35	34	33	0.25	0.44	44	8,700
B	42	18	11	11	0.18	0.25	21	6,200
C	28	13	11	11	0.24	0.61	—	22,000
D	27	15	14	14	0.16	0.55	2.9	21,000
E	20	11	6.6	6.0	—	—	3.8	9,400
F	17	8.8	6.2	6.2	0.12	0.37	4.8	4,400
Average	34	17	14	14	0.19	0.44	15	12,000
Range	17–69	9–34	6–34	6–33	.12–.25	.25–.61	2.9–44	4,400–22,000

[*]kg – kilogram; kkg – thousand kilogram.
COD = Chemical Oxygen Demand; BOD = Biological Oxygen Demand; SS = Suspended Solids (Total);
VSS = Volatile Suspended Solids; Total P = Total Phosphorus; TKN = Total Kjeldahl Nitrogen;
L/kkg = Liters per thousand kilograms.

late night or early morning cleanup shift. Additional cleanup takes place at changes of shifts, changes of items being processed, or products being prepared, and as a result of spills. Washing, rinsing, and blanching of vegetables represent other sources of wastewater. Frying, breading, and cooking represent other sources. Tables 10-5 and 10-6 present "average" values of wastewater characteristics obtained from several prepared frozen foods processing plants. These tables have been reproduced from a report on a study conducted by the AFFI during the 1980s. These results are not presented as

"average" or even typical for the industry as a whole, since there is such great variability from one plant to another regarding raw materials taken in, processes taking place, frequency of cleanup, items produced, and plant maintenance procedures. Tables 10-5 and 10-6 are presented here simply to illustrate the wastewater characteristics observed at six individual prepared frozen food plants.

In general, it can be said that the wastewaters from prepared frozen dinner plants, as represented by the six plants that were the subjects of Tables 10-5 and 10-6, were relatively strong, were organic in nature, and had

Table 10-6 Prepared Dinners, Average Wastewater Characteristics

	Concentration (mg/L)						
Plant Code	COD	BOD	SS	VSS	Total P	TKN	Grease and Oil
A	7,900	4,000	3,900	3,800	29	51	5,100
B	6,800	2,900	1,800	1,700	30	34	3,400
C	1,300	620	530	510	11	28	—
D	1,300	720	680	650	7.6	26	140
E	2,100	1,240	700	640	—	—	400
F	3,800	2,000	1,400	1,400	28	85	1,100
Average	3,900	1,900	1,500	1,500	210	45	2,000
Range	1,300–7,900	620–4,000	530–3,900	510–3,800	7.6–30	26–85	140–5,100

every indication of being biodegradable. For instance, Table 10-6 shows that for the six plants included in this portion of the report, the concentration of BOD_5 varied from a low of 620 mg/L (plant C) to a high of 4,000 mg/L (plant A), with the average concentration for all six plants equaling 1,900 mg/L. This compares to the BOD_5 for domestic wastewater of 250 to 350 mg/L.

Frozen Bakery Products

Compared with frozen dinners, frozen bakery product processing plants produce wastewaters that are higher in fats, oils, and greases, higher in carbohydrates, and lower in protein. The high fats, oils, and greases result from the use of butter, shortening, and cooking oils. The high carbohydrates result from the use of starch (flour) and sugar.

Again, there is a tendency for the bulk of the waste load to be generated by cleanup activities, and, again, there is every indication that the substances in the wastewater would respond well to biological treatment.

Tables 10-7 and 10-8 are also reproduced from the AFFI study mentioned earlier. These data show that the wastewaters from the preparation of frozen bakery products, as represented by two plants studied by the AFFI during the early 1980s, can be charac-

terized as "strong." For instance, the concentration of BOD_5 in the wastewater from the two plants was about 2,100 mg/L for one of the plants, and 4,300 mg/L for the other plant, compared with about 250–350 mg/L for domestic wastewater.

Italian Specialties

Prepared frozen foods included in the category of "Italian specialties" include frozen spaghetti, lasagna, ravioli, pizza, and sauces for Italian foods. The raw materials for these products include tomatoes, cheese, and flour (starch). Meat and seasonings are included as well.

Tables 10-9 and 10-10 are reproductions of tables presented in the AFI study mentioned above. The data included in Tables 10-9 and 10-10 indicate, among other things, the characteristic variability of wastewater from one processing plant to another. For instance, the BOD_5 reported for the wastewater from plant Q was 200 mg/L. The BOD_5 from plant R was 690 mg/L. These results compare to the BOD_5 from domestic wastewater of 250 to 350 mg/L. A wastewater treatment plant designed on the basis of the characteristics reported for plant Q would appear to be severely overloaded if placed into operation at plant R. This observation is

Table 10-7   Frozen Bakery Products, Average Pollutants Contained in Wastewater per Unit Production

| | Constituent (kg/kkg finished product) | | | | | | | |
Plant Code	COD	BOD	SS	VSS	Total P	TKN	Grease and Oil	Volume (L/kkg)
G	52	23	14	14	0.082	0.30	11	11,000
H				No production information provided				

Table 10-8   Frozen Bakery Products, Average Wastewater Characteristics

| | Concentration (mg/L) | | | | | | |
Plant Code	COD	BOD	SS	VSS	Total P	TKN	Grease and Oil
G	4,600	2,100	1,300	1,200	7.8	27	940
H	9,300	4,300	3,100	3,000	5.7	45	690
Average	7,000	3,200	2,200	2,100	6.8	36	820

Table 10-9   Italian Specialties, Average Pollutants Contained in Wastewater per Unit Production

| | Constituent (kg/kkg product) | | | | | | | Volume |
Plant Code	COD	BOD	SS	VSS	Total P	TKN	Grease and Oil	(L/kkg)
O	39	19	14	13	0.79	0.59	—	80,000
P	—	3.3	—	—	—	0.12	—	9,800
Q	8.8	5.2	3.4	3.1	0.052	0.15	4.7	26,000
R	2.6	1.1	0.65	0.59	0.011	0.061	—	1,800
Average	17	7.2	6.0	5.6	0.28	0.23	4..7	29,000

Table 10-10   Italian Specialties, Average Wastewater Characteristics

| | Concentration (mg/L) | | | | | | |
Plant Code	COD	BOD	SS	VSS	Total P	TKN	Grease and Oil
O	500	240	180	150	10	7.6	—
P	—	340	—	—	—	11.8	—
Q	340	200	130	120	2.0	5.6	180
R	1,500	690	360	330	6.0	34	—
Average	780	370	220	200	6.0	15	180

contradicted by the results presented in Table 10-9. When expressed as units of BOD$_5$ per thousand units of product, plant Q discharged 5.2 mg/kkg, but plant R discharged only 1.1 pounds. From these data, it would appear that the wastewater treatment plant designed on the basis of data obtained at plant Q would be severely underloaded if placed into operation at plant R.

Chinese Foods and Mexican Foods

Prepared frozen Chinese and Mexican foods result in the production of waste characteristics that are somewhat stronger in terms of conventional pollutants (BOD, COD, TSS, O&G) than domestic wastes and are expected to be biodegradable. Both types of plants use vegetables, some poultry, and some meats. Both use rice to a significant degree, and neither of the two uses large amounts of oils or fats. It is common for producers of both prepared frozen Chinese foods and prepared frozen Mexican foods to receive, as raw materials, at least some fresh vegetables, and to preprocess them by cleaning and peeling.

Tables 10-11 and 10-12 present the characteristics of wastewater from two different plants that produced frozen Chinese foods. Again, these tables are reproductions from the AFFI study discussed above. There were no data presented for prepared frozen Mexican foods in the AFFI study. In lieu of data on prepared frozen food production, the AFFI presented data on Plant "X," which produced canned Mexican foods. Since all of the processes are the same up to the point of either freezing or canning, and since neither the freezing nor canning processes should produce significant wastes, solid, airborne, or waterborne, it appears to be a good assumption to consider that wastes generated during the preparation of frozen Mexican foods should be similar in characteristics to wastes generated during the preparation of canned Mexican foods.

As shown in Tables 10-11 and 10-12, the BOD$_5$ of the two plants that prepared frozen Chinese foods was 370 to 450 mg/L, reasonably close to what would be expected for normal domestic wastes. The BOD$_5$ of the wastewater from the preparation of canned

Table 10-11 Frozen Chinese and Mexican Foods, Average Pollutants Contained in Wastewater per Unit Production

Plant Code	Constituent (kg/kkg finished product)							Volume (L/kkg)
	COD	BOD	SS	VSS	Total P	TKN	Grease and Oil	
V	12	6.3	2.4	2.2	0.084	0.36	1.2	14,000
W	12	6.7	4.0	3.8	0.041	0.27	4.7	18,000
X	12	7.8	1.9	1.2	0.29	0.21	—	8,900
Average	12	6.9	2.8	2.4	0.14	0.28	3.0	14,000

Table 10-12 Frozen Chinese and Mexican Foods, Average Wastewater Characteristics

Plant Code	Concentration (mg/L)						
	COD	BOD	SS	VSS	Total P	TKN	Grease and Oil
V	830	450	170	160	6.0	26	85
W	670	370	220	210	2.3	15	260
X	1,300	900	210	140	34	22	—
Average	930	570	200	170	14	21	170

Mexican foods was about 900 mg/L (see Table 10-12), significantly higher than what would be expected in domestic wastes (250–350 mg/L).

Breaded Frozen Products

Major products from the breaded frozen products category are breaded "fish sticks," onion rings, mushrooms, and shellfish. Some plants purchase fresh raw material, such as fresh picked onions and mushrooms. Some plants purchase "raw material" already processed and frozen, such as frozen processed fish or shellfish. These plants thaw the frozen raw material, process it further, then freeze it again. Tables 10-13 and 10-14 are, again,

Table 10-13 Breaded Frozen Products, Average Pollutants Contained In Wastewater per Unit Production

Plant Code	Constituent (kg/kkg finished product)							Volume (L/kkg)
	COD	BOD	SS	VSS	Total P	TKN	Grease and Oil	
Y	40	15	23	23	0.12	0.33	1.2	3,300
Z	66	37	30	29	0.58	4.8	—	92,000
Average	53	26	26	26	0.35	2.6	—	48,000

Table 10-14 Breaded Frozen Products, Average Wastewater Characteristics

Plant Code	Concentration (mg/L)						
	COD	BOD	SS	VSS	Total P	TKN	Grease and Oil
Y	12,000	4,500	7,100	7,100	37	100	360
Z	720	400	330	320	6.3	52	—
Average	6,400	2,400	3,700	3,700	22	76	—

reproductions of tables presented in the report of the AFFI study mentioned earlier. The data shown in these tables illustrates, again, that there is a great deal of variability in wastewater characteristics from one plant to another and that at least some of these plants can be expected to have relatively strong wastes. For instance, Table 10-14 shows that one of the prepared frozen food plants producing breaded products, plant Z, was discharging wastewater characterized by a BOD_5 of 400 mg/L, slightly on the strong side compared with the strength of domestic wastewater. Another plant was discharging wastewater with a concentration of BOD_5 of 4,500 mg/L. This plant's wastewater was very strong compared with domestic wastewater.

Wastes Minimization

Most of the wastes that require management in the form of containment, treatment, and disposal from prepared frozen food industrial plants have been shown to have their principal source in cleanup operations, spills, and leaks. The most important steps that can be taken to reduce to a minimum the quantity of wastes generated are the following:

- Pursue an aggressive, ongoing, daily program to prevent accidental spills. Spills are the most prevalent of the preventable occurrences that add to the quantity and strength of discharged wastes in food processing plants of all types.
- Pursue an aggressive preventive maintenance program to eliminate the occurrence of leaks of water or wastewater from anywhere in the plant.
- Aggressively employ technologies for recovering lost raw materials, as well as separating and removing contaminants. The general objective should be to treat wastes as closely as possible to the source (individual manufacturing process) and to reuse as much material as possible.
- Limit plant cleanup occurrences to as few as possible on any given day. Plan the processing of different products so as to change what is being produced on each line as few times per day as is feasible.
- During plant cleanup, use dry methods such as brooms, air squeegees, and shovels to remove as much of what needs to be cleaned up as possible. This material becomes solid waste, which can probably be treated and disposed of, or used as animal feed via a rendering facility, at much less cost than for wastewater treatment.
- Never let water run at wash stations not in use. Water should be turned on by hand-, knee-, or foot-actuated valves and should turn off automatically when the user is finished.
- Never leave water running in hoses that are not in use.
- Make sure ingredients of all detergents are known and scrutinized to ensure compatibility with wastewater treatment processes.

The AFFI study found that the attitude of plant management was an extremely important factor regarding the quantity of wastes that ultimately required handling and treatment. One specialty food processing plant employed continuous monitoring to keep track of the loss of valuable product. When waste quantity increased to 1% of the product produced, an investigation was initiated to determine and correct the cause.

Other factors that have been shown to influence the quantity of wastes generated at prepared frozen foods processing plants include plant size, number of shifts, relative amounts of ingredients preprocessed at other locations, the cost of water, the cost of waste disposal, the age of the plant, and the age of the individual processes and equipment. There definitely appears to be an economy of scale regarding waste generated (proportional to product loss). Larger plants have been shown to generate less waste material (equivalent to "lose less product") than smaller plants.

Treatment and Disposal of Wastes

Two prominent characteristics of both solid and waterborne wastes from most processing plants engaged in the production of prepared frozen foods are that they are amenable to biological methods of treatment and that they tend to be relatively strong. For these reasons, candidate treatment technologies that would appear to hold promise are composting, in the case of solid wastes, and anaerobic or aerobic biological treatment, for waterborne wastes.

Solid Wastes

A significant portion of the solid waste stream from prepared frozen foods plants consists of normal industrial plant trash, such as packaging and shipping material (associated with both incoming and outgoing material), construction debris from remodeling, plant expansion and regular maintenance, and equipment and appurtenances that are no longer usable. In fact, because packaging and shipping are such major activities for this industry, the quantity of waste packaging and shipping material, such as cardboard boxes, paper boxes, paper wrapping, and strapping material, is considerable. Since much of this material is recyclable, a large portion of the solid wastes disposal problem can be avoided.

Success in solving waste disposal problems by recycling is best enhanced by proper setup of storage facilities, proper arrangements for transportation to a recycling facility, and an aggressive program for keeping recyclable materials cleaned up around the plant and placed in appropriate containers. Nonrecyclable wastes can usually be landfilled, since hazardous materials are not normally included in wastes from food processing plants of this type.

Other than packaging and shipping wastes, the principal solid waste stream from prepared frozen foods plants consists of trimmings, rejected raw material, and other portions of the organic (edible, for the most part) wastes from foods. Composting, therefore, presents itself as a potential disposal method that can accept all of this portion of the plant's solid waste stream and result in a useful product, namely soil conditioner and fertilizer. Another possibility for disposal of organic solid wastes is by direct application on land, followed by tilling into the soil to avoid problems with odors and pests such as rodents and flies.

Waterborne Wastes

Three characteristics that strongly influence selection of candidate treatment technologies for wastewaters from prepared frozen foods industrial plants are:

1. The wastes are amenable to biological treatment.
2. In the case of most plants, the liquid waste streams are relatively strong in terms of concentration of BOD_5, TSS (largely organic solids), and sometimes oil and grease.
3. The wastewaters tend to come in slugs, such as the high quantity of flow during the occurrence of plant cleanup and the emptying of processing facilities when a product line is changed.

The characteristic being generated in slugs, as opposed to being generated at a steady rate, strongly indicates that flow equalization should be one of the first components in the wastewater treatment system. The flow equalization device should be preceded by bar racks, screens, or both, to remove large objects and screenable materials that would settle to the bottom of the equalization device. A grease trap or other oil and grease removal device should also be placed upstream of the equalization device. As discussed in Chapter 7, if the equalization device is a variable-depth holding basin, it should be equipped with surface skimming equipment as well as bottom scraping and sludge removal. The high degree of putrescibility of food processing wastes requires that there be no opportunity for solids that are either lighter than water or heavier than water to collect and remain for any length of time before being removed and processed.

The characteristics of amenability to biological treatment, along with relatively high strength, strongly suggest anaerobic biological treatment as a candidate treatment technology, especially one of the high-rate anaerobic technologies, such as the upflow anaerobic sludge blanket (UASB) (a suspended growth system) or one of the fixed-film systems, such as the fluidized reactor or the expanded bed reactor. These technologies are presented and discussed in Chapter 7. There are two major advantages of anaerobic treatment technologies, as opposed to aerobic systems. The first is a far lower requirement for electrical energy. The second is a much smaller quantity (20% to 30%) of biological solids generated that must be managed.

Notwithstanding the apparent advantages of anaerobic treatment technologies, the most prevalent methods of wastewater treatment that have been used in the past have been aerobic systems. Aerobic lagoons, extended aeration activated sludge, and land disposal have been used extensively. In many cases, wastewaters have been discharged untreated to municipal sewer systems where they were treated by means of the POTW. The principal advantage of having wastewaters treated by means of a POTW is convenience. The principal disadvantage is cost. POTW treatment normally involves a substantial surcharge penalty for wastes with BOD and TSS concentrations significantly greater than normal domestic wastewater.

Although in general food processing wastewaters are amenable to wastewater treatment by nearly all of the available aerobic and anaerobic treatment technologies, prepared foods sometimes involve the use of sugar, flour, starch, and other high-carbon, low-nitrogen, and/or phosphorus substances. Prepared frozen foods sometimes fall into that category. The report that resulted from the AFFI study referenced earlier contained a table summarizing, by way of averaging, the BOD_5/TKN/phosphorus ratios found for the five categories of pre-

Table 10-15   Carbon:Nitrogen:Phosphorus Ratios

Category	BOD	TKN	Phosphorus
Prepared Dinners	100	2.4	1.1
Frozen Bakery Products	100	1.1	0.2
Italian Specialties	100	4.0	1.6
Chinese and Mexican Foods	100	3.7	2.5
Breaded Frozen Products	100	3.2	0.9

pared frozen food products, reproduced here as Table 10-15.

Figure 10-6 presents a schematic of a wastewater treatment facility that was in use when the AFFI study was conducted. As shown in Figure 10-6, this particular wastewater treatment system, which was said to have evolved over a considerable period of time, was complex, had many sources of solids or "sludge," and was obviously a relatively expensive system to operate. It is almost certain that a single anaerobic system, such as a UASB or AAFEB, followed, possibly, by an aerobic polishing system, would be more reliable, efficient, and economical.

Wastepaper De-inking

Historically, paper has been produced in the United States primarily from virgin wood pulp. However, since the last half of the 20th century there has been an ever-increasing shift to produce paper from recycled fibers. A common process used in the production of recycled pulp (and subsequently paper) is de-inking.

De-inking wastes are regulated under *40 C.F.R. Part 430, Subpart,* pertaining to the De-inking Division of the Secondary Fibers subcategory of the Pulp, Paper, and Paperboard point source category. The Secondary Fibers subcategory includes all recycled paper. The De-inking division includes those secondary fiber processes where ink is removed prior to production of white (recycled) paper. The objective of the de-inking process is to remove ink in order to brighten the pulp and to remove other noncellulosic substances such as pigments, fillers, and

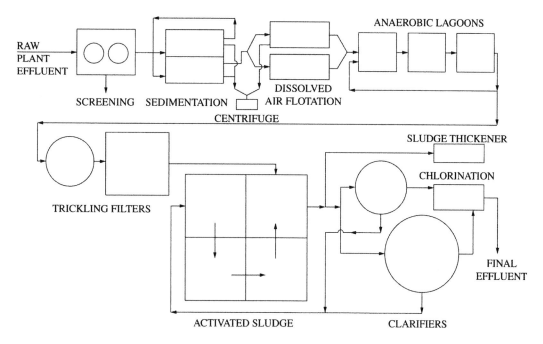

Figure 10-6   Wastewater treatment facility.

coatings. Therefore, wastes from de-inking processes will contain all of these substances plus a portion of additional substances added in the de-inking process.

For purposes of federal regulation, the de-inking division is divided into three subdivisions that correspond to differences in final product production requirements, as well as differences in wastewater characteristics. Those de-inking facilities that produce pulp for tissue paper generally have the highest pollutant load, in terms of daily flow, BOD_5, and TSS. Those that produce pulp for newsprint generally have the lowest pollutant loads, and those that produce fine papers such as office stationery, copier paper, and computer printout paper have pollutional loads that are less than those of the tissue paper mills and more than those of the newsprint mills.

The De-inking Process—Wastes Generation

Wastepapers to be reused as secondary fiber products must first be sorted and classified as to suitability for final product. Newer secondary fiber processing facilities are capable of handling a much wider range of recycled paper types, and are equipped to separate out wire, bottles, and a wide variety of unwanted foreign objects. Older mills are not as well equipped. Once the "used" paper has been sorted, the following processes convert it to a pulp product that is ready for the paper mill. In some cases, the recycled pulp is mixed with virgin pulp. In many cases, it proceeds directly to the paper-making process by itself.

There are ten basic steps in the de-inking process, and they are as follows:

1. Pulping
2. Prewashing, heat and chemical loop
3. Screening (coarse and fine)
4. Through-flow cleaning (or reverse cleaning)
5. Forward cleaning
6. Washing
7. Flotation
8. Dispersion
9. Bleaching
10. Water recirculation and makeup

Different de-inking facilities employ different numbers and different sequences of these ten steps depending on the requirements of the final product and the characteristics of the wastepaper. A brief description of each follows.

Pulping

Waste paper is loaded into a pulper where it is mixed with hot water, alkali (pH = 9 to 11), and various solvents, detergents, and dispersants. This mixture is "cooked," which produces a "stock" of the resulting pulverized paper. The added chemicals dissolve and disperse adhesives, fillers, sizes, ink pigments, binders, and coatings, all of which eventually end up in one or another waste stream from the process. A built-in coarse screen allows smaller solids and liquids to continue on. Those solids that do not pass the screen are either returned for another pass through the pulper or enter the waste stream.

Prewashing

Gross amounts of ink, clay, and other materials are removed by prewashing, which consists of fine screening, partial dewatering, dissolved or dispersed air flotation, and/or settling.

Screening

The prewashed stock is next subjected to both coarse and fine screening. The fine screens are sometimes operated under pressure.

Through-flow Cleaning

Also called "reverse cleaning," this process is typified by a counter-current washing process. In one form, the stock flows down an inclined screen with several intermediate barriers. The stock is sprayed with water at each barrier, which washes substances such as ink particles through the screen. Clean water is applied at the lowest barrier and recycled. Progressively dirtier water is applied at progressively higher barriers. Due to the relatively large amount of water used, this process is a significant source of wastewater.

Forward Cleaning

Heavy contaminants that pass through the through-flow and fine screening processes are the target pollutants for the forward cleaning process. This process operates in a multistage sequence similar to that of the through-flow process. However, the stock is much more dilute (less than 1% solids). Large amounts of water are used. This water is cleaner than that used for through-flow cleaning.

Washing

The washing process makes use of counter-current flow washing to remove ink from the stock that has not yet been successfully removed. Equipment includes sidehill screens, gravity deckers, and dewatering screws.

Flotation

Those colloidal substances, including inks that are resistant to screening and washing processes, are the target substances for the flotation process. Flotation does not make use of added water but may use coagulation chemicals, including organic polymers. In some instances, the flotation process is located ahead of the washing process. The high pH from the pulping process sometimes aids significantly in flotation.

Dispersion

Those quantities of inks that are not removed by screening, through-flow cleaning, forward cleaning, washing, and flotation are dispersed in order to make them undetectable in the finished paper.

Bleaching

Bleaching of the recycled pulp is highly specific to each individual mill. Bleaching can be done in the pulper, just after prewashing, or after flotation and dispersion. Bleaching chemicals can include chlorine, chlorine dioxide, peroxides, and/or hydrosulfites.

Water Recirculation and Makeup

While water recirculation and makeup are not a "step," they are inherent to each of the processes previously discussed.

The ten processes discussed previously are diagrammed in Figure 10-7. One of many alternatives to the processing sequence shown in Figure 10-7 is presented in Figure 10-8. In the alternative process, some of the steps shown in Figure 10-8 have been eliminated or combined, leaving seven of the major processing steps in five processes.

Wastes Generation and Wastes Minimization

De-inking facilities produce significant solid wastes in the form of wastepaper sorting rejects, screening rejects, and sludges from flotation and sedimentation. Air pollutants are relatively few and are treated by use of wet scrubbers, which creates more wastewater and solid waste (sludge).

De-inking facilities use very large quantities of water, which can be reduced by aggressive application of recycle. The different processes, sequence of processes, and chemicals used result in varying wastewater loads and characteristics from one de-inking plant to another.

The de-inking process illustrated in Figure 10-7 shows that the major source of wastewater is blow-down from recycle of rinsewaters and process water makeup. To this blow-down water is added contaminated water from leaks and spills, which becomes incorporated in the second major source of wastewater, plant washdown water. The plant must be washed down periodically because of the leaks and spills.

Consequently, aggressive preventive maintenance to prevent leaks and careful management of spills and cleanup become among the most important waste management activities.

Contaminants in de-inking wastewater are those substances extracted from the waste paper, plus a portion of the detergents, dispersants, coagulants, and other chemicals added during the de-inking process. A listing of the major pollutants of concern includes adhesives, starches, clays, ink particles and carriers, sizing, fillers, detergents, dispersants, coagulants, lost fiber, solvents, and bleaching chemicals.

Wastewater Characteristics

Table 10-16 presents average values of characteristics of wastewater from the de-inking industry, as published in the EPA's Development Document for the pulp, paper, and paperboard point source category.

Toxic Pollutants

Pentachlorophenol (PCP) and trichlorophenol are among the most common toxic pollutants in de-inking wastewater. The source of these substances is the slimicides and biocides used to keep the growth of troublesome biological growths under control. This fact amounts to strong indications of biological treatment as a candidate treatment process for de-inking plant wastewaters.

Wastes Minimization

Minimization of wastes from the de-inking industry is best accomplished by the following:

- Substitution of nontoxic chemicals for toxic chemicals
- Aggressive pursuit of good housekeeping
- Preventive maintenance to eliminate leaks
- Equipment modifications to prevent spills
- Reuse of water to the maximum extent feasible
- Recovery of usable fiber
- A continuous program to reduce usage of water

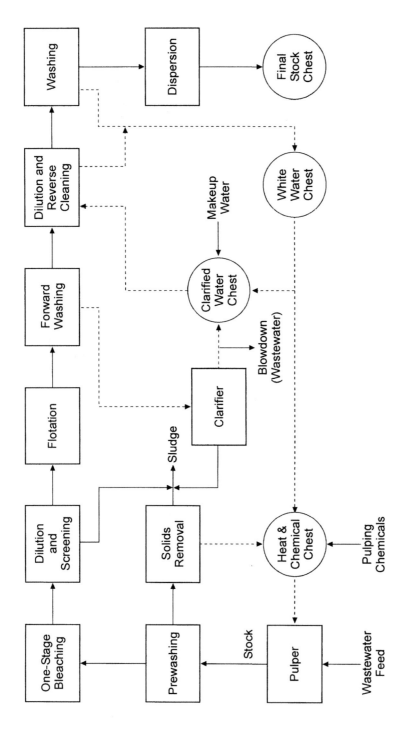

Figure 10-7   Ten-step de-inking process flow diagram.

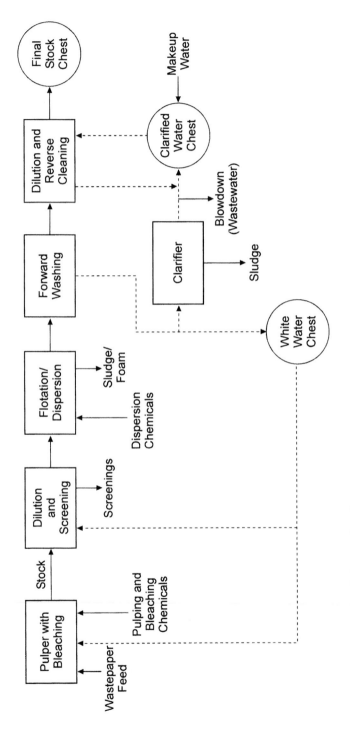

Figure 10-8 Seven-step de-inking modified process flow diagram.

Table 10-16   De-inking Wastewater—Average Values of Untreated Wastewater Characteristics by Subdivision*

	Flow		BOD₅		TSS	
	kL/kkg	*kgal/ton*	*kg/kkg*	*lb/ton*	*kg/kkg*	*lb/ton*
Fine Papers	88.1	21.1	37.3	74.6	174.1	348.1
Tissue	136.9	32.8	87.2	174.3	251.0	501.9
Newsprint	67.5	16.2	15.9	31.7	96.8	193.5

Toxic Pollutants—De-inked Pulp for Tissue Papers

2,4,6— Trichlorophenol, *ug*/L	Pentachlorophenol, *ug*/L	PCB-1242, *ug*/L
8.8	4.8	21.3

*All values obtained from the *Development Document for Effluent Limitations, Guidelines, and Standards* for the Pulp, Paper, and Paperboard Point Source Category, Tables V-14 and V-33.

Wastewater Treatment

After substitution of nontoxic materials for toxic materials, the most advantageous technology for treatment of wastewaters from the de-inking industry has often been determined to be aerobic or anaerobic biological treatment. When preceded by appropriate primary treatment—for instance, pH neutralization, screening, flow equalization, and primary clarification—biological secondary treatment has garnered a history of success. Because of low nutrient levels, it is necessary to add sources of nitrogen, phosphorus, and some trace nutrients. The following biological treatment technologies have been used.

Aerated Lagoons

Aerated lagoons have been used extensively for the treatment of wastewaters from many pulp- and paper-related industries, including de-inking wastes. Two attractive features of aerobic lagoons for these and other industrial wastes are (1) the low-stressed nature of the system, owing to relatively low organic loading rates, and (2) the fact that a significant fraction of aeration is atmospheric, as opposed to mechanical, or by way of diffusers. The low organic loading rates result in the ability to absorb shock or "spike" loads and require relatively low intensity of operator attention, and the atmospheric aeration results in lower costs for power. As is the case with all lagoons, aerated lagoons must be lined to protect the groundwater.

Oxidation Basins

Oxidation basins have been used in southern regions, where higher temperatures favor higher biological reaction rates and more intense sunlight provides more energy for oxygen generation by algae. These lagoons should be preceded by primary treatment, have excellent buffering capacity against shock loads, and have no mechanical requirements. These facilities must be lined to protect the groundwater.

Activated Sludge

Modifications of the activated sludge process that have been used with success to treat wastewaters from de-inking include complete mix, conventional, tapered aeration, step aeration, contact-stabilization, and pure oxygen. A two-stage activated sludge process having a detention time of four hours and an integrated selector was discussed in the development document as being particularly effective for the treatment of wastewaters from de-inking.

Anaerobic Contact Filter

The anaerobic contact filter is essentially a nonaerated trickling filter. This technology has proved to be successful at several de-inking facilities. Detention times of up to three days have been used.

Die Casting: Aluminum, Zinc, and Magnesium

Die casting is one of the oldest methods used to shape metals. The metal or metal alloy is melted, then poured into a prepared mold and allowed to cool. The molded piece is then removed from the mold and processed further by one or more of a great variety of processes. The mold, depending on the process used, may be used again as-is, may be rebuilt to varying degrees and used again, or may be completely destroyed during the process of removing the shaped piece. Molds are normally made of a metal or metal alloy that has a significantly higher melting point than the metal being molded. There are three principal elements to all die casting machines: (1) a casting machine to hold the die into which the molten metal to be cast is injected, (2) the mold itself that receives the molten metal and is capable of ejecting the solidified product, and (3) the casting metal or alloy. First, the metal is melted and any desired additives are added. Then, a source of hydraulic energy impacts a high velocity to the molten metal, causing it to rapidly fill the die. The die must absorb the stresses of injection and dissipate the heat from the molten metal.

Two types of die casting machines are in common use. The first is an air-operated machine. Compressed air forces the molten metal (or metal alloy) into the die by exerting high pressure on the surface of the molten metal in a special ladle referred to as the "goose." The second type of die casting machine has a cylinder and piston submerged in the molten metal to force the molten metal into the die.

There are three primary variations of the die casting process: (1) the hot chamber process, used for lower-melting metals such as zinc and magnesium; (2) the cold chamber process, used for higher-temperature melting metals such as aluminum; and (3) the direct injection process. In the hot chamber process, the hydraulically actuated cylinder and piston are submerged in the molten metal. In the cold chamber process, the molten metal is fed to the cylinder and piston from a reservoir. In the direct injection process, nozzles directly inject molten metal into the dies. Large amounts of noncontact cooling water are normally associated with the die casting process. Also, lubricants, referred to as "die lubes," are used to prevent adherence of the casting to the die. Selection of the die lube is governed, first, by wastewater treatment and discharge permit considerations and, second, by its performance in providing the casting with a better finish (allowing the metal to flow into all cavities of the die) and handling characteristics. Die lubes that were used historically contained complex phenolic compounds, and even PCBs have been replaced with die lubes having a vegetable oil base.

Aluminum Die Casting

Figure 10-9 presents a schematic of a typical aluminum die casting process. Aluminum die castings are used in automobiles and many other products. The raw material for aluminum die castings is largely recycled aluminum cans and other articles. The first step in an aluminum die casting operation is to crush, shred, and sort the raw material. Then, the raw material is melted, by use of coreless and channel induction furnaces, crucible and open-hearth reverberatory furnaces fired by fuel oil or natural gas, or electric resistance and electric radiation furnaces. Air flows from the furnaces to wet scrubbers. Furnace temperatures are in the range of 425°F to 600°F.

Next, salts are added to remove oxides from the melt and then hydrogen, which

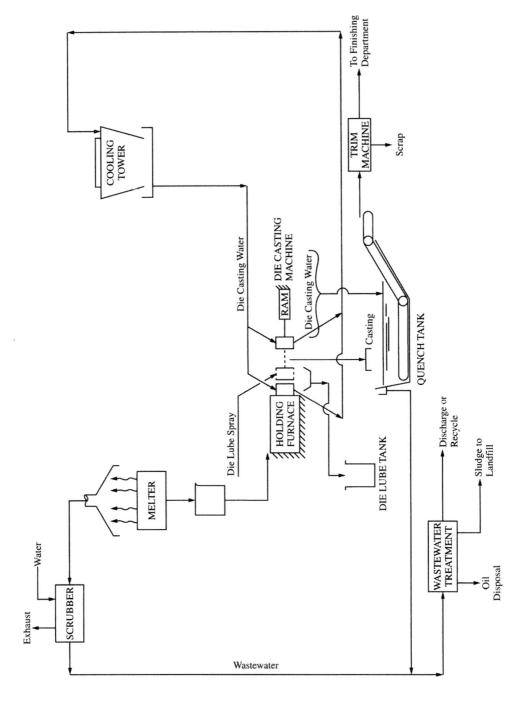

Figure 10-9 Schematic of an aluminum die casting process (from the *EPA Development Document*, 1980).

causes voids in the product casting, is removed by addition of dry, chemically pure nitrogen, argon, or chlorine.

The next step is to lubricate the die, which is normally done by automatic spray. Then, the molten metal is injected into the mold. Cold chamber processing is typically used for die casting aluminum. Processes referred to as gating and risering are sometimes used to minimize shrinkage and to produce directional solidification.

Quenching in a water bath is then used to rapidly cool the casting to room temperature. Oil, salt baths, and various organic solutions are alternative quenching solutions, but water is most often used. After quenching, the casting is sometimes "heat treated" by holding the casting at a temperature of 95°C to 260°C. This process is sometimes referred to as "aging." Finally, the casting is cleaned by use of alkaline and/or acid solutions.

Zinc Die Casting

Figure 10-10 presents a schematic diagram of a typical zinc die casting process. The first step is to crush the scrap metal and prepare it for melting. Then, the metal is melted at temperatures between 325°F and 475°F. Next, the molten metal alloy, having a typical mixture of 1% copper, 3.9% aluminum, 0.06% magnesium, and the rest zinc, is maintained at the desired temperature within ±6 degrees in the holding furnace. After the die has been lubricated, the molten zinc alloy is injected into the mold. The hot chamber process is used. After solidifying, the casting is extracted, trimmed, then dropped into a quenching tank. Finishing may include texturing by acid-etching or other process, electroplating, or polishing.

Magnesium Die Casting

Magnesium die casting follows approximately the same procedure as that shown for zinc die casting in Figure 10-11. Magnesium alloy typically contains zinc plus aluminum,

beryllium, nickel, and copper. Oxygen may be released from the molten alloy by use of a flux (magnesium chloride, potassium chloride, or sodium chloride) or by use of a flux-less process that uses air/sulfur-hexafluoride. The temperature of the molten alloy is held in the range of 475°F to 525°F by use of a crucible furnace. Before it is injected into the die, the molten alloy is surface skimmed to remove oxides, and the die is lubricated very lightly (or, in some cases of magnesium die casting, not at all).

The molten alloy is injected using the hot chamber procedure. Relatively low pressure is used. After solidification the casting is extracted, quenched, and then finished by use of grinding.

Waste Streams and Waste Management

Waste streams from each of aluminum, zinc, and magnesium die casting processes have similar sources, and the waste substances are characteristic of the alloys used and the state of cleanliness of the raw material. The following is an itemization of sources of wastes common to all three die casting processes, with comments as to treatment.

Heating Furnace

Off-gases are normally treated by wet scrubbers. Sludge from metal residuals develops in the bottom of the furnace and can be recovered.

Scrubbers

Wet scrubbers are used extensively at die cast facilities. Ammonia, cyanide, magnesium, phenols, sulfide, copper, iron, and zinc, as well as total suspended solids (TSS) and oil and grease, are waste substances that are common to the scrubber blow-down from all three types of facilities. In addition, aluminum and nickel are found in scrubber blow-down from aluminum die casting facilities. Copper is found in scrubber blow-down from magnesium die casting facilities, and copper is a normal ingredient of scrubber blow-down from zinc die casting facilities.

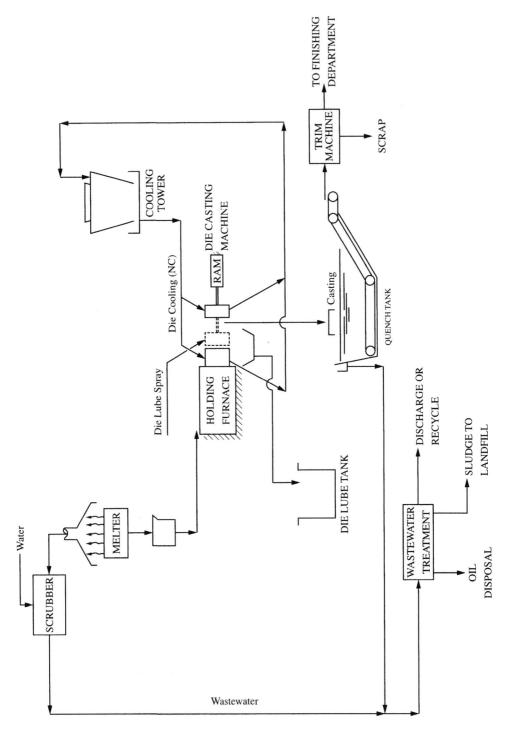

Figure 10-10   Schematic of a zinc die casting process (from the *EPA Development Document*, 1980).

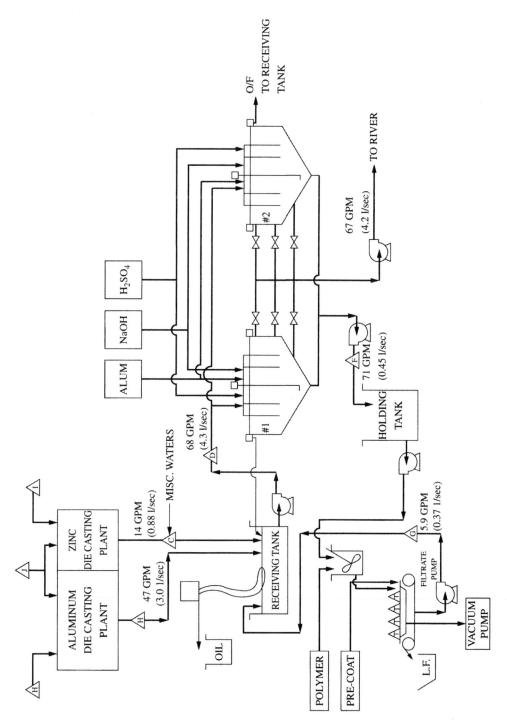

Figure 10-11   Schematic of a magnesium die casting process (from the *EPA Development Document*, 1980).

Treatment of scrubber blow-down after recycle is carried out to the maximum extent it can be by precipitation, coagulation, sedimentation, filtration, and ion exchange.

Quench

The molded product is allowed to cool just enough to solidify in the die. It is then ejected into a water bath, or "quench tank." The water bath is continually renewed to prevent the buildup of impurities to levels that would adversely affect the casting and to keep it cool. The overflow goes to wastewater treatment. The principal objective of water management in the die casting industry is to recycle as much of the treated wastewater as possible. A certain fraction must always be bled, or "blown-down," from the system and an equal amount of clean water replaced, or "made up," thus eliminating the possibility of complete recycle. The fraction that must be blown down is a function of the maximum concentration allowable in the quench tank of the substance that is least well removed in the wastewater treatment system. Blow-down from the quench tanks contains metal residuals, plus substances used to lubricate the dies. Biological treatment has been used successfully where the die lube has been derived from vegetable oils. Metals have been a problem and have resulted in the necessity of disposing of the dewatered sludge as hazardous waste.

Die Casting

Surface cooling sprays and machine and floor wash-down waters, together with leaks from hydraulic systems, leaks from noncontact cooling water piping, and, in some cases, intermingling of all of the above with quench water, produce a waste stream that must be managed. Where all substances in this waste stream have been carefully selected to be biodegradable, including the hydraulic oils and the detergents used for machine, floor, and general plant cleanup, biological treatment has been used successfully.

Finishing

Waste characteristics differ between the three die casting processes. In the case of aluminum die casting, trim pieces are knocked off mechanically, as well as by hand. The solid waste stream that results is recycled back to the scrap aluminum crusher at the head of the process, and associated oils, greases, and "dirt" thus become a portion of the waste discussed earlier. In the case of magnesium and zinc die casting, trimming is done by grinding, which creates dust. The wastes (dust) from this grinding operation are captured by scrubbers. The scrubber blow-down then becomes the principal waste stream from the magnesium die casting finishing process. This waste stream can be treated by chemical precipitation, coagulation, sedimentation, and filtration. Recycle and reuse of the clarified effluent as quench water, cooling water, and/or plant wash-down water can significantly reduce overall waste discharges from the plant. Since lead and zinc are listed as toxic pollutants, all steps in the waste treatment and recycle system must be carefully managed.

Additional Waste Management Considerations

Several substances that have normally been used in aluminum, magnesium, and zinc die casting operations are listed as hazardous. These substances include degreasing solvents, such as perchloroethane, perchloroethylene, trichloroethane, and other chlorinated solvents, as well as cyanide, copper, lead, zinc, and nickel. In some cases, phenolic compounds are present. The following discussion is arranged such that categories of pollutant substances are grouped, regardless of the individual sources, and treatment techniques are presented.

Solids

As discussed in Chapter 7 under Physical Treatment Methods, suspended (TSS) and settleable solids are normally removed from wastewater streams by sedimentation, followed by, or along with, chemical coagulation. Plate or tube settlers are often used. These processes can be followed by filtration. Dissolved solids, such as dissolved metals, can be removed by alkaline precipitation followed by sand or other filtration. Pressure filtration has been used with success. Other filtration processes that have been used prior to recycle and reuse or direct discharge include vacuum filtration and ultrafiltration. Resulting sludges and filter cakes must be disposed of as hazardous waste, unless they fall in a category that has been "delisted."

Dissolved Organics

Dissolved organics in wastewaters from die casting processes can be successfully treated by biological processes, if care is taken to ensure that all products used in the manufacturing process contain only biodegradable organic substances.

Dissolved Inorganics

The principal dissolved inorganics in wastes from die casting processes include the metal being cast, plus those added as alloys, and their salts. Other inorganic substances include sodium, calcium, compounds of sulfur, nitrogen, and chloride and those additional inorganics included in proprietary products. In general, alkaline, carbonate, sulfide, or phosphate precipitation, followed by sedimentation and filtration, are used to remove the metals. As discussed in Chapter 7, the solubility characteristics of the various metal hydroxides, carbonates, sulfides, and phosphates must be carefully considered when designing a metal removal process. In addition, the toxic or other hazardous characteristics of the treatment residuals are a major factor regarding treatment and disposal of these wastes.

Oil and Grease

The two principal methods for removing substances that are measured and reported as "oil and grease" are mechanical skimming for the nondissolved fraction and chemical coagulation (sometimes with just pH adjustment), often with dissolved air flotation, for the dissolved and/or suspended (including emulsified) fraction. Ultrafiltration or reverse osmosis can be used as polishing steps, or, in some cases, as the only method for removal of oil and grease.

Destruction of Phenolics

Every effort should be made to exclude phenolic substances from die casting processing and cleaning materials, and thus from the wastewater. Where unavoidably present, simple phenolic compounds can quite easily be removed by biological treatment. Some complex phenolics must be removed by activated carbon adsorption, which must be preceded by sand (or other) filtration.

Anodizing and Alodizing

Anodizing and alodizing are industrial processes that enhance the property that some metals have of forming a protective coating of the metal oxide on their surfaces. This layer of metal oxide is quite stable and protects the metal from further contact with oxygen, hydrogen ions, and other substances that would otherwise cause further corrosion.

Anodizing is an electrochemical process; alodizing is a strictly chemical process. Both processes act to produce a thicker, more even, and more predictable coating than would be formed naturally. In addition, the anodized or alodized metals have a strong affinity for paints and other organic coatings.

The processes by which anodizing or alodizing coatings form are complex. Basically, oxygen from the hydrolysis of water, in the case of anodizing, reacts with the metal itself to produce two results. One is the formation of pits on the surface of the metal; the other is the formation of metal oxide molecules and ions. The next step is an attachment of the metal oxide species to the newly formed surfaces of the pits. As this process progresses, the local areas on the metal surface where the process has taken place become nonreactive. Consequently, those local areas that have not yet been coated become somewhat more reactive, until the entire surface of the metal becomes coated.

In nature, this progressive process is seldom complete. There are almost always local areas on the metal surface that are not completely coated and are thus exposed to additional corrosion. The controlled industrial processes have the objective of taking the progressive self-coating process to completion.

The thickness of the protective metal oxide coating can be increased by increasing the intensity of the electrochemical (in the case of anodizing) or chemical (in the case of alodizing) process and the time over which it is allowed to take place. Thus, there is control over the product in terms of money spent (energy, chemicals, and time).

Anodizing

When aluminum, for example, is placed in an electrolytic (good conductor of electric current) solution and is made the anode, oxygen from the electrolysis of water reacts with the aluminum at its surface, as illustrated by the following simplified reactions:

$$2H_2O + Elect \rightarrow O_2 + 2H_2 \qquad (10\text{-}2)$$

$$2Al + 3O_2 \rightarrow 2AlO_3 \qquad (10\text{-}3)$$

As aluminum atoms are extracted from the surface of the metal, pits are formed. As well, the surfaces of the pits are particularly reactive, probably because of the availability of electrons for covalent bonding. The aluminum oxide that is in the process of forming also has available electrons; therefore, aluminum oxide bonds with the "raw" metal surface of the pits, forming a relatively stable complex.

The desired properties of the electrolyte are that it conducts electric current efficiently and is a good solvent for the metal species (aluminum ions, partially formed aluminum oxides) involved in the electrolysis process. It should not, however, be a solvent for the final coating product, which is the aluminum oxide–aluminum metal complex.

The electric current intensity, electrolyte characteristics, temperature, and process duration all influence the characteristics of the pitting and the thickness and integrity of the coating. Higher applied voltage increases the speed of oxidation, the size of the pits, and the thickness of the coating. Longer duration increases coating thickness. Temperature and electrolyte characteristics influence the rate of dissolution of the metal surface. Electrolyte characteristics affect pore density. For instance, use of sulfuric acid will result in more than twice the number of pores generated, compared with use of chromic acid.

The size and density of the pores affect abrasion resistance as well as the capacity of the coating to absorb dyes for coloring and paints for desired surface characteristics. Generally, coatings with a higher density of smaller pores have higher resistance to abrasion.

Alodizing

Alodizing produces a protective oxide coating on metal surfaces similar to that produced by electrolysis, but the process is purely chemical. Often, the metal to be coated is "dipped" in an acid solution containing chromate, phosphate, and fluoride ions. Alternatively, coatings can be applied by brushing or swabbing. The acid acts to dissolve metal from the surface, which then reacts with oxygen in the alodizing bath to form the metal oxide, which then reacts with the newly exposed metal surface to result in the protective metal oxide coating. The coating formed by alodizing is characteristically thinner and has less abrasion resistance than does the coating formed by anodizing. Alodizing is an economical alternative to anodizing if resistance to abrasion is not required.

Processing Steps and Wastes Generation

The anodizing and alodizing processes have similar steps that produce wastes. The basic production steps are: cleaning, rinsing, deoxidizing, rinsing, etching, anodizing or alodizing, rinsing, coloring, rinsing, and sealing. All steps produce wastes, described as follows.

Cleaning

The cleaning of metal objects to be processed produces various wastes. If degreasing is required, an important waste that must be managed is the waste solvent, as well as the substance removed. In general, the most appropriate way to manage the waste solvents is to regenerate for reuse, then dispose of the residuals. For instance, if a chlorinated solvent is used for degreasing, regeneration is accomplished by use of a still. The still bottoms are then treated and disposed of by one of the methods described in Chapter 7.

Rinsing

In general, rinsing after any of the five processing steps (cleaning, deoxidizing, etching, anodizing or alodizing, and coloring) produces wastes that are simply dilute forms of the wastes produced directly by those processing steps. Often, rinsing wastes are best managed by operating the rinsing process in a counter-current mode, where two or more rinsing baths are used for each of the processing steps. Clean makeup water is continually added to the final rinse tank, which overflows in the rinse tanks that precede it in the processing steps, and so on, if more than two rinse tanks are used. The overflow from the first rinse tank after each processing step, anodizing, for instance, is then used as makeup water for the process itself.

Deoxidizing and Etching

Deoxidizing and etching are both done with either caustic or acid solutions. Deoxidizing may use both, in series, if the metal to be anodized or alodized is badly oxidized. The purpose of deoxidizing is to remove oxides that have formed naturally but in an undesirable manner. The purpose of etching is to expose a clean, fresh metal surface for the anodizing or alodizing process. Wastes contain spent acid or caustic solutions and ions of the metal being coated.

Anodizing or Alodizing

The spent acid baths from the anodizing or alodizing process constitute the major waste from the coating process. These solutions must be maintained above a certain quality for the coating processes to be satisfactory. Build-up of metal salts (from the metal being coated) must be kept below a level where they interfere with the anodizing or alodizing process. Also, as the solution is weakened by drag-in from the previous rinse process, the active ingredients must be made up to maintain a required minimum concentration. In continuous flow anodizing or alodizing operations, continuous makeup of water and active ingredients can maintain successful operation for a period of time; however, as the acid or alkali bath solutions continually attack the metal being coated, eventually the bath must be dumped and the process restarted with new anodizing or alodizing solution. These dumps of spent solutions, which normally occur about once per month, represent a major waste stream.

Coloring

Coloring, if included in the process, produces wastes that are specific to the coloring process being used.

Tables 10-17 and 10-18 present typical wastewater sources with typical flows and characteristics from an aluminum anodizing operation, which includes acid polishing prior to the caustic etch. The process also includes a coloring process.

Wastes Minimization

As discussed previously, wastes from rinsing can be minimized and sometimes eliminated by use of counter-current rinsing and by

Table 10-17 Wastewater from an Actual Anodizing Plant (anodizing capacity of the plant: 2,500 tons aluminum per month)

Description	Alkaline Wastewater	Acidic Wastewater	Rinsewater
pH	Strong alkaline solution	Strong acidic solution	4–5
Liquid Temp. (°C)	15–60	15–60	15–25
NaOH (%)	5–7	—	—
H_2SO_4 (%)	—	15	—
Al^{3+} (%)	5.0–9.0	1.5–1.8	—
SS* (mg 1)	—	—	100–200
Flow Rate	56 cu m/day	60 cu m/day	250 cu m/hr

*Primarily insoluble $Al(OH)_3$.

using the most concentrated rinse wastes as makeup water for baths that precede that particular rinse. Other waste minimizing alternatives include use of air squeegees to minimize dragout and use of either static rinses or maintaining makeup water flow-through at as low a rate as possible. Strict adherence to the rule of turning water off

Table 10-18 Wastewater Sources

Extrusion	
Extrusion Press Cooling Tower Blow-down	Negligible
Extrusion Press Die Quelch	1,300
Caustic Die Cleaner	1,300
Fabrication	Negligible
Finishing	
Tap Water Rinses	52,000
Tap Water Sprays	2,400
Chiller Cooling Water Rinse	10,000
Deionized Water Rinse	2,000
Demineralized Water Rinse	15,000
Rectifier Cooling Water Rinse	15,000
Periodic Tank Dumping	3,500
Miscellaneous	
Air Compressor Cooling Water	12,500
Regeneration of Water Deionizer Demineralizer and Softener Units	Negligible
Boiler Water Blow-down	Negligible
Total	101,000

when the rinse line is not in use is an absolute requirement. Regarding the processing baths, which may include cleaning baths, deoxidizing, caustic etching, acid etching, anodizing, alodizing, and coloring, all solutions should be rebuilt and/or maintained at working strength according to need based on work done rather than by a regular timetable.

In some cases usable and even marketable products have been prepared from spent acid etch solutions. When aluminum is the metal being coated, sodium hydroxide can be recovered using the Bayer reaction. This can be accomplished by operating the caustic soda etch tank (where applicable) so as to favor movement of the following chemical equilibrium to the right:

$$NaAlO_2 + 2H_2O \rightarrow Al(OH)_3 + NaOH \quad (10\text{-}4)$$

When operation is conducted in such a manner, it is said that the etching bath is operated in a metastable range. However, crystallization nuclei are needed for sodium aluminate to decompose to sodium hydroxide and aluminum hydroxide. A process patented by Alcoa introduces the waste etch solution to a reactor that contains 300 to 500 g/L aluminum hydroxide. Precipitated aluminum hydroxide is separated from the sodium hydroxide solution by filtration, and the sodium hydroxide solution is returned to the caustic etch process. The aluminum hydroxide can be marketed as a waste treatment chemical.

Another method that combines waste treatment (of the caustic etch wastes) with wastes minimization is to precipitate and recover calcium aluminate from spent caustic etching solutions. High-calcium lime that contains aluminum ions at a high pH is added to the spent etch solution. The recovered calcium aluminate can be sold or given to a cement manufacturer to be used as an additive. Spent caustic etch solutions have been processed as shown by the following chemical reaction:

$$NaAlO_2 + 2SiO_3Na_2 + 9H_2O$$
$$\rightarrow 4NaOH + Al_2O_3 \bullet Na_2O \bullet 2SiO_2$$
$$(10\text{-}5)$$

The product, nephelin hydrate, can be marketed as a toilet-cleaning product.

Regarding spent acid etch solutions, a strong base ion exchange resin can be used to separate the acid from its salts.

Waste Treatment and Disposal

Treatment and disposal of wastes, including solid wastes and liquid wastes, are best accomplished by recovering substances for reuse wherever feasible and then treating remaining waste solutions and solids using the procedures presented in Chapter 7 so as to produce as few residuals for disposal as possible. For instance, spent acid and caustic solutions can be combined to produce water and nonhazardous salts such as sodium sulfate. Precipitated metal salts can be treated and conditioned so as to enable disposal as ordinary nonhazardous waste.

Production and Processing of Coke

Coke is a dark gray, porous solid that is produced when pulverized soft coal is heated in an oxygen-deprived atmosphere. It contains 87% to 89% carbon and burns with intense heat and very little smoke. Coke is used as fuel in blast furnaces for the manufacture of iron and steel.

The Coking Process

The process of producing coke involves heating coal to about 2,000°F. Many of the organic substances that make up coal volatize at that temperature, leaving the "coke" behind. The volatilized gas is then subjected to sequentially lower temperature condensing chambers (as shown in Figure 10-12), which capture tar (a mixture of many relatively heavy organic compounds), oils, light oils, and then low-molecular-weight gases. The separated coke is then used as fuel or as a component of steel. Often, some of the coke is used as fuel in the same coke ovens where it was produced. Production of coke is one of the major processes in an integrated steel mill.

As illustrated in Figure 10-12, the first compounds (highest condensing temperature) to be recovered are tars. Some coke plants recover several high-molecular-weight materials from several separate high temperature condensers. Others simply use only one high-temperature condenser, which collects all of the high-molecular-weight compounds together as the tar. The next successively lower-temperature condensers collect oils, then light oils, and finally low-molecular-weight (e.g., methane, ethane, propane, etc.) gases.

The heating of the coal is done in narrow, rectangular, silica brick ovens. The ovens stand in groups of 10 to 100 or more, called batteries. After about 18 hours of "cooking," the remaining substance, "coke," which amounts to 1,300 to 1,550 pounds for each ton of coal heated, is pushed into a quenching car that transports the coke to quenching towers. Here, the coke is sprayed with water to lower the temperature. About 35% of this water evaporates and leaves the system as steam or water vapor. The remaining water drains to a settling basin where the coke fines are removed.

The vaporized organics produced in the ovens are withdrawn during cooking by exhausters and then sprayed with water in the first of several condensers. The sprayed water saturates and cools the gas, causing condensation of the tar. The mixture of "flushing liquid" or "flushing liquor" and tar flows to a separator where the water is decanted. The remaining gas flows on through a tar extractor, which is an electrostatic precipitator, which removes most of the remaining tar.

Ammonia can also be recovered from the flushing liquid, using one of two methods: the semidirect or the indirect processes. In the semidirect process, ammonia is recovered by use of an ammonia absorber or saturator after the tar extractor. The gas is brought into contact with 5% to 10% sulfuric acid solution, causing ammonium sulfate to precipitate. The crystals of ammonium sulfate are dried and sold. In the indirect process, some of the ammonia dissolves into the flushing liquor. More ammonia is scrubbed from the gas with water and mixed with the flushing liquor. The flushing liquor is then distilled, dephenolized, and the recovered ammonia is marketed.

The phenol that is removed from the ammonia liquor is recovered as sodium phenolate. Light oil is recovered as the gas passes through a scrubber that uses an absorbent known as straw oil. The straw oil absorbs 2% to 3% of its weight of light oil. The remaining gas is then used for fuel, either on site or marketed. Steam distillation is then used to strip the oils from the straw oil that was previously the absorbent. The wash oil is then cooled and returned to the scrubbers as illustrated in Figure 10-12. Hydrogen sulfide is removed from the coke oven gas after the light oil scrubbers.

Sources of Wastes

The production of coke gives rise to considerable solid, airborne, and waterborne wastes. Phenolic compounds, which make up a considerable portion of coal, are found in wastes discharged to all three media. It is essential, then, to determine the status regarding hazardous nature of all wastes as part of wastes management.

Solid Wastes

Solid wastes from the production of coke include reject coal, reject coke, sludges from clarification of quench water, as well as normal packaging and shipping wastes. These are in addition to the normal solid waste stream from most "typical" industrial facilities, which includes construction debris from plant maintenance, repair, and expansion. Also, there are normally items of broken or worn out equipment that must be managed.

Airborne Wastes

Air pollution is a major concern at coke production facilities. Blowing coal dust must be controlled by containment. Since the coking process is a heating and vaporization process, emissions from the oven stacks must be controlled. Also, there is the need to contain the vapors to prevent loss via fugitive emissions or directly out of one or more stacks. Electrostatic precipitators, bag houses, and wet scrubbers are used in conjunction with aggressive preventive maintenance to avoid leaks.

Waterborne Wastes

The principal sources of wastewater from the production of coke are as follows:

- Excess flushing liquor
- Final cooling water overflow
- Light oil recovery wastes
- Condenser wastes from the crystallizer
- Gas stream desulferization
- Sludges from air pollution control equipment
- Coal pile runoff
- Overflows from sumps, including the quench sump

The major source of wastewater is flushing liquor. The quantity of this wastewater varies

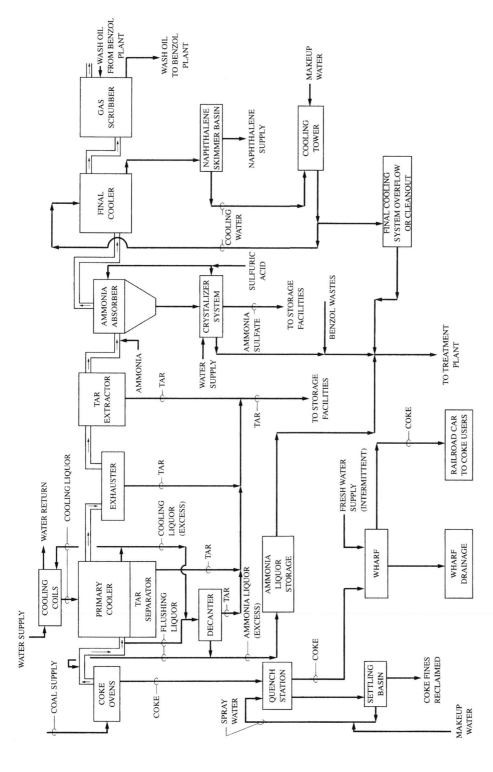

Figure 10-12   The coking process for producing coke and byproducts (from the *EPA Development Document*, 1973).

with the moisture content of the coal and with the process used for recovery of ammonia. Since it the first water to come into contact with the coke oven gases, it has the highest concentrations of pollutants. Contaminants include ammonia, phenol, sulfides, thiocyanates, and cyanides.

Final cooling water overflow results from spraying the gas stream with water to remove remaining water-soluble components and naphthalene crystals. After the naphthalene has been removed from the stream, the water can be used for cooling. Some process wastewaters have been recycled by using them for quenching; however, caution must be exercised. As the water is evaporated during quenching, pollutants remaining in the process wastewater may be volatilized, causing air pollution. Those contaminants that are not volatilized stay with the quenched coke and thus represent a potential problem during subsequent use of the coke.

Wastes Minimization

Cover and containment are extremely important in order to minimize wastes generated at a coke production facility. If it is at all possible, the coal pile should be covered to prevent contamination of stormwater. Also, it is important to cover the quenching system and to contain the quench water; that is, to not allow quench water sumps to overflow. In addition, the following wastes minimization measures should be strictly adhered to:

- Drips must be contained and returned to the source.
- Aggressive maintenance must be practiced to eliminate the occurrence of leaks or other "accidents" that could lead to noncontainment of chemicals and other substances.
- Dry methods of cleanup, including brooms, shovels, and dry vacuuming, should be used to the maximum extent.
- Purchasing should be guided by aggressive selection of raw materials to obtain the cleanest possible materials.

- Purchasing should be guided to demand that the packaging of materials delivered to the plant be recyclable or otherwise of low solid waste nature.
- There should be a constant and consistent program to substitute less polluting and nonpolluting substances for those that require expensive treatment and expensive disposal. For instance, low-sulfur coal should be used to the extent that financial feasibility allows to minimize the quantity of sulfides and other sulfur compounds in the waste air and water.

Air Pollution Control

Control of blowing coal dust and coke fines can only be accomplished by covering the sources. Covering has the dual benefit of reducing wastewater by preventing contamination of stormwater.

Scrubbers, both wet and dry, are used extensively at coke production facilities. Also, bag houses are used in conjunction with systems that have the purpose of maintaining clean air within the work areas of the production facilities. The blow-down from the wet scrubbing systems is a significant source of wastewater that must be managed.

Wastewater Treatment

Biological treatment systems have been used to successfully treat wastewaters from the coking process, even though these wastewaters normally contain significant amounts of toxic substances. Two procedures are used to overcome the toxic effects of these substances. The first is to remove some of them by use of a pretreatment step. The second is to employ the process of gradual acclimation of the biological treatment system to the wastewaters.

The removal step involves combining wastewaters from light oil recovery, final cooling, air pollution control blow-down, and excess flushing liquor. This combined wastewater stream is then passed through the free leg of the ammonia still. The waste from this process is then increased in pH by adding lime. It is then passed through the fixed leg of the ammonia still, where ammonia is removed and recovered. The effluent from this process is then combined with wastewater from the crystallizer (see Figure 10-12) and held in a storage tank for a period of time. The pH of the effluent from this holding tank is then lowered with acid to the neutral range. The pretreated wastewater is now ready for treatment by a properly acclimated biological system.

The acclimation step involves subjecting the biological treatment system, activated sludge, for instance, to gradually increasing levels of the toxic substances in the full strength wastewater, until full-strength wastewater is being treated. One way to accomplish acclimation is to develop the activated sludge system using domestic wastewater until it is fully functional as a wastewater treatment system. Then, the pretreated industrial wastewater is added to the domestic wastewater starting with a mix of about 10% pretreated industrial wastewater and 90% domestic wastewater. When it has been established that the treatment system is working well on the 10%/90% mixture, the proportion of pretreated industrial wastewater is increased, and so on, until the treatment system is receiving and operating well on full-strength pretreated wastewater. What happens during the acclimation procedure is that the microorganisms develop the capability to produce enzymes that can metabolize the toxic materials in the pretreated industrial wastewater and that can grow and flourish in the activated sludge (or other biological treatment medium). Those that cannot gradually die off are replaced over time. The result is an "acclimated" biological treatment system that is able to successfully treat the pretreated industrial wastewater.

An acclimated biological treatment system consisting of three extended aeration basins in series has been developed. Phenols are removed in the first basin, oxidation of both ammonia and cyanide takes place in the second basin, and the third basin (nonaerated) is used to remove nitrogen via the denitrification process. It is necessary to maintain favorable concentrations of other nutrients, such as phosphorus, in each of the biological treatment basins.

Chemical–physical treatment systems have been used with some success. One version of chemical–physical treatment has been to use chemical oxidation to destroy organics, ammonia, and cyanide and then to use activated carbon to remove unreacted substances as well as partially treated substances. Oxidation is accomplished by adding chlorine to the pretreated wastes in an aeration basin. Oxygen from air and chlorine are the oxidizing agents. Spent activated carbon can then be recovered (partially) by the process of heating (incinerating).

The Wine Making Industry

Production of wine is one of the oldest endeavors of human industry. The historical record shows that the Egyptians as well as the Assyrians were making wine from grapes by 3500 B.C. The basic process has remained unchanged for centuries; however, some new wine products have been developed during more modern times.

In the year 2000, approximately 12% of the world's production of wine was taking place in the United States. About 80% of the world production was in Europe. California produced more than 20% of U.S. wine. Other wine producing states, in approximate order of quantity, were New York, Washington, Pennsylvania, and Oregon.

The most widely used grape for wine production is *Vitis vinifera*, known as the European grape. It is grown throughout Europe, the United States, Australia, Chile, and in regions of Asia. Other grapes used for wine production include *Vitis rotundifolia* and *Vitis labrusca*, but it is widely agreed that superior wines are produced from *vinifera* varieties.

The Wine Production Process

The basic, age-old process for producing wine includes six steps: destemming, crushing, pressing, fermentation, racking, and bottling. While some destemming takes place before crushing, destemming is also accomplished simultaneously with the crushing process.

When the grapes are ripe they are picked by hand or by use of mechanical harvesters. They are transported to the winery, which is typically close by, and are destemmed and then crushed. The amount of destemming that takes place before crushing depends on the type of wine to be made. The stems impart tannins to the wine and thus influence the color and flavor. The product of the grape crushing process becomes what is called "must," sometimes before and in some cases after the solids are removed from the juice. The must is what then enters the fermentation process. If white wine is to be produced, the solids, including the skins, seeds, and the quantity of stems that remain, are removed before fermentation. If red wine is to be produced, the solids are considered part of the must. Most of the pigments in grapes are located in the skins. Depending on the characteristics desired in the wine product, varying degrees of care are exerted to avoid breaking the seeds during the crushing process.

Sulfur dioxide is often used to treat the must before fermentation. In some cases, sufficient sulfur dioxide treatment is used to kill all of the yeast and other microorganisms naturally present in the must. Then the desired species of yeast is added. Other effects of treating the must with sulfur dioxide include the following: settling characteristics for solids removal are improved, thus improving clarification (desired for some white wines); the color of red wines is changed somewhat; and, in some cases, storage characteristics are improved due to inhibition of undesirable enzyme activity.

After treatment, if any, the must is pumped into tanks for fermentation. Typically, some of the juice is pumped over the top of the "cap" that forms as a result of skins and other solids collecting on the surface. This is done to increase the extraction of pigments from the skins. As ethanol increases in amount during the fermentation process, it increases the rate of extraction of pigments.

When the desired intensity of color has been achieved, the partially fermented must is pressed to separate solids from the juice, as is done prior to fermentation for white wines. Many types of presses are used, but all have the common objective of gently squeezing juice from the skins, seeds, pulp, and other solids without extracting undesirable substances. At this point, the solids become part of the waste stream, while the juice is returned to tanks to complete the fermentation process. When fermentation is complete, the process called "racking" is initiated. The objective of racking is as follows: after the grapes have been pressed to remove the skins, stems, and seeds, and most of the pulp and the juice has been returned to vats for completion of fermentation, a layer of sediment called "the lees," composed of dead yeast cells and bits of grape fragments, forms on the bottom of the vat. If the developing wine is allowed to remain in contact with the lees, off-flavors develop from decomposition of the lees. Consequently, the developing wine is racked: drawn off and placed in clean vats.

When fermentation has proceeded to completion, various substances are often added, which are referred to as "fining agents." Fining agents act to remove colloidal solids and include bentonite clay, egg whites, and gelatin. After clarification by sedimentation, the wine is normally filtered and then aged in wooden vats or bottled.

Variations on the wine process include the processes for making brandy, dessert wines, sparkling wines, and champagne. Brandy is made by first distilling the alcohol from the waste materials such as the solids, or pomace, left after pressing. Then, this alcohol is added to the wine to raise the total alcohol content to about 20%. In some cases, brandy is added to other wines, along with sugar and possibly other substances, to produce dessert wines.

Sparkling wines are made by allowing the fermentation process to take place in a closed container. The CO_2 produced during the fermentation process dissolves into the wine under pressure. In some cases compressed CO_2 is added to enhance the natural CO_2 content of the sparkling wine.

Champagne is normally produced by adding sugar and more yeast to finished wine. The yeast ferments the added sugar to produce more alcohol and CO_2. The process takes place in a closed container in order to retain the CO_2, as explained previously for sparkling wines.

Winemaking Wastes

The six major steps in wine production all produce wastes, either solid or liquid, or both. Air pollution is not a normal problem in winemaking, unless unusual circumstances result in a problem with emissions from boilers or in the production of odors. Figure 10-13 presents a schematic diagram of an example winery, where production of both wine and brandy take place. As illustrated in Figure 10-13, plant cleanup operations produce plant washdown wastes from essentially all the wine making steps.

The destemming step produces waste stems and "dirt" that can be land-applied. This can usually be done without causing pollution problems. The next step, crushing, results in normal plant cleanup wastes. The waste lees that result from the pressing step represent varying quantities of waste material depending on whether or not by-product recovery is practiced. Fermentation also results in normal plant cleanup wastes, plus the dregs from the fermentation process itself (dead yeast cells and various settled or filtered solids from the grapes themselves).

Racking results in sediment in the bottom of the vats as well as normal plant washdown wastes. Finally, bottling results in more plant washdown wastes. All of the above wine making steps, except for destemming, con-tribute wastes in the form of lost product, including the final step, bottling, where lost wine product from spillage, overfilling, bottle breaking, and other unintentional releases contribute to the BOD level of the overall plant waste stream.

Wastes Minimization

As is the case with many food processing industries, by-product recovery can reduce the quantity of wastes that require treatment. One example is the recovery of tartrates, salts of tartaric acid that occur in lees, pomace, and on the surfaces of wine storage tanks. Tartrates are rinsed off the surfaces of storage tanks and extracted from pomace using water and are then precipitated with calcium carbonate or lime. The dried precipitates are used in cooking as cream of tartar. Another example is the recovery of an edible oil from grape seeds.

Pomace is sometimes used for animal feed. Otherwise, it can be mixed with the stems and other solids and returned to the vineyards for use as a soil conditioner and fertilizer.

Treatment of Winery Wastes

The principal liquid waste from wineries is the wash water from general plant cleanup, as well as from vat cleaning, bottle cleaning, and product loss. In general, biological treatment processes have been used with success, including anaerobic as well as aerobic processes. Farmer, Friedman, and Hagin (1988) reported on a pilot plant project in which an upflow anaerobic sludge blanket (UASB) and an anaerobic contact process were used to treat high-strength wastes (Chemical Oxygen Demand, COD, of about 15,000 mg/L) from a winery. Both processes were successful in removing 98% of the soluble COD. Following the processes with aerobic treatment increased overall COD removal to more than 99%. As expected, the principal advantages of the anaerobic processes over using only aerobic treatment were that the anaerobic

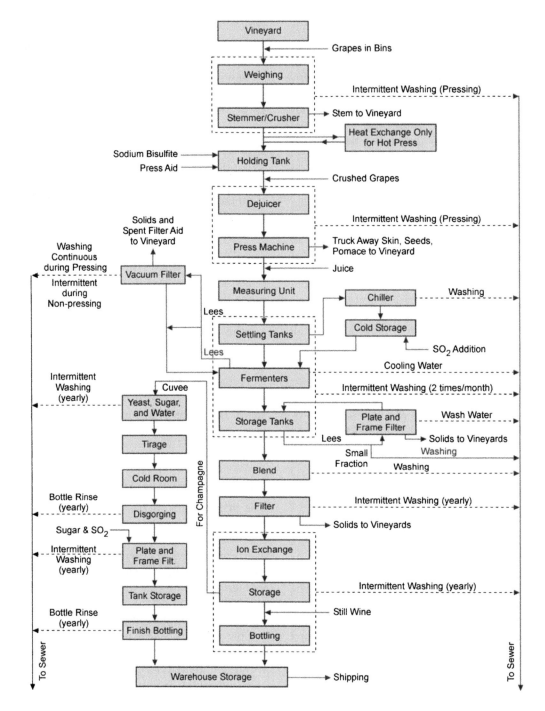

Figure 10-13   Production of wine and brandy.

processes produced about 80% less sludge and occupied a significantly smaller footprint.

One of the most popular methods for treating winery wastes has been that of land disposal, especially the ridge and furrow method. Russell et al. (1976) reported on a land application system that treated up to 50,000 gallons of wastewater per day in California. The operation was called an "unqualified" success, with negligible contamination of groundwater.

Ryder (1973) reported on the successful use of an aerobic lagoon to treat winery wastes. Ryder emphasized the value of the aerated lagoon effluent as irrigation water for the vineyards.

An aspect of winery waste treatment that is different from most other industrial wastes is that the destruction of any waste containing alcohol must be in accordance with regulations of the Bureau of Alcohol, Tobacco, and Firearms (BATF). The law specifically regulates disposal of lees, requiring that wine must be completely pressed or drained from the lees before it can be used for animal feed or production of cream of tartar or other uses.

The Synthetic Rubber Industry

There are many different types of synthetic rubber produced today and a host of products made from them. Annual (world) production in the year 2000 exceeded 20 million tons, accounting for well over 80% of all rubber products produced, including natural rubber.

Nearly 300 years after Columbus discovered Haitian children playing with what would later be known as rubber balls, Joseph Priestly, an English chemist, discovered that the gum from certain trees grown in South America, Haiti, and nearby islands would rub out pencil marks. Hence, the name "rubber" was born. During the next 100 years, use for this remarkable substance, produced from the sap of *Hevea brasiliensis*, which was still confined to the jungles of South America and nearby islands, developed slowly. In 1876, the British transplanted thousands of these trees in southeastern Asia and began large-scale production of rubber, now known as natural rubber. The milky sap from which natural rubber is made has, since that time, been called latex, from the Latin word for milky, *lacteus*.

Shortages during the war years, from World War I through the early 1950s, prompted the development of synthetic rubber. During World War II, the effort put forth by the United States to develop its synthetic rubber manufacturing capability was second only to its efforts to develop the atomic bomb. The principal synthetic rubber produced during that time period was called GR-S, short for "Government Rubber—Styrene," which is still produced as SBR, Styrene Butadiene Rubber.

The many different types of synthetic rubber produced in the early 2000s include silicone rubbers, used in the aerospace industry for their usefulness over wide ranges of temperature; butyl rubber, used for the manufacture of inner tubes; chloroprene rubber, used to line tanks to prevent corrosion; latex (foam rubber), used for the manufacture of cushions, padding, and other things; nitrile rubber used in the manufacture of oil-resistant and heat-resistant hoses, gaskets, and other things; and styrene butadiene rubber (SBR), the most commonly used type of synthetic rubber. SBR is used in the manufacture of automobile tires, conveyor belts, noise and vibration insulators, foul weather gear, and many other things. More than 500 different types of SBRs are produced, most for the automobile industry.

For purposes of regulation, the EPA has divided the rubber manufacturing industry into two categories: the Tire and Inner Tube Industry and the Synthetic Rubber Industry. Both categories make extensive use of SBR. Because the raw materials from which SBR is manufactured are derived from petroleum, the synthetic rubber industry tends to be located within two clusters. The plants that produce synthetic rubber are clustered in the oil producing states, mainly Louisiana and Texas, while those that produce products made from synthetic rubber are located in industrialized areas, especially those associated with the automobile industry.

Production of Synthetic Rubber

So-called tire rubber, also used for many other products, is produced as a substance known as "crumb rubber." There are two principal methods for producing crumb rubber: solution crumb production and emulsion crumb production. Solution crumb production involves mixing the raw materials in a homogenous solution, wherein polymerization takes place. Emulsion crumb production involves producing an emulsion of the raw materials, resulting in bulk polymerization of droplets of monomers suspended in water. The mechanisms of solution polymerization are those of ionic interaction. The mechanisms of emulsion polymerization require sufficient emulsifier, in the form of a soap solution, to maintain a stable emulsion, and proceed as explained below.

Emulsion Crumb Production

Production of crumb rubber by emulsion polymerization has been the traditional process for production of synthetic rubber. It is still the most commonly used process, accounting for 90% of the world's production of SBR. Figure 10-14 presents a schematic diagram of the emulsion crumb rubber production process.

As illustrated in Figure 10-14, raw materials in the form of monomers, produced by the petroleum industry, is delivered to tank farms. The monomers include styrene and butadiene. Because production facilities such as these operate 24 hours per day, 365 days per year, the tank farm always maintains a constant supply. Other materials delivered and stored in tank farms include soap (or

detergent), activator, catalyst, modifier, extender oil, and carbon black. Butadiene, as well as certain other polymers, is delivered with polymerization inhibiters mixed in to prevent premature polymerization during delivery and storage.

The production process begins with removal of the polymerization inhibiters, by passing the monomer through a caustic scrubber. The monomers to be polymerized are mixed with soap solution, deionized water, catalyst, activator, and modifiers prior to entering the first of a series of reactors. The purpose of the soap solution (or detergent) is to hold the entire mixture in a stable emulsion throughout the polymerization process. Rosin acid soap or fatty acid soap is typically used. The purpose of the catalyst is to generate free radicals to initiate and maintain polymerization. Typical catalysts used are hydroperoxides or peroxysulfates.

The function of the activator is to assist the catalyst in generating free radicals, as well as to enable reaction at a lower temperature. The modifier acts to control the length of polymerized chain and, consequently, the size (and molecular weight) of the polymer. The process is operated as either "cold" (40°F to 45°F at 0–15 psig) or "hot" (122°F at 40–60 psig), to produce cold SBR or hot SBR, respectively. Cold SBR processes are stopped at 60% polymerization. Hot SBR processes, the older of the two, are allowed to proceed to near completion of the polymerization process.

"Cold rubbers" have improved properties, compared with hot rubbers but require more extensive process management. The emulsified mixture resulting from the initial mixing of monomers and additives must be kept cool by means of an ammonia refrigerant prior to entering the reactors. Also, a "short-stop" solution must be added to the solution exiting the reactor, in order to halt the polymerization process at 60% completion. Then, the unreacted monomers, as well as the catalysts, activators, modifiers, water, and emulsifiers, must be separated from the poly-

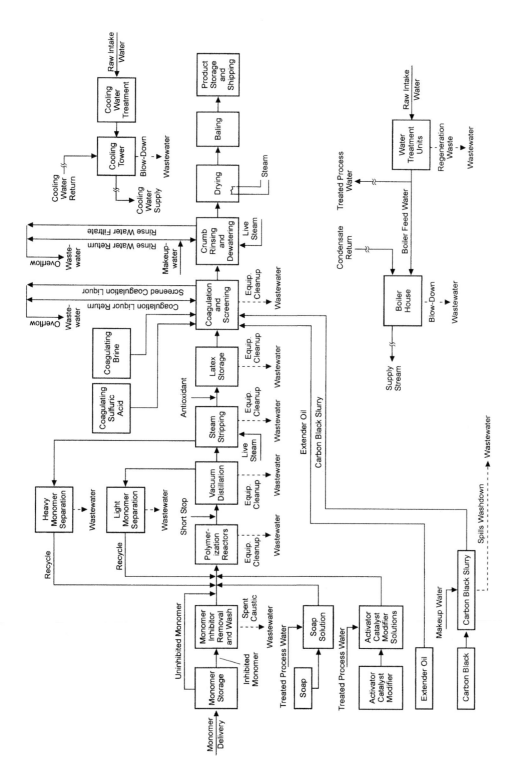

Figure 10-14   General water flow diagram for an emulsion polymerized crumb rubber production facility.

merized material, reconditioned, and returned for continued use.

The shortstop solution is made up of sodium dimethyl dithiocarbonate and hydroquinone. The 60% reacted mixture, at the point of being "stopped," is milky-white in appearance and is called "latex." Although it resembles the sap from the natural rubber tree in appearance, the two are not to be confused with each other.

The next step is recovery of unreacted butadiene by use of a vacuum flash tank. Recovery of styrene is accomplished by use of perforated plate stripping columns. The stripped latex must be protected from oxidation by oxygen or ozone. This is accomplished by addition of an antioxidant in a blend tank. The polymerized material is then separated from the remaining latex by coagulation, which is brought about by the addition of dilute sulfuric acid (pH 4 to 4.5) and sodium chloride (brine). At this point, the product is called "crumb." Carbon black and various oils are added to the crumb to produce desired properties, including color.

Separation of the coagulated crumb is accomplished on a shaker screen. The liquor is then returned for reuse, after reconditioning by addition of fresh acid and brine. The crumb is washed by resuspension in water in a reslurry tank. Gravity separators, called crumb pits, are used as clarifiers to recover floatable crumb rubber from both the liquor that passed through the shaker screen and the overflow from the reslurry tank.

The filtered and washed crumb rubber is dried by use of hot air, then weighed, baled, and stored for shipping.

Solution Crumb Production

Production of crumb rubber by the solution crumb process allows the use of stereospecific catalysts that are able to produce polymers nearly identical to natural rubber. In the proper organic solvent, as opposed to an emulsion, as used in emulsion polymerization, the cis structure can be obtained in an amount up to 98% of the total.

Figure 10-15 presents a schematic diagram of a solution polymerized crumb rubber production facility. As can be seen, there are similarities to the emulsion polymerization process, but there are several important differences. Solution polymerization requires that the monomers be of a very high degree of purity. As well, the solvent in which the monomers, catalyst, and modifiers are dissolved must be absolutely anhydrous. The polymerization process proceeds to more than 90% completion, in contrast to the emulsion polymerization process, which is stopped by the introduction of the shortstop solution at the point of 60% completion.

Figure 10-15 shows that the stored monomers are pumped from the tank farm through caustic soda scrubbers to remove polymerization inhibitors. The monomers proceed to fractionater-drying towers where water is removed. These towers are also used to remove water from recycled as well as fresh solvents. The monomers and solvents are mixed to produce "mixed feed," and catalysts are added. This begins the polymerization process, which is exothermic. For this reason, the reaction vessels must be cooled, typically by an ammonia refrigerant. The shortstop solution is added after the reaction has reached 90% or more completion.

After the polymerization reaction has been halted, additional substances are added, including antioxidants and oil for oil extension. The mixture is then transferred to a coagulation vessel where the polymerized material precipitates as the crumb. Carbon black is typically added at this point.

The mixture is now ready for separation of the crumb from the liquor and separation of some of the components of the liquor for reuse. A series of strippers is used to strip off solvents as well as unreacted monomers. The vapors of monomers and solvents are condensed and sent to a decant system. The organic portion is returned to the fractionator. Decant is discharged as wastewater. The stripped crumb slurry is separated further and proceeds to a vibrating screen where it is washed with water. Most of the washwater

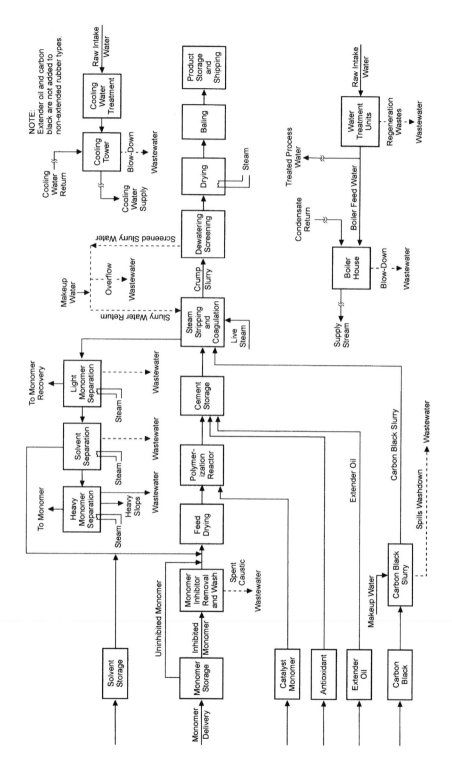

Figure 10-15   General water flow diagram for a solution polymerized crumb rubber production facility.

recycles to the coagulation stage; the rest is discarded as blow-down. The crumb is dried, baled, and stored for shipment to a processor.

Specialty Products

Far less important in terms of annual production totals, but important in terms of pollutant generation, are the specialty rubbers. These products have more diverse composition as well as end uses. The largest of these, in terms of production volume, are the butadiene rubbers, which are sold by producers in a latex form as opposed to the crumb rubber form.

Production of latex is the same as emulsion crumb production, except that the coagulation step for separation of the crumb rubber from the liquor is not carried out. Another difference, important in terms of pollutants generated, is that in latex production polymerization is allowed to proceed essentially to completion. This is in contrast to the production of crumb rubber in which the polymerization process is stopped at the point of 60% completion. Consequently, recovery of unpolymerized monomer is not feasible. Figure 10-16 presents a schematic of the latex production process.

Wastes Generation

The generation of wastes at manufacturing facilities engaged in the production of synthetic rubber is dependent to an unusually high degree on how well the entire facility is managed to prevent such generation. Raw materials are delivered as bulk liquids, and the delivery process can be pollution-free if there are no spills and fumes are contained. The entire process, except for clarification and decant steps, is contained within tanks, pumps, piping systems, and mixing vessels; therefore, the only opportunities for air pollution are from leaks and sloppy handling. The only solid material involved in the production process that is of significant volume is the product rubber itself. Any loss of this material to solid waste is a loss of valuable product.

Solid Wastes

Solid wastes requiring management at a typical synthetic rubber production facility should be only nonproduction-related wastes, such as packaging and shipping wastes and construction debris from plant maintenance, modifications, expansions, and periodic facility upgrade projects. In addition, sludges from wastewater treatment and waste resins from process water deionization require management.

Airborne Wastes

The many scrubbers and strippers used at synthetic rubber production facilities are potential sources of air pollutants. Vents on tank farm storage facilities are also potential sources.

Waterborne Wastes

Figures 10-14 through 10-16 show principal sources of wastewater from (1) the emulsion crumb production process, (2) the solution crumb production process, and (3) the emulsion latex production process, respectively. As shown in these figures, essentially every major processing unit is a source during the normal processing schedule. For instance, Figure 10-14 shows that as the monomers are being transferred from storage to the polymerization reactors, they are passed through the caustic scrubbers to effect removal of the polymerization inhibitors. There is a blow-down of spent caustic wash solution, contaminated with the polymerization inhibitor material, entering the wastewater stream. What is not shown is that each of these processing units is a source of wastewater as a result of periodic washdown. However, these figures do show that additional areas of the plant are at least potential sources of wastewater as the result of spills or overflows. When these largely unnecessary events occur, they must be managed properly to avoid damage to the environment.

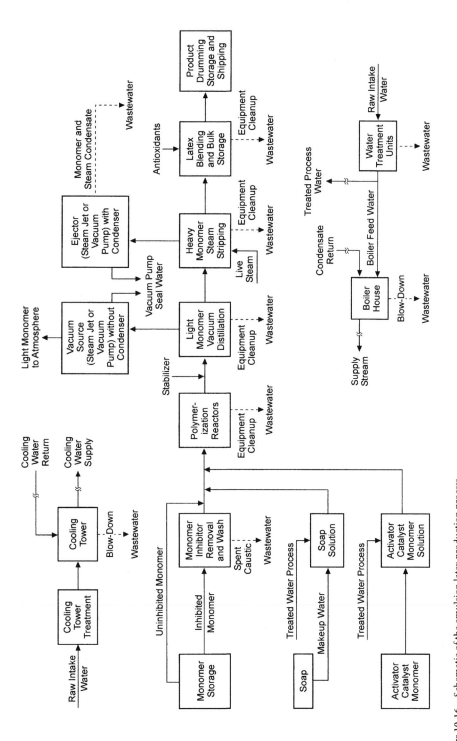

Figure 10-16   Schematic of the emulsion latex production process.

Table 10-19   Summary of Potential Process-associated Wastewater Sources from Crumb Rubber Production via Emulsion Crumb Production

Processing Unit	Source	Nature of Wastewater Contaminants
Caustic soda scrubber	Spent caustic solution	• High pH, alkalinity, and color
		• Extremely low average flow rate
Monomer recover	Decant water layer	• Dissolved and separable organics
Coagulation	Coagulation liquor overflow	• Acidity, dissolved organics, suspended and high dissolved solids, and color
Crumb dewatering	Crumb rinse water overflow	• Dissolved organics and suspended and dissolved solids
Monomer strippers	Stripper cleanout and rinse water	• Dissolved organics and high suspended and dissolved solids
		• High quantities of uncoagulated latex
All plant areas	Area washdowns	• Dissolved and separable organics and suspended and dissolved solids

The sources of wastewater from the emulsion crumb production process, as illustrated in Figure 10-14, are presented in Table 10-19.

The sources of wastewater from the solution crumb production process, as illustrated in Figure 10-15, are presented in Table 10-20.

The sources of wastewater from the emulsion latex production process, as illustrated in Figure 10-16, are presented in Table 10-21.

Wastes Minimization

Maintaining a very high level of attention to containing fumes and preventing spills, overflows, leaks, and other unintended discharges is of paramount importance. Containment of the ammonia refrigerant, used to prevent the exothermic polymerization reactions from attaining too high a temperature, is important.

Several of the processing steps in all three of the production procedures involve separation of product from liquor by screening, gravity separation, or combinations of these processes. These steps are followed by liquor renovation and recycle for reuse, integrated with a blow-down. Close attention to the characteristics of the blow-down to ensure that it is not a larger fraction of the total quantity of liquor than it needs to be is an important wastewater minimization procedure.

In all of the variations (emulsion and solution) for production of crumb, there is a procedure for washing and rinsing the prod-

Table 10-20   Summary of Potential Process-associated Wastewater Sources from Crumb Rubber Production via Solution Polymerization Processing

Processing Unit	Source	Nature of Wastewater Contaminants
Caustic soda scrubber	Spent caustic solution	• High pH, alkalinity, and color
		• Extremely low average flow rate
Solvent purification	Fractionator bottoms	• Dissolved and separable organics
Monomer recovery	Decant water layer	• Dissolved and separable organics
Crumb dewatering	Crumb rinse water overflow	• Dissolved organics and suspended and dissolved solids
All plant areas	Area washdowns	• Dissolved and separable organics and suspended and dissolved solids

Table 10-21   Summary of Potential Process-associated Wastewater Sources from Latex Production via Emulsion Polymerization Processing

Processing Unit	Source	Nature of Wastewater Contaminants
Caustic soda scrubber	Spent caustic solution	• High pH, alkalinity, and color
		• Extremely low average flow rate
Excess monomer stripping	Decant water layer	• Dissolved and separable organics
Tanks, reactors, and strippers	Cleanout rinse water	• Dissolved organics, suspended and dissolved solids
		• High quantities of uncoagulated latex
Tank cars and tank trucks	Cleanout rinse water	• Dissolved organics, suspended and dissolved solids
		• High quantities of uncoagulated latex
All plant areas	Area washdowns	• Dissolved and separable organics and suspended and dissolved solids

uct crumb with water. The spent wash water enters the wastewater stream. Development of countercurrent washing and rinsing procedures, by which fresh water is used only for the final rinse and maximum recycle of all effluents is practiced, is imperative. For instance, the effluent from the first rinse should be used as makeup for the wash water used in conjunction with the vibratory screen separation process, discussed above.

There are success stories involving waste minimization by use of substitution in the synthetic rubber production industry. For instance, coagulation of latex with acid-polyamine rather than brine solution results in significantly lower TDS in the wastewater. As another example, the use of a steam grinding technique for the addition of the carbon slurry has resulted in a significant reduction in spillage of carbon black. This, in turn, has resulted in significant reduction in washdown and runoff wastewaters contaminated with carbon black. As still another example, the use of alum, on the spot, to coagulate crumb material from spilled latex solution can reduce the amount of washdown water required. Finally, use of dual crumb pits can effectively avoid resuspension of colloids during cleaning.

Wastewater Treatment

In general, all of the raw materials used in the production of synthetic rubber are biode-gradable, and all of the wastewaters are amenable to biological treatment. Also, as a general rule, wastewaters generated by emulsion crumb and emulsion latex production procedures require chemical coagulation (to break the emulsions) prior to chemical treatment. This is not true of the solution polymerized production wastewaters. Following appropriate preliminary treatment, conventional primary treatment by gravity clarification, possibly aided by settling aids; secondary treatment by any of the variations of the activated sludge processes, with appropriate addition of nitrogen and phosphorus as nutrients; and additional treatment as required for compliance with receiving water classification requirements, have been successful. The additional treatment technologies have included sand filtration and mixed media filtration.

An equalization system preceding the wastewater treatment system can effectively even out the flows and loads and allow construction and operation of a smaller treatment system. Also, pH control in conjunction with the equalization system has been effective in improving the performance of primary clarification. Settling aids such as cationic polymers may improve the performance of primary clarification.

Wastewater characterization parameters that have been of significance are the conventional parameters, namely, pH, TSS, BOD_5, COD, and FOG. Other parameters have included TDS, acidity, alkalinity, surfactants, color, and temperature. The COD-to-BOD ratio is typically high, due to COD demand of inorganic constituents. COD values have ranged from about 9 pounds per 1,000 pounds of product for wastewaters from the solution crumb process to about 35 pounds per 1,000 pounds of product for the emulsion latex process. BOD values have ranged from about 1 pound per 1,000 pounds of product (solution crumb) to about 5 pounds per 1,000 pounds of product in the case of the emulsion crumb process.

Wastewaters from emulsion crumb and latex plants are characterized by high values of TSS, which are usually due to uncoagulated latex. In general, TSS values have ranged from about 3 pounds per 1,000 pounds of product, for solution crumb wastewaters, to about 6 pounds per 1,000 pounds of product for emulsion latex wastewaters. Total dissolved solids are typically attributable to carbonates, chlorides, sulfates, phosphates, and nitrates of calcium, magnesium, sodium, and potassium. Traces of iron and manganese can also be present. A major source of TDS from the emulsion crumb process has been the liquor from the coagulation process used to separate the crumb. High TDS values are also characteristic of wastewaters from the solution crumb and emulsion latex processes.

FOG in wastewaters from all synthetic rubber production processes has been attributable to undissolved monomers and extender oils. Leaked or spilled machine lubricating oils and hydraulic fluids are also potential sources.

The Soft Drink Bottling Industry

In the late 1800s, the then-fledgling soft drink industry in the United States was party to the introduction of a new drink made from an extract from the African kola nut, an extract from cocoa, and water saturated with carbon dioxide. The first cola drink thus began what is now one of the world's major industries. This industry is still experiencing significant growth in the early 2000s.

Production of Bottled Soft Drinks

The soft drink industry's products (in the United States) are divided into two categories: soda water and colas, also known as peppers. Soda water is simply water saturated with carbon dioxide. Saturation is defined as that amount of carbon dioxide dissolved, at equilibrium, under a pressure of one atmosphere (about 15 psig) at a temperature of 60°F. This quantity is referred to as one volume of carbon dioxide. A cola (or pepper) must contain caffeine from the kola nut or from extracts of other natural substances. In the United States the quantity of caffeine is not allowed to exceed 0.02% by weight. The exact formulas for most commercial beverages are closely guarded; however, the following two formulas presented in Table 10-22 are said to be representative.

Table 10-22 Typical Soft Drink Formulas

Carbonated Cola Beverage	
Sugar syrup, 76 Brix	305.00 ml
Phosphoric Acid, 85%	1.25 ml
Caffeine solution, 4%*	10.00 ml
Caramel color, double strength	4.00 ml
Natural cola flavor	2.00 ml
Water	177.75 ml
Sodium benzoate	5%
Water	91%
Caffeine	4%
Pale Dry Ginger Ale	
Pale dry ginger ale flavor	2.00 ml
Citric acid solution, 50%	9.20 ml
Sodium benzoate solution, 18%	5.00 ml
Carmel color 2X, 20% solution	0.30 ml
Sugar syrup, 76 Brix	248.00 ml
Water	235.50 ml

*Caffeine solution, 4% (wt/wt).

Table 10-23   Types of Soft Drinks and Their Ingredients

		Soft Drink Ingredients (General)			
Soft Drink	Flavors	Color	Sugar (%)	Edible Acid	Carbon Dioxide (CO^2) (volume of gas)
Cola	Extract of kola nut, lime oil, spice oils, and caffeine	Caramel	11–13	Phosphoric	3.5
Orange	Oil of orange and orange juice	Sunset yellow FCF with some Tartrazine	12–14	Citric	1.5–2.5
Ginger Ale	Ginger root oil or ginger and lime oil	Caramel	7–11	Citric	4.0–4.5
Root beer	Oil of wintergreen, vanilla, nutmeg, cloves, or anise	Caramel	11–13	Citric	3

Table 10-23 presents a general summary of the ingredients of various types of carbonated soft drinks.

Sweeteners include dry or liquid sucrose, invert sugars, dextrose, fructose, corn syrup, glucose syrups, sorbitol, or an artificial substitute. Flavorings are often used in a carrier such as ethanol, propylene glycol, or glycerin, or they may be in the form of fruit juice extracts or the juices themselves, dehydrated or full strength, or in the form of extracts from bark, vegetables, roots, or leaves. Edible acids used include acetic, citric, fumaric, gluconic, lactic, malic, tartaric, or phosphoric. This statement of ingredients is by no means intended to be complete; rather, it is intended to illustrate the variety of substances the environmental engineer may be confronted with in situations involving spills, leaks, or sloppy management of the production plant.

Figure 10-17 presents a flow diagram of a typical soft drink bottling industrial facility. The term bottling is used generically to include packaging, under a pressure of about one atmosphere, in bottles, cans, plastic containers, kegs, or other bulk containers. The first significant activity is delivery of raw materials to the receiving and storage facilities, usually a relatively small tank farm. Also, water is stored, after treatment to the specifications of the bottling company and the requirements of applicable agencies and federal, state, and local regulations. Carbon dioxide may be delivered to the site or may be generated on site.

Treatment of the water may include simple filtration or activated carbon adsorption, or it may include extensive treatment such as coagulation, flocculation, and sedimentation, followed by filtration, activated carbon adsorption, disinfection, and dechlorination. In addition to high-quality water from the standpoint of health, the concerns of the bottling company are any substances in the water that may affect taste or cause high carbonate hardness, which can cause the beverages to go flat in a short amount of time. Also, certain heavy metal ions can cause rapid loss of carbonation. Desirable characteristics of the water to be used in soft drink production include complete absence of bacteria, turbidity, dissolved carbon dioxide, and chlorine.

Preparation of the bottles (or other containers) includes washing and sterilization. As will be discussed later, this represents one of the most important, and largely unavoidable (but not amenable to minimization), sources of wastewater. Reusable bottles require washing in a hot caustic solution. They then are subjected to the washing and rinsing procedures used on new bottles.

The mixing of ingredients by transferring them from storage to blending systems is one of three major operations in the soft drink

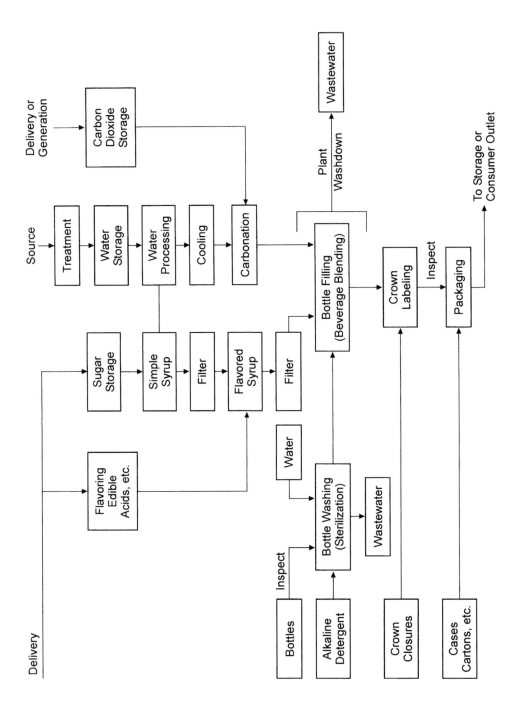

Figure 10-17 Flow diagram of standard manufacturing process: bottled and canned soft drinks.

production process. However this is accomplished—whether by means of pumps and pipes, by use of transfer vessels, by cleanup operations at the end of each processing day and in some cases throughout the processing day—represents a major source of waste generation.

The next steps after mixing are carbonation and bottling, usually done concurrently. Here, again, no true wastewater is generated, except unintentionally as from spills, leaks, broken bottles, or other containers, or, in very rare instances, the need to discard a "bad batch" because of contamination. Daily washdown of equipment, of course, is an ever-present major source of wastewater.

After bottling, the filled and sealed containers are labeled and then packaged for storage and/or shipment.

Wastes Generation

Except for boiler operation, the bottled soft drink industry is not of concern as a significant source of air pollutants. Both solid wastes and waterborne wastes, however, are of major concern to plant and environmental managers.

Solid Wastes

There are several significant sources of solid wastes from the bottled soft drink industry. The first, using the flow-through of materials from receipt and storage to shipment of final product as the approach, is the discarded (and not recycled) material used for packaging and shipping raw materials to the facility. There are also the materials intended for recycle that are found to be unacceptable for recycling. Next, there are the residuals from treatment of the water to be used in the product. Then, there are the substances cleaned from the processing facility floor during cleanup operations. This can include broken bottles or other objects, lost caps, and discarded personal items from employees. The residuals from wastewater treatment

constitute a major source of solid wastes. Finally, there are the items discarded from labeling and packaging the product for storage and/or shipment.

Nonprocess-related solid wastes include construction debris from plant upgrades, facility expansions, general facility repair and upkeep, and incidental wastes, such as garbage and trash from the cafeteria. In general, there are no hazardous wastes from these facilities.

Waterborne Wastes

The major sources of wastewater from soft drink bottling operations include plant washdown wastewater and the wastewater from washing bottles (and other containers), as shown in Figure 10-17. Plant washdown water includes the water to clean the machines, pumps, pipes, and mixing and bottling equipment, as well as washing the floors and the general work area.

Wastes Minimization

Wastes minimization at soft drink bottling facilities is best accomplished by the following:

- Installing overflow warning devices on all storage tanks, mixing tanks, blending tanks, and holding tanks
- Using biodegradable detergents
- Containing drips and returning them to the source
- Practicing aggressive maintenance to eliminate the occurrence of leaks or other accidents that could lead to non-containment of sweeteners, finished product, and other substances
- Using dry methods of cleanup, including brooms, shovels, and dry vacuuming, to the maximum extent
- Rinsing bottles after cleaning in a counter-current manner, with respect to fresh water makeup and spent rinse water overflow

Wastewater Treatment

Because there are no nonbiodegradable substances used at soft drink bottling facilities, all wastewaters are amenable to biological treatment. As a general rule, no pretreatment other than screening and pH adjustment is required before secondary treatment by either an on-site biological treatment system or a POTW. Adjustment of pH is required because of the caustic used for cleaning bottles, pipes, tanks, floors, and mixing and bottling machinery.

A two-stage aerated lagoon has been used with success at a bottling plant in New England. The layout and characteristics of this system are as follows:

- The first lagoon is operated as a sequencing batch reactor.
- The second lagoon is operated as a conventional aerobic lagoon.
- Potassium is added to both lagoons as a nutrient.

The primary reasons for selecting the two-lagoon system were as follows:

- Lagoons, because of their relatively large volume, are capable of withstanding large variations in flows and loads, pH, and concentrations of chlorine.
- The sandy nature of the soil in the area of the plant would allow percolation of liquid into the ground and to the groundwater for removal from the area.
- There was ample distance to the groundwater for complete treatment (renovation of the wastewater) before reaching the groundwater.
- The sandy nature of the soil allowed free movement of oxygen into the soil to support the in situ, on-site biological treatment.
- It was calculated that sludge would have to be removed only once every five to ten years.

Pepsi Corp. has patented a wastewater treatment system with the following layout and operational characteristics:

- Waste from the returned bottle prerinse was directed to a trickling filter operated as a roughing filter.
- The effluent, along with the sloughings from the roughing filters was directed to an oxidation tank, which also received flow from the bottling area in the processing plant, the plant washdown water, and the wastewater from a solids separator used for the process water filter backwash.
- The oxidation tank was aerated with air for biological treatment as well as ozone for chemical oxidation.
- The effluent from the oxidation tank flowed to an aerated mixing tank, where it mixed with wastewater from the final bottle rinse.
- This mixture flowed to an activated sludge aeration tank followed by a final clarifier and then a disinfection tank, where ozone was added as the disinfectant and the pH was given a final adjustment before discharge.

Production and Processing of Beef, Pork, and Other Sources of Red Meat

In the United States, over 90% of the protein consumed by the population was derived from animals in the year 2000. This compares to about 55% in Western Europe, about 40% in Eastern Europe, about 40% in Japan, and about 20% in Africa. As a general rule, people appear to increase their consumption of animal-derived protein as their standard of living increases. It is therefore expected that as so-called third-world countries develop, the raising, slaughtering, processing, and consumption of animals will increase.

Processing animals for use as food results in very large quantities of wastes that must be managed. Many of the processes discussed in this section are applicable, with some modification of actual characteristics, to the processing of goats, sheep, and other animals processed throughout the world for the purpose of producing edible protein.

The Production and Processing of Beef

Cattle are grown, processed, and eaten around the world. Table 10-24 presents a summary of the leading producers of cattle and the per capita consumption of beef in those countries. As a general rule, cattle are processed for market in facilities owned and operated by parties other than those that breed and raise cattle. Beef processing plants have receiving and holding facilities to which cattle are delivered by truck or by rail car, generally on the day they will be slaughtered and processed.

Plants engaged in the processing of beef range from those that only slaughter (slaughterhouses) to those that perform many operations: killing, sticking, bleeding, dressing, trimming (hide removed), washing, processing, and packaging (processing plants). Many plants perform all of the functions of a processing plant, plus engage in what is known as further processing, which includes cooking, curing, smoking, pickling, canning, and many other processes (packinghouses).

Figure 10-18 presents a flow diagram of a "typical" packinghouse. As shown, a typical slaughterhouse, processing plant, or packinghouse has facilities at which animals are received and held in pens. It is here that wastes are first generated, consisting of excrement and washdown water.

The slaughtering operation begins with stunning, usually with electric shock or a plastic bullet to the brain, followed by hanging, sticking, and bleeding. Blood is collected on the killing floor. The animals are then dressed (disemboweled), trimmed or skinned, washed, and then hung in cooling rooms.

Table 10-24   Leading Producers of Cattle and Per Capita Consumption of Beef: World Beef and Veal Production in 1998

The four major areas of production are:

North America

   35.5% of world production (14.7 million tons)

   United States: 11.7 million tons; Mexico: 1.8; Canada: 1.0

South America

   19.1% of world production (7.88 million tons)

   Brazil: 4.96 million tons; Argentina: 2.55; Uruguay: 0.37

Western Europe

   18.0% of world production (7.46 million tons)

   France: 1.640 million tons; Germany: 1.438; Italy: 1.000; UK: 0.740; Ireland: 0.500; The Netherlands: 0.4900; Spain: 0.485; Belgium-Luxembourg: 0.325; at less than 0.200: Austria, Denmark, Sweden, Portugal, and Greece

Asia

   12.6% of world production (5.22 million tons)

   China: 4.400 million tons; Japan 0.485; Korea 0.233

Miscellaneous

   Russia: 2.633 million tons

   Australia: 1.775 million tons

   New Zealand: 0.616 million tons

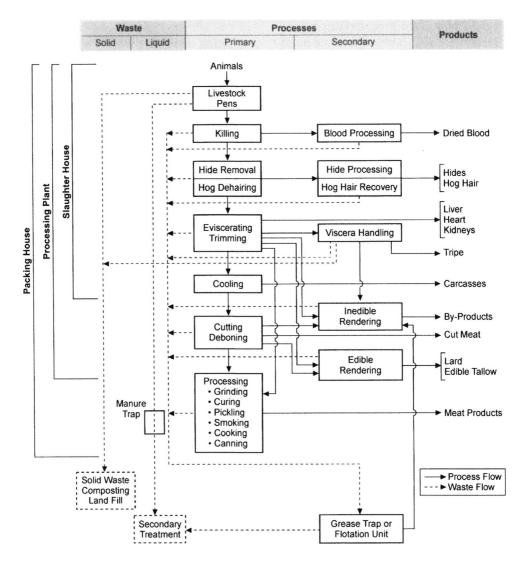

Figure 10-18   Flow chart for a packinghouse (from North Star Research & Development Institute, "Final Report, Industrial Waste Study of the Meat Products Industry," EPA Contract No. 68-01-0031).

The finished product from slaughterhouses consists of the cooled carcasses, plus hearts, livers, and tongues. The hides are salted, folded, piled, and then shipped to a tannery. Viscera, feet, and head bones are either sent to a rendering facility or, in many cases, processed in an on-site rendering facility.

The next series of operations is referred to as processing. The carcasses are cut into smaller sections or into individual cuts. At some plants, curing, smoking, canning, and possibly many additional activities take place.

Packinghouses are also capable, typically, of performing many by-product processing operations, such as processing blood collected just after the animals are killed and rendering the nonedible materials. Rendering refers to separating the fats and proteinaceous materials by use of heat and sometimes pressure. Fats are harvested as lard and other products. Protein material is harvested as animal feed supplement.

The Production and Processing of Pork

As is the case with the processing of beef, discussed above, plants engaged in the processing of swine range from those that only slaughter (slaughterhouses) to those that perform many operations: killing, scalding, dehairing, washing, eviscerating, processing, and packaging (processing plants). Many plants perform all of the functions of a processing plant plus engage in what is known as further processing, which includes cooking and producing bacon, sausage, and many other products (packinghouses). There are also many specialty plants that receive processed swine and produce items such as summer sausage and specialty meats, all of which can be categorized as "further processing."

Production Processes

In the slaughterhouse portion of the operation the animals are first stunned, usually by electric shock, then hung over a bleeding trough. Movement of the animals is continuous through the next several steps, enabled by a moving conveyor to which the hooks that hang the animals are attached. As the animals move over the bleeding trough they are stuck (the jugular vein is cut), and the draining blood is collected in the trough. The animals are then passed through a scalding tank where very hot water softens the hair follicles. After emerging from the scalding tank the animals are dehaired, usually by abrasion from some type of rotating rubber fingers in a machine that becomes extremely loaded with hair, bits of hide, and blood throughout the processing day. An open flame to remove the last traces of hair singes the carcasses, and then they are sprayed with water for the purposes of cooling and washing.

At an increasing number of slaughtering facilities, the animals are skinned, which eliminates the scalding and dehairing. However, the hides will eventually have to be cleaned and dehaired, which amounts to transferring the waste management problem to another location.

The next step is to open the carcasses (by handheld knives) for disemboweling, referred to as the eviscerating procedure. The carcasses are cut in half and hung in a cooler for 24 hours or more. They are then cut into smaller sections or into individual cuts, and it is at this point that major differences occur from one packinghouse to another. Swine (usually called hogs in the United States) are cut into hams, sides, shoulders, and loins. At some plants they are cured. Loins, however, are usually packaged without curing.

The heart, liver, and kidneys are washed and processed as edible meat products. Meat for sausage is ground and emulsified and then blended with herbs, spices, nitrite, and nitrate preservatives. The sausage mix may then be packaged or extruded into casings that have been removed from the outside of hog or other animal intestines. Cellulose materials are also used for sausage casings. The sausage may then be cooked, smoked, or packaged for sale as fresh. Smoking is accomplished by hanging the meat or meat product in an (above 137°F to prevent botulism) atmosphere of smoke-generated by burning hardwood sawdust. The creosols from this burning process are responsible for the characteristic flavor. An alternative process is to soak or inject the meat or meat product in a solution of salt, sugar, and natural and/or artificial flavorings.

Wastes Generation

Animal slaughtering operations result in large quantities of solid, airborne, and waterborne wastes. Even when all possible materials are contained and processed as by-products, there are large quantities of solid wastes from paunch manure and sludges from wastewater treatment to be managed. There are many airborne substances in the form of odors and very large amounts of strong waterborne wastes from blood, plant washdown, and by-product operations such as on-site rendering facilities.

Solid Wastes

Generation of solid wastes begins in the receiving and holding pen area. Wastes include excrement, wash water, and often rainfall runoff. Unless the holding pen area is covered and has a well-engineered, constructed, and maintained stormwater management system, the runoff can become heavily contaminated and represent a significant management problem. Well managed, the excrement can remain as solid waste and be treated and disposed of by use of composting (see Chapter 9) or applied directly to farmland.

Hides are major products of animal slaughtering facilities. Although the hides themselves are not solid waste, they typically contain large amounts of excrement and dirt. This material must be removed and managed as waste material somewhere, either at the slaughterhouse or at the tannery.

Airborne Wastes

Air pollution from animal processing facilities is typically a significant issue only if the release of odors is not controlled. If a rendering plant is operated on site, it is a major issue. Rendering facilities are almost always sources of generation of nuisance odors. Typically, containment and treatment by use of wet scrubbers can keep the problem under control.

Waterborne Wastes

Almost every operation involving slaughtering and processing animals results in wastewater. Certain operations, such as the scalding of hogs as part of the cleaning and hair removal process, produce large volumes directly. Other operations, such as butchering, produce most of the wastewater during cleanup activities.

Components of wastewater are blood, paunch manure, fat solids, meat solids, grease, oil, and hair. In-house laundry operations also contribute. Each beef stomach contains 55 to 80 pounds of paunch manure. One hundred gallons or more of water are typically required to wash out a single paunch. This quantity can be reduced by using dry removal and handling systems.

Blood, a major contributor to the high strength of wastewater, has upwards of 150,000 mg/L of BOD_5. Sources of blood within an animal processing facility begin at the killing and bleeding facility and continue on through the processing facility. Normally, each successive processing operation at a (complete) packinghouse will be a less important source of blood than the one before it. However, it is the usual practice at animal processing facilities to combine wastewater from all sources before treatment.

Cutting and packaging operations (butchering or further processing) include additional operations that contribute to wastewater. Intestinal casings require squeezing or pressing to remove the contents. The casing used for production of edible foods, such as sausages and hot dogs, has to be removed from the outside of the intestine. The intestine itself is sent to the rendering facility. The intestine casing is salted and then drained. This operation, as a whole, results in wastewater that is high in grease, as well as sodium chloride.

Scalding is used to process tripe, the muscular part of an animal's stomach. This process results in wastewater containing grease and TSS. Again, all processing stations are washed down each processing day, some several times each day. The washdown water is a major source of wastewater from any animal processing facility.

In the case of slaughtering and processing hogs, the freshly killed and bled carcasses are scalded by immersion in a tank of near-boiling water to wash and prepare them for removal of hair. The scalding tanks overflow at a certain constant rate to prevent solids and dissolved materials (pollutants) from building to an unacceptable level. This overflow constitutes a major source of wastewater from the processing of hogs. It is comparable to the scalder overflow encountered in the processing of chickens, turkeys, and ducks. The BOD_5 of any of these scalder overflow wastewaters is typically in the range of 2,000 to 5,000 mg/L.

Wastes Minimization

Washing the cattle, hogs, or other animals before delivery to the slaughtering facility can best minimize solid wastes that have the holding pens as the source. A second major source of solid materials to be managed is the sludges from wastewater treatment and from air pollution control; however, if the processes that create these residuals are managed properly, they can be used as raw material to the rendering facility. The result will be twofold: a very large decrease in solid wastes to be disposed of and an increase in the animal feed supplement that is the product of the rendering facility.

Paunch manure should be handled by dry methods, such as dry conveyors and dry collection systems. In addition, daily (or more frequent) plant cleanup should begin by thorough sweeping, squeegeeing, air blast, or other feasible dry method. All of the material thus removed from the killing, bleeding, and processing equipment and floors can be processed into animal feed supplement in an on-site or off-site rendering facility.

Regarding airborne wastes and, more specifically, nuisance odors, the best minimization strategy is to maintain a scrupulously clean facility, in order to minimize putrefaction of organic substances. The products of biodegradation such as fatty acids, amines, amides, and reduced sulfur compounds such as hydrogen sulfide and mercaptans, are responsible for bad odors from meat processing facilities.

Treatment and Disposal

There is considerable similarity between the management of wastes from all types of red meat production and processing facilities. Beef, pork, and most other red meat source animals result in solid, airborne, and waterborne wastes that are appropriately managed in very much the same ways.

Solid Wastes

Most solids from meat packing facilities are organic. As such, they have a potential use. The excrement and sometimes the paunch manure have been successfully processed in composting facilities and thus converted to useful material such as fertilizer or soil improvement agent. Another use for paunch manure is as feed to appropriate rendering facilities.

All other solid materials resulting from the slaughtering and processing of animals can be managed as feed for an appropriate rendering facility.

Airborne Wastes

The treatment of nuisance odors, the only air pollutants of consequence at red meat processing facilities, is done by containment and wet scrubbing. Containment usually means prevention of the release of fugitive emissions. If there is a rendering facility on site, containment definitely includes maintenance of a negative pressure in side buildings (compared with the atmospheric pressure outside the buildings). During windy periods, a greater differential pressure is required. Blowing significant quantities of air from the inside of the buildings to the outside, at a constant rate, creates the negative pressure. Control of nuisance odors is accomplished by exhausting the blowers through wet scrubbers. This technology is discussed more completely in the section titled "Rendering of By-Products from the Processing of Meat, Poultry, and Fish."

Waterborne Wastes

Treatment of wastewaters generated at slaughterhouses, processing plants, and/or packinghouses, where red meat is produced and processed, typically consists of biological treatment preceded by screening and primary sedimentation. In some instances, dissolved air flotation (DAF) has either replaced primary sedimentation or has been placed between primary sedimentation and biological treatment. The solids harvested from the

DAF process (flot) can be used as raw material for rendering facilities, resulting in production of a valuable animal feed supplement.

Primary Treatment

Screening should be used to the maximum extent feasible, since the solids captured by screening are excellent raw material for the rendering process. Many different types of screens have been used successfully, including shaking screens, tangential screens, conveyor screens, rotary screens, and static screens. It may be necessary to place a grease trap ahead of the screening system to prevent blinding.

Flow equalization has been found to be valuable, and even necessary, in some cases to enable screening systems, gravity clarifiers, and DAF systems to operate successfully. There are many occurrences during the processing day that cause significant change in both rate of wastewater flow and wastewater strength. These occurrences include periodic plant washdown, daily plant washdown, spills, and dumps of unacceptable product or intermediate.

Flow equalization systems must be equipped to cope with floating and settling solids. The variable quantities of fats, oils, and settleable solids can cause debilitating problems if they are not so equipped.

Secondary Treatment

Wastewaters from the processing of red meat are well suited to biological treatment, due to their very high organic content. Either anaerobic or aerobic methods can be used; however, the high strength of these wastes, in terms of BOD_5, TSS, and FOG, make them particularly well suited to anaerobic treatment. The elevated temperature of these wastewaters is also a factor. Higher temperatures decrease the oxygen transfer efficiency of aeration equipment, but tend to increase the rate of anaerobic treatment. These considerations are discussed more fully in Chapter 7.

As discussed in Chapter 7, the more recently developed anaerobic treatment technologies, such as the upflow anaerobic sludge blanket (USAB) and the anaerobic contact process, have had significant success treating wastewaters from the dairy industry. Although there has not been significant success reported for the USAB process for the meat packing industry, it appears obvious that significant potential exists.

Anaerobic contact systems have been used with success. These systems consist of a conventional anaerobic digester, with mixing equipment and a clarifier portion. With detention times of only 6 to 12 hours, 90% BOD_5 removals have been typical. Using the 6- to 12-hour detention time, solids have been separated in the clarifier portion and, for the most part, returned to the active digester. It has been necessary to waste only a relatively small fraction of the biological solids. In order to achieve good separation, it is necessary to degas the solids as they are transferred from the active digester to the settling component.

There is a long history of use of anaerobic lagoons for treating wastewaters from the processing of cattle and hogs. The most successful of these treatment systems has been designed with low surface-to-volume ratios to conserve heat and to minimize aeration. Depths of 12 to 18 feet have been used. Covers are required for odor control. In some cases, the natural cover formed by floating greases has sufficed, but it has typically been necessary to place covers made of Styrofoam or other floating material. Polyvinyl chloride and nylon-reinforced hypalon have been used with success. These lagoons, of course, must be lined to prevent contamination of the groundwater.

BOD_5 loading rates of 15 to 20 pounds of BOD_5 per 1,000 cubic feet have been common. Detention times of 5 to 10 days have resulted in effluent BOD_5 concentrations of 50 to 150 mg/L, amounting to removals of 70% to 85%. Additional treatment, usually by means of an aerobic stabilization pond or other aerobic biological treatment system, is generally required to enable compliance with EPA standards. Ammonia is another issue. Anaerobic lagoons, as described here, typically discharge effluents having up to 100 mg/L of ammonia.

It has been found advantageous in several instances to place the inlet to the anaerobic lagoon near the bottom of the lagoon. This arrangement allows the incoming wastewater to immediately contact the sludge blanket, which contains active microorganisms.

Aerobic lagoons have been used for secondary treatment; however, because of the high cost of providing the large quantity of oxygen required by these high-strength wastewaters, they are most often used as a step following another technology, such as an anaerobic lagoon or anaerobic contact system. Aerated lagoons used for treatment following anaerobic systems have operated with detention times of two to ten days. Depths of 8 to 18 feet have been used with mechanical or diffused air systems.

In appropriate climates, oxidation ponds have been used as a final (polishing) treatment step, following either anaerobic or aerobic biological treatment systems. Oxidation ponds with depths of four to eight feet, sized for organic loadings of 20 to 40 pounds of BOD_5 per acre-day and having detention times of one to six months, have worked with success.

When preceded by effective equalization, the activated sludge process has successfully treated wastewaters from red meat processing facilities. Various modifications have been used, including tapered aeration, step aeration, extended aeration, and contact stabilization. Extended aeration appears to have had the most success.

Trickling filters have been used as roughing filters and as components of nitrogen management systems. As roughing filters, the trickling filters receive a relatively heavy load of BOD_5 for a relatively short period of time. In this mode, trickling filters are capable of removing 25% to 40% of the BOD_5, at a relatively low cost per unit of BOD_5 removed. As components of nitrogen management systems, trickling filters are appropriate media for the nitrification step, preceding denitrification.

Rendering of By-Products from the Processing of Meat, Poultry, and Fish

Rendering is a process for separating fat from animal tissue. Heat is very often used, and sometimes one or more chemicals, pressure, and vacuum are used as well. Rendering serves the invaluable function of converting what would otherwise be waste materials to useful products, including animal, poultry, or fish feed supplements, and oils for industrial and household use, such as soaps and rustproofing paints. A material made from rendering feathers obtained from the processing of poultry has long been used to manufacture a foam product used in firefighting. Edible rendering produces edible lard, chicken fat, or specialty fats or oils.

The rendering process itself results in considerable, sometimes difficult-to-manage wastes. In total, however, wastes resulting from rendering are only a small fraction of the wastes it makes use of. A huge number of rendering plants operate in the United States and around the world. Virtually all facilities that manufacture foods derived from animal, poultry, or fish produce by-products that can be used as raw material at a rendering plant.

The Rendering Process

The rendering process is customized depending on the products and the raw materials. However, the following discussion is applicable to the rendering process in general. The wastes that result from most rendering operations are remarkably similar. Figure 10-19 presents a generalized schematic of a "typical" rendering plant. There are both "wet" and "dry" rendering processes. In the wet rendering process, live steam is injected into the rendering tank, along with the material being rendered. In dry rendering, steam is

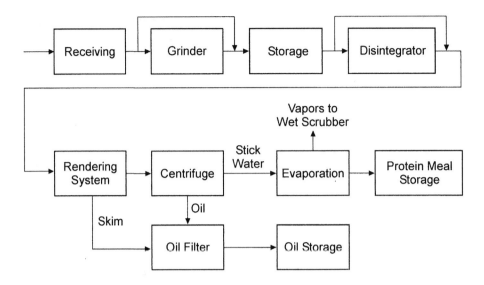

Figure 10-19 Schematic of a "typical" rendering plant.

confined in a jacket that surrounds the tank containing the material being rendered.

As illustrated in Figure 10-19, rendering plants have facilities for receiving and storing the raw material. The materials are taken from storage and are usually ground and blended. The ground and blended material is then pumped, conveyed, or placed by use of a bucket loader, into a tank in which the rendering process takes place (the rendering system). Live steam at 40 to 60 psig is introduced, and the rendering process, not to be confused with cooking, takes place. In some cases, pH is adjusted. Sometimes, other chemicals are introduced. As the rendering process proceeds, fats and oils are drawn off the top of the tank. The remaining liquid, called "stick water," contains the protein material. The stick water is typically evaporated and added to animal feed.

Dry rendering typically involves placing the materials to be rendered in a steam-jacketed rendering system that is placed under vacuum. Screening and centrifugation are used to separate the fats and oils from the solids, which contain the protein matter.

A large rendering plant in the northeast rendered a variety of materials, including by-products (the mostly inedible substances remaining after slaughtering and processing) from beef, poultry, and fish packing operations. Table 10-25 presents a list of materials rendered by this facility. Fat and bone, poultry by-products (heads, feet, entrails, blood, and carcasses of rejected birds), fish, and fish by-products were converted to high protein meal by use of a continuous dry rendering process. In this system, diagrammed in Figure 10-20(a), a slurry of recycled fat (from restaurants, butcher shops, and beef packing plants) and ground-up bones, meat, and other substances was dehydrated in a multiple effect evaporator. Vapors were vented to a barometric condenser. Expellers (centrifuges) were used to separate the hot fat product from the protein meal. The process is often referred to as, "cooking"; however, temperatures were maintained at 140°F or lower.

Feathers from chicken processing plants were handled in a separate system. A schematic diagram of the feather processing is shown in Figure 20(b). The feathers were hydrolyzed using high temperature and pressure and then dried in a steam tube rotary dryer.

A variation of the rendering process is low-temperature rendering, in which the raw

Table 10-25   Material Received and Produced by Rendering Facility

Raw Material	Mode of Delivery	Processing Steps	Final Product
Fat, bone viscera	Drum and truckload	Carver-Greenfield	Beef meal*
Poultry by-product	Trailer truck	Carver-Greenfield	Poultry meal*
Sludge from treatment plant (poultry processing)	Trailer truck	Carver-Greenfield	Poultry meal*
Fish by-products	Trailer truck	Carver-Greenfield	Poultry meal*
Whole fish	Trailer truck and boat	Carver-Greenfield	Fish meal*
Menhaden	Trailer truck and boat	Carver-Greenfield	Fish meal*
Feathers	Trailer truck	Feather cooker and dryer	Feather meal* for fire extinguisher foam
Restaurant grease	Drum	Steam table and filtration	Poultry feed supplement

*Poultry feed supplement.

materials are heated to just above the melting point of the fat. Centrifugation is used to separate the fats and oils from the protein matter.

Wastes Generation

Solid, airborne, and waterborne wastes are generated at rendering plants in considerable quantities. Some of these are relatively diffi- cult to manage, largely because of their nuisance odor–causing potential.

Solid Wastes

Solid wastes are generated at the receiving and storing area in the form of containers that are no longer usable and broken pallets. Also, construction debris from plant mainte- nance, modifications, expansions, and peri-

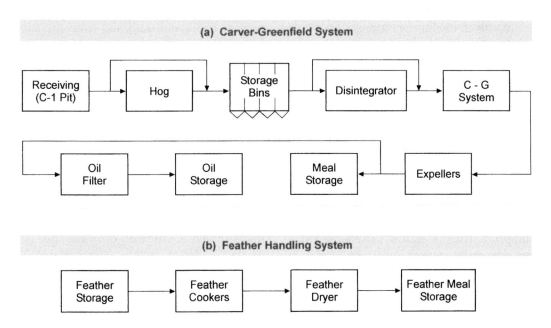

Figure 10-20   Schematic of rendering systems.

odic facility upgrade projects contributes to the solid wastes that must be managed.

Airborne Wastes

Airborne wastes are an ever-present potential at typical rendering plants. All objects with which the raw material for the rendering plant comes in contact are subject to becoming the source of a nuisance odor problem. The process of anaerobic biodegradation of the easily degraded animal flesh proteins causes the odors. This process is referred to as "putrefaction," and it produces foul-smelling amines, amides, indole, skatole, various volatile acids, and reduced sulfur compounds such as hydrogen sulfide and methyl as well as ethyl mercaptans.

Waterborne Wastes

Primary sources of wastewaters from rendering plants are as follows:

- Blood and other "juices" that drain from the animal, poultry, and fish by-products that are the raw material for the rendering facility
- Plant washdown water, which is constantly in use for the purpose of controlling air pollution by foul odors
- Blow-down from wet scrubbers used for air pollution control

Plant washdown, as part of housekeeping practices implemented to control nuisance odors, is the major source of wastewater in terms of volume. Constant washdown of the plant equipment and floors is necessitated by the nature of the equipment as well as the nature of the material handled. Most rendering plants produce nothing for human consumption. For this reason, the equipment is not the easily cleaned, relatively expensive type used in edible food processing plants. The material handling equipment typically consists of screw conveyors and open impellor pumps. This equipment handles poultry and animal entrails, fish and fish parts, and sludge from wastewater treatment plants (both on-site treatment plants and treatment plants located at off-site meat, poultry, and fish processing plants). This combination of sloppy material, along with the types of handling equipment, leads to almost continuous dripping and spilling.

In addition, oil vapors within the processing areas continually condense upon surfaces, resulting in a biodegradable film that will get thicker as time goes on if it is not regularly washed off. The consequence is that plant washdown must continue in a never-ending fashion. As soon as an area is cleaned, it begins to downgrade. As well, wet scrubbers must be constantly in operation to control odors.

Wastes Minimization

Notwithstanding the fact that plant washdown water creates most of the wastewater that requires treatment, good housekeeping practices are still the most important wastes minimization technique for rendering facilities. The key is to make use of drip pans and other containment devices to decrease as much as possible the quantity of material required to be cleaned from unconfined areas such as floors and walls and from the surfaces of the machinery.

Dry methods of cleaning must always be used to the maximum extent feasible. Dry methods of cleaning, using brooms, shovels, vacuum cleaners, and air blowers, are an important waste minimization technique in any industrial setting. It is especially important in the setting of a meat, poultry, or fish by-product rendering facility, because of the ever-present "juices" of very high organic content.

Whenever possible, liquid drainings and juices should be disposed of before they get to the rendering facility. This, of course, involves a tradeoff, since these liquids must be managed somewhere. From the point of view of the rendering facility, the less of this material the better.

One example of this type of waste minimization occurred at a trash fish ("nonedible" fish, such as menhaden) rendering facility on the east coast of the United States. Before the wastes minimization measure was instituted, fishing boats would catch fish by use of the "purse-seine method" and load them aboard boats. The boats would transport them to shore, where they were transferred to trucks by use of a vacuum pump. During the earlier days of this industry, the vacuum pump routinely pumped the fish, as well as the "fish juice," out of the boat. Because of the high water content of the fish juice, it was sent overboard into the harbor, rather than put into the trucks and carried to the rendering facility. Evaporation of this water was too costly. Eventually, however, it became illegal to dump this fish juice in the harbor. At this point, the juice was pumped into the trucks and then dumped on the receiving room floor at the rendering facility. This caused a massive wastewater problem at the rendering facility. The solution that was eventually worked out was to handle and dispose of this fish juice at dockside in a nonpolluting manner.

In the case of wet rendering, there may be excess tank water at the end of the rendering process. This water, which typically has a BOD_5 in the range of 50,000 mg/L and a TSS value in the range of 2%, should be evaporated and blended into animal feed supplement.

Treatment and Disposal

Treatment and disposal of wastes from meat, poultry, and fish by-product rendering facilities is a matter of minimizing contact between organic material and the solid wastes, in order to control odors from the solid wastes, and handling and disposal of the solid wastes as ordinary (nonhazardous) waste. Regarding air pollutants, the key is containment and treatment by means of wet scrubbers. Wastewaters are entirely organic, with the possible exception of wet scrubber blow-down and chemicals used for pH adjustment and for cleaning the plant. Bio-

logical treatment should, therefore, be appropriate.

Solid Wastes

The principal solid wastes from meat, poultry, and/or fish by-products consist of packaging and transport materials, such as 55-gallon drums, broken pallets, steel strapping, and nonrecyclable containers previously used for chemicals, detergents, lubricants, and laboratory supplies. After aggressive wastes minimization and recycling, these materials can be disposed of as ordinary solid wastes. There should be no organic materials in the solid waste stream, since any such wastes should simply have been added to the feed to the rendering system.

Airborne Wastes

For many rendering facilities, the most troublesome problem is the nuisance odors from the anaerobic biodegradation of protein matter. The solution is to rigorously contain any odors so generated and to treat them by the oxidative action of the chemical solutions in wet scrubbers.

The foul odors that require containment, collection, and treatment emanate from handling and storage of the raw material, as well as from the handling and processing equipment on the processing floor. The rendering process itself, as well as subsequent processes such as pressing and drying, are generally agreed to be the sources of strongest odors. These "high-intensity" sources emit varying concentrations of odors, depending on the type and age of the raw material, and the temperature of the raw material during the hours or days prior to its placement in the rendering system. In general, the older the raw material and the higher its temperature, the more concentrated, or stronger, the bad odors are.

Odors from rendering operations have been generally described as those of ammonia, ethylamines, and hydrogen sulfide. Skatole, other amines, sulfides, and mercaptans also contribute to the blend of rendering plant odors.

The types of equipment normally used for controlling odors in the rendering industry are afterburners, condensers, adsorbers, and wet scrubbers. Gases from wet rendering processes are almost always condensed. Removing condensable odors results in as much as a tenfold volume reduction. The condensed gases can then be directed to additional treatment.

Flame incineration (afterburning) offers a positive method of odor control. However, the cost of fuel makes afterburner operation expensive. Scrubbing with chemical solutions is the primary method for controlling odors. Activated carbon has also been used for removing odors by the mechanism of adsorption; however, its application is limited to treatment of relatively cool and dry gases.

The typical situation at rendering plants of large size is to have from two to four wet scrubbers. Air from any location within the plant must go through one of these scrubbers before reaching the outside. Figure 10-21 is an illustrative diagram of a typical scrubbing system used for this purpose. The system shown in Figure 10-21 consists of a venturi (fixed throat opening) section followed by a packed tower. A large induced-draft fan pulls air from within the building that houses the rendering system and forces it through the venturi scrubber system shown in Figure 10-21. Process gases are also forced to pass through a venturi scrubber system, similar to the system shown in Figure 10-21.

The wet scrubber portion of the venturi scrubber system is typically a tower packed with plastic tellerettes, which provide a very large surface area. A solution of hypochlorite, with the pH adjusted for maximum effectiveness, is sprayed onto the top of the tellerettes. This solution flows down over the (very large) surface area of the packing. At the same time, gases are forced up through the packing by the action of the fan. As the gases contact the scrubbing solution, the oxidizing action of the hypochlorite changes the chem-

ical nature of the odor-causing substances. Also, many of these substances become dissolved in the scrubbing solution.

One of the most important objectives of the odor control system described above is maintaining a negative pressure within the rendering plant building. If the barometric pressure inside the building is less than the barometric pressure outside the building, it is more difficult for fugitive emissions to pass from inside to outside. This negative pressure (negative inside relative to outside) is maintained simply by keeping all the doors and windows closed while operating the wet scrubber systems. It is necessary for the buildings to have good structural integrity, that is, to be free of leaks in the walls, and to have good, reasonably tight fits regarding doors and windows. If the building is relatively sound, it is a simple matter to maintain a negative pressure and to maintain effective control over fugitive emissions. Occasional opening and closing of doors for human and truck traffic can be tolerated.

Waterborne Wastes

Wastewaters from rendering plants using red meat, poultry, and/or fish by-products as raw material are amenable to biological treatment, since these wastewaters are almost entirely organic. The only sources of inorganic chemicals are the blow-down wastes from the air pollution control systems and possibly inorganic agents used for cleaning.

Anaerobic as well as aerobic technologies are valid potential treatment technologies. The normally very high strength character of these wastewaters, as well as elevated temperatures, makes them particularly well suited to anaerobic treatment technologies. However, aerobic treatment has been used with success as well. A large rendering facility in New England successfully treated wastewaters by making use of chemical coagulation and dissolved air flotation (DAF) to remove most of the protein, fats, and oils (which accounted for about 80% of the BOD_5), followed by activated sludge for removal of the dissolved fraction of waterborne pollutants.

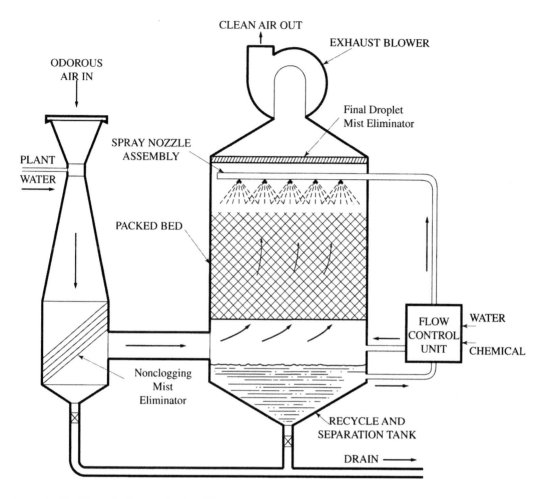

Figure 10-21   Schematic diagram of wet scrubber system.

This rendering facility used by-products from red meat processing, poultry processing, fish processing, and sludge from wastewater treatment as raw material. Two rendering systems were used. One was a continuous flow dry rendering system that operated at about 140°F and under a vacuum. The second was a batch-type dry rendering system. On a typical day the continuous flow system was started up on poultry (offal, heads, feet, and blood) and operated at the rate of 30,000 lb/hr for seven hours. It would be switched to fat and bone and run at the rate of 23,000 lb/hr for 2½ hours. It would then be switched to fat, bone, and meat scraps from red meat processing and butcher shops, as well as

grease from restaurants in the area. It would then be switched to by-products from fish processing plants and operated at 50,000 lb/hr for three hours, or so. Finally, it would be switched to menhaden, a particularly oily fish, for four to ten hours, depending on supply.

The batch rendering system would be started up on feathers from poultry processing plants on Monday afternoons and run at a rate of about 3,000 pounds per hour until the supply of feathers was exhausted on Saturday.

The characteristics of the combined wastewaters from this plant, including plant washdown water, were as follows:

- Flow   96,000 gpd
- BOD_5 85,000 mg/L   (68,000 lb/day)
- TSS   6,000 mg/L   (4,800 lb/day)

Chemical coagulation in combination with DAF removed about 80% of the BOD_5 and about 90% of the TSS. About 95% of the oil and grease was also removed by DAF. All of the sludge (flot) from this system was used as raw material in the continuous feed dry rendering system.

The aerobic, activated sludge system that followed the DAF system was operated by use of MLVSS concentrations between 9,000 and 11,000 mg/L and a hydraulic retention time of about three days. An integral clarifier was attached directly to the aeration tank, and sludge was wasted from the aeration tank. Sludge age was maintained at ten days. All of the waste sludge was used as raw material in the continuous flow dry rendering system.

Overall BOD_5 removal for the combined DAF–complete-mix activated sludge system consistently exceeded 99%. TSS and oil and grease removals exceeded 95%.

The Manufacture of Lead Acid Batteries

A battery, as the word is used in this section, is a modular source of electric power in which all or part of the fuel is contained within the unit and the electrical power is generated by a chemical reaction within the unit. The primary components are the anode, the cathode, and the electrolyte. The function of the anode and cathode (the electrodes) is to convert chemical energy into electrical energy. If an electrical circuit outside the unit is connected between the anode and the cathode, electric current will be caused to flow through the circuit.

The EPA has divided the manufacturing of batteries into eight categories: cadmium, calcium, lead, leclanche, lithium, magnesium, nuclear, and zinc. The lead subcategory is the largest, in terms of number of manufacturing plants and volume of batteries produced.

Products that use batteries included in the lead subcategory include automobiles, portable hand tools, lanterns, and various implements used in industry and the military.

There are four common types of lead acid batteries: wet-charged, dry-charged, damp, and dehydrated. Wet-charged batteries are shipped after manufacture with electrolyte. All others are shipped without electrolyte.

Lead Battery Manufacture

Only production of the anode, the cathode, and ancillary devices are considered part of battery manufacture. Production of the structural components, such as the cases, terminal fittings, electrode support grids, seals, separators, and covers, are all included in other manufacturing categories. However, any or all of them may be manufactured at the same plant that manufactures the electrodes.

Anodes are metals when in their fully charged state in a battery. In the case of most lead acid batteries, anodes are manufactured by applying a paste of lead oxide to a support. The paste-support structure is allowed to dry. Cathode active materials are typically metal oxides. The lead oxide substance used to make cathodes for lead acid batteries is called "leady oxide" within the battery manufacturing industry. This substance is a specific oxidation state of lead oxide that is 24% to 30% lead free. It is used for the manufacture of both the anodes and cathodes in lead acid batteries and is manufactured by the so-called "Barton process" or by a ball mill process.

Cathodes for use in lead acid batteries are manufactured by applying a paste of the leady oxide to a structural grid. The grid must be able to carry the desired electrical current as well as be strong enough to support the leady oxide. Thus, the fabrication of anodes and cathodes for use in lead acid batteries is very similar. There is a difference, however, in that cathodes remain in the lead peroxide state.

Ancillary operations are operations of the battery manufacturing process other than fabrication of the anodes and cathodes. Included are battery assembly; production of leady oxide; battery washing; and washing of floors, manufacturing equipment, and personnel.

Figure 10-22 presents a flow diagram for the process of manufacturing wet-charged lead acid batteries using a closed formation process. It is more or less typical of all lead acid battery manufacturing processes, as far as wastes generation is concerned. As illustrated in Figure 10-22, there are two "initial" processes. One is the production of leady oxide to be used in fabricating anodes and cathodes. The other is casting of the grids to be used for this same purpose.

Leady Oxide Production

Finely divided metallic lead is mixed with lead oxides to produce the active materials used for manufacturing the battery electrodes or plates. So-called "leady oxide" is produced by placing high-purity lead particles in a ball mill. The friction within the ball mill generates heat, and a forced flow of air provides oxygen. The result is particles of red lead letharge and a certain percentage of unoxidized metallic lead. Noncontact cooling, along with regulation of the air flow, governs the speed of oxidation of the lead.

An alternative process, the Barton process, produces leady oxide by feeding molten lead into a pot and vigorously agitating it to break the lead into small droplets. Oxygen from a stream of air oxidizes a certain percentage of the lead into a mixture of yellow lead, red letharge, and metallic lead.

Grid Manufacturing

Grids are fabricated by casting lead alloys such as lead-antimony and lead-calcium. Trace amounts of arsenic, cadmium, selenium, silver, and tellurium are also added.

Paste Preparation and Pasting

The anodes are fabricated by applying a paste of lead oxides mixed with binders and other substances to the grids. The paste is prepared by mixing leady oxide and granular lead or red lead. The cathodes are prepared in the same manner, except that the paste is prepared by mixing leady oxide, lead, sulfuric acid, water, and expanders such as lampblack, barium sulfate, and various organic materials. One of the principal objects in fabricating the electrodes is to create a very large surface area by preparing the paste so as to have a porous, very rough texture. The paste is often applied by hand. Some facilities use mechanized equipment.

Curing

A principal objective of the curing process is to induce the electrodes to obtain proper porosity and strength. To this end, the curing process is strictly controlled. The plates are flash-dried, stacked, and covered or placed in a humidity-controlled room for several days. Small crystals of tribasic lead convert to lead peroxide. One technique is to soak the plates in sulfuric acid to enhance sulfation and improve mechanical properties. This soaking is done in the battery case or in a separate tank.

Stacking and Welding

The cured plates are stacked for convenient access at the assembly line. Separators are placed so as to prevent short-circuiting. Separators are made from plastic, rubber, fiberglass, or paper.

Assembly

The assembly process involves placing the stacked, alternating anode-cathode plates within the battery case, welding the connecting straps, and installing the covers and vents. Also, the connections are made to the battery posts.

There are two types of assembly: open formation and closed formation (formation refers to charging the battery as a result of

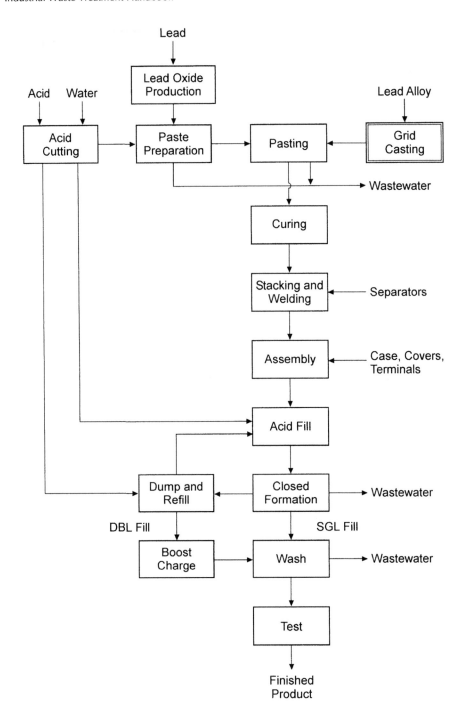

Figure 10-22 Production of wet-charged lead acid batteries using the closed formation process.

acid reacting with substances on the electrodes). Figure 10-22 is appropriate to the closed formation process. In the closed formation process, the top and vents of the battery are installed before the acid (electrolyte) is added. The opposite is true for the open formation process, which allows visual inspection of the plates, separators, strap welds, and posts after the electrolyte has been added.

Electrolyte Preparation and Addition (Acid Fill)

The electrolyte solution for the battery is prepared by diluting concentrated sulfuric acid with water. This acid-cutting process is typically performed in two steps: The first cut is dilution to 45% acid, by weight. This acid solution is used for preparation of the paste, described earlier. The second acid cut is dilution to 25% to 30%. This acid solution is used for battery formation and filling.

With respect to the quantity of wastewater generated, there are several different methods used for filling lead acid batteries. One is to fill each battery to overflowing to ensure that filling is complete. The other is to immerse the batteries in a tank of the electrolyte, after which the batteries are rinsed. A third is to make use of sensing and relay equipment to automatically stop the filling process at the proper time.

Formation

The active substance for the cathode is lead peroxide, although it is not a component of the paste that is applied to the cathode grids during fabrication of the cathodes. The formation of lead peroxide takes place after the newly fabricated cathodes are immersed in the electrolyte and an electric current is applied. In the case of the closed formation battery manufacturing process, this process takes place just after the acid fill step.

The forming process itself consists of applying an electric current between the anodes within a newly filled battery and the cathodes. When this is done, lead oxide and sulfate are converted to lead peroxide on the cathode, and lead oxide is converted to lead on the anode. The final composition on the cathode is 85% to 95% lead peroxide, and the composition on the anode is greater than 90% lead.

The formation process takes time. Several methods are used to ensure proper formation. Two of them are as follows: in one method, each battery is filled with full-strength electrolyte, and the battery is ready for shipment after one to seven days, depending on the type of battery and the characteristics desired. The second method involves filling each battery with a more dilute electrolyte and then emptying and refilling with full-strength electrolyte before shipment. A boost charge is applied prior to shipment. The entire process takes place in one day, or less.

Battery Washdown

Typically, batteries are washed before shipment. This washdown has the purpose of removing electrolyte and other materials from the outside surfaces of the batteries, resulting from filling and forming. Detergents are usually used to remove oils and dirt.

Battery Testing and Repair

Testing batteries seldom generates wastes; however, when testing determines that a battery is faulty, it is taken apart for repair. When this is done, faulty parts are replaced, and the faulty parts become solid waste. Electrolyte drained from the battery becomes waterborne waste.

Wastes Generation

Lead acid battery manufacturing plants are highly variable in terms of wastes generated. For instance, work done in preparation for producing the *EPA Development Document* for the battery manufacturing point source category determined that total plant discharges of wastewater ranged form 0 to 390,000 gpd, with a median of 22,000 gpd.

Solid Wastes

Solid wastes are generated from packaging and shipping wastes and construction debris from plant maintenance, modifications, expansions, and periodic facility upgrade projects. In addition, sludges from wastewater treatment and waste resins from process water deionization require management. Lead acid battery manufacturing plants also generate solid wastes as a result of reject batteries and faulty battery parts, including cases, tops, posts, electrode grids, straps, and separators.

Airborne Wastes

The major source of air pollutants from lead acid battery manufacturing plants is acid fumes. Typically, these substances are controlled by use of scrubbers. There is always a discharge, owing to the less than 100% efficiency of the scrubbers; therefore, an aggressive operation and maintenance program, including aggressive preventative maintenance, is necessary.

Scrubbers are located throughout the processing plant, including the processes carrying out the following production steps: leady oxide production, grid manufacturing, electrolyte preparation and addition, formation, battery washdown, battery testing and repair, and any acid storage and dispensing facilities.

Waterborne Wastes

Wastewater generated from the production of lead acid batteries has three primary sources: the paste preparation and pasting operations, the electrode forming operations, and the washing of finished batteries. However, there are wastewaters from each of the manufacturing steps in the production of lead acid batteries, as illustrated in Figure 10-21.

Leady Oxide Production

Leady oxide production results in wastewaters from cooling and leakage from ball mills, contact cooling during grinding of oxides, and from wet scrubbers used for air pollution control.

Paste Preparation and Pasting

Paste preparation and pasting results in wastewaters from washdown of the equipment as well as the production area. These washdown wastewaters typically contain high concentrations of suspended solids and lead, as well as additives used in making the paste.

Electrolyte Preparation and Addition (Acid Fill)

There is almost always dripping, spilling, and overflowing of acid in this area. Therefore, washdown and rinse waters are typically highly acidic. Also, wet scrubbers are normally used in this area because of the fumes from the dripped, leaked, spilled, or overflowed acid.

Formation

The formation production area is subject to acid spills, similar to the fill area. The wastewaters generally contain significant concentrations of various metals. Copper and iron in the wastewater result from corrosion, caused by the acid, of process equipment and charging racks. Oil and grease also result from equipment washdown. Wet scrubbers in this area generate significant wastewater.

Battery Washdown

This battery washdown process is the site of considerable wastewater generation. Wet scrubbers, which generate significant wastewater, are used in this area as well.

Battery Testing and Repair

When faulty batteries are detected, they are often drained. The drained electrolyte can be a major source of wastewater. Wet scrubbers are often used in this area.

Wastes Minimization

There are many opportunities for wastes minimization at lead acid battery manufacturing facilities. Most fall within the category of good plant housekeeping. As is the case with many other industries, the key to minimizing the cost for eventual waste treatment and disposal is to aggressively implement the following:

- Whenever possible, nontoxic substances should be used for degreasing and cleaning.
- If toxic substances, for instance, chlorinated volatile organics, must be used for degreasing or other cleaning process, containment, recycle, and reuse must be practiced to the maximum extent possible.
- Biodegradable detergents should be used.
- Drips must be contained and returned to the source.
- Aggressive maintenance must be practiced to eliminate leaks and "accidents" that could lead to noncontainment of chemicals and other substances.
- Reconstitution of cleaning baths, acid baths, alkali baths, and plating baths should be done on an as-needed basis according to the work performed, rather than on a regular timing or other schedule.
- Water used for the first stages of battery washdown should be recycled water from other processes.
- Dry methods of cleanup, including brooms, shovels, and dry vacuuming, should be used to the maximum extent.
- Techniques such as treating wastewater from washdown of the pasting area should be treated in multistage clarifiers. At some plants, this technique has enabled total reuse of the water. The clarified water has been used for washdown, and the solids have been reused in new batches of paste.

- Rinsing should be counter-current, with respect to freshwater makeup and spent rinsewater overflow.
- Purchasing should be guided by aggressive selection of raw materials in order to obtain the cleanest possible materials.
- There should be a constant and consistent program to substitute less polluting and nonpolluting substances for those that require expensive treatment and expensive disposal.
- In concert with the aforementioned, there should be a constant and consistent program for replacing process controls, including sensors, microprocessors, and hardware, with the objective of decreasing waste and maximizing retention, containment, recycle, and reuse of all substances. For instance, the battery filling system must be constantly maintained and upgraded to keep overflowing to a minimum.
- Technologies for recovering and regenerating acids, lead, and other chemicals as well as separating and removing contaminants should be aggressively employed. Ultrafiltration to effect removal of oils from alkaline cleaning solutions is an example. Centrifugation has also been used for this purpose. Activated carbon can be used to remove organic impurities.

Both filtration and centrifugation produce concentrated impurities that offer the possibility of recovery. If recovery of substances is not feasible, the concentrated impurities are in a form more easily disposed of.

Wastewater Treatment

Wastewaters from the manufacture of lead acid batteries are typically acidic, as the result of contamination by the sulfuric acid used as the electrolyte. In addition, dissolved lead and suspended solids containing lead are major components. Differences from one plant to another regarding the presence of pollutants other than acid and lead depend on whether or not leady oxide is produced at the plant, whether or not the electrode grids are produced at the plant, and the type of grids fabricated (i.e., antimony alloy, pure lead, or calcium alloy). Other factors include differences in the method of plate curing, forming, and assembly. Parameters that are regulated, and therefore have been found in wastewaters from plants manufacturing lead acid batteries, include antimony, cadmium, chromium, copper, lead, mercury, nickel, silver, zinc, iron, oil and grease and TSS. pH value is also regulated. Some plants pretreat and discharge to a POTW. Others treat and discharge on site. Either way, the wastewater treatment system must be capable of pH control and removal of multiple metals.

In general:

- Wastewaters containing significant quantities of oil and grease should be isolated, treated for removal of oil and grease (for instance, by reverse osmosis, or RO), and then combined with other wastes for removal of metals.

- Wastewater, including equipment washdown water, from the paste makeup and pasting areas should be isolated and treated on site by multiple clarifiers, and both the clarified water and the solids reused. The water should be used for paste makeup and for acid dilution in the electrolyte makeup area. The solids should be returned to paste makeup. As usual, a certain amount of blow-down will probably be required.

- The final wastewater stream from a typical lead acid battery manufacturing facility, after wastes minimization, wastes segregation and pretreatment, and combining of the pretreated flows, can best be treated by use of precipitation for removal of metals, sedimentation for removal of precipitated metals and other suspended solids, and chemical coagulation, sedimentation, sand, mixed media filtration, final pH adjustment, and ion exchange as necessary to achieve required levels of metals, TSS, pH, and

oil and grease. It may be necessary to perform the metals removal operations in two or more stages in series, using different values of pH for each stage, to remove all the metals (some of which have different values of pH for lowest solubility). These wastewater treatment techniques are discussed in detail in Chapter 7.

Bibliography

"A Report on Bottled and Canned Soft Drinks, SIC 2086, and Flavoring Extracts and Syrups, SIC 2087." Associated Water and Air Research Engineers, Inc., August 1971.

ABMS Non-Ferrous Metal Data, 1988.

Amerine, M. A., H. W. Berg, R. E. Kunkee, C. S. Ough, V. L. Singleton, and A. D. Webb. *The Technology of Wine Making*. 4th ed. Westport, CT: AVI Publishing, 1980.

Anonymous. "California Wine: On the Grapevine." *Economist* 307, No. 7547 (April 23, 1988).

Apotheker, S. "Plumbing Lead from the Waste Stream." *Resource Recycling* (April 1992): 34.

Azad, H. S. *Industrial Waste Management Handbook*. New York: McGraw-Hill, 1976.

Bagot, B. "Brand Report No. 161-Domestic Wine: Wine Tastings." *Marketing & Media Decisions* 24, No. 5 (May 1989).

Baker, D. A., and J. E. White. "Treatment of Meat Packing Waste Using PVC Trickling Filters." *Proceedings of the Second National Symposium on Food Processing Wastes*. U.S. Environmental Protection Agency. Denver, CO: 1971.

Bethea, R. M., B. N. Murthy, and D. F. Carey. "Odor Control for Rendering Plants." *Environmental Science and Technology* 7 (June 1973).

Bishop, P. I. *Pollution Prevention: Fundamentals and Practice*. New York: 2000. p. 426.

Blanc, F. C., and S. H. Corr. "Treatment of Beef Slaughtering and Processing Waste Waters Using Rotating Biological Contactors." *Proceedings of the 38th Industrial Waste Conference.* Purdue University, 1983. pp. 133–140.

"Bottling Breakthrough at 1000 to 2000 BPM." *Food Engineering* (March 1969).

Bramer, H. C. *Industrial Wastewater Management Handbook.* New York: McGraw-Hill, 1975.

Brooks, C. S. "Metal Recovery from Waste Sludges." *Proceedings of the 39th Purdue Industrial Waste Conference.* Purdue University, 1985. pp. 545–553.

Carron, H. *Modern Synthetic Rubbers.* New York: Van Nostrand Reinhold, 1937.

Chadwick, T. H., and E. D. Schroeder. "Characterization and Treatability of Pomace Stillage." *Journal of the Water Pollution Control Federation* 45 (1978).

Considine, D. M. *Chemical and Process Technology Encyclopedia.* New York: McGraw-Hill, 1974.

Considine, D. M., and G.D. Considine. *Foods and Foods Production Encyclopedia.* New York: Van Nostrand Reinhold, 1982.

Crandall, C. J., and J. R. Rodenberg. "Waste Lead Oxide Treatment of Lead Acid Battery Manufacturing Wastewater." *Proceedings of the 29th Purdue Industrial Waste Conference.* Purdue University, 1974. pp. 194–206.

Crow, D. R., and R. F. Secor. "The Ten Steps of De-inking." *TAPPI Journal* 70, No. 7 (July 1987).

Current Industrial Reports. U.S. DOC. Bureau of the Census. 1986, 1989.

Cushnie, G. C., Jr. *Removal of Metals from Wastewater: Neutralization and Precipitation.* Park Ridge, NJ: Noyes Publications, 1984.

Eldridge, E. F. "Michigan Engineering Experiment Station Bulletin #60." 1934.

Ellis, J., and R. W. Williams. "Recovery of Tin from Electroplating and Rinsewaters." *Effluent and Water Treatment Journal* 18, No. N8 (1978): 389.

Etzel, J. E. Private conversation. March 1989.

Farmer, J. K., A. A. Friedman, and W. C. Hagen. "Anaerobic Treatment of Winery Wastewaters." *Proceedings of the 43rd Industrial Waste Conference.* Purdue University, May 1988.

"Frozen Foods." *Supermarket Business* 44, No. 9 (September 1989): 175.

Fukuyama, Yoji, Y. Misaka, and K. Kato. "Recovery of Aluminum Hydroxide from Fabricating Plant of Aluminum Products." *Proceedings of the 29th Industrial Waste Conference.* Purdue University, May 1974.

Gillies, M. T. "Soft Drink Manufacture." *Food Technology Review* 8 (1973).

Graham, A. Kenneth. *Electroplating Engineering Handbook.* New York: Van Nostrand Reinhold, 1971.

Gudger, C. M., and J. C. Bailes. "The Economic Impact of Oregon's Bottle Bill." Corvalis, OR: Oregon State University, March 1974.

Hamilton, J. O. C. Abstract No. 88-17491. "What Slump? It's a Vintage Year for Premium California Wines." *Business Week* (Industrial Technology Edition) 2947 (May 1986).

Hunt, G. E. et al. "Cost-Effective Waste Management for Metal Finishing Facilities." *Proceedings of the 39th Industrial Waste Conference.* Purdue University, 1985. pp. 545–553.

"Increasing Wastewater Treatment." *Food Engineering* 59, No. 8 (August 1987): 97.

Jones, D. D. "Biological Treatment of High-Strength Coke-Plant, Wastewater." *Proceedings of the 38th Industrial Waste Conference.* Purdue University, 1983. pp. 561.

Kanicki, D. P. "Casting, Advantages, Applications, and Market Size." In *45M Handbook V. 14: Castings.* ASM International, 1988.

Krofta, M., and L. K. Wang. "Total Closing of Paper Mills with Reclamation and De-inking Installations." *Proceedings of the 43rd Industrial Waste Conference.* Chelsea, MI: Lewis Publishers Inc., May 1988.

Labella, S. A., I. H. Thaker, and J. E. Tehan. "Treatment of Winery Waste by Aerated Lagoon, Activated Sludge and Rotating Biological Contactor." *Proceedings of the 27th Industrial Waste Conference.* Purdue University, May 1972.

Loehr, R. C. *Pollution Control for Agriculture.* 2nd ed. Orlando, FL: Academic Press, 1984.

Lowenheim, F. A. *Electroplating.* Sponsored by the American Electroplaters' Society. New York: McGraw-Hill, 1978.

Macauley, M. N., W. T. Stebor, and C. L. Berndt. "Anaerobic Contact Pretreatment of Slaughterhouse Waste." *Proceedings of the 42nd Industrial Waste Conference.* Purdue University, 1987. pp. 647–655.

Manzione, M. A., D. T. Merrill, M. McLearn, and W. Chow. "Meeting Stringent Metals Removal Requirements with Iron Adsorption/Coprecipitation." *Proceedings of the 44th Industrial Waste Conference.* Purdue University, 1989.

Maxwell, J.C., Jr. "Coolers Have Chilling Effect on Wine Sales." *Advertising Age* 59, no. 27 (June 27, 1988).

Medbury, H. (ed.) *The Manufacture of Bottled Carbonated Beverages.* Washington, DC: American Bottlers of Carbonated Beverages, 1945.

Mermelstein, H. H. "Water Purification for Beverage Processing." *Food Technology* (February 1972).

Metal Finishing 55th Guidebook Directory 85, No. 1A (Mid-January 1987).

Metals Handbook. 9th ed. ASM International, 1988.

MHR-VIANDES. *World Beef and Veal Production in 1996: Meat Industry Statistics.* France: Rond Point de l'Europe, 1998.

Modern Pollution Control Technology, Vol. 1, Air Pollution Control. New York: Research and Education Association, 1978.

Morton, M. *Rubber Technology.* American Chemical Society. New York: Van Nostrand Reinhold, 1973.

Nemerow, N. L. *Liquid Waste of Industry.* Reading, MA: Addison-Wesley, 1971.

Neveril, R. B., J. U. Price, and K. L. Engdahl. "Capital and Operating Costs of Selected Air Pollution Control Systems IV." *Control Technology News* 28 (November 1978).

Niles, C. F., and H. Gordon. "Operation of an Anaerobic Pond on Hog Abattoir Wastewater." *Proceedings of the 26th Industrial Waste Conference.* Purdue University, 1971.

Norcross, K. L., S. Petrice, R. Bair, and G. Beaushaw. "Start Up and Operation Results from SBR Treatment of a Meat Processing Wastewater." *Proceedings of the 42nd Industrial Waste Conference.* Purdue University, 1987.

Osag, T. R., and G. B. Crane. *Control of Odors from Inedibles Rendering Plants.* EPA 450/1-74-006. July 1974.

"PAC-Activated Sludge Treatment Coke-Plant Wastewater." *Proceedings of the 40th Industrial Waste Conference.* Purdue University, 1985.

Pandey, R. A. "Wastewater Characteristics of LTC Process of Coal." *Journal of Environmental Science and Health* A24, no. 6 (1989): 633.

Parsons, W. A. *Chemical Treatment of Sewage and Industrial Wastes.* Bulletin No. 215, National Lime Association. Washington, DC: 1965.

Patrick, G. C. et al. "Development of Design Criteria for Treatment of Metal-Containing Wastewaters at Oak Ridge National Laboratory." *Proceedings of the 42nd Industrial Waste Conference.* Purdue University, 1988. pp. 819–830.

Patterson, J. W. *Industrial Wastewater Treatment Technology.* Boston: Butterworth-Heinemann, 1985.

Patterson, J. W. "Metals Speciation, Separation, and Recovery." *International Symposium on Metals Speciation, Separation, and Recovery.* Chelsea, MI: Lewis Publishers, 1987.

Perry, R. H. *Perry's Chemical Engineers' Handbook.* 6th ed. New York: McGraw-Hill, 1984.

Peters, R. W., and Young Ku. "The Effect of Citrate, a Weak Chelating Agent, on the Removal of Heavy Metals by Sulfide Precipitation." In *Metals Speciation, Separation, and Recovery.* Chelsea, MI: Lewis Publishers, 1987.

"Pilot Study of Upgrading of Existing Coke Oven Waste Treatment Facility With Trickling Filter." *Proceedings of the 41st*

Industrial Waste Conference. Purdue University, 1986. pp. 586.

Porges, R. and E. J. Struzeski. "Wastes from the Soft Drink Bottling Industry." *Journal of the Water Pollution Control Federation* 33 (1961): 167.

Potter, N. N. *Food Science.* 3rd. ed. Westport, CT: AVI Publishing, 1978.

Prokop, W. H. "Plant Operations." *Renderer Magazine* (October 1979).

Prokop, W. H. "Wet Scrubbing of High Intensity Odors from Rendering Plants." Presented at the APCA Specialty Conference: *State-of-the Art of Odor Control Technology II.* Pittsburgh, PA. March 10–11, 1977.

Pulp and Paper Manufacture, Vol. II: Control, Secondary Fiber, Structural Board, and Coating. Joint Textbook Committee of the Paper Industry, 1969.

Ramierez, E. R., and O. F. D'Alessio. "Innovative Design and Engineering of a Wastewater Pretreatment Facility for a Metal Finishing Operation." *Proceedings of the 39th Industrial Waste Conference.* Purdue University, 1985. pp. 545–553.

Robertson, W. M., J. C. Egide, and J. Y. C. Huang. "Recovery and Reuse of Waste Nitric Acid from an Aluminum Etch Process." *Proceedings of the 35th Industrial Waste Conference.* Purdue University, May 1980.

Rudolfs, W. *Industrial Wastes: Their Disposal and Treatment.* New York: L.E.C. Publishers, 1961.

Russell, J. M., and R. N. Cooper. "Irrigation of Pasture with Meat-Processing Plant Effluent." *Proceedings of the 42nd Industrial Waste Conference.* Purdue University, 1987. pp. 491–497.

Russell, L. L., J. N. DeBoice, and W. W. Carey. "Disposal of Winery Wastewater." *Proceedings of the 7th National Symposium on Food Processing Wastes.* EPA 600/2-76-304. December 1976.

Ryder, R. A. "Winery Wastewater Treatment and Reclamation." *Proceedings of the 28th Industrial Waste Conference.* Purdue University, May 1973.

Sawyer, C. N., and P. L. McCarty. *Chemistry for Environmental Engineering.* 3rd ed. New York: McGraw-Hill, 1978.

Sell, N. J. *Industrial Pollution Control.* New York: Van Nostrand Reinhold, 1981.

Sirrine, K. L., P. H. Russell, Jr., and J. Makepeace. *Winery Wastewater Characteristics and Treatment.* EPA 600/2-77-102.

Snoeyink, V. L., and D. Jenkins. *Water Chemistry.* New York: John Wiley & Sons, 1980.

Stander, G. T. "Treatment of Wine Distillery Waste by Anaerobic Digestion." *Proceedings of the 22nd Industrial Waste Conference.* Purdue University, May 1967.

Steffan, A. J., and M. Bedker. "Operations of a Full Scale Anaerobic Contact Treatment Plant for Meat Packing Wastes." *Proceedings of the 16th Industrial Waste Conference.* Purdue University, 1971.

Steffan, A. J. *In-Plant Modifications to Reduce Pollution and Pretreatment of Meat Packing Wastewater for Discharge to Municipal Systems.* Environmental Protection Agency. West Lafayette, IN: 1973.

Stevenson, T. *Southby's World Wine Encyclopedia.* Boston: Little Brown, 1988.

Supermarket Business 44, No. 3 (March 1989): 85.

Synthetic Rubber: The Story of an Industry. New York: International Institute of Synthetic Rubber Production, 1973.

"The Looming Battle for Center of the Plate." *Advertising Age* 60, No. 49, S10 (November 13, 1989).

Title 40. *Code of Federal Regulations.* Effluent Guidelines and Standards. Part 430. Subpart Q De-ink Subcategory. January 30, 1987.

U.S. Bureau of the Census. *Statistical Abstract of the United States: 1988.* Washington, DC: 1987.

U.S. Bureau of the Census. *Statistical Abstract of the United States: 1990 (110th Annual Edition).* January 1990.

U.S. Department of Energy. *Quarterly Coal Report.* October–December 1989.

U.S. Department of Health, Education, and Welfare. *An Industrial Guide to the Meat Industry.* Publication No. 386. Washington, DC: U.S. Government Printing Office, 1954.

U.S. Environmental Protection Agency. *Biological Removal of Carbon and Nitrogen Compounds from Coke Plant Wastes*. April 1973.

U.S. Environmental Protection Agency. *Development Document for Effluent Limitations, Guidelines, and New Source Performance Standards for the Red Meat Processing Segments of the Meat Products Point Source Category*. Washington, DC: U.S. Government Printing Office, 1974.

U.S. Environmental Protection Agency. *Development Document for Effluent Limitations, Guidelines, and Standards for the Foundries (Metal Molding and Casting)*. EPA 440/1-80/070a. 1980.

U.S. Environmental Protection Agency. *Development Document for Effluent Limitations, Guidelines, and Standards for the Iron and Steel Manufacturing Point Source Category* vol. II. December 1980.

U.S. Environmental Protection Agency. *Development Document for Effluent Limitations, Guidelines, and Standards of Performance for the Battery Manufacturing Point Source Category*. EPA 440/1-82/067-b. Washington, DC: U.S. Government Printing Office, 1982.

U.S. Environmental Protection Agency. *Development Document for Effluent Limitations, Guidelines, and Standards of Performance, Rubber Processing Industry*. EPA 440/1-74-013-A. Washington, DC: U.S. Government Printing Office, 1974.

U.S. Environmental Protection Agency. *Proceedings 8th National Symposium on Food Processing Wastes*. Washington, DC: U.S. Government Printing Office, August 1977.

U.S. Environmental Protection Agency. *Treatability Manual*. EPA 600/2-82-001b. 1982.

U.S. Environmental Protection Agency. *Upgrading Meat Packing Facilities to Reduce Pollution*. Washington, DC: U.S. Government Printing Office, 1973.

U.S. Environmental Protection Agency. *Wastewater Characteristics for the Specialty Food Industry*. Washington, DC: U.S. Government Printing Office, 1974.

U.S. Environmental Protection Agency. Office of Water. *Development Document for Effluent Limitations, Guidelines, and Standards for the Pulp, Paper and Paperboard Point Source Category*. EPA 440/1-82/025. October 1982.

U.S. Environmental Protection Agency. Office of Water Regulation and Standards, Effluent Guidelines Division. *Proposed Development Document for Effluent Limitations, Guidelines, and Standards for the Copper Forming Point Source Category*. EPA 44011-82/074-b. October 1982.

Van Nostrand, N. J. *Van Nostrand's Scientific Encyclopedia*. New York: John Wiley & Sons, 1983.

Varley, P. C. *The Technology of Aluminum and Its Alloys*. Cleveland, OH: CRC Press, 1970.

Ward, R. C. "Aluminum Anodizing Wastewater Treatment and Reuse." *Proceedings of the 36th Industrial Waste Conference*. Purdue University, May 1981.

Westra, M. A., and B. L. Rose. "Oil and Grease Removal from a Concentrated Source in the Metal Finishing Industry." *Proceedings of the 44th Industrial Waste Conference*. Purdue University, 1989.

Whitby, G. S. *Synthetic Rubber*. New York: John Wiley & Sons, 1954.

Glossary & Acronyms

aerobic—an environment in which oxygen is used as the terminal electron acceptor

agglomerate—to cluster together

anaerobic—an environment in which no oxygen, either as free O_2 or in the form of nitrite or nitrate are available and the shift is toward less efficient electron acceptors like SO_4^{2-}, Fe^{3+}, etc.

anode—the positively charged electrode of an electrolytic cell, or a negatively charged terminal

anoxic—an environment in which there is limited free oxygen (O_2) present and the shift is toward nitrite and nitrate the terminal electron acceptors

anion—a negatively charged ion

bag house—a filtration system for air pollution control which utilizes fabric bags as the filter media

batch-type—occurring in discreet units from start to finish of process, as opposed to a continuous process

binary fission—division into two parts

blow-down—removal of a portion of the process stream to prevent the concentration of a contaminant from exceeding a setpoint. For example, in a cooling tower contaminants are concentrated due to evaporation of the water being cooled. Blow-down occurs when the a variable (typically conductivity) exceeds a setpoint. A sidestream is removed and replaced with fresh make up water to control the desired concentration.

cathode—the negatively charged electrode of an electrolytic cell, or a positively charge terminal

cation exchange capacity—the ability of a soil to exchange cations on the soil particle for cations that may be introduced via percolation of a waste stream.

caustic—a substance of pH greater than 7, also referred to as alkaline or basic

chelating agent—a chemical used to sequester or hold metal ions in solution over a broad range of process conditions, including pH and ORP.

clarification—removal of impurities by way of gravity settling

colloid—a particle that does not settle to a practical extent (for waste treatment puposes) in a medium under the influence of gravity

condenser—equipment designed to liquefy gases via cooling

corona—ionized gas molecules created by high energy electrons in a strong electrical field.

denitrification—the biological conversion of nitrate to nitrogen gas (requires an anoxic environment)

diatomaceous earth—skeletal remains of diatoms (small invertebrates) used in waste treatment as filtration media

divalent—an ion with a charge (negative or positive) of two

drag-out—the liquid that is carried out of a plating batch by the pieces being plated

entropy—the measure of the disorder or randomness in a system

evapotranspiration—water loss via evaporation and through plant leaves

exothermic—a reaction that gives off heat

F/M ratio—food to microorganism ratio

facultative—able to function in aerobic, anoxic and anaerobic environments

floc—an agglomeration of small particles to form a larger one flocculation—the process of gently agitating particles causing them to collide with one another and join together to form larger particles

freeboard—the distance between the intended water level and the top of a tank

flux—1) a mineral added to the charge of a furnace to drive out oxygen and prevent the formation of oxides. 2) A rate of flow of a substance expressed as mass or volume per unit time

mesophilic—an environment with a temperature range of 20° to 45°C that favors bacteria with optimal growth in this range (mesophiles)

monomer—a single unit in a polymer chain of similar chemical units

monovalent—an ion with a charge (negative or positive) of one

nitrification—the biological conversion of ammonia to nitrate (requires an aerobic environment)

osmotic pressure—pressure created by a difference in concentration of a solute on either side of a semi-permeable membrane

polymerization—a chemical process where two or more monomers join to form a polymer

psychrophilic—an environment with a temperature less than 20°C that favors bacteria with optimal growth in this range (psychrophiles)

putrescible—capable of biological decomposition under anoxic or anaerobic conditions

reverse osmosis—the application of pressure to a semi-permeable membrane to create a concentrated solution on one side of the membrane and a clean, or less concentrated solution on the other.

sequencing batch reactor—a single tank reactor for treating wastewater using a series of timed steps

stoichiometric—the molar ratio of products and reactants in a chemical reaction

surface loading—the rate of flow per unit area of surface

thermophilic—an environment with a temperature range of 45° to 90°C that favors bacteria with optimal growth in this range (thermophiles)

trivalent—an ion with a charge (negative or positive) of three

Acronyms

AA	atomic adsorption
AAFEB	anaerobic fixed-film expanded bed
ABC	activity-based costing
AFFI	American Frozen Food Institute
APC	air pollution control
APMP	Air Pollution Management Plan
A/S	air to solids ratio
ASCE	American Society of Civil Engineers
ASTs	above-ground storage tanks
ASTM	American Society for Testing Materials
BATF	Bureau of Alcohol, Tobacco and Firearms
BET	Brunauer-Emmett-Teller equation
BMPs	best management practices
BOD	biochemical oxygen demand
CAA	Clean Air Act
CBOD	carbonaceous BOD
CERCLA	Comprehensive Environmental Response, Compensation, and Liability Act
C.F.R.	Code of Federal Regulations
CMR	completely mixed mode
COD	chemical oxygen demand
CSOs	combined sewer overflows
CWA	Clean Water Act
DAF	dissolved air flotation

DMRs	discharge monitoring reports	MLVSS	mixed liquor volatile suspended solids
DNA	deoxyribonucleic acid	MSGP	Multi-Sector General Permit
DO	dissolved oxygen	MSWLFs	municipal sold wastes landfill facilities
DOT	Department of Transportation		
EDTA	ethylenediaminetetraacetic acid	MTBE	methyl-tert-butyl ether
EPA	U.S. Environmental Protection Agency	MWCO	molecular weight cutoff
		NAAQS	National Ambient Air Quality Standards
EPCRA	Emergency Planning & Community Right-to-Know Act	NBOD	nitrogenous BOD
ESPs	electrostatic precipitators	NESHAP	National Emission Standards for Hazardous Air Pollutants
FF	fixed film systems		
F/M	food-to-microorganism ratio	NF	nanofiltration
FOG	fats, oils, and greases	NLR	nonlinear regression
FWPCA	Federal Water Pollution Control Act	NOI	Notice of Intent
		NPDES	National Pollutant Discharge Elimination System
FWS	free water surface		
GACTs	generally available control technology	NRC	National Resource Council
		NSPS	new source performance standards
GCL	geosynthetic clay layer		
GPD	gallons per day	NTUs	Nephelometric Turbidity Units
GR-S	Government Rubber—Styrene	O&G	oil and grease
HAPs	hazardous air pollutants	O&M	operation and maintenance
HDPE	high-density polyethylene	OIT	operator interface terminal
HRT	hydraulic retention times	ORP	oxidation-reduction potential
HSWA	Hazardous and Solid Waste Amendments	OTE	oxygen transfer efficiency
		OWPS	oily wastewater pretreatment system
HVAC	heating, ventilation, and air conditioning		
		OWTP	oily wastewater treatment plant
LDRs	land disposal restrictions	P&M	process and maintenance
LFG	landfill gas	PACT	powder activated carbon technology
MACTs	maximum achievable control technology		
		PCBs	polychlorinated biphenyls
MF	microfiltration	PCP	pentachlorophenol
MGD	million gallons per day	PE	polyethylene
MTBE	methyl-tert-butyl ether	PFLT	Paint Filter Liquids Test
ML	mass loading	PLC	programmable logic controller
MLSS	mixed liquor suspended solids		

POTW	publicly owned treatment works		**SIP**	state implementation plan
ppb	parts per billion		**SOTR**	standard oxygen transfer rate
ppm	parts per million		**SP3**	Stormwater Pollution Prevention Plan
PQL	Practical Quantitation Limit		**S/S**	solidification and stabilization
PRP	potentially responsible party		**TCE**	trichloroethylene
RBC	rotating biological reactor		**TCLP**	Toxic Characteristics Leaching Procedure
RCRA	Resource Conservation and Recovery Act		**TDS**	total dissolved solids
RDF	refuse-derived duel		**tpy**	tons per year
RMF	Runoff Management Facility		**TRS**	total reduced sulfer
RNA	ribonucleic acid		**TSD**	treatment storage and disposal
RO	reverse osmosis		**TSS**	total suspended solids
RPP	rinsewater pretreatment plant		**TVSS**	total volatile suspended solids
SARA	Superfund Amendments and Reauthorization Act		**UASB**	upflow anaerobic sludge blanket
SBR	sequence batch reactor		**UF**	ultrafiltration
SBR	Styrene Butadiene Rubber		**UV**	ultraviolet light
SDWA	Safe Drinking Water Act		**VOCs**	volatile organic compounds
SFS	subsurface flow system		**WTE**	waste-to-energy
SIC	Standard Industry Classification			

Index